MORPHOGENESIS IN PLANT TISSUE CULTURES

MORPHOGENESIS IN PLANT TISSUE CULTURES

Edited by

WOONG-YOUNG SOH

and

SANT S. BHOJWANI

Springer-Science+Business Media, B.V.

A C.I.P. Catalogue record for this book is available from the Library of Congress.

ISBN 978-90-481-5206-3 ISBN 978-94-015-9253-6 (eBook)
DOI 10.1007/978-94-015-9253-6

Printed on acid-free paper

Originally published by Kluwer Academic Publishers in 1999.
Softcover reprint of the hardcover 1st edition 1999

Contents

Introduction

Morphogenesis, or developmental morphology, is a dynamic process and a fascinating field of investigation. Since the beginning of this century plant morphologists (Eames, 1961), anatomists (Eames and Macdaniels, 1947) and embryologists (Maheshwari, 1950) have studied the processes of development and differentiation by observing whole plants and their histological preparations and have generated a wealth of information on the structural changes associated with various developmental stages in the life cycle of a large number of plant species. Around 1940 plant morphogenesis became an experimental field when plant physiologists initiated studies on the effect of treatments, such as application of physiologically active compounds or growth regulators, irradiation, exposure to different day-length, temperature and injury, on morphological and structural changes. During the past two decades geneticists and molecular biologists have become interested in plant morphogenesis and extensive work is being done to understand the regulation of gene expression during morphogenesis and how the products of genetic information control the developmental processess.

All sexually reproducing organisms begin as a single-celled zygote, which by growth and cell divisions gives rise to the complex body of the mature organism. Since the divisions involved are mitotic, every cell in the plant body of an organism should have the same genetic material as in the zygote. Yet the organism comprises a vast array of cell, tissue and organ types which are structurally and functionally very different. This process of differentiation of genetically identical cells into various different types of tissues has intrigued biologists for a very long time.

As early as 1868, Hofmeister, in his publication "General Morphology of Growing Things" had raised a very basic question, as to how during growth does the observed form or structure of an organism comes to be as it is and what factors are involved. A similar question was asked by Haberlandt in his famous address to the German Academy in 1902. The classic experiments of Vochting on polarity in cuttings, carried out in 1878, had clearly demonstrated that all cells along the stem length are capable of forming shoots as well as roots but it was Haberlandt who introduced the concept of cellular trotipotency and suggested that terminally differentiated plant cells, as long as they contain the normal complement of chromosomes, should be capable of regenerating whole plants. He made a novel approach to prove this hypothesis in which single cells were isolated from a highly differentiated tissue of the plant body and cultivated in nutrient solution. Although Haberlandt did not succeed in his experiments due to technical problems at that time, his ideas attracted many scientists to pursue this line of investigation, and in 1939 ability of plant cells for unlimited growth was demonstrated by Gautheret, Nobecourt and White, independently. Further progress in the cultivation of plant cells, tissues and organs on artificial medium, under controlled environmental conditions, gave birth to a large number of aseptic techniques, collectively called Plant Tissue Culture. These laboratory techniques have proved to be a powerful approach to manipulate morphogenesis under highly controlled conditions and to understand the role of various extrinsic factors and correlative effects of various tissues in morphogenesis. The three important aspects of morphogenesis to which significant contributions have been made with the aid of plant tissue culture techniques are: (1) cytodifferetiation, (2) organogenic differentiation and (3) somatic embryogenesis. These studies are not only of academic interest but of considerable practical importance.

1 CYTODIFFERENTIATION

In vitro tracheary element differentiation has long been used to study cytodifferentiation. A variety of experimental systems, such as internodes, storage tissues, suspension cultures, single cells and protoplasts, have been used in these studies. One of the most significant advances was the optimization of the *Zinnia* mesophyll cell system by Fukuda and Komamine (1980) and Church and Galston (*see* Church, 1993). In this case 50-60% of the cells directly (without a cell division) and fairly synchronously differentiate into tracheary elements within 72 h (Church, 1993).

2 ORGANOGENIC DIFFERENTIATION

As early as 1939, White (1939b) had reported controlled differentiation of shoots from tissue cultures of tobacco. In 1957 Skoog and Miller put forth the concept of hormonal control of organ formation, according to which root and shoot differentiation is a function of the auxin-cytokinin ratio; relatively higher concentration of cytokinin promotes shoot formation and relatively higher auxin concentration favours root formation. This concept proved to of wide applicability. In 1965 it was shown for the first time that full plants can be obtained starting from a single isolated cell (Vasil and Hildebrandt, 1965). By emperically manipulating the medium and other environmental factors it has been possible to induce shoot differentiation from explants, calli, cells and/or protoplasts of over 1000 species (Thorpe, 1993). It is now well established that plant regeneration can be achieved from all living cells irrespective of their nature of specialization and ploidy level. In 1964 Guha and Maheshwari first time reported organogenic differentiation from haploid microspores. The technique of anther and isolated microspore culture to produce androgenic haploid plants is now well established. Even the triploid endosperm cells have been shown to be totipotent (Johri and Bhojwani, 1965).

3 SOMATIC EMBRYOGENESIS

In 1958 an important contribution was made to the field of experimental morphogenesis when Steward (working in the USA) and Reinert (working in Germany) observed the differentiation of embryos in somatic cell cultures of carrot. The development and appearance of these somatic embryos were comparable to the zygotic embryos of this plant. These reports attracted the attention of several developmental botanists because until that time it was believed that embryo formation is the monopoly of the zygote and a few other ovular cells, and embryo development requires a special environment available only inside the female gametophyte or embryo sac. During the last three decades somatic embryogenesis has dominated the research activities in the field of plant morphogenesis not only because of its curiosities but because of its potential role in basic and applied aspects of plant sciences. Until recently, plant embryogenesis had been studied intensively to describe morphological and anatomical changes that characterize embryo development in different species (Natesh and Rau, 1984; Raghavan, 1986; Bhojwani and Bhatnagar, 1999); almost nothing was known about the molecular and genetic control of embryogenesis because of the practical problems of inaccesibility to the zygotic embryo which develops under several cell layers. The progress in the field of somatic embryogenesis during the last two decades has been such that in systems such as carrot (Terzi et al., 1985) gram quantities of practically pure fractions of embryos (histologically and biochemically similar to zygotic embryos) at different developmental stages can be obtained. It has, therefore, now become possible to study genetic and molecular control of morphogenesis during embryogenesis. Using this system about 21 "embryo specific" or "embryo enhanced" genes have been cloned from somatic embryos (Zimmerman, 1993). Investigations related to cellular and molecular aspects of organogenesis and somatic embryogenesis would not only increase our knowledge and understanding of the processes involved but is also expected to enhance our capacity to increase the level of regeneration in presently recalcitrant systems.

4 PRACTICAL IMPORTANCE

The value of in vitro morphogenesis as a commercially viable procedure for plant propagation and as an essential step in genetic improvement of crops through the modern methods of biotechnology has considerably promoted research in this area. Mass production of genetically identical plants by regeneration from vegetative tissues *via* organogenesis or somatic embryogenesis is of considerable interest to the horticulturists and foresters. Regeneration *via* embryogenesis is of particular interest as embryos are bipolar structures bearing both root and shoot meristems and may allow certain degree of automation during production and their field planting. Several groups are investigating large scale production of somatic embryos in bioreactors and their encapsulation to produce artificial seeds to facilitate their mechanized planting in the field.

Regeneration of haploid plants from pollen grains has become a popular technique in plant breeding as it enables to achieve homozygosity in a single step, as against several years required to achieve it by the traditional method of recurrent selfing. Androgenesis is also serving as a novel sourse of useful genetic variability (Gametoclonal variation). Plant regeneration from protoplasts, cells and tissue explants is extremely important in harnesing the potential values of somatic hybridization and genetic engineering in crop improvement.

Thus, in vitro morphogenesis is of considerable importance in basic and applied areas of growth and differentiation. However, most of the tissue culture books published during the last two decades (over 100) mainly deal with the basic aseptic techniques and their practical applications with very little or no emphasis on morphogenesis. The only book which has discussed in vitro morphogenesis to some extent is "Plant Tissue Culture and Morphogenesis", written in 1964 by Butenko. On the other hand, the book entitled "Plant Morphogenesis", by Sinnot (1960), only deals with in vivo morphogenesis. The substantial advances made in the field of in vitro morphogenesis during the last 25 years was the driving force to bring out this volume on Morphogenesis in Tissue Cultures. It covers all aspects of morphogenesis at cellular, tissue and organ levels. This book includes 16 chapters written by some internationally renowned scientists with considerable experience in their fields. Each chapter discusses upto-date literature in the field, and the text is amply supported by tables and illustrations. The chapters have been classified into two sections, one dealing with articles on the basic aspects and the other on applied aspects of morphogenesis. We feel that the students and teachers of plant sciences would find this book useful.

We are indeed very grateful to the contributors for their ready co-operation without which it would have not been possible to bring out this volume. We would also like to thank Professor Duck Yee Cho, Department of Biology, WooSuk University, Chonbuk and our students Mrs. Sil Yoon, Ec Youb Kim, Jong Cheun Lee, Hak Soo Kim, Ki Shik Moh and Jae Dong Lee, and Misses. Byung Sook Kang, Ae Ryeong Cho, Eun Kyung Lee, Sun Ah Joo, Bo Rib Jang, and Jung Soon Shin at the Department of Biological Sciences, Chonbuk National University, Korea for their help in various ways in the preparation of this book for press.

Woong -Young Soh
Sant S. Bhojwani
Editors

PART I

BASIC STUDIES

Chapter 1

DIFFERENTIATION OF VASCULAR ELEMENTS IN TISSUE CULTURE

G. Paul BOLWELL[1] and Duncan ROBERTSON[2]
[1]Division of Biochemistry, School of Biological Sciences, Royal Holloway, University of London, Egham, Surrey TW20 0EX, and [2]Department of Biological Sciences, Durham University, Durham DH1 3LE, UK.

1 INTRODUCTION

It remains a major goal of modern biology to understand the molecular events involved in the differentiation of specialized cells. This chapter considers to what extent the experimental induction of vascular differentiation in cell and tissue cultures can be achieved and what insight such studies can generate into similar processes in the intact plant. The lack of complexity, relative homogeneity and isolation of events within individual or small groups of cells has great potential for study of aspects of signal transduction leading to the modulation of genes responsible for structural and other changes during differentiation. Although such systems cannot reconstitute cell lineages and position effects they could, in principle, help define the role of extracellular signals and other factors and their modulation by a largely empirical approach involving optimization of their effects under various nutrient and growth regulator requirements. Furthermore, biochemical studies may be able to identify aspects of metabolic regulation that are not accessible in the intact plant. Finally, tissue culture systems can be used to understand intracellular signal transduction pathways leading to changes in gene expression.

Differentiation of plant tissues gives rise to a relatively small number of cell types. Vascular cells are early discernible products of commitment and determination of the meristematic cells of the stem apex, root tip and vascular cambium. As such, their formation is tied into the developmental programming that is a feature of multicellular organisms, where networks of information serve to determine the fate of each meristematic cell following division with such precision and positioning that it can only be perturbed by extreme environmental effects or mutation within the developmental pathway. Vascular differentiation exhibits many of the central features of plant cell morphogenesis, signal perception, cell lineages, differential gene expression and programmed cell death.

In tissue cultures, there is much constraint on the multiplicity and types of signals possible and the duplication of positioning effects. Of course, these may be preserved to some extent in organ culture but cell cultures lack many of these features and cannot imitate many of the aspects of developmental control in the whole plant. Nevertheless, they have proven useful in understanding some of the fundamental changes at the cellular level. Additionally, such studies offer a relatively unique feature, in that, much of the differentiation involves biosynthesis and modification of the cell wall (Fukuda, 1992, 1994; Bolwell, 1993). This is especially true for the differentiation of vascular tissue from cambium. Study of the regulation of wall biosynthesis at the gene and macromolecular assembly levels allows the investigation of defined targets (Northcote, 1995) and insights at all levels to the processes of differentiation through signal transduction and gene activation to fundamental changes in cell biology. Wall changes during differentiation have been thoroughly characterized as cessation of pectin synthesis, change in the nature of hemicelluloses and increased cellulose synthesis before the major onset of lignin deposition. There appears to be a programmed phased change in biosynthesis. This is also reflected in the cell biology where secondary wall is laid down evenly to the inside of the primary wall in phloem and xylem and in the latter more specifically, mediated by microtubule alignment, in the secondary thickenings that give rise to the scalariform structures of the tracheary elements. There is a lack of knowledge about the regulation of gene expression associated with these changes in the developing cambium leading to the characteristic wall structures of xylem and phloem. Tissue culture systems set out to model such processes.

2 SYSTEMS OF STUDY

A number of cell culture systems have been developed to study aspects of vascular differentiation. Some of these, especially the *Zinnia* system, afford an opportunity for studying the cell biology and changes in gene expression in developmental process that is closest to vascular differentiation in higher plants even though the vascular tissue is not derived from vascular cambium. Others, such as the French bean system, do not suffer from restrictions in the amount of cells available and afford an opportunity for detailed study of biochemical changes associated with secondary wall synthesis in cells that develop to resemble fibre cells. These complementary systems can be used for producing molecular probes, cDNA and antibodies specific for components of vascular differentiation which would not be readily accessible from the intact plant. However, there are many instances now where such probes derived from studies of these systems were found to be specific for vascular differentiation in the intact plant using *in situ* hybridization or immunolocalization. Furthermore, comparative studies with other model systems using such probes are revealing that similarities in gene expression exist between vascular differentiating systems and others such as wounding, elicitation of suspension cultured cells and cells undergoing somatic embryogenesis. Such comparative studies will be important in understanding signal transduction and how disparate signals give rise to the induction of similar gene products in the overall plasticity of plant responses.

2.1 Xylogenesis

2.1.1 *Zinnia* Mesophyll cells

The most striking example of vascular differentiation in cultured cells yet available concerns the induction of xylogenesis in isolated mesophyll cells isolated from the leaves of *Zinnia elegans* and put into suspension culture (Fukuda, 1992, 1994; Fukuda et al., 1993;Church, 1993). After a period of enlargement, the cells differentiate directly and relatively synchronously into tracheary elements in the presence of auxin (naphthalene acetic acid, 0.1 $mg l^{-1}$, cytokinin (6-benzyladenine, 1.0 $mg\ l^{-1}$) and sucrose ($10g\ l^{-1}$) as inducing agents. Between 40-60 % of the cells differentiate by 96 h, one of the highest recorded in tissue culture. The tracheids are terminally differentiated and show the composition of secondary wall (Ingold et al., 1988). Because of the high degree of synchrony it has been possible to map the crucial times of commitment and determination in cytological terms, by manipulation of the culture conditions and by use of agents that can block further development (Fukuda, 1992; Fukuda et al., 1993;Church, 1993). The precise time of this commitment has been delineated by transfer experiments between inductive and non-inductive medium and show that after 48 h the cells will differentiate even after transfer back into non-inductive medium (Church and Galston, 1988; Church, 1993) and this observation has been repeated by others subsequently. All these groups have combined to demonstrate that there are clear and demarcated early, intermediate and late changes during differentiation. Furthermore, such physiological and cytological studies have been complemented by a growing number of biochemical and molecular studies that give an extensive description of the series of events that take place. As a result, a number of detailed maps relating changes in gene expression and the appearance of immunodetectable epitopes and enzyme activities which correlate to cytological events have become available, giving an exciting insight in the processes of vascular differentiation.

The role of the cytoskeleton in these changes has been investigated extensively both with respect to actin filaments and microtubules by a number of workers (Fukuda, 1992). The organization of the microtubules is now known to change from a rare scattered distribution through a longitudinal orientation of dense bundles which then resolve into several thick transverse structures underlying the forming secondary cell wall bands (Kobayashi et al., 1988; Fukuda et al., 1993). These changes in microtubules required synthesis and turnover. Three tubulin genes were identified that were differentially regulated, the expression of two of which was associated with the onset of secondary thickening between 24 and 48 h (Fukuda et al., 1993). Depolymerization occurred after 48 h. Coodinated changes in actin filaments occurred which changed from cytoplasmic to intercalating between the microtubules at the onset of secondary wall formation (Kobayashi et al., 1988; Fukuda et al., 1993).

There appears to be a requirement for some DNA reorganization during this early stage (Sugiyama et al., 1994). Cell expansin has also been investigated in the early stages (Roberts and Haigler, 1994) and is promoted by maintenance of the medium at pH 5.5 or reduced osmotic potential. Interestingly, cross-linking of wall components which presumably leads to rigidification as seen in the elicitation response is promoted by alkalinization of the apoplast and maintenance of the cells at pH 5.5 prevents hydrogen peroxide production (Bolwell et al., 1995; Wojtaszek and Bolwell, 1995). Expansion of *Zinnia* cells can be also induced by conditioned medium and an expansion inducing factor has been subjected to preliminary characterization (A. Roberts and C. H. Haigler personal communications). Prolonging the expansion phase delays the differentiation of tracheary elements and the development of their characteristic morphology.

The intermediate phase is characterized by the laying down of secondary walls and thickenings. During this period there is increased cellulose and xylan deposition (Figure 1A). At this stage, many of the organelles are still intact. However, terminal differentiation resembles xylogenesis *in planta*, in that complete autolysis occurs. Disruption of the vacuole occurs several hours after secondary wall formation and is followed by rapid loss of the cell contents (Fukuda,1992). Thus, overall, the *Zinnia* system is the best available for studying differentiation in tissue culture. However, it is not absolutely perfect since the mesophyll cell is not initially meristematic and the product of differentiation itself. Other tissue culture systems may resemble cambial cells to a better extent and while there is no particular strong evidence to substantiate this, their continued use may also contribute to the understanding of vascular differentiation.

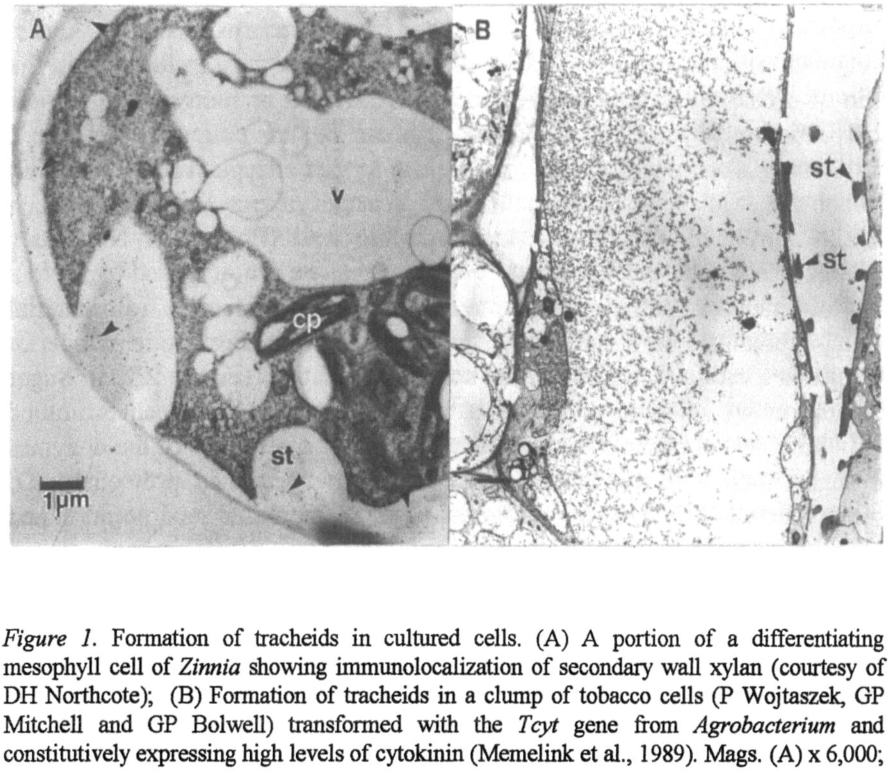

Figure 1. Formation of tracheids in cultured cells. (A) A portion of a differentiating mesophyll cell of *Zinnia* showing immunolocalization of secondary wall xylan (courtesy of DH Northcote); (B) Formation of tracheids in a clump of tobacco cells (P Wojtaszek, GP Mitchell and GP Bolwell) transformed with the *Tcyt* gene from *Agrobacterium* and constitutively expressing high levels of cytokinin (Memelink et al., 1989). Mags. (A) x 6,000; (B) x 1,000. cp = chloroplast; v = vacuole; st = secondary thickening. In A arrows indicate gold deposits. Scale bars (A) 1 μm; (B) 5 μm

2.1.2 French bean cells and other suspension cultures

Before the development of the *Zinnia* system and its wide adoption as the model system of choice, an extensive range of callus and suspension cultures were investigated for induction of vascular differentiation and shown to exhibit low levels of xylogenesis following manipulation of auxin, cytokinin and sucrose levels (Bolwell, 1985; Fukuda, 1992). There has not been such widespread use of these systems in more recent times as the *Zinnia* system has quite rightly found favour, but they can still prove useful for a number of biochemical studies because many of the cells undergo changes with the biochemical characteristics of xylogenesis. They have one particular advantage in that much more material is available for conventional purification purposes than is possible with *Zinnia* cells although in this case the amount of available cells may be improved with use of a leaf disk system (Church and Galston, 1989). With respect to the other systems, only relatively recent work is discussed; earlier work is discussed in much greater detail elsewhere (Fukuda, 1992).

The French bean system exhibits typical metabolic changes seen in response to elevated levels of exogenous or endogenous cytokinin levels. Treatment of cell suspension cultures of French bean with an increase in the ratio of cytokinin to auxin and in the concentration of sucrose also models differentiation to some extent. These cells can be manipulated to produce a wall having a composition reasonably close to that of typical secondary wall (Robertson et al. 1995a). The cells show changes of enzyme activity which can be correlated with these changes in the wall (Bolwell & Northcote, 1983a,b; Robertson et al., 1995a). Wall changes are characterized by up to a five fold increase in thickness due to the laying down of extra wall material which intercalates between the middle lamella and probably the remainder of the primary wall although this may not remain distinct (Figure 2A,B). Sugar flux following labelling of the cells with [^{14}C]-sucrose has been examined during the period of maximum extractable catalytic activities of the enzymes of sugar nucleotide conversion (e.g. Fig. 3A) determined previously (D. Robertson and G. P. Bolwell, unpublished data). Increased incorporation and secretion was observed in response to the transfer to xylogenic medium and this was observed in or all major groups of polysaccharides but most particularly the cellulosic fraction. Furthermore, analysis of the sugars in the hemicellulosic fraction indicated that newly synthesized polysaccharide was xylan, a secondary wall polysaccharide rather than xyloglucan (D. Robertson and G.P. Bolwell, unpublished data). Xylan was also detected by immunolocalization in these walls but not those of the control (Figure 2D). This contrasted with the pectin fraction which was seen to be relatively much less in the induced cells. This treatment also increased incorporation into the wall of other components characteristic of secondary walls since they also contain a thirty fold increase in phenolics and increased levels of hydroxyproline indicative of increased deposition of structural cell wall proteins (Robertson et al., 1995a). Similar increases in wall thickness in tracheids have been observed in cell walls of suspension cultured cells (Figure 1B) derived from transgenic tobacco transformed with the T*cyt* gene (Memelink et al., 1989).

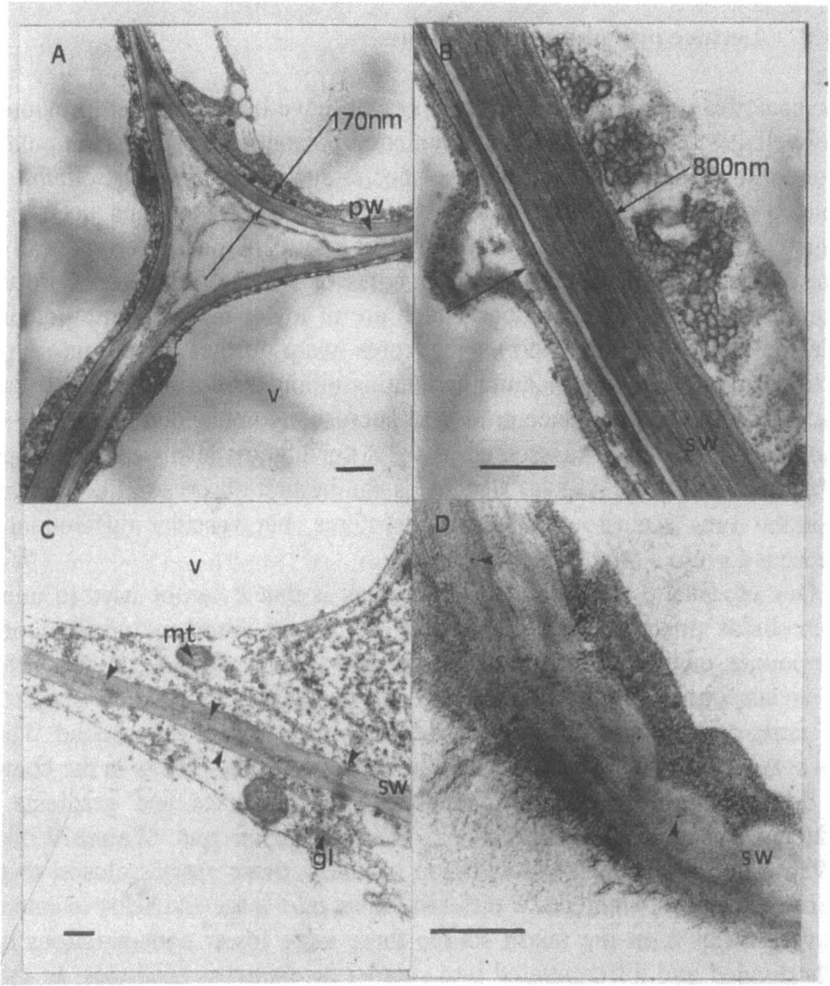

Figure 2. Secondary cell wall formation in French bean cells. (A) Primary walls of cells before transfer to xylogenic medium; (B) Thick walls formed after five days in xylogenic medium; (C) Immunolocalization (arrowed) showing deposition of a specific secondary wall glycoprotein (Wojtaszek and Bolwell, 1995); (D) Immunolocalization of secondary wall xylan in induced bean cells. Mags. (A) x 15,000, (B) x 20,000 ; (C) x 12,000; (D) x 35,000. gl= Golgi apparatus; mt = mitochondrion; pw = primary wall; v = vacuole; sw = secondary wall. In (C) and (D) arrows indicate gold particles. Scale bars 0.5 μm.

Few other studies as extensive as those carried out in French bean have been reported for other systems in more recent times although suspension cultured cells such as tobacco have been used to follow lignification (Nagai et al., 1994). Carrot cultures have been used to study somatic embryogenesis (De Vries et al., 1994) and, together with *Catharanthus*, cell expansion (Masuda et al., 1984; Suzuki et al., 1990).

2.1.3 Lettuce pith and other explants

The basic features of the lettuce pith system have been reviewed previously (Bolwell,1985). Sections of undifferentiated parenchyma can be manipulated to undergo xylogenesis from diverse species such as *Helianthus, Sambucus*, tobacco and *Coleus*, while the most widely used system has been from lettuce. This is because large amounts of tissue of one cell type can be excised from the sub-apical region as cores of 3-5 mm diameter. Care was taken to avoid the lower regions which are of lower morphogenic potential (Warren-Wilson et al., 1991). The cores were further sliced into 2mm sections and placed on a medium containing indole acetic acid (10mg l^{-1}) and zeatin (0.1mg l^{-1}). The concentration of sucrose recommended previously (2-3%) has recently been reassessed as 0.2% for highest levels of xylogenesis (Warren-Wilson et al.,1994a). Tracheids can be first observed after 3-5 days, a similar time scale to that in other systems, but vascular differentiation proceeds for up to 14 days.

One advantage of the lettuce pith system is that it can be used to mimic intercellular polar transport of signals. By taking 6mm cylindrical cores, compounds can be administered at one end (Warren-Wilson et al., 1982), unlike suspension cultured systems where presumably all cells are subject to the same concentration of stimuli, although they do not all respond. Thus, this system may be closer to modelling those positional effects in the control of differentiation which may be dependent on fluxes and gradients of inducing substances (Sachs, 1991; Warren-Wilson and Warren-Wilson, 1993). Upon unilateral administration of auxin, those vessels closest to the source underwent immediate differentiation into large tracheary elements. Moving away from the auxin source there were lower concentrations and cells divided and differentiated into smaller isodiametric tracheids. In more distant regions tracheary strands could be observed where auxin transport had become canalized. Similarly, application of sucrose and zeatin resulted in spatial arrangements of cell division and xylogenesis which correlated with the mobility and concentration of each factor. Sucrose was found to be freely mobile whereas zeatin, although slightly less mobile than auxin, also formed a steep concentration gradient in relation to the source (Warren-Wilson et al., 1994b).

2.2 Phloem differentiation in culture

In addition to xylogenesis in tissue cultures, the induction of phloem sieve elements can also be used to model differentiation and also as a source for the acquisition of molecular probes. Cultures of *Stepanthus tortuosus* can be induced to form phloem sieve elements. By substituting 0.1 mg l^{-1}

naphthalene acetic acid in the general maintenance medium with 5 mg l^{-1} 2,4-D in the presence of 2% sucrose and 1 mg l^{-1} kinetin, the cells could be readily induced to differentiate (Toth et al., 1994). A phloem-enriched preparation can be obtained by preferential digestion of parenchyma cell walls and collecting the undigested intact sieve elements by centrifugation on percoll gradients. This is an exciting development since there are likely to be many phloem-specific gene products (Fisher et al. 1992; Sakuth et al., 1993) and since, like xylem, mature phloem elements are enucleate and there are likely to be novel regulatory aspects to protein synthesis. Although study of phloem exudates has given some insight into the range of proteins, use of tissue cultures with their bulk potential will allow a deeper study in defining individual components belonging to this complex physiological and biochemical system.

3 MARKERS FOR VASCULAR DIFFERENTIATION

Plants appear to respond to developmental and environmental cues by changes in gene expression. An early goal in studying vascular differentiation in plants has been to identify specific targets for regulation during xylogenesis (Northcote, 1995). Many of these have been related to lignification and secondary wall formation but more recently novel induced genes have been identified. Such programmes as differential screening of libraries (Demura and Fukuda, 1993, 1994; Ye and Varner, 1993, 1994) and more recently differential display (L. T. Koonce and C. H. Haigler, personal communications) have identified additional types of cDNAs coding for genes with no known function or with sequence similarity to regulatory proteins in other organisms.

3.1 Polysaccharide biosynthesis

Polysaccharide biosynthesis and metabolism can be measured by the flow and accumulation of radioactivity from labelled sugar substrates into specific polysaccharides and cellular compartments. Cell cultures and in particular suspension cultures can be manipulated in such a way that they offer an excellent model to study polysaccharide biosynthesis. Alternatively, another approach to gauging the rates of cell wall biosynthesis can be through assaying component enzymes and correlating their specific activities with pathway fluxes in relation to developmental events.

The actual synthesis of cell wall polysaccharides is a Golgi-based process with the exception of the glucans, cellulose and callose which are synthesized at the plasmalemma. All these glycosyl transferases remain rather poorly described but progress is being made in their characterization (Gibeaut and Carpita, 1994). When cells of French bean are induced, arabinosyl transferase activity catalyzed by an M_r 70,000 Golgi-localized enzyme is reduced indicating accession of pectin synthesis (Bolwell and Northcote, 1983 a,b; Rodgers and Bolwell 1992). Similarly there was a loss of polygalacturonic acid synthase in sycamore cells on differentiation (Bolwell et al., 1985). On the other hand, xylosyl transferase activity involved in xylan synthesis, which is probably catalyzed by an Mr 40,000 protein (Rodgers and Bolwell, 1992), is seen to rapidly increase several fold in cells grown in induction medium (Bolwell and Northcote, 1983 a,b). This increase in xylan synthase activity reflects that seen in differentiating sycamore (Bolwell et al., 1985) and for enzymes responsible for the synthesis of other hemicelluloses such as glucomannan (Dalessandro et al., 1986). In cultures the kinetics of appearance of these examples of glycosyl transferase activities correlates to the changes observed between the induced and non-induced cell walls. The cell walls found in the French bean system which have undergone xylogenic-like changes in composition are also reminiscent of those found in differentiated *Zinnia* cells (Ingold et al, 1988). Similarly, in the *Zinnia* system, xylosyl transferase activities have also been correlated to the increase in xylan synthesis (Suzuki et al, 1991). Furthermore, these observations made in French bean cells have also been made in developing bean hypocotyls which would tend to confirm that cell culture systems provide a valid model for xylogenic differentiation (Bolwell and Northcote, 1983b). Other systems such as flax which accumulates high levels of xylan may be more amenable to study the biosynthesis of these hemicelluloses (N Carpita, personal communication). Cellulose synthesis has not been measured directly due to the difficulty in measuring this enzyme activity but there is known to be increased cellulose deposition in *Zinnia* (Taylor et al., 1992; Suzuki et al., 1992).

Although it is generally assumed that the major controlling factor in the qualitative production of cell wall polysaccharides resides in the complement of the membrane bound synthases (Bolwell, 1993), the underlying enzymes involved in the supply of UDP-sugars, as substrates for the synthases, may affect the overall balance of cell wall polysaccharide biosynthesis. Recently, an intimate association has been demonstrated between sucrose synthase, thought to be the major enzyme in the provision of UDP-glucose from sucrose, with plasmalemma and with cellulose and callose synthesis (Amor et al., 1995). Immunolocalization demonstrates an upregulation of sucrose synthase in differentiating tracheary elements of *Zinnia* (C. H. Haigler, personal communication). In French bean cells, however, sucrose synthase declined in activity and it seems more likely that in these cells starch is a major source for wall polysaccharides (Robertson et al., 1995a). This work also showed that UDP-glucose dehydrogenase increased in activity (Figure 3A) which correlated with the xylogenic-like changes found in cell walls of induced French bean cells.

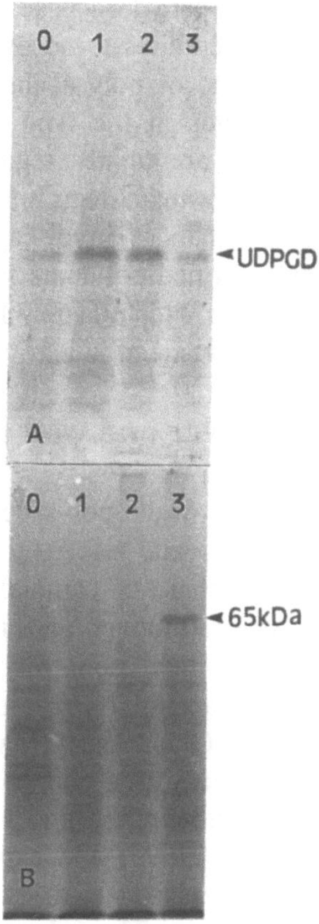

Figure 3. Examples of the induction of specific proteins in cells forming secondary wall in French bean. (A) Western blot of UDP-glucose dehydrogenase involved in the provision of UDP sugars for polysaccharide synthesis. Soluble cell extracts from French bean induced by cytokinin treatment at days 0, 1, 2 and 3 were subjected to SDS/PAGE and immunodecorated with anti-(UDP-glucose dehydrogenase) serum (Robertson et al., 1996). Induction of the protein is seen at days 1 and 2 which coincides with the appearance of maximum enzyme activity (Robertson at al., 1995a). (B) $CaCl_2$ extracts (Wojtaszek et al., 1995) of cell wall proteins from the same cells as in (A), were separated by SDS/PAGE and stained with Coomassie blue. Note appearance of a prominent polypeptide of M_r 65000 at day 3. This protein is an N-linked glycoprotein (cf Wojtaszek and Bolwell, 1995) and may be a member of a small family of glycoproteins responsible for the high affinity of secondary walls for wheat germ agglutinin (Benhamou and Asselin, 1989).

This soluble enzyme, which localises to vascular tissue in the plant (Robertson et al., 1996) is thought to be committal, since the reaction is irreversible. The product, UDP-glucuronic acid can readily be converted to UDP-xylose by the action of a decarboxylase. This enzyme also appeared to be under developmental control but to a lesser extent than the dehydrogenase (Robertson et al., 1995a). UDP-xylose can exert negative feedback control on the activity of the purified dehydrogenase (Robertson et al., 1996). Therefore, by manipulating cell cultures to become xylogenic it is possible to envisage how different aspects of polysaccharide biosynthesis interact. Not only are there gross changes in the type of cell wall carbohydrate polymers synthesized but these are tightly regulated with respect to the metabolism of the necessary UDP-sugars required. Moreover this is timed to coincide with the cytoskeletal changes involving microtubules which are required to deposit the newly formed polymers into the forming secondary wall. This is especially so in cells which display highly architectured arrays of secondary thickenings characteristic of xylem vessels and tracheids.

3.2 Polysaccharide turnover and wall assembly

Turnover and modification of wall polysaccharides is a feature of developmental change and this is true also in tissue cultures (Carpita and Gibeaut, 1993). The expansion phase of cambial development is an event that probably involves polymer turnover and rearrangement of existing structures in addition to a massive increase in biosynthesis and deposition of new material. A number of extracellular enzymes and proteins have been implicated in these processes. Glucanases (Maclachlan and Carrington, 1991), xyloglucan endotransglycosylases and endoxyloglucan transferases (Fry, 1995; Fry et al., 1992; Hetherington and Fry, 1993; Potter and Fry, 1994; Xu et al., 1995; Nishitani, 1995) and expansins (Cosgrove and Li, 1993; Li et al., 1993; McQueen-Mason et al., 1992, 1993) have all been championed as the major influence. The role of these have not been as studied extensively in cultures although hydrolases have been demonstrated in many tissue cultures (Fry, 1995). However, an extensive study is commencing to explore the complex changes in the glycan structures of the wall in *Zinnia* (Stacey et al., 1996). Changes in the secretion and turnover of pectins, xyloglucan and the arabinogalactan epitopes on AGPs, show that a rhamnogalacturonan appears around the time of determination, while a specific AGP appears later and accumulates in secondary thickenings. The fucose-containing epitope on xyloglucan disappears just before the onset of secondary thickening. Such studies may reveal which polysaccharides are targetted for the generation of possible modulatory signals. A fucosidase appears to be one hydrolase that is activated. There may some involvement in the production of oligosaccharin signals as feedback or monitoring signals for the cells as to the stage on the differentiation pathway that has been reached. Indeed, induction of other hydrolases such as chitinase appears to be necessary for somatic embryogenesis in carrot cells where cell expansion is a process that appears inversely related to this phenomenon (De Vries et al., 1994). In *Zinnia,* expansion also appears to need completion before differentiation (Roberts and Haigler, 1994). Elongation has also been modeled in a carrot system (Masuda et al., 1984) and in *Catharanthus* (Suzuki et al., 1990). Differences were seen in the arabinose and galactose containing components and while these may reflect changes in neutral pectins, arabinogalactan proteins have also been implicated. Arabinogalactan proteins (AGPs) have been characterized from carrot lines and have been shown to influence somatic embryogenesis (Baldwin et al., 1993; Kreuger and van Holst, 1993). Mapping of these and their influence on the differentiation of *Zinnia* mesophyll cells is underway and may lead to a deeper understanding of the role of these proteins in morphogenesis (Stacey et al., 1996).

The polymers which are exocytosed into the extracellular compartment have to be deposited at defined points along the plasmalemma to give rise to the characteristic architecture of the secondary cell wall during xylogenesis. Use of the *Zinnia* system in the presence of inhibitors of cellulose biosynthesis suggests that the secreted polymers assemble in a self perpetuating cascade (Taylor and Haigler, 1993). Normal secondary cell wall thickenings contain cellulose, xylan and lignin as well as specific proteins. When cells were treated with either 2,6 dichlorobenzonitrile or isoxaben at concentrations that inhibit cellulose biosynthesis at the sites of secondary thickening then xylan and glycine-rich proteins could not be detected immunologically. At lower inhibitor concentrations where some cellulose synthesis occurred, xylan and glycine-rich proteins could be detected between thickenings but were not assembled indicating that a whole population of components were required to allow self assembly (Taylor and Haigler, 1993). Once the thickenings are established it is only then that lignin deposition occurs.

3.3 Lignin biosynthesis

There is now extensive knowledge of the components of lignification, where most proteins and genes encoding them have been isolated and characterized from a number of species (Sederoff et al., 1994; Whetten and Sederoff, 1995). All the proteins, with the exception of *p*-coumarate-3-hydroxylase have been purified for the phenylpropanoid pathway together with the enzymes of the lignin branch pathway, cinnamoyl CoA reductase (CCR) and cinnamyl alcohol dehydrogenase (CAD) and the probable enzymes of polymerization, lignin peroxidase and laccase.

Induction of phenylalanine ammonia-lyase (PAL) activity and the appearance of mRNA has been determined in *Zinnia* during differentiation where the gene was activated (Lin and Northcote, 1990). Similarly, PAL has been shown to be induced during hemicellulose production in cells of French bean stimulated to produce thick walls by cytokinin and sucrose treatment (Bolwell and Northcote, 1983b; Bolwell and Rodgers, 1991; Fig 4). PAL was also induced in tobacco cells in response to cytokinin (Nagai et al., 1994). The second enzyme of the pathway, cinnamate-4-hydroxylase (C4H), which is a cytochrome P450 (Bolwell et al., 1994) designated CYP73, is induced by elicitor treatment of suspension cultured cells of Medicago (Fahrendorf and Dixon, 1994).

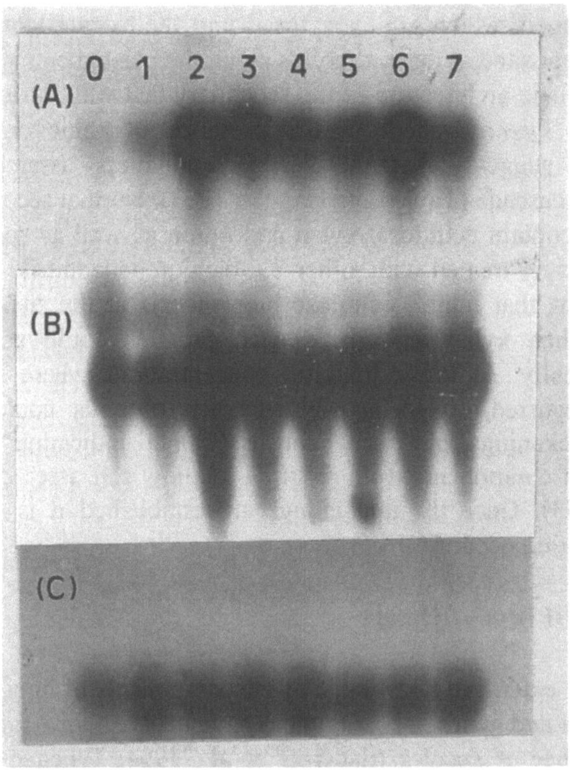

Figure 4. Transcriptional activation of lignin biosynthesis genes. mRNA was isolated from French bean cells at 0, 1, 2, 3, 4, 5, 6 and 7 days after subculture into xylogenic medium. They were northern blotted and probed with (A) French bean pPAL5 cDNA; (B) cDNA coding for cinnamate-4-hydroxylase from French bean (Jupe SC and Bolwell GP, unpublished data); (C) cDNA coding for cinnamyl alcohol dehydrogenase, a kind gift of A. O'Connell and W. Schuch

We have recently shown that it is also induced by cytokinin in a coordinate manner with PAL (S. C. Jupe and G. P. Bolwell, Fig. 4). Clones for *Zinnia* C4H have been reported in the Genebank data base and presumably a similar coordinate induction will be demonstrated. *p*-Coumarate-3-hydroxylase is as yet unequivocally identified (Bolwell, 1993). However, there are two types of methyl transferases, one working at the level of the free acid and the other on the CoA ester have been characterized and cloned from a number of sources. The latter appears to be the major species involved in lignification in differentiating *Zinnia* cells (Ye et al., 1994; Ye and Varner, 1995). The caffeic acid O-methyl transferase appears to have a more pronounced role in stress responses. Ferulate-5-hydroxylase has been localized to lignification and is a cytochrome P450 designated CYP84 (Meyer et al., 1995) and a second methylation takes place (Bolwell, 1993). The intermediates for lignin, *p*-coumarate, ferulate and sinapic acid are activated by CoA ligase for which a xylem specific form has been identified and cloned from Loblolly pine (Voo et al., 1995). This enzyme has been shown to increase in activity during the later stages of differentiation in *Zinnia* just prior to the onset of the major burst of lignification (Church and Galston, 1988; Sudibyo and Anderson, 1993). Cinnamoyl CoA reductase (CCR) has been purified (Goffner et al., 1994) and cloned (O'Connell A, Boudet AM and Schuch W, personal communications). Cinnamyl alcohol dehydrogenase (CAD) has been extensively characterized from a number of species (Sederoff et al., 1994; Grima Pettenati et al., 1994; Whetten and Sederoff, 1995), cloned (Grima-Pettenati et al., 1993; Van Drosselaere et al., 1995) and subjected to genetic manipulation (Halpin et al., 1994). Promoter analysis has been carried out (Walter et al., 1994; Feuillet et al, 1995) and has identified regulatory elements. Fig 4 shows the induction of CAD mRNA in bean cells in response to cytokinin and sucrose.

Two types of enzyme have been associated with the polymerization process, the copper oxidase, laccase and the heme-protein, peroxidase. Laccase has been purified from suspension cultured cells of sycamore where it was abundantly secreted (Driouich et al., 1992). This proved to be a glycoprotein with Mr 66,000. However, this form was not detected in the intact plant and the major species of Mr 56,000 was confined to the epidermis. Other examples of laccase were found to be vascular (Bao et al., 1993) and a laccase has been detected in *Zinnia* cells (Liu et al., 1993; Dean et al, 1995). Laccase may be involved in polymerization of lignin in special areas while the bulk of the lignin polymerization is carried out by peroxidase (Bolwell, 1993).

There is much controversy about the identification of a lignin peroxidase. A number of claims have been made but there are likely to be many possible candidates since, for example, a PCR approach has identified some 16 separate peroxidases in bean (S. C. Jupe and G. P. Bolwell, unpublished data). There has been an in depth study of the peroxidase isoforms expressed in differentiating *Zinnia* cells (Sato et al., 1993, 1995). Five isoforms in particular were identified of which the expression of two correlated with differentiation and lignification. All these forms were found capable of efficiently oxidizing coniferyl alcohol whereas only one form had an absolute requirement for calcium. These were anionic. A likely candidate for the lignin peroxidase in French bean is an M_r 46,000 cationic isoform which localizes to secondary thickenings and is present in walls of the suspension cultured cells (Smith et al., 1994; Zimmerlin et al., 1994). The substrate specificity, which was particularly high for ferulate residues, also suggests an extracellular role (Zimmerlin et al, 1994; Smith et al, 1994). This cationic peroxidase was implicated in developmental as well as pathogenic responses (Bolwell et al., 1995) and may well be involved in the lignification process in xylogenic cells.

3.4 Other wall proteins

In addition to the extracellular hydrolases, peroxidases and laccases already discussed there are a number of subsets of families of wall proteins that are considered specific for the primary cell wall but as yet the number of wall proteins that are specific to the secondary wall is small. When tissue cultured cells are induced to differentiate, especially to form tracheary elements, the most striking visible changes are associated with the cell wall. It can be assumed that a large part of cell metabolism is directed to providing the necessary carbohydrate polymers, phenolic materials and proteins which either form part of the secondary wall or are involved in the biosynthesis and deposition of these materials in the extracellular compartment.

As far as cell wall proteins which may be involved in xylogenesis are concerned, there exists little information at the cellular level regarding the positioning of structural or enzymic proteins. There is, however, a myriad of information regarding the hydroxyproline-rich glycoproteins (HRGPs), proline-rich proteins (PRPs), glycine-rich proteins (GRPs) and some lipid transfer proteins (LTPs) which have been based upon studies using tissue printing, *in situ* hybridizations and differential screening of cDNA libraries (Ye and Varner, 1991, 1993; Keller,1993; Taylor and Haigler, 1993; Bao et al., 1992; Demura and Fukuda, 1994; Showalter, 1993).These observations were made in *Zinnia* cells or developing seedlings which were undergoing differentiation. The means of unequivocable localization at the cell level were not always tested but the available evidence would suggest that almost all these proteins are found in the cell wall compartment. However, it is difficult to assess which cell wall proteins are specifically involved in differentiation because in the case of these structural proteins at least, they exist as members of multigene families and are implicated in other developmental events as well as defense responses (Showalter, 1993).

The only examples of specific secondary wall proteins localized directly other than inferred from cDNA sequence, have been shown in Loblolly pine (Bao et al., 1992), hypocotyl of French bean (Wojtaszek and Bolwell, 1995) and in differentiating *Zinnia* cells (Stacey et al 1996). The French bean protein proved to be hydroxyproline-poor, glycosylated, (being recognized by wheat germ agglutinin) and localized to tracheary elements, xylary and phloem fibres and can be localized in secondary walls induced in bean cultures (Fig. 2C). The epitope conferring recognition by wheat germ agglutinin is a particular feature of secondary thickenings (Benhamou and Asselin, 1989). Another epitope found in secondary thickenings of *Zinnia* cells was recognized by a monoclonal antibody (JIM 13) which is specific for an arabinogalactan protein. There are probably AGPs specifically expressed in xylem (Loopstra and Sederoff,1995). As yet there is no defined function for these proteins. Fig. 3B shows another novel M_r 65,000 wall protein of unknown function that is induced in French bean tissue cultures.

One of the difficulties in defining which proteins are up- or down-regulated in response to cytokinin treatment is illustrated by an example of a tentative extracellular protein which was found to be "repressed" in cytokinin treated tobacco suspension cultures and was identified as a β1-3 glucanase (Bauw et al, 1987). Although the authors did not define the exact nature of the phenotypic changes in their cytokinin-treated cells, it may be that the enzyme which is a defense protein and induced by constitutive cytokinin production (Memelink et al., 1989), is immobilized. The glucanase was quantified by measuring the intensity achieved on two dimensional electrophoretograms and positively identified by obtaining the N-terminal sequence. Although the protein appeared to be down regulated it may also have become recalcitrant to the method of extraction. By analogy to systems used to model defense responses, where several subsets of wall proteins are known to be immobilized in response to a pathogen derived elicitor (Bradley et al, 1992; Wojtaszek and Bolwell, 1995) a similar situation may exist in differentiating cells. For example, defense responses share some similarities to xylogenic development. Both show an increase in the amount of hydroxyproline-rich glycoproteins and phenolics accumulated compared to the relevant control cells. These *de novo* synthesized materials are also cross linked forming hydrophobic impervious barriers for the purpose of defense or as water conducting reinforced strengthening in the case of xylogenesis. It may be that during the polymerization phase other proteins as well as some carbohydrate polymers become entrapped or covalently-linked in the process. For this reason it can be difficult to characterize structural wall proteins involved in xylogenesis.

Some wall proteins by their very nature may elude detection by normal aqueous chemical methods, an example being the glycine-rich proteins of French bean which are specifically localized in the vascular tissue and other lignified cells of this plant. To date these very hydrophobic proteins have been largely studied through molecular techniques including the generation of an antibody raised against a fusion protein (Keller et al., 1988, 1989). Subsequent tissue printing confirmed the tissue specific localization (Ye and Varner, 1991).

Various roles have been assigned to the classical cell wall proteins ranging from the strengthening needed in the absence of lignin, which is deposited at later stages in the developmental process to potential nucleation sites for the deposition of lignin. They have also been implicated in the formation of cross-linked networks either through dityrosine or isodityrosine residues. In the case of arabinogalactan proteins a role in determination of development has even been suggested. However the precise function of many of these proteins remains to be determined. Bearing in mind that most cell wall proteins belong to multigene families and that their functional roles may overlap then defining a role for any one protein will not be an easy task but may be aided by the use of tissue culture systems.

3.5 Lipid metabolism

Changes in lipid metabolism are not usually considered features of differentiated cells but are probably worth examination. In fact a revaluation of the lipid content of cell wall which is a major component of suberised cells is probably timely. Hydroxy-fatty acid-phenolic esters especially with ferulic acid may contribute to the hydrophobic sealing of vessels. Furthermore, cells undergoing somatic embryogenesis induce lipid transfer proteins which are probably involved in exporting lipids to the extracellular matrix (De Vries et al., 1994). Interestingly, one of the clones that was differentially screened from *Zinnia* cDNA libraries proved to have high sequence similarity to a lipid transfer protein (Ye and Varner, 1993).

3.6 Miscellaneous

Autolysis is a feature of the development of vascular cells. There is however evidence for continued biochemical activity after breakdown of organelles. Certainly enzymes of lignin biosynthesis can be detected immunologically in vascular cells at this time (Smith et al., 1994). It is also of significance because it involves programmed cell death or apoptosis, a feature shared by the hypersensitive response in which subsets of proteins are produced that are similar to those induced during vascular differentiation while others are unique to defense reactions.

During the phase leading up to programmed cell death and autolysis there would be an expected battery of degradative enzymes would be produced and employed in providing changes to the cell shape, i.e. breaking down cross walls and also in breaking down other macromolecular cell components. In addition, a number of enzymes would be involved in recycling released metabolites. In *Zinnia*, for example, cDNAs which are specific for differentiating cells have been isolated which code for adenylate kinase and a protease (Ye and Varner 1993). Both cysteine and serine proteases have been cloned from *Zinnia* (Ye and Varner,1996). Expression patterns link the cysteine protease with events during xylogenesis while the serine protease showed an additional involvement in autolysis. A cysteine protease has also been partially purified from *Zinnia* cells where it showed maximum activity between 48 and 60 h (Minami and Fukuda, 1995) when the number of defined tracheary elements increases rapidly. The M_r of this protein, 30,000, indicated that it may be a different isoform to that of 20,000 predicted for the cloned species. The involvement of proteases may be even more complex (E. Beers, personal communication). Four proteinases have also been characterized and it was possible to distinguish between those induced by sucrose starvation during growth of the cultures from thosc genuinely part of programmed cell death. These appear to have different M_r to those reported by the other workers. The role of ubiquitin in protein turnover during vascular differentiation (Bachmair et al., 1990; Stephenson et al., 1996) is also probable in *Zinnia* (E. Beers, personal communication).

Fragmentation of nuclear DNA is also a feature of terminal differentiation of xylem in the developing plant (Mittler and Lam, 1995). The timing of nuclear degradation in *Zinnia* has been determined (A. Jones, personal communication). Differentiation-specific nucleases specific for both DNA and RNA have also been characterized. A nuclease from *Zinnia* was one of the earliest identified markers for xylogenesis (Thelen and Northcote, 1989). Similarly, two cloned RNAses (Ye and Droste, 1996) showed differential expression in a number of developmental and stress situations. However one, an M_r 24,000 ribonuclease was expressed in differentiating mesophyll cells.

3.7 Phloem-specific proteins

There are likely to be many phloem-specific proteins. The best characterized is the P-protein of the slime which is involved in the sealing of wounded sieve tubes (Northcote, 1995).The filamentous proteins of the slime are also thought to bind to and physically restrict the progress of pathogens (Sjolund et al., 1993). Monoclonal antibodies have been raised to the P-proteins from differentiating cultures of *Strepanthus tortuosus* (Toth et al., 1994) by initially raising a panel of monoclonals against total phloem proteins and screening for P-protein specific species by western blotting and immunolofluorescence localization. A second phloem-specific protein was also characterized from *Strepanthus* cultures as B-amylase (Wang et al., 1995). This protein was purified using a monoclonal antibody as a means of detection and the target protein was identified by protein sequence and enzyme activity and the phloem localization confirmed by immunofluorescence on *Arabidopsis* seedlings. The amylase may be involved in mobilizing starch for differentiation. A similar observation was made in the bean system where amylase rather than sucrose synthase was induced by cytokinin treatment (Robertson et al., 1995a).

Other phloem-specific proteins are known. Callose synthase is also likely to be induced during phloem differentiation since callose lines the pores at the sieve plate. Callose is also found in the cell plates of newly divided cells. Fig. 5 shows the immunolocalization of a probable subunit of callose synthase localizing to a sieve plate (C. G. Smith, B. MacCormack, & G. P.

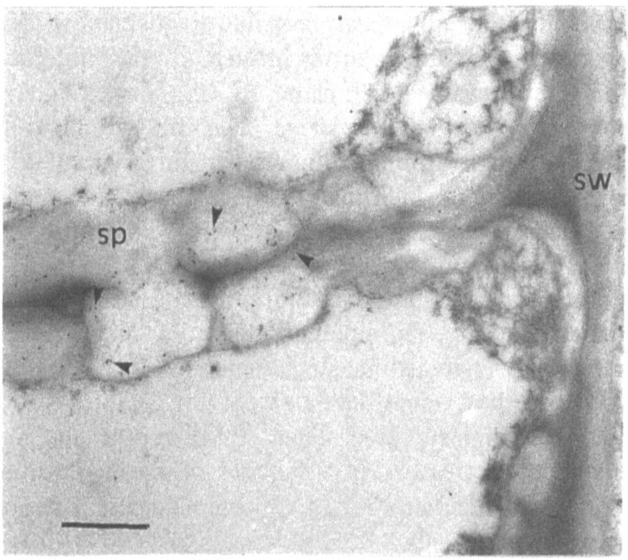

Figure 5. Localization of a subunit of callose synthase to sieve plates of phloem. Antiserum was raised against an M_r 70000 subunit of callose synthase (B. McCormack and G. P. Bolwell, unpublished data). Immunolocalization is shown to the callose-rich sieve plate of a phloem cell (photograph, C. Smith and G. P. Bolwell). Note entrapment of the callose synthase (arrowed) in the callose surrounding the plate. Mag. X20000. sp = sieve plate; sw = secondary wall. Scale bar 0.5 μm.

Bolwell, unpublished data) and we are presently examining possible induction of callose synthase in cytokinin-treated bean cells similar to that seen in elicitor-treated cells (Robertson et al., 1995b).

4 ANALYSIS OF TRANSCRIPTIONAL REGULATION OF VASCULAR GENE EXPRESSION

The most studied gene system related to vascular differentiation is that for PAL. It is interesting to compare induction of this enzyme with its induction in elicitation systems. There has now been an extensive analysis of the involvement of the PAL promoter in conferring vascular-specific expression (Bevan et al., 1989; Liang et al., 1989; Lois et al., 1989; Leyva et al., 1992) in addition to stress related expression. The regulatory box structure is complex and contains defined stress and environmental responsive elements such as G-boxes, H-boxes and a P-box. In contrast, the sequences involved in conferring vascular expression are less well understood. A study has been made of the promoter sequences that confer vascular expression in gPAL2 of French bean (a second PAL gene gPAL3 showed additional expression patterns). Deletion analysis indicated that vascular expression was dependent on combinatorial negative and positive elements (Leyva et al., 1992). The negative element suppresses a cryptic *cis* element that promotes phloem expression and is found in a number of PAL genes from various sources. 4-Coumarate ligase (Hauffle et al., 1993) was also revealed to have a negative element that suppresses expression in the phloem and appeared to be responsible for restricting vascular expression to the xylem. Study of the PAL gene family in *Zinnia* is underway (Fukuda et al. 1993). Although most of these expression systems have been studied through the use of GUS fusions in transgenic plants, *a priori* it is to be expected that similar promoter structures are involved in regulation of transcriptional activity in tissue culture systems subjected to auxin and cytokinin treatment. They may have a future use in transient expression systems and in elucidating the signal transduction coupled to these changes in gene expression and identifying the transcriptional factors involved. H-box binding factors such as the *myb* family are already well characterized (Sablowski et al., 1994; Solano et al., 1995) while their exact relationship to the H-box binding factors defined as KAP-1 and KAP-2 (Yu et al., 1993) is unclear. Binding factors (BPF-1) for the P-box found in PAL and 4-coumarate ligase promoters appear to be involved in stress expression (da Costa e Silva et al., 1993), whereas G-box factors have a wide pleiotropic involvement in responses to many stimuli (Menkens et al., 1995). While these systems may participate to some cooperative extent in regulating vascular expression, factors specific for only vascular expression are yet to be identified. These may become apparent when the genes of other vascular proteins are studied in as great detail. Glycine-rich glycoproteins (Keller et al., 1988, 1989; Keller and Baumgartner, 1991) are other vascular expressed genes whose promoters have been extensively studied (Hauf et al., 1995) and share some promoter sequences in common with xylem specific PRP-like proteins in loblolly pine (Loopstra and Sederoff, 1995). The promoters of the *Eucalyptus* CAD (Feuillet et al., 1995) and CCR genes (Boudet, 1995) are also currently under investigation. However, even these may not clarify the situation as they also show some responses to stress. The answer may lie in investigating the expression of genes that only show xylem or phloem expression when these are identified.

5 SIGNALS

5.1 Extracellular signals

The most common method of inducing differentiation-related changes is to increase the ratio of cytokinin to auxin and to raise the level of sucrose in the medium. In fact, manipulation of many of these factors can induce morphological changes in cell and tissue cultures including shape, such as elongation and adhesion and clump size as in cytokinin constitutive cultures of tobacco. Tracheids can also be found spontaneously in these cultures (Fig. 1B). There is still an absolute requirement for auxin in all systems studied (reviewed by Fukuda, 1992). This requirement for auxin may be sufficient to initiate tracheary formation in the absence of cytokinin and in such systems the level of endogenous cytokinin production does not appear to be a limiting factor. Both auxin and cytokinin administration to cell cultures stimulate synthesis *de novo* of a large number of proteins (Bolwell, 1993). In cells of sycamore and *Haplopapus gracilis* low levels of auxin fed to cultures in pulses are more effective in promoting differentiation than a single addition (Kuternozinska et al., 1991). This differentiation in *Halopapus* was also dependent on mitochondrial protein synthesis (Kuternozinska and Pilipowicz, 1993).

Sucrose has a known influence on the expression of a number of genes (Jang and Sheen, 1994). These include proteinase inhibitors (Johnson and Ryan, 1990; Kim et al., 1991), chalcone synthase (Tsukaya et al., 1991), aminopeptidases and pathogenesis related proteins (Herbers et al., 1995). In some cases, a sugar response element in the promoters has been identified and a common mechanism of sugar-sensing resulting in the repression of photosynthetic genes and the activation of stress-related and pathogenesis related genes has been proposed (Jang and Sheen, 1994). The interrelatedness of the response to various stimuli is compounded when the effect of constitutive cytokinin expression in inducing stress-response genes is considered (Memelink et al., 1989) and suspension cultured calli derived from these lines develop tracheids (Fig. 1B). Perturbation of sugar metabolism by viruses or in transgenic plants is accompanied by the induction of stress-response genes (Herbers et al., 1995). In this study on the effect of the accumulation of soluble sugars in cells, ACC oxidase was also induced. Carbohydrates have been reported to induce ethylene production (Philisoph-Hadas et al., 1985) and this has been investigated with respect to sucrose concentration in the lettuce pith system (Warren-Wilson et al., 1994a). Ethylene has been implicated in vascular differentiation in a number of studies (Fukuda, 1992). The ethylene pathway of signal transduction is now one of the best understood especially in relation to stress (Ecker, 1995). It is striking how those genes classically turned on by wounding, β 1-3 glucanase, PR proteins, chitinase and HRGPs are also induced by the expression of the Tcyt gene in transgenic tobacco leading to high endogenous levels of cytokinin (Memelink et al., 1989). Lignification genes are also stimulated by stress (Bolwell, 1993) and, for example, a major anti-fungal hybrid chitin binding protein of French bean is expressed in developing xylem at the plasmalemma wall interface (Millar et al., 1992). The analogy between the wounding response and xylogenesis has often been raised. It may well be that some of the transduction components first identified in the ethylene response have pleiotropic effects and also participate in xylogenesis.

Other low molecular mass compounds have also been implicated as signals working in the developmental pathway. Experiments with the *Zinnia* system using conditioned medium on the cessation of cell expansion before the differentiation phase suggest an involvement of oligosaccharins (A.W. Roberts and C. H. Haigler, personal communication). Oligosaccharins have been implicated in modulating growth patterns in a large number of systems (Aldington and Fry, 1993). Most have been characterized as regulators of expansion growth while others affect development including floral initiation. However, oligogalacturonides induce stress-related lignification (Bruce and West, 1989) while xylanase-treatment induces ethylene production (Fuchs et al., 1989). Transformed tobacco cells undergoing tracheid formation (Fig 1B) have an extractable xylanase present in their cell walls which is absent from control cultures and could be involved in generating such morphogenic signals (G.P. Mitchell and G.P. Bolwell, unpublished data). Experiments involving the use of conditioned medium augurs well for the identification of modulatory signals in vascular differentiation. In comparison, characterization of these would be extremely difficult in whole plants.

Experiments using conditioned medium have been particularly successful in studies of somatic embryogenesis and have indicated an involvement for wall-derived signals in the initiation of formation of other cell types related to their relative position. Nevertheless, some of these signals will influence vascular differentiation (De Vries et al., 1994). There is some evidence that *nod*-like factors may be involved. This was based on the observation that addition of an M_r 32,000 chitinase could release globular embryos arrested in development and that this could also be mimicked by the addition of Rhizobial *nod*-factors (De Jong et al., 1992, 1993). More recent work suggests that putative endogenous *nod*-factors may affect the balance between the action of auxin and cytokinin (Schmidt et al., 1994) and may thus influence vascular differentiation also. An enhanced production of chitinase is required (De Jong et al., 1995) and this is a feature of transformed cells expressing high endogenous cytokinin levels (Memelink at al., 1989; G. P. Mitchell and G. P. Bolwell, unpublished data). Certainly, the occurrence of potential substrates for chitinases is highest in secondary walls (Benhamou and Asselin, 1989; Wojtaszek and Bolwell, 1995).

A number of other signals such as brassinolides have also been implicated (Fukuda, 1992). Since phenolic conjugates have also been considered signalling molecules (Bolwell, 1993; Dixon and Paiva, 1995), it is probably worth investigating further since elaborate mechanisms exist for their synthesis and hydrolysis (Whetten and Sederoff, 1995). They may prove to be effective in modulating single cell vascular differentiation since some circumstantial evidence exists for the continued synthesis and transport of lignin precursors from xylem parenchyma to differentiated xylem (Bevan et al., 1989; Smith et al., 1994; Whetten and Sederoff, 1995).

5.2 Intracellular signals

Little is known of the initial signal transduction events that trigger responses to cytokinin, although many of the target genes for regulation are known. Identification of transcription factors that are *specific* to vascular differentiation and their isolation may allow the use of newer technology such as the two hybrid system (Allen et al., 1995) which may be used to delineate the signal transduction chain. Use of signal transduction modulators indicate a role for Ca^{2+}/calmodulin in the differentiation process (Roberts and Haigler, 1990; Kobayashi and Fukuda, 1994). This signalling system also has a role in responses to light which includes regulation of the genes of phenolic metabolism (Bowler and Chua, 1994; Menkens et al., 1995). Use of calcium antagonists also indicate the requirement for calcium fluxes in differentiation in hypocotyls of French bean (Soumelidou et al., 1994). There is little evidence for other signalling systems at present and indeed for some systems such as cAMP there is still a debate whether it has a significant role in plants at all (Bolwell, 1995). However, methylxanthines, which could potentially target a cAMP signalling system, were only effective in modulating differentiation before the involvement of the Ca^{2+} signal and acted on premature calcium mobilization (Roberts and Haigler, 1992). In further studies use of cAMP analogues were interpreted as acting as cytokinins, however, since this work was carried out, there has accumulated evidence that cAMP may be involved in opening ion channels so it could have a role in cross talk (Bolwell, 1995). Phosphoinositides are also probably an area of useful further investigation due to their relationship with calcium signalling and cross-talk between the systems.

6 CONCLUSIONS

One of us (GPB) last reviewed this field in 1985 when the application of recombinant DNA technology was first being made to systems of vascular differentiation. Like many areas of plant biology there has been astonishing progress in identifying vascular specific genes, the lignification pathway is known in its entirety, a number of wall specific proteins have been identified and cloned and progress is being made on the regulation of polysaccharide biosynthesis. In this latter case, once one glycosyl transferase has been cloned it will probably open up the identification of others by PCR using primers to conserved sequences, since a similar phenomenon happened with plant cytochrome bs such as peroxidases and cytochrome P450s (Bolwell et al., 1994) once the first cDNAs were isolated. Once the genes are cloned then promoter analysis can identify the transcription factors involved and allow elucidation of signal transduction pathways coupled with biochemical studies using modulators of signalling components. Tissue culture systems will continue to contribute to this understanding since they allow modelling aspects that are not easily accessible in the intact plant. Coupled with studies of developmental mutants it is anticipated that vascular differentiation will continue to be one of the better understood developmental processes in plants.

7 ACKNOWLEDGEMENTS

The authors are indebted to Hiroo Fukuda, Candace Haigler, Keith Roberts, Zheng-Hua Ye, Alison Roberts, Eric Beers, Kazuhiko Nishitani, Nick Carpita, Dianne Church and Alan Jones for providing reprints/preprints and their thoughts in the preparation of this manuscript.

8 REFERENCES

Aldington S & Fry SC (1993) Oligosaccharins. Adv. Bot. Res. 19: 1-10.

Allen JB, Walberg MB, Edwards MC & Ellege SJ (1995) Finding prospective partners in the library: the two-hybrid system and phage display find a match. TIBS 20: 511-516.

Amor Y, Haigler CH, Johnson S, Wainscott M & Delmer DP (1995) A membrane-associated form of sucrose synthase and its potential role in synthesis of cellulose and callose in plants. Proc. Natl. Acad. Sci. USA 92: 9353-9357.

Bachmair A, Becker F, Masterson RV & Schell J (1990) Perturbation of the ubiquitin system causes leaf curling, vascular tissue alteration and necrotic lesions in a higher plant. EMBO J. 9: 4543-4549.

Baldwin T, McCann TC & Roberts K (1993) A novel hydroxyproline-deficient arabinogalactan protein secreted by suspension cultured cells of Daucus carota. Plant Physiol. 103: 115-123.

Bao W, O'Malley DM & Sederoff RR (1992) Wood contains a cell-wall-structural protein. Proc. Natl. Acad. Sci. USA 89: 6604-6608.

Bao W, Whetten R, O'Malley D & Sederoff, R. (1993) A laccase associated with lignification in loblolly pine xylem. Science 260: 672-674

Bauw G, De Loose M, Inze D, Van Montagu M & Vanderkerchove J (1987) Alterations in the phenotype of plant cells studied by NH_2-terminal amino acid sequence analysis of proteins electroblotted from two dimensional gel-separated total extracts. Proc. Natl. Acad. Sci. USA 84: 4806-4810

Benhamou N & Asselin A (1989) Attempted localization of a substrate for chitinases in plant cells reveals abundant N-acetyl-D-glucosamine residues in secondary walls. Biol. Cell 67: 341-350.

Bevan M, Shufflebottom D, Edwards K, Jefferson R & Schuch W (1989) Tissue- and cell-specific activity of a phenylalanine ammonia lyase promoter in transgenic plants. EMBO J. 8:899-1906.

Bolwell GP (1995) Cyclic AMP, the reluctant messenger in pants. Trends Biochem. Sci. 20: 492-495.

Bolwell GP & Northcote DH (1983a) Arabinan synthase and xylan synthase activities of *Phaseolus vulgaris*; subcellular localization and possible mechanism of action. Biochem. J. 210:497-507.

Bolwell GP & Northcote DH (1983b) Induction by growth factors of polysaccharide synthases in bean cell suspension cultures. Biochem. J. 210: 508-515.

Bolwell GP (1985) Use of tissue cultures for studies on vascular differentiation :In Dixon RA (ed.) Plant Tissue Culture: a practical approach. 1st edition. (pp.107-126), IRL Press Oxford, Washington.

Bolwell GP (1993) Dynamic aspects of the plant extracellular matrix. Int. Rev. Cytol. 146: 261-324.

Bolwell GP, Bozak KR & Zimmerlin A (1994) Plant Cytochrome P450. Phytochemistry 37: 1491-1506.

Bolwell GP, Butt VS, Davies DR & Zimmerlin A (1995) The origin of the oxidative burst in plants. Free Rad. Res. 23: 517-532.

Bolwell GP, Dalessandro G & Northcote DH (1985) Loss of polygalacturonic acid synthase during xylem differentiation of the vascular cambium of sycamore. Phytochemistry 24: 699-702.

Bolwell GP, Rodgers MW (1991) L-Phenylalanine ammonia-lyase from French bean; characterization and differential expression of antigenic multiple M_r forms. Biochem. J. 279: 231-236.

Boudet AM (1995) Lignification genes in plants. 7th Cell Meeting. Santiago. Spain. Aug 25-29,1995.
Bowler C & Chua N-H (1994) Emerging themes of plant signal transduction. Plant Cell 6: 1529-1541.
Bradley DJ, Kjellbom P & Lamb CJ (1992) Elicitor- and wound-Induced oxidative cross-linking of a proline-rich cell wall protein: a novel rapid defense response. Cell 70: 21-30.
Bruce RJ, West CA (1989) Elicitation of lignin biosynthesis and isoperoxidase activity by pectic fragments in suspension cultures of castor bean. Plant Physiol. 91: 889-897.
Carpita NC & Gibeaut DM (1993) Structural models of primary cell walls in flowering plants: consistency of molecular structure with the physical properties of walls during growth. Plant J. 3: 1-30.
Church DL & Galston AW (1988) 4-coumarate:coenzyme A ligase and isoperoxidase expression in Zinnia mesophyll cells induced to differentiate into tracheary elements. Plant Physiol. 88: 92-96.
Church DL & Galston AW (1989) Hormonal induction of vascular differentiation in cultured *Zinnia* leaf disks. Plant Cell Physiol. 30: 73-78
Church DL (1993) Tracheary element differentiation in Zinnia mesophyll cell-cultures. Plant Growth Regul. 12: 179-188.
Cosgrove DJ & Li Z-C (1993) Role of expansin in cell enlargement of oat coleoptiles. Plant Physiol. 103: 1321-1328.
da Costa e Silva O, Klein L, Schmeltzer E, Trezzini GF & Hahlbrock K (1993) BPF-1, a pathogen-induced DNA-binding protein involved in the plant defense response. Plant J. 4: 125-135.
Dalessandro G, Piro G & Northcote DH (1986) Glucomannan synthase activity in differentiating cell of *Pinus sylvestris* L. Planta 169: 564-574.
De Jong AJ, Cordewener L, Lo Schiavo F, Terzi M, Vanderkerkove J, Van Kammen A & De Vries SC (1992) A carrot somatic embryo mutant is rescued by chitinase. Plant Cell 4: 425-433.
De Jong AJ, Heidstra R, Spaink HP, Hartog MV, Meijer EA, Hendricks T, Lo Schiavo F, Terzi M, Bisseling T, Van Kammen A & De Vries SC (1993) *Rhizobium* lipooligosaccharides rescue a carrot somatic embryo mutant. Plant Cell 5: 615-620.
De Jong AJ, Hendriks T, Meijer EA, Penning M, Lo Schiavo F, Terzi M, van Kammen A & De Vries S (1995) Transient reduction in secreted 32kD chitinase prevents somatic embryogenesis in the carrot (*Daucus carota* L.) variant ts11. Develop. Genet. 16: 332-343
De Vries SC, de Jong AKJ & van Engelen FA (1994) Formation of embryogenic cells in plant cell tissue culture. Biochem. Soc. Symp. 60: 43-50
Dean JFD, LaFayette PR, Tristram AH, Hoopes JT & Eriksson K-EL (1995) Laccase involvement in lignification. Abs 7th Cell Wall Meeting. Santiago. Spain. Aug 25-29, 1995.
Demura T & Fukuda H (1993) Molecular cloning and characterization of cDNAs associated with tracheary element differentiation in cultured Zinnia cells. Plant Physiol. 103: 815-821.
Demura T & Fukuda H (1994) Novel vascular cell-specific genes whose expression is regulated temporally and spatially during vascular system development. Plant Cell 6: 967-981.
Dixon RA, Paiva NL (1995) Stress-induced phenylpropanoid metabolism. Plant Cell 7: 1085-1097.
Driouich A, Laine A-C, Vian B & Faye L (1992) Characterization and localization of laccase forms in stem and cell cultures of sycamore. Plant J. 2: 13-24.
Ecker J (1995) The ethylene signal transduction pathway in plants. Science 268: 667-675.

Fahrendorf T & Dixon RA (1993) Stress responses in alfalfa (*Medicago sativa* L.) XVIII Molecular cloning and expression of the elicitor-inducible cinnamic acid 4-hydroxylase cytochrome P450. Arch. Biochem. Biophys. 305:509-516.

Feuillet C, Lauvergeat V, Deswarte C, Pilate G, Boude A-M, Van Montagu M & Inze D (1995) Tissue-and cell-specific exppression of a cinnamyl alcohol dehydrogenase in transgenic poplar plants, Plant Mol. Biol. 27: 651-667

Fisher DB, Wu Y & Ku SB (1992) Turnover of soluble proteins in the wheat sieve tube. Plant Physiol. 100: 1433-1441.

Fry SC (1995) Polysaccharide-modifying enzymes in the plant cell wall. Annu Rev. Plant Physiol. Plant Mol. Biol. 46: 497-520

Fry SC, Smith RC, Renwick KF, Martin DJ, Hodge SK & Matthews KJ (1992) Xyloglucan endotransglycosylase, a new wall-loosening enzyme activity from plants. Biochem. J. 282: 821-828.

Fuchs Y, Saxena A, Gamble HR, Anderson JD (1989) Ethylene biosynthesis inducing protein from Cellulysin is an endoxylanse. Plant Physiol. 89: 138-143.

Fukuda H (1992) Tracheary element formation as a model system of cell differentiation. Int. Rev. Cytol. 136: 289-332.

Fukuda H (1994) Redifferentiation of single mesophyll cells into tracheary elements. Int. J. Plant Sci. 155: 262-271.

Fukuda H, Sato Y, Yoshimura T & Demura, T (1993) Molecular mechanisms of xylem differentiation. J. Plant Res. 3: 97-107.

Gibeaut DM & Carpita NC (1994) Biosynthesis of plant cell wall polysaccharides. FASEB J 8: 904-915.

Goffner D, Campbell MM, Campargue C, Clastre M, Borderies G, Boudet A & Boudet A-M (1994) Purification and characterization of cinnamoyl-coenzyme A: NADP oxidoreductase in *Eucalyptus gunii*. Plant Physiol. 106: 625-632.

Grima-Pettenati J, Campargue C, Boudet A & Boudet A-M(1994) Purification and characterization of cinnamyl alcohol dehydrogenase isoforms from *Phaseolus vulgaris*. Phytochemistry 37: 941-947.

Grima-Pettenati J, Feuillet C, Goffner CD, Borderies G & Boudet A-M (1993) Molecular cloning and expression of a *Eucalyptus gunii* cDNA clone encoding cinnamyl alcohol dehydrogenase. Plant Mol. Biol. 21: 1085-1095.

Halpin C, Knight ME, Foxon GA, Campbell MM, Boudet AM, Boon JJ, Chabbert B, Tollier M-T & Schuch W (1994) Manipulation of lignin quality by down regulation of cinnamyl alcohol dehyrogenase. Plant J. 6: 339-350.

Hauf G, Ringli C & Keller B (1995) Purification and biochemical characterization of glycine-rich proteins and analysis of gene regulation. 7th Cell Meeting. Santiago. Spain. Aug 25-29, 1995.

Hauffle KD, Lee SP, Subramaniam R & Douglas CJ (1993) Combinatorial interactions between positive and negative cis-acting elements control spatial patterns of 4CL expression in transgenic tobacco. Plant J. 4: 235-253.

Herbers K, Monke G, Badur R & Sonnewald U (1995) A simplified procedure for the subtractive cDNA cloning of photoassimilate-responding genes: isolation of cDNAs encoding a new class of pathogenesis-related proteins. Plant Mol. Biol. 29: 1027-1038.

Hetherington PR & Fry SC (1993) Xyloglucan endotransglycosyase activity in carrot cell suspensions during cell elongation and somatic embryogenesis. Plant Physiol. 103: 987-992.

Ingold E, Munetaka S & Komamine A (1988) Secondary cell wall formation: changes in cell wall constituents during the differentiation of isolated mesophyll cells of *Zinnia elegans* to tracheary elements. Plant Cell Physiol. 29: 295-303.

Jang JC & Sheen J (1994) Sugar sensing in higher plants. Plant Cell 6: 1665-1679.
Johnson R & Ryan CA (1990) Wound-inducible potato inhibitor II genes: enhancement of expression by sucrose. Plant Mol. Biol. 14: 527-536.
Keller B & Baumgartner C (1991) Vascular-specific expression of the bean GRP 1.8 gene is negatively regulated. Plant Cell 3: 1051-1061.
Keller B (1993) Structural cell wall proteins. Plant Physiol. 101: 1127-1130.
Keller B, Sauer N & Lamb, CJ (1988) Glycine-rich cell wall proteins in bean: gene structure and association of the protein with the vascular system. EMBO J. 7: 3625-3633.
Keller B, Templeton TD & Lamb CJ (1989) Specific localization of a plant cell wall glycine-rich protein in protoxylem cells of the vascular tissue. Proc. Natl. Acad. Sci. USA 86: 1529-1533.
Kim ER, Costa MA & An G (1991) Sugar response element enhances wound response of potato proteinase inhibitor II promoter in transgenic tobacco. Plant Mol. Biol. 17: 973-983.
Kobayashi H & Fukuda H (1994) Involvement of calmodulin and calmodulin-binding proteins in the differentiation of tracheary elements in Zinnia cells. Planta 194: 388-394.
Kobayashi H, Fukuda, H & Shibaoka H (1988) Interrelationship between the spatial disposition of actin filaments and microtubules during differentiation of tracheary elements in cultured *Zinnia* cells. Protoplasma 143: 29-37.
Kreuger M & van Holst G-J (1993) Arabinogalactan proteins are essential in somatic embryogenesis of *Daucus carota* L. Planta 189: 243-248.
Kuternozinska W & Pilipowicz M (1993) The effect of chloramphenicol on the growth and xylogenesis in callus of *Haplopappus gracilis*. Biol. Plant. 35: 307-309.
Kuternozinska W, Pilipowicz M & Korohoda W (1991) Transitory pulse -like treatment with IAA solution effectively induces xylogenesis in callus-culture than permanent presence of the auxin. Biol. Plant. 33: 433-438.
Leyva A, Liang X, Pinto-Toro JA, Dixon RA & Lamb CJ (1992) *Cis*-Element combinations determine phenylalanine ammonia-lyase gene tissue specific expression patterns. Plant Cell 4: 263-271.
Li Z-C, Durachko DJ & Cosgrove DJ (1993) An oat coleoptile wall protein that induces wall extension in vitro and that is antigenically related to a similar protein from cucumber hypocotyls. Planta 191: 349-356.
Liang X, Dron M, Cramer CL, Dixon RA & Lamb CJ (1989) Developmental and environmental regulation of a PAL-β-glucuronidase gene fusion in transgenic tobacco plants. Proc Natl. Acad Sci USA 86: 9284-9288.
Lin Q & Northcote DH (1990) Expression of PAL gene during tracheary-element differentiation from cultured mesophyll cells of *Zinnia elegans* L. Planta 182: 591-598.
Liu L, Dean JFD, Friedman WE & Eriksson KEL (1993) A laccase-like activity is correlated with lignin biosynthesis in *Zinnia elegans* stem tissues. Plant J. 6: 213-224.
Lois R, Dietrich A, Hahlbrock K & Schultz W (1989). A phenylalanine ammonia-lyase gene from parsley: structure regulation and identification of elicitor-and light-responsive *cis*-elements. EMBO J. 8: 1641-1648.
Loopstra CA & Sederoff RR (1995) Xylem-specific gene expression in Loblolly pine. Plant Mol. Biol. 27: 277-291.
Maclachlan G, Carrington S (1991) Plant cellulases and their role in plant development. In: CH Haigler & PJ Weimer (eds) Biosynthesis and Biodegradation of Cellulose. (pp. 599-621) Dekker, New York.
Masuda H, Ozeki Y, Amino S & Komamine A (1984) Changes in cell wall polysaccharides during elongation in a 2,4 D free medium in a carrot suspension culture. Physiol. Plant 62: 65-72.

McQueen-Mason S, Drachko DM & Cosgrove DJ (1992) Two endogenous proteins that induce cell wall extension in plants. Plant Cell 4: 1425-1433.

McQueen-Mason SJ, Fry SC, Drachko DM & Cosgrove DJ (1993) The relationship between xyloglucan endotransglycosylase and *in-Vitro* cell wall extension in cucumber hypocotyls. Planta 190: 327-331

Memelink J, Hoge JHC and Schilperoort R (1989) Cytokinin stress changes the developmental regulation of several defense-related genes in tobacco EMBO J. 6: 3579-3583.

Menkens AE, Schindler U & Cashmore AR (1995) The G-box: a ubiquitous regulatory element in plants bound by the GBF family of bZIP proteins. Trends Biochem. Sci. 20: 506-510.

Meyer K, Cusumano J, Somerville C & Chapple C.(1995) Cloning of a cytochrome P-450 dependent monoxygenase required for syringyl lignin biosynthesis in Arabidopsis. 7th Cell Meeting. Santiago. Spain. Aug 25-29, 1995.

Millar DJ, Allen AK, Sidebottom C, Smith CG, Slabas AR & Bolwell GP (1992) A major stress-inducible M_r 42,000 wall glycoprotein of French bean (*Phaseolus vulgaris* L.) Planta 187: 176-184.

Minami A & Fukuda H (1995) Transient and specific expression of a cysteine endopeptidase associated with autolysis during differentiation of *Zinnia* mesophyll cells into tracheary elements. Plant Cell Physiol. 36: 1599-1606.

Mittler R & Lam E (1995) In situ detection of nDNA fragmentation during the differentiation of tracheary elements in higher plants. Plant Physiol. 108: 489-493.

Nagai N, Kitauchi F, Okamoto K, Kanda T, Shimosaka M & Okazaki M (1994) A Transient increase of phenylalanine ammonia-lyase transcript in kinetin-treated tobacco callus. Biosci. Biotech. Biochem. 58: 558-559.

Nishitani K (1995) Endo-xyloglucan transferase, a new class of transferase involved in cell wall construction. J. Plant Res. 108: 137-148.

Northcote DH (1995) Aspects of vascular differentiation in plants: parameters that may be used to monitor the process. Int. J. Plant Sci. 156: 245-256.

Philosoph-Hadas S, Meir S & Aharoni N (1985) Carbohydrates stimulate ethylene in tobacco leaf discs. Plant Physiol. 78: 139-143.

Potter I & Fry SC (1994) Changes in xyloglucan endotransglycosylase (XET) activity during hormone-induced growth in lettuce and cucumber hypocotyls and spinach suspension cultures. J. Exp. Bot. 45: 1703-1710.

Roberts AW & Haigler CH (1990) Tracheary-element differentiation in suspension cultures of Zinnia requires uptake of extracellular Ca^{2+}. Experiments with calcium-channel blockers and calmodulin inhibitors. Planta 180: 502-509.

Roberts AW & Haigler CH (1992) Methylxanthines reveribly inhibit tracheary-element differentiation in suspension-cultures of *Zinnia elegans* L. Planta 186: 586-592.

Roberts AW & Haigler CH (1994) Cell expansion and tracheary element differentiation are regulated by extracellular pH in mesophyll cultures of *Zinnia elegans* L. Plant Physiol. 105:699-706.

Robertson D, Beech I & Bolwell GP (1995a) Regulation of the enzymes of UDP-sugar metabolism during differentiation of French bean (*Phaseolus vulgaris* L.). Phytochemistry 39:21-28.

Robertson D, MacCormack B & Bolwell GP (1995b) Cell wall polysaccharide biosynthesis and related metabolism in elicitor-stressed cells of French bean (*Phaseolus vulgaris* L.). Biochem. J. 306: 745-750.

Robertson D, Smith CG & Bolwell GP (1996) Inducible UDP-glucose dehydrogenase from French bean (*Phaseolus vulgaris* L.) locates to vascular tissue and has alcohol dehydrogenase activity. Biochem. J. 313: 311-317.

Rodgers MW & Bolwell GP (1992) Partial purification of arabinosyl-transferase and two xylosyl-transferases from French bean. Biochem. J. 288: 817-822.

Sablowski RWM, Moyano E, Culianez FA, Schuch W, Martin C & Bevan M (1994) A flower-specific Myb protein activates transcription of phenylpropanoid biosynthetic genes. EMBO J. 13: 128-137.

Sachs, T. (1991) Pattern formation in plant tissues. Cambridge University Press. Cambridge

Sakuth T, Schobert C, Pecsvaradi A, Eichholz A, Komor E & Orlich G (1993) Specific proteins in the sieve tube exudate of *Ricinus communis* L. seedlings: separation, characterization and *in vivo* labelling. Planta 191: 207-213.

Sato Y, Sugiyama M, Gorecki RJ, Komamine A & Fukuda H (1993) Interrelationship between lignin deposition and the activities of peroxidase isoenzymes in differentiating tracheary elements of *Zinnia*. Planta 189: 584-589.

Sato Y, Sugiyama M, Komamine A & Fukuda H (1995) Separation and characterization of the isoenzymes of wall-bound peroxidase from cultured Zinnia cells during tracheary element differentiation. Planta 196: 141-147.

Schmidt EDL, de Jong AJ & de Vries SC (1994) Signal molecules involved in plant embryogenesis. Plant Mol. Biol. 26: 1305-1313.

Sederoff R, Campbell M, O'Malley D & Whetten R (1994) Genetic regulation of lignin biosynthesis and the potential modification of wood by genetic engineering in Loblolly pine. In: Ellis BE et al. (eds) Genetic Engineering of Plant Secondary Metabolism. (pp. 313-355) Plenum, New York.

Showalter AM (1993) Structure and function of plant cell wall proteins. Plant Cell 5: 9-23

Sjolund RD, Shih CY & Jensen KG (1993) Freeze fracture analysis of phloem structure in plant tissue culture. III P-protein, sieve area pores and wounding. J. Ultrastructure Res. 82: 198-211.

Smith CG, Rodgers MW, Zimmerlin A, Berdinando D & Bolwell GP (1994) Tissue and subcellular immunolocalization of enzymes of lignin synthesis in differentiating and wounded hypocotyl tissue of French bean (*Phaseolus vulgaris* L.). Planta 192: 155-164.

Solano R, Nieto C, Avila J, Canas L, Diaz I & Pax-Ares J (1995) Dual DNA binding specificity of a petal epidermis-specific MYB transcription factor (MYB.Ph3) from *Petunia hybrida*. EMBO J. 14: 1773-1784.

Soumelidou K, Li H, Barnett JR, John P & Battey NH (1994) The effect of auxin and calcium-antagonists on tracheary element differentiation in *Phaseolus vulgaris* L. J. Plant Physiol. 143:717-721.

Stacey NJ, Roberts K, Carpita NC, Wells B & McCann MC (1995) Dynamic changes in cell surface molecules are very early events in the differentiation of mesophyll cells from *Zinnia elegans* into tracheary elements. Plant J. 8: 891-906.

Stephenson P, Collins BA, Reid PD & Rubinstein B (1996) Localization of ubiquitin to differentiating vascular tissues. Am. J. Bot. 83: 140-147.

Sudibyo AS & Anderson JW (1993) Xylogenesis and phenylpropanoid metabolism in cultured *Zinnia* mesophyll cells. Phytochemistry. 34: 1245-1250.

Sugiyama M, Fukuda H & Komamine A (1994) Characterization of the inhibitory effect of aphidicolin on transdifferentiation into tracheary elements of isolated mesophyll cells of *Zinnia elegans*. Plant Cell Physiol. 35: 519-522.

Suzuki K, Amino S, Takeuchi Y & Komamine A (1990) Differences in the composition of the cell walls of two morphologically different lines of suspension-cultured *Catharanthus roseus* cells. Plant Cell Physiol. 31: 7-14.

Suzuki K, Ingold E, Sugiyama M & Komamine A (1991) Xylan synthase activity in isolated mesophyll cells of Zinnia elegans during differentiation to tracheary elements. Plant Cell Physiol. 32: 303-306.

Suzuki K, Ingold E, Sugiyama N, Fukuda H & Komamine A (1992) Effects of 2,6-dichlorobenzonitrile on differentiation to tracheary elements of isolated mesophyll cells of *Zinnia elegans* and formation of secondary cell walls. Physiol. Plant. 50: 43-48.

Taylor JG & Haigler CH (1993) Patterned secondary cell-wall assembly in tracheary elements occurs in a self perpetuating cascade. Acta. Bot. Neerland. 42: 153-163.

Taylor JG, Owen TP, Koonce LT & Haigler CH (1992) Dispersed lignin in tracheary elements treated with cellulose synthesis inhibitors provides evidence that molecules of the secondary-wall mediate wall patterning. Plant J. 2: 959-970.

Thelen MP & Northcote DH (1989) Identification and purification of a nuclease from *Zinnia elegans* L: a potential molecular marker for xylogenesis. Planta 179: 181-195.

Toth KF, Wang Q & Sjolund RD (1994) Monoclonal antibodies against phloem P-protein from plant tissue cultures. 1. Microscopy and biochemical analysis. Am. J. Bot. 81: 1370-1377.

Tsukaya H, Oshima T, Naito S, Chino M & Komeda Y (1991) Sugar-dependent expression of the CHS-A gene for chalcone synthase from petunia in transgenic *Arabidopsis* Plant Physiol. 97: 1414-1421.

Van Drosselaere J, Baucher M, Feuillet C, Boudet A-M, Van Montagu M & Inze D (1995) Isolation of cinnamyl alcohol dehydrogenase cDNAs from two important economic species: alfalfa and poplar. Demonstration of high homology of the gene within angiosperms. Plant Physiol. Biochem. 33: 105-109.

Voo KS, Whetten RW, O'Malley DM & Sederoff RR (1995) 4-Coumarate-Coenzyme A ligase from Loblolly pine xylem: isolation, characterization and complementary-DNA cloning. Plant Physiol. 108: 85-97.

Walter M, Schaaf J & Hess D (1994) Gene activation in lignin biosynthesis: patterns of promoter activity of a tobacco cinnamyl-alcohol dehydrogenase gene. Acta Hortic. 381: 162-168.

Wang Q, Monroe J & Sjolund RD (1995) Identification and characterization of a phloem-specific ß-amylase. Plant Physiol. 109: 743-750.

Warren-Wilson J & Warren-Wilson PM (1993) Mechanisms of auxin regulation of structural and physiological polarity in plants, tissues, cells and embryos. Aust. J. Plant Physiol. 20: 555-571.

Warren-Wilson J, Keys WMS, Warren-Wilson PM & Roberts LW (1994b) Effects of auxin on the spatial distribution of cell division and xylogenesis in lettuce pith explants. Protoplasma 183: 162-181.

Warren-Wilson J, Roberts LW, Gresshoff PM & Dircks, SJ (1982) Tracheary element differentiation induced in isolated cylinders of lettuce pith: a bipolar gradient technique. Ann. Bot. 50: 605-614.

Warren-Wilson J, Roberts LW, Warren-Wilson PM & Gresshoff PM (1994a) Stimulatory and inhibitory effects of sucrose concentration on xylogenesis in lettuce pith explants: possible mediation by ethylene biosynthesis. Ann. Bot. 73: 65-73.

Warren-Wilson J, Warren-Wilson, PM & Walker, ES (1991) Patterns of tracheary differentiation in lettuce pith explants: positional control and temperature effects. Ann. Bot. 68: 109-128.

Whetten R & Sederoff R (1995) Lignin biosynthesis. Plant Cell 7: 1001-1013.

Wojtaszek, P & Bolwell GP (1995) Secondary cell-wall-specific glycoprotein(s) from French bean hypocotyls. Plant Physiol. 108: 1001-1012.

Xu W, Purugganan MM, Antosiewicz DM, Polisensky DH & Fry, SC (1995) *Arabidopsis* TCH4, encoding an XET that modifies plant cell wall xyloglucans, shows diverse regulation of expression. Plant Cell 12: 1555-1567.

Ye ZH & Droste, D. L. (1996) Isolation and characterization of cDNAs encoding xylogenesis-associated and wounding-induced ribonucleases in *Zinnia elegans* . Plant Mol. Biol. 30: 697-709.

Ye ZH & Varner JE (1991) Tissue-specific expression of cell wall proteins in developing soybean tissues. Plant Cell 6: 1427-1439.

Ye ZH & Varner JE (1993) Gene-expression patterns associated with in-vitro tracheary element formation in isolated single mesophyll cells of *Zinnia elegans*. Plant Physiol. 103: 805-813.

Ye ZH & Varner JE (1994) Expression of an auxin- and cytokinin-regulated gene in cambial region in *Zinnia* . Proc. Natl. Acad. Sci. USA 91: 6539-6543.

Ye ZH & Varner JE (1995) Differential expression of two O-methyltransferases in lignin biosynthesis in *Zinnia elegans*. Plant Physiol. 108: 459-467.

Ye ZH & Varner JE (1996) Induction of cysteine and serine proteases during xylogenesis in *Zinnia elegans* . Plant Mol. Biol. 30: 1233-1246.

Ye ZH, Kneusel, RE, Matern U & Varner JE (1994) An alternative methylation pathway in lignin biosynthesis in *Zinnia* . Plant Cell 6: 1427-1439.

Yu LM, Lamb CJ & Dixon RA (1993) Purification and biochemical characterization of proteins which bind to the H-box *cis*-elements implicated in transcriptional activation of plant defense genes. Plant J. 3: 805-816.

Zimmerlin A, Wojtaszek P & Bolwell GP (1994) Synthesis of dehydrogenation polymers of ferulic acid by a purified cell-wall peroxidase from French bean (*Phaseolus vulgaris* L.). Biochem. J. 299: 747-753.

Chapter 2

PLANT REGENERATION FROM CULTURED PROTOPLASTS

Zhi-Hong XU and Hong-Wei XUE
National Laboratory of Plant Molecular Genetics, Shanghai Institute of Plant Physiology, Chinese Academy of Sciences, Shanghai 200032, China

1 INTRODUCTION

The protoplasts isolated from various plant species, organs or tissues, can regenerate cell wall, divide and form cell colonies and calli, followed by plant regeneration via different pathway of morphogenesis, under proper culture conditions. Since the first report of plant regeneration from mesophyll protoplasts of tobacco (Nagata and Takebe, 1971; Nitsch and Ohyama, 1971), the number of species with successful plant regeneration from protoplasts reached about 100 by mid 1980's (Binding, 1985; Maheshwari et al.,1986). According to Roest and Gilissen (1989), this number increased to 212 during the next 4-5 years. Unfortunately, these authors missed several Chinese contributions which appeared in Chineses language (see Loo and Xu, 1986; Xia, 1992). According to our recent survey (Xu and Xue, 1995), there are 368 species or subspecies of higher plants, covering 161 genera of 46 families, for which plant regeneration from protoplasts has been achieved. Among them, the Solanaceous members appear to be most responsive followed, in the declining order, by Leguminosae, Gramineae, Compositae, Cruciferae, Umbelliferae and Rosaceae (See Table 1).

Surprisingly in the Orchidaceae, one of the three largest families of higher plants, there is still no convincing demonstration of plant regeneration from protoplasts, even callus formation from orchid protoplasts is quite difficult (Yasugi,1989; Liu et al., 1995).

Table 1. List of families and number of genera and species or subspecies in which plant regeneration from protoplast has reported

Family	Number		Family	Number	
	Genera	Species or subspecies		Genera	Species or subspecies
Actinidiaceae	1	3	*Linaceae*	1	8
Apocynaceae	1	1	*Magnoliaceae*	1	1
Araceae	1	1	*Malvaceae*	2	2
Asclepiadaceae	3	3	*Moraceae*	2	2
Campanulaceae	1	1	*Musaceae*	2	2
Caprifoliaceae	1	1	*Oleaceae*	1	1
Cariceae	1	1	*Passifloraceae*	1	3
Caryopyllaceae	3	9	*Pinaceae*	6	9
Chenopodiaceae	1	1	*Platanaceae*	1	1
Compositae	15	30	*Polygonaceae*	1	1
Convolvulaceae	1	2	*Ranuncluaceae*	2	3
Cruciferae	12	28	*Resedaceae*	1	5
Cucurbitaceae	1	4	*Rosaceae*	5	12
Euphorbiaceae	1	1	*Rubiaceae*	1	1
Generiaceae	1	1	*Rulaceae*	3	9
Gernmaceae	1	3	*Salicaceae*	1	8
Gramineae	19	35	*Santalaceae*	1	1
Hypericaceae	1	2	*Sapindaceae*	1	1
Iriclceae	1	1	Scrophularliaceae	5	6
Labiatae	1	1	Solanaceae	11	81
Leguminosae	23	55	Umbelliferae	13	16
Lillaceae	6	8	Urticaceae	1	1
Total 46 families, 161 genera and 368 species or subspecies					

During the past 15 years, substantial progress has been made in regenerating plants from protoplasts of cereal crops and grasses (Table 2; see also Vasil and Vasil,1992), grain legumes and forage legumes (Table 3; see also Xu and Wei,1993; Li and Xu,1995), woody plants and fruit trees (Table 4), and a number of other economic plants, such as cotton (*Gossypium hirsutum*, Chen et al.,1989), *Linum usitatissimum* (Barakat and Cocking, 1983; Ling and Binding, 1987), sunflower (Burrus et al., 1991; Fischer et al., 1992), sugar beet (Krens et al.,1990; Hall et al., 1993), and various vegetable crops (see Yarrow and Barsby,1988). These achievements have laid basis for the genetic improvement of the important crops by using protoplast system (somatic hybridization, gene transfer, etc.), and provide a valuable system for fundamental studies in plant sciences.

2 REGENERATION PROCESS IN PROTOPLAST CULTURE

The whole regeneration process from protoplast to plant can be divided into several phases: cell wall regeneration, initiation of sustained divisions cell colony and callus formation, and plant regeneration from these tissues. EM observation and staining by Calcofluor demonstrated that under optimal culture conditions cell wall regeneration occurs within 24 h. Experiments using radioactive precursors showed that cellulose synthesis might start within 10-20 min of culture (Klein et al., 1981). In early studies, some investigators emphasized that proper cell wall regeneration is a prerequisite for normal divisions of the protoplast-derived cells. However, inhibition of cell wall synthesis, by 2,6-dichlorobenzonitrile does not effect nuclear divisions thus indicates that the two processes can be separated in space and time (Meyer and Herth, 1978; Galbraith and Shields, 1982. also see.

The time required for the first cell division in protoplast cultures varies with plant species genotypes, protoplast source, procedure of protoplast isolation, protoplast viability, medium composition and culture conditions. Nagata and Takebe (1971) reported that in the culture of mesophyll protoplasts isolated from leaves of tobacco plants grown in pot in green house first division occurred after 2-5 days. This time could be shortened to 2-3 days, when the leaves were derived from aseptic shoot cultures. Generally, the protoplasts isolated from seedling tissues (hypocotyl, cotyledon and radicle), embryogenic calli, suspension cultures immature cotyledons, and shoot tips of *in vitro* grown plantlets divide more readily (within 24-48 h of culture) than those from cells. Mahaeshwari et al., 1986).

*Abbreviations used in Table 2-4 are as follows:

Protoplast source: C-Callus; Cot-Cotyledon; Cot(im)-Immature cotyledon; EC-Embryogenic callus; Epic-Epicotyl; ESC-Embryogenic suspension culture; FE-Fertilized egg; Hyp-Hypocotyl; L-Leaf; LSC-Leaves from shoot culture or aceptic seedlings; NC-Nucellar callus; R-Root; SC-Suspension cultured cells; SC(h)- Suspension cultured cells (haploid); ShT-Shoot tips; St-Stem; Sh(cul)-cultured shoots

Regeneration type: CP-Plants regenerated from calli through shoot formation; CPg-Shoot(s) from callus need to be grafted to form whole plants(s); CRP-Callus forms root(s) first, then plants in regenerated from root through bud formation; CS-Only shoot(s) form from calli; E-Somatic embryo or embryo-like structure; EP-Plant regenerated from somatic embryos; Epa-Omly albino plants regenerated from somatic embyos; Epg-Shoots from somatic embryo need to be grafted to form whole plants(s).

Table 2. List of the species of plants regenerated from protoplasts in cereals and grasses

Plant species	Protoplast Source	Regeneration type	Reference
Agrostis alba	*ESC*	*EP*	*Asano & Sugiura, 1990*
A. palustris	*ESC*	*CP*	*Terakawa et al., 1992*
Bromus inermis	*SC*	*Epa* *CP*	*Kao et al., 1973* *Power & Berry, 1979*
Dactylis glomerata	*ESC*	*EP*	*Horn et al., 1988*
Festuca arundinacea	*ESC* *ESC* *SC*	*EP/Epa* *CP/Cpa* *CP*	*Dalton, 1988* *Wang et al., 1994* *Tajanuzo et al., 1990*
F. pratensis	*ESC*	*CP/CPa*	*Wang et al., 1994*
F. rubra	*ESC*	*CP*	*Wang et al., 1995*
Haynaldia villosa	*SC*	*CE*	*Zhou A et al., 1994*
Hordeum vulgare	*ESC* *ESC* *FE*	*Cpa* *CP* *EP/Epa* *EP*	*Luhrs & Lorz, 1988* *Yan Q et al., 1990* *Jahne et al., 1991* *Holm et al., 1994, 1995*
H. murinum	*?*	*EP*	*Wang et al., 1994*
Leymus racemosus- L.giganteus	*EC*	*CP*	*Zhang et al., 1993*
Lolium multiflorum	*ESC*	*Cpa* *EP/Epa*	*Dalton, 1988* *Wang et al., 1994*
L. perenne	*SC* *ESC*	*Epa* *Cpa* *EP/Epa*	*Creemers-Molenaar et al., 1988* *Wang et al., 1994*
L. × boucheanum	*ESC*	*EP/Epa*	*Whang et al., 1995*
Oryza glumaeptura	*ESC*	*CP*	*Yan Q et al., 1995*
O. granulata	*ESC*	*CP*	*Baset et al., 1993*
O. rufipogon	*ESC*	*CP*	*Baset et al., 1991*
O.sativa subsp. Indica	*ESC*	*CP*	*Kyozuka et al., 1988* *Lee et al., 1989*
	SC	*EP*	*Datta et al., 1990* *Yin YH et al., 1993*
	SC	*CP*	*Sun et al., 1989* *Yang et al., 1992*

O. sativa subsp. Japonica	*SC*	*CP*	*Fujimura et al., 1985*
			Toriyama et al., 1986
			Yamada et al., 1986
			Lei et al., 1989
	SC	*EP*	*Abdullah et al., 1986*
			Thompson et al., 1986
	SC,C	*EP*	*Kyokuka et al., 1987*
	C	*CP*	*Coulibaly & Demarly, 1986*
			Wang & Xia, 1986, 1987
O. sativa(AA) × O. minuia(BBCC)	*C*	*CP*	*Zhang et al., 1994*
Panicum maxinum	*ESC*	*EP*	*Lu et al., 1981*
P. miliaceum	*SC*	*EP/Epa*	*Heyser, 1984*
Paspalum dilatatum	*ESC*	*EP*	*Akashi & Adachi, 1992*
Pennisetum Americanum	*ESC*	*EP*	*Vasil & Vasil, 1980*
P. purpureum	*ESC*	*EP*	*Vasil et al., 1993*
Poa pratensis	*SC*	*Csa*	*Van der Valk & Zaal, 1988*
	ESC	*EP*	*Nielsen et al., 1993*
Polypogon fugax	*ESC*	*EP*	*Chen & Xia, 1987*
Saccharum officinarum	*ESC*	*EP*	*Srinivasan & Vasil, 1985*
	SC	*CP*	*Chen et al., 1988*
	EC,SC	*CP*	*Liao et al., 1994*
	C	*EP*	*Gupta et al., 1994*
Setaria italica	*ESC*	*EP*	*Dong & Xia, 1989, 1990*
	EC	*EP*	*Ren et al., 1990*
Sorghum vulgare	*ESC*	*CP*	*Wei & Xu, 1989, 1990*
Triticum aestivum	*ESC*	*EP*	*Harris et al., 1988*
	ESC		*Vasil et al., 1990*
	EC	*CP/EP*	*Wang et al., 1989*
	FE	*EP*	*Ren et al., 1989*
			Guo et al., 1991
		EP	*Holm et al., 1995*
T. aestivum L. × Haynaldia villosa	*SC*	*CP*	*Zhang et al., 1994*
T. dvum	*SC*	*EP*	*Yang YM et al., 1993*
	SC(h)	*CP*	*Lu TG & Vasil, 1985*
T. durum × Elytrigia intermedium	*ESC*	*CP*	*Wang & Qian, 1990*
Zea mays	*EC*	*EP*	*Cai et al., 1987*
	ESC	*CE*	*Vasil & Vasil, 1987*
	ESC	*EP*	*Kamo et al., 1987*
			Rhodes et al., 1988
			Shillito et al., 1989
			Sun JS et al., 1989
	ESC	*CP*	*Petersen et al., 1992*

Table 3. List of the species of plants regenerated from protoplasts of grain and forage legumes*

Plant species	*Protoplast Source*	*Regeneration type*	*References*
Arachis hypogaea	*Cot(im)*	*CP*	*Wei et al., 1993* *Xu & Wei, 1993*
A. paraguariensis	*SC*	*CP*	*Li et al., 1993*
Astragalus gurnbovis	*C*	*CP*	*An et al., 1994*
A. hyangheensis	*Cot,Hyp*	*CP*	*Luo et al., 1991*
A. melilotoides	*LSC*	*CP*	*Zhang et al., 1993*
A. tennis	*L*	*CP*	*An et al., 1991*
Canavallia ensiformis	*LSC*	*CP*	*Yan et al., 1991*
Clianthus formosus	*Sh(cul)*	*CP*	*Binding et al., 1983*
Coronilla varia	*Cot*	*EP/CP*	*Lu et al., 1987*
Crotalairia juncea	*Cot*	*CP*	*Rao et al., 1982*
Dolichos biflorus	*SC*	*CP*	*Sinha & Das, 1986*
Glycine canescens	*Cot(im)*	*EP*	*Grant, 1984*
	Hyp	*CP*	*Newell & Liu, 1985*
	Cot	*CP*	*Hammatt et al., 1987*
	SC	*CS*	*Myers et al., 1989*
G. clandestina	*Cot*	*CP*	*Hammatt et al., 1987*
	SC	*CS*	*Myers et al., 1989*
G. max	*Cot(im)*	*CP*	*Wei & Xu, 1988, 1990* *Luo et al., 1990* *Dhir et al., 1991*
	Cot(im)	*EP*	*Zhang et al., 1993*
G. soja	*SC*	*CE*	*Gamborg et al., 1983*
	Cot(im)	*CP*	*Wei & Xu, 1990*
G. tabacina	*SC*	*CE*	*Gamborg et al., 1983*
Hedysarum coronarium	*Cot*	*CP*	*Arcioni et al., 1985*
Lotus corniculatus	*R,Cot, Hyp, SC*	*CP*	*Ahuja et al., 1983*
	Cot, Hyp	*CP*	*Lu et al., 1986*
	Root hair	*CP*	*Rasheed et al., 1990*
	Cot, Hyp	*CP*	*Picirilli et al., 1988*
L. tenuis	*L,R*	*CP*	*Mariotti et al., 1984*
Medicago arborea	*L*	*CP*	*Arcioni et al., 1985*
M. coerulea	*L,SC*	*EP*	*Arcioni et al., 1982*
M. difalcata	*Cot*	*EP*	*Gilmour et al., 1987*
M. falcata	*Cot*	*EP*	*Gilmour et al., 1987*
M. glutinosa	*L, SC*	*EP*	*Aricioni et al., 1982*
	Cot	*EP*	*Gilmour et al., 1987*
M. hemicycla	*Cot*	*EP*	*Gilmour et al., 1987*
M. sativa	*L*	*EP*	*Kao & Michayluk, 1980* *Santos et al., 1980*
	R	*EP*	*Xu et al., 1982, 1984*
	Cot,R, L	*EP*	*Lu et al., 1982, 1983*
	C, SC	*EP*	*Mezentsev, 1981*

Plant species	*Protoplast Source*	*Regeneration type*	*References*
M. varia	*Cot*	*EP*	*Gilmoudr et al., 1982*
Onobrychis viciaefolia	*L*	*CP*	*Ahuja et al., 1983*
	C	*CP*	*Jia et al., 1989*
	L,C	*CP*	*Pupilli et al., 1989*
	SC	*CP*	*Zhao et al., 1990*
Oxytropis leptophylla	*C*	*CP*	*Zhang et al., 1994*
Phaneolus angularis	*L*	*CP*	*Ge et al., 1983*
Pisum sativum	*Epic, L*	*CS*	*Puonti-Kaevla & Erksson,1988*
	ShT	*EP*	*Lehminger-Mertena & Jacobsen, 1989*
Pithecellobium dulee	*LSC*	*CP*	*Saxena & Gill, 1987*
Psophocarpus tetragonolobus	*SC*	*CP*	*Zakri, 1984* *Wilson et al., 1985*
Sesbania bispinosa	*Cot*	*CP*	*Zhao et al., 1992, 1995*
S. formosa	*Cot*	*CP*	*Zhao et al., 1992*
S. grandiflora	*Cot*	*CP*	*Zhao et al., 1993*
S. sesban	*Cot*	*CP*	*Zhao et al., 1993*
Stylosanthes guyanensis	*SC*	*CP*	*Meijer & Steinbiss, 1983*
	Cot	*CP*	*Vieira et al., 1990*
S. macroce phala	*Cot*	*CP*	*Vieira et al., 1990*
S. scabra	*Cot*	*CP*	*Vieira et al., 1990*
Trifolium hybridum	*L, R*	*CP*	*Webb et al., 1984, 1986*
T. lupinasier	*SC*	*CP*	*Zhao & Luo, 1990*
T. pratense	*Cot, L*	*EP*	*Davey & Power, 1988*
T. repens	*SC*	*CP*	*Gresshoff et al., 1980*
	L	*CP*	*Ahuja et al., 1983; White, 1983*
T. rubens	*L, SC*	*EP*	*Grosser & Collines, 1984*
Trigonella cormiculata	*L*	*EP*	*Lu et al., 1982a*
	L,SC	*EP*	*Santos et al., 1983*
T. foenum-graecum	*L*	*CS*	*Shekhawat & Galston, 1983*
Vicia faba	*Cot(im)*	*CP*	*Wei & Xu, 1993* *Xu & Wai, 1993*
V. narbonensis	*ShT, Epic*	*CPg*	*Tegeder et al., 1995*
	ShT, Epic	*Epg*	*Tegeder et al., 1995*
Vigna aconitifolia			*Shekhawat & Galston, 1983*
	L	*EP*	*Krishnamurthy et al., 1984*
	C	*CP*	*Gill & Eapen, 1986*
V. mungo (Phaseolus mungo)	*L*	*CE*	*Sinha et al., 1983*
V. radiata (Phaseolus aurreus)	*L*	*CE*	*Shinha, 1982*
V. sinensis	*L*	*CE/CS*	*Davey et al., 1974*
	Cot(im)	*EP*	*Li XB et al., 1993*
V. sublobata	*Hyp*	*CP*	*Bhadra et al., 1994*

*Abbreviations: see footnotes in Table 2

Table 4. List of the fruit trees and woody species in which plant regeneration from protoplast has been achieved*

Plant species	*Protoplast Source*	*Regeneration type*	*Reference*
Actinidiaceae			
Actinidia chinensis	*SC*	*CP*	*Mii & Ohashi, 1988*
	C	*CP*	*Shang et al., 1992*
			Tsai, 1988
A. deliciosa	*C*	*CP*	*Oliveria & Pais, 1991*
			Shang et al., 1992
A. eriantha	*L*	*CP*	*Zhang et al., 1995*
Anocnaaceae			
Rauvolfia vonitoria	*LSC*	*EP*	*Guiller & Chenieux,1991*
Cariceae			
Carica papaya × *C.eauliflora*	*ESC*	*EP*	*Chen & Chen, 1992*
Magnoliaceae			
Liriodendron tulipifera	*ESC*	*EP*	*Merkle & Sommer, 1987*
Malvaceae			
Hibiscus syriacus	*C*	*CP*	*Zhao & Yao, 1993*
Moraceae			
Broussonetia kazinoki	*LSC*	*CP*	*Oka & Lhyama, 1985*
Morus alba	*LSC*	*CP*	*Wei et al., 1992, 1994*
Oleaceae			
Forsythia × intermedia	*LSC*	*CP*	*Ochatt, 1994*
Pinaceae			
Abies alba	*SC*	*E*	*Hartmann et al., 1992*
Cathaya argyrophylla	*ESC*	*EP*	*Lang & Kohlenbach, 1989*
Larix × *europeolepis*	*EC,SC*	*EP*	*Klimaszewska, 1989*
L. × *europeosis*	*ESC*	*EP*	*Klimaszewska, 1989*
Picea abies	*ESC*	*EP*	*Gupta et al., 1990*
P. glauca	*ESC*	*EP*	*Attree et al., 1987, 1989*
Pinus caribaea	*ESC*	*EP*	*Laine & David, 1990*
			Klimaszewska, 1989
P. lacda	*ESC*	*EP*	*Gupta & Durzan, 1987*
Pseudotsuga menziesii	*ESC*	*E/EP*	*Gupta et al., 1988*
Platanaceae			
Platanus orientalis	*LSC*	*CP*	*Wei et al., 1991*
Rosaceae			
Malus × *domestica*	*L*	*CP*	*Patartt-Ochatt et al., 1988*
			Wallin & Johansson, 1989
	L	*CP*	*Perales & Schieder, 1993*
	Sh(Cul)	*CP*	*Patat-Ochatt, 1994*
M. pumila cv 'Starkrimsou'	*SC*	*CP*	*Ding et al., 1994*
Prunus avium × *P. pseudocerasus*	*LSC*	*CP*	*Ochatt et al., 1987*
P. cerasifera	*LSC*	*CP*	*Ochatt et al., 1992*
P. cerasus	*L*	*CRP*	*Ochatt & Power, 1988*

Plant species	Protoplast Source	Regeneration type	Reference
P. erasus	C	CP	Ochattd, 1990
P. spinosa	LSC	CP	Ochatt, 1992
Pyrus communis	LSC	CP	Ochatt & Power, 1988
P. communis var. Pyraster	L	CP	Ochatt & Caso, 1986
Rubiaceae			
Coffea canephora	SC	EP	Schopke et al., 1987
Rutaceae			
Citrus aurantium	NC	EP	Vardi & Spiegel-Roy, 1982
C. limon	NC	EP	Vardi & Spiegel-Roy, 1982
C. mitis	ESC	EP	Sim et al., 1988
C. paradisi	NC	EP	Vardi & Spiegel-Roy, 1982
C. reticulata	NC	EP	Vardi & Spiegel-Roy, 1982
C. sinensis	NC	EP	Vardi et al., 1975 Vardi & Spiegel-Roy, 1982 Kobayashi et al., 1983 Deng et al., 1988
Foetunella hindsii	NC	EP	Deng et al., 1989
Microcitrus australis × *M. australasica*	NC	EP	Vardi et al., 1986
Microcitrus sp.	NC	EP	Vardi et al., 1983
Salicaceae			
Populus alba × *P. grandidentata*	L	CP	Russell & McCown, 1986, 1988
P. deltoides × *P. simonii*	LSC	CP	Wang Y et al., 1995a
P. nigra × *P. maximowiczii*	LSC	CP	Park & Son, 1992
P. nigra × *P. trichocarpa*	L	CP	Russell & McCown, 1986, 1988
P. simonii	LSC	CP	Wang Y et al., 1995b
P. tomentosa	LSC	CP	Wang SP et al., 1991
P. tremula	L	CP	Russell & McCown, 1986, 1988
P. tremula × *P. simonii*	L	CP	Wang et al., 1995
Santalaceae			
Samtahum album	C	EP	Bapat et al., 1985
	SC	EP	Rao & Ozias-Akins, 1985
Sapindaceae			
Dimocarpus longan	EC	EP	Wang F et al., 1992
Scrophulariaceae			
Paulownia fortunet	LSC	EP	Wei et al., 1991
Solanaceae			
Solanum dulcamara	L	CP	Binding & Nehls, 1977 Binding et al., 1980,1981 Binding & Mordhorst,1984
	SC	CP	Chand et al., 1990

*Abbreviations: see footnotes in Table 2.

It is still very difficult to induce and sustain cell divisions in the cultures of mesophyll protoplasts of cereals and other monocots, and so far plant regeneration could not be achieved from mesophyll protoplasts of these plants, making people wonder if the mesophyll protoplasts of these plants are totipotent. Generally, the more difficult thing to induce callusing from protoplast is its slow starting division. In many species pretreatment of donor plants or explants under specific light and/or temperature conditions, or preculture of leaves or other proper tissues on specific medium (Donn, 1978; Kao & Michayluk, 1980; Xu et al., 1981, 1984) has been found beneficial for enhancing cell division, colony formation and even plant regeneration in protocal culture.

Among the media constitution, growth regulator is the most critical for protoplast culture. In order to induce sustained divisions, an auxin-type growth regulator and a cytokinin are generally required; the type and concentration of the growth regulators vary with plant species. Recently, Ferreira et al. (1994) reported that in tobacco mesophyll protoplast culture high level of expression of the gene *cycl* At, (a cyclin gene) related to mitosis, was observed only in the presence of both NAA and BA. But, Meyer and Cooke (1979) demonstrated that the cytokinin was required later (20-24 h before division starts) than the auxin. The concentration of plant hormones could be very critical, especially in low plating density cultures. NAA in the concentration range generally used in protoplast culture inhibited cell division after a few days of culture at a low plating, but such an inhibition did not occur in high density cultures of tobacco protoplasts (Caboche, 1980). In the latter case, free NAA level reduced, and the combined NAA increased in the medium, which may be one of the reasons that a higher plating density is needed at the initial culture stage. The protoplasts isolated from embryogenic cultures of Citrus are an exception. They do not require an exogenous auxin in the culture medium; even a very low level of exogenous auxin was toxic for these caused the cultured protoplasts dying (Vardi et al.,1975; see also Deng et al,, 1988). Another interesting exception is the protoplasts of crown gall tissue of Parthenocissus, that required exogenous hormones at the initial stage of culture, although the crown gall callus could grow in their absence (Scowcroft et al., 1973).

Besides plant hormones, the type and concentration of sugars also have significant affect on protoplast culture. In many cases glucose seems to be better carbon source and osmoticum than other sugar (Xu et al., 1981, 1984; Wang Y et al., 1995; see also Eriksson, 1985). Some other media constituents that have been reported to have beneficial effects on protoplast division and/or colony formation are m-inositol (Xu et al., 1981, 1984;

Attree et al., 1989), organic acids (Kao & Michayluk, 1975; Negrutiu & Mousseau, 1980) and some amino acids (Donn, 1978).

Use of conditioned medium and feeder cells has been found promotable for cell division and colony formation in protoplast cultures of many species (see Roset and Gilissen, 1989). It is generally believed that some factors stimulating cell division might be present in the conditioned medium, or released from the feeder cells. The beneficial effect of initial high density culture may also be related to such stimulating substances released into the medium. The beneficial effect of the feeder cells is more apparent in single protoplast culture. With tobacco mesophyll protoplast-derived cells as feeder cells, the plating efficiency in the cultures of individual mesophyll protoplasts of tobacco could be raised to 90%. Microcolony formation has also been achieved from individual protoplasts of *Brassica napus* by using feeder cells from suspension cultures (Eigel and Koop, 1989). Recently feeder cells were used for successful culture of *in vitro* fused sperm and egg protoplasts, and of the protoplasts of fertilized egg. Plants were regenerated from the fusion products obtained by electrofusion of sperm and egg protoplasts of maize, by coculture with cells from non-morphogenic suspension cultures of maize (Kranz and Lorz, 1993). When the protoplasts from *in vivo* fertilized eggs of barley were cultured with androgenic microspores as feeder cells, about 75% of the protoplasts formed embryo-like structures and 50% of them produced fertile plants (Holm et al., 1994, 1995). Plants have also been raised from the protoplasts of fertilized eggs of wheat following a similar method(Holm et al., 1995).

Various protoplast culture techniques have been developed so far. They also apparently affect regeneration in protoplast cultures. The most frequently used techniques are: liquid culture, liquid over agar (or agarose) culture and protoplast-embedded method(solid or semisolid culture). Since plating in agarose was found to stimulate cell division and colony formation in protoplast culture of a number of plant species (Shillito et al., 1983), this method has been widely applied in protoplast culture. In some plant species, such as *Helianthus giganteus* (Krasnyanski et al., 1992) and Brassica (Yamashita & Shimamoto, 1989; Wei & Xu, 1990; Xue et al., 1994; Yang et al., 1994) cell division started earlier and the frequency of division was higher when the protoplasts were cultured in liquid medium. However, the colonies formed in liquid medium generally grew worse than those formed in agarose-embedding method. By combining the advantages of both liquid and solid culture methods, initial culturing protoplast in liquid medium for several days to 1-2 weeks and then embedding protoplast-derived cells or small colonies in agarose stimulated cell division and the development of colonies and microcalli (Lillo & Shahin, 1986; Kirti, 1988; Puonti-Kaerlas & Eriksson, 1988; Liu et al., 1995). Adding Ficoll in the liquid medium further enhanced early protoplast division and cell colony formation in oil seed rape (Millam et al., 1988). Other improved agarose embedding methods have also been developed, in which protoplasts are cultured in agarose droplets or beads surrounded by liquid medium. Among them, streaky culture technique, in which streaks of agarose droplets with protoplasts embedded at high density are covered with liquid medium. Protoplast division was induced in some recalcitrant species, and organogenesis accellerated in nearly all the species tested that susceptible to the treatment for shoot formation (Binding et al., 1988).

In order to stimulate sustained divisions and colony formation, it is necessary to gradually reduce osmotic pressure and the plating density by adding fresh medium with lowered levels or devoid of the osmoticum. Sustained divisions result in colony and callus formation. When protoplast-derived calli are transferred onto differentiation medium, plant regeneration can be achieved through organogenesis or embryogenesis (see Binding, 1985; Maheshwari et al., 1986; Roest and Gilissen, 1989).

3 PLANT REGENERATION VIA ORGANOGENESIS

According to our survey (Xu and Xue, 1995), in more than 72.26% of the plant species that have been reported to regenerate plants from protoplasts, especially those of belonging to the most successful families, such as Solanaceae, Compositae, Cruciferae and Leguminosae, regeneration occurred via organogenesis. In this process the protoplast-derived calli are induced to differentiate shoots , and the shoots are rooted to obtain whole plants. Thus, even in a simple procedure for plant regeneration from protoplasts, at least three media, usually different in type and concentration of auxin and cytokinin, and level of osmoticum, must be used: (1) Protoplast culture medium, used for cell wall regeneration, sustained divisions, and formation of colony and small callus. This medium generally contains an auxin (2,4-D or NAA) and a cytokinin (BA, KT or Zeatin), and is of high osmolarity (maintained by mannitol, sorbitol, or glucose). In order to stimulate the growth of cell colonies and microcalli, it is necessary to add fresh medium of low osmolarity to the culture, or sometimes to transfer the calli to a fresh medium, for reducing osmolarity and plating density; (2) Differentiation medium, for shoot formaion: This medium is usually devoid of osmoticum and contains a higher level of cytokinin and lower level of auxin. If protoplasts are cultured in liquid or soft agar(or agarose) medium, the colonies or microcalli formed should be transferred a medium solified with agar, agarose or Gelrite at this stage; and (3) Rooting medium for inducing the shoots to form roots and obtaining whole plants that can be transplanted into soil for further growth and development. This medium lacks a cytokinin, but contains lower level of auxin (IAA, NAA, or IBA). For some plant species, the rooting medium does not need any growth regulator (see also Maheshwari et al., 1986).

The process of shoot formation is more or less same as in calllus cultures. Prior to the differentiation of shoot primordia, dark green areas or spots generally appeared on the protoplast-derived calli transferred onto differentiation medium. In some cases, nodular calli are formed. However, when shoot tips are used as the source of protoplasts, shoot primordia may differentiate directly from the protoplasts. In such cases, shoot regeneration could occur within a short time; it took only 14 days from protoplast to shoot formation in *Petunia hybrida* and *Solanum nigrum* (see Binding, 1985). Some investigators demonstrated that shoot formation is readily induced, when protoplast-derived calli are transferred onto differentiation medium at an early stage. For example, delay in transfer of protoplast calli onto differentiation medium reduced the incidence of shoot formation in *Solanum nigrum* (Binding et al., 1982) and cauliflower (Yang et al., 1994).

3.1 Effects of plant materials

Besides the time required for protoplast-division and callus formation, organogenesis or embryogenesis in protoplast cultures also varies with the plant species varieties, or tissues. Aseptic shoot cultures, seedling tissues (cotyledon, hypocotyl, root, etc.), immature cotyledons and in many cases, suspension cultures or calli have proved to be good sources of protoplasts with a high potential of plant regeneration.

We can take the Genera *Brassica* as example. The genetic relationships of the 6 economically important species of *Brassica* have been revealed: *B.juncea* (2n=36, AABB), *B.napus* (2n-38, AACC) and *B.carinata* (2n=34, BBCC) are the natural amphidiploids of the three diploid species, vi2. *B.campestris* (2n=20, AA), *B.nigra* (2n=16, BB) and *B.oleracea* (2n=18, CC). Significant genotypic differences in shoot regeneration response were observed in cotyledonary cultures of this species. *B.campestris* is the lowest regenerating species and *B.nigra* and *B.oleracea* regenerate with high frequencies; the presence of *B.campestris* component in *B.napus* reduces its shoot regeneration frequency below that of *B.oleracea*. Regeneration capacity of *B.carinata* is less than the parental diploid species, while additive effect of combining genomes was found in *B.juncea* (Narasimhulu & Chopra, 1988). More or less the same tendency has been observed in protoplast cultures of various species of *Brassica*. Many investigators have demonstrated that *B.campestris* and other *Brassica* species with AA genome

(2n=20) are rather recalcitrant, and *B.oleracea*, *B.nigra* and *B.napus* good for plant regeneration from protoplasts (see Yang & Xu, 1995).

Robertson and Earle (1986) reported that in broccoli (*Brassica oleracea var. italica*) the division frequency of mesophyll protoplasts from the plants regenerated from hypocotyl explants showed higher plating efficiency and subsequent shoot differentiation from the protoplast-derived calli (60-70% and 77%, respectively) than those from plants grown directly from commercial hybrid seeds (22% higher and 15%, respectively). Higher morphogenetic capacity of protoplasts from plants regenerated in tissue culture has also been observed in *B.napus* (Kohlenbach et al., 1982; Glimelius, 1984). These results indicate that plants regenerated through a tissue culture cycle might be a better source of protoplasts with higher potentials for cell division and plant regeneration. Robertson et al. (1988) further observed that cotyledon or mesophyll protoplasts from selfed progeny of regenerated plants produced more vigorous calli and more shoots than protoplasts from hybrid seeds. It demonstrated that the increased totipotency of the protoplasts from regenerated plants was apparently caused by a genetic factor.

3.2 Effects of media constituents

The composition of the used medium plays a critical role in regulation of organogenesis. In a rich protoplast culture medium (such as KM8P, V-KM), meristemstic cells are readily formed. Bud formation occurs soon after these colonies with meristematic cells are transferred to low osmotic medium. In some cases, such as *Orychophragmus violaceus*, a number of buds or embryo-like structures were formed even in protoplast culture medium, during repeated dilutions with fresh medium of low osmolarity or no osmoticum at all (Xu and Xu, 1988).

Among the media constituents, the type species and concentrations of plant hormones or growth regulators are most critical for shoot formation. In order to induce bud formation a cytokinin is usually required in the medium at a concentration of 0.2-5 mgl^{-1}, and rarely, up to 10 mgl^{-1}. BA, KT and zeatin (or zeatin riboside) are most frequently used. In recent years, thidiazuron (TDZ) has been found with very strong cytokinin activity. In poplar protoplast cultures, low level of TDZ (0.1 μM) could significantly stimulate bud formation (Rusell and McCown, 1988). Such a stimulating effect of TDZ on shoot bud differentiation is also reported in apple (Ding et al., 1994) and *Vicia faba* (Tegeder et al., 1995). The most effective cytokinin and its optimum concentration would vary with plant species. It should be noted that at higher concentrations of cytokinin vitrified shoots could be frequently formed.

In the cultures of protoplasts form immature cotyledons of several important grain legumes, viz. soybean, peanut and faba bean (*Vicia faba*), combination of two or three cytokinins proved more effective for shoot formaiton than any one of them (Wei and Xu, 1988, 1990b). Besides cytokinin, one auxin-type growth regulator at lower concentration is also required. It is generally added in the range of 0.01-0.2 mgl^{-1}. Most frequently, NAA and IAA have been used.

In soybean, immature cotyledon protoplast-derived microcalli required transfer to a proliferation medium with 0.05-0.1 mgl^{-1} 2,4-D and 0.5 mgl^{-1} BA for further growth to form compact and nodular calli, before shoot formation on the differentiation medium containing 0.15 mgl^{-1} NAA, and 0.5 $mg\ l^{-1}$ each of BA, KT and zeatin. If the growth medium contained only 2,4-D or NAA (even as high as 10-20 mgl^{-1} 2,4-D, but without BA), the calli grew rapidly, and became soft. Such calli did not show embryogenesis or organgenesis after transfer to the differentiation medium. The results showed that hormone type and concentration in the proliferation medium have a significant effect on the subsequent morphogenic response of the protocalli (Wei and Xu, 1988). Proliferation culture also proved to be an important step in shoot formation in some woody species, such as *Populus* (Wang SP et al., 1991; Park & Son, 1992; Wang Y et al., 1995) and *Morus alba* (Wei et al., 1994), etc.

There are only a few studies dealing with the effects of other types of plant hormones (ABA, ethylene and GA). ABA was reported to stimulate shoot formation in potato protoplast culture (Shepard, 1980) and, as described in the next section (4.2), in some cases it also promoted embryogenesis from protoplast-derived colonies or calli. In the cultures of hypocotyl protoplasts of *Brassica juncea*, Pua (1990) observed that embryogenic capacity of protoplast-derived calli of two varieties tested was very different. For the variety "Leaf Heading", which is more difficult to induce embryogenesis, a number of shoots and plantlets were regenerated, when protoplasts-derived embryogenic calli were transferred onto the differentiation medium supplemented with 10 μM BA, 2-7 μM NAA and 30 μM $AgNO_3$. In the medium without $AgNO_3$, none or only a few shoots or plantlets were formed. GA is added to differentiation medium for protoplast culture of some plant species, such as cassava (Shahin and Shepard, 1980), peanut (Wei et al., 1993), and potato (Bokelmann and Roest, 1983). In sunflower, shoot buds differentiated only in the presence of 0.1 mgl^{-1} GA, in addition to NAA and BA (Burrus et al., 1991). It was reported that GA could reduce vitrification of protoplast-derived shoots of *Brachycome* (Malaure et al., 1990). GA frequently stimulated shoot growth and elongation. For instance, although removing GA from differentiation medium in pear protoplast culture had no effect on bud formation, the shoots formed were dwarf, and difficult for multiplication and rooting. Thus, shoots had to be transferred to 1/2 MS medium with BA and GA three times (for 4 weeks each time) before transfer to rooting medium for obtaining whole plantlets (Ochatt and Power, 1988). In Chinese cabbage (*Brassica pekinensis*), addition of spermidine to protoplast culture medium not only enhanced cell division, but also to be beneficial to plant regeneration from protoplast-derived calli (Li SJ et al., 1992).

3.3 Effects of physical factors

Although several studies have been carried out on the effects of different physical factors (light, temperature, pH, etc.) on protoplast culure (see Maheshwari et al., 1986; Roest and Gilissen, 1989), but only a few of them deal with the effect of these factors on plant regeneration from protoplast-derived tissues. Light condition plays an important role in shoot formation and growth. Thus, protoplast cultures are required to be returned to a stronger light regime after initial period of culture in dark or dim light. It is interesting to find that electroporation not only enhanced protoplast division, but also stimulated growth and plant regeneration from protoplast-derived calli in colt cherry (*Prunus avium* × *pseudocerasus*) (Ochatt et al., 1988) and *Solanum dulcamara* (Chand et al., 1988).

3.4 Other types of organogenesis

Besides the most frequent model of plant regeneration via shoot formation, there are several other types of organogenesis in protoplast cultures. For example, in *Rudbeckia laciniata* (Compositae), roots were formed from protoplast-derived tissues, and similarly, plants were regenerated from pieces of these roots (Al-Atabee et al., 1990) in *Callistephus chinensis*, another member of the Compositae, plants differentiated regenerated when decapitated roots formed from protoplast calli were transferred to differentiation medium (Pillai et al., 1990). Bulblet formation followed by plant regeneration was reported in onion mesophyll protoplast culture (Wang et al., 1986). It was observed that shoots or plantlets regenerated from protoplast-derived tissues readily formed flower buds or even flowered *in vitro* in *Nicotiana rustica* (Gill et al., 1979; Li and Wang, 1979; Li, 1987), *Panax ginseng* (Chen et al., 1990), *Callistephus chinensis* and *Centaurea cyanus* (Pillai et al., 1990).

3.5 Rooting of protoplast-derived shoots

Rooting of protoplast-derived shoots is not much different from routine procedures in tissue culture. Normally, it is achieved by excising the shoots from the protoplast-derived calli and transferring them to an auxin containing or hormone-free medium. The most frequently used auxins are IAA, NAA and IBA. Reducing the level of sugar and/or minerals (e.g. using 1/2 MS strength), or adding activated charcoal are beneficial for root formation in some cases. Where it is difficult to root the shoots or to establish the plantlets in the soil, grafting may be tried to obtain healthy and complete plants (Schieder, 1980; Tegeder et al., 1995).

4 PLANT REGENERATION FROM PROTOPLAST VIA EMBRYOGENESIS

Plant regeneration from protoplasts via embryogenesis mostly occurs in species belonging to the families Gramineae, Umbelliferae, Rutaceae, Cucurbitaceae and Leguminosae (especially forage legumes). Pinaceae of gymnosperms is also included in the list (Xu and Xue, 1995). In the Solanace*ae*, plant regeneration from protoplast via somatic embryogenesis, as reported in *Nicotiana, N,glutinosa* (Liu and Xu, 1988).

4.1 Effects of plant materials

In most of the cereal crops and grasses, plant regeneration from occurs through somatic embryogenesis, due to using embryogenic callus or cell suspension culture derived from immature inflorescence, young or mature embryos as protoplast source (Table 2, see also Vasil and Vasil, 1992; Vasil, 1995). This is also true many species of Umbelliferae and gymnosperm were reported to be similar to it. It should be noted that embryogenic cell suspension culture frequently reduce and eventually lost embryogenesis potential after repeated subcultures. For solving this problem, cryopreservation of embryogenic calli can be applied, and new cell suspensions, initiated from the cryoperservated or fresh tissues at regular intervals, can provide a reliable source material of totipotent protoplasts. In fact, a protoplast population, even from the same batch of cultured cells, is generally heterogeneous, and the protoplasts are physiologically different. Klimaszewska (1989) recovered somatic embryos and plantlets from protoplasts of hybrid larch (*Larix* × *eurolepis*) isolated from two embryogenic callus (L1) and cell suspension (L2) lines. The protoplasts were centrifuged on a discontinuous medium/Percoll density gradient (0-30%).

The fraction, collected at the 10-20% Percoll interface, contained small, densely cytoplasmic uninucleate protoplasts (80%). Protoplast division frequency of this fraction at day 8 of culture was higher for L1 (28-39%) than that for L2(18-20%). However, the L2 protoplasts formed quite a lot of early somatic embryos (25-33 for each petri dish) after 23-28days, while the L1 protoplasts grew as a culture composed of loosely associated long cells and aggregates of small cells after 4-5 weeks, and only a few embryos formed (1-3 for each dish). The protoplasts collected at the 0-10% Percoll interface were large and vacuolated, did not regenerate cell walls and survived only for a short time in culture. In the fraction at 20-30% Percoll interface, protoplasts were mainly multinucleate, and only divided occassionally to form multicellular aggregates, which did not yield callus and embryoids recovered. Attree et al. (1989) noted that newly established suspension cultures of white spruce (*Picea glauca*) gave lower protoplast yields and plating efficiencies than the long term established suspensions. Even the somatic embryos formed by protoplasts from young cultures was poorer latter. It was suggested that changes in the morphology and texture of embryos in older suspensions may be favourable for readily release of protoplasts. However, such older suspension cultures also suffer with the problem, that of gradual loss of embryogenic capacity after repeated subcultures. Lu et al. (1983) compared protoplasts from leaves, seedling cotyledons and seedling roots of *Medicago sativa*, and found that the time from protoplast isolation to embryogenesis under identical culture conditions varied with the source of protoplasts root protoplast took the shortest time (30-35 days), possibly due to their more meristematic nature. In *Citrus sinensis*, plant regeneration was observed in the cultures of protoplasts isolated from nucellus-derived embryogenic calli or suspension cultures (Kobayashi et al., 1983; Deng et al., 1988). However, cell clusters were formed by the protoplasts from stem calli (Deng et al., 1988). From the results described above, it can be concluded that the source material very much affects growth and somatic embryogenesis in protoplast cultures.

4.2 Effects media constituents

2,4-D is the most important component of culture medium for somatic embryogenesis even in protoplast culture. In order to induce embryogenesis, the initial culture of protoplasts and cell colonies derived from them needs to be carried out in the presence of 2,4-D. Subsequent reduction is the 2,4-D level or its omission from the medium leads to the formation and development of embryos and plantlets. In some cases such as *Brassica napus* (Li and Kohlenbach, 1982), cabbage (Fu et al., 1985), *Lycopersicon peruvianum* (Zapata and Sink, 1981), *Medicago sativa* (Gilmour et al.,

1987), wheat (Guo et al., 1990; Li et al., 1992) and some of other cereals (see Vasil and Vasil, 1992), embryos are regenerated directly from the protoplast-derived cell colonies without an apparent unorganized callus phase. NAA alone induced embryo formation in *Atropa* protoplast cultures (Gosch et al., 1975). Although plant regeneration from the protoplasts of immature cotyledons of soybean was reported via organogenesis by several laboratories (Wei and Xu, 1988, 1990; Dhir et al., 1991), Zhang et al. (1993) achieved it via embryogenesis when protoplast-derived embryogenic calli were transferred to a medium supplemented with higher concentration of NAA (10 mg l^{-1}) and lower sucrose level (0.5-1%), than those in the embryo induction medium. In protoplast cultures of melon (*Cucumis melo*), somatic embryogenesis was induced in the medium containing 1 mg l^{-1} 2,4-D and 0.1mg l^{-1} BA, while shoot formation occurred in the medium with 0.5 mg l^{-1} each of BA and kinetin (Debeaujon and Branchard, 1992). Thus, plant regeneration could be achieved through different pathways of morphogenesis in the same species by manipulation of the exogenous auxin/cytokinin balance in the differentiation medium. In rice, protoplasts of high quality could be isolated from suspension cultures grown in ABA containing medium. Embryogenesis was also promoted by ABA (Yang et al., 1989; Yin et al., 1993). Beneficial effect of ABA on the formation of embryo-like structures in protoplast-cultures was also observed in apple (Kouider et al., 1984) and Carica (Chen and Chen, 1992). In the gymnospermous species. *Abies alba* and *Picea glauca*, ABA was shown to stimulate the development and maturation of somatic embryos (Attree et al., 1989; Hartmann et al., 1992). In some cases, GA also accelerated the growth and development of somatic embryos (Davey and Power, 1988; Deng et al., 1988; Lehminger-Mertens and Jacobsen 1989).

It has been observed that the method used for protoplast culture may influence somatic embryogenesis in the differentiation medium., Krasnyanski et al.(1992) observed that colonies formed from mesophyll of *Helianthus giganteus* protoplasts cultured in agarose solidified medium, consisted of spherical cells rich in cytoplasm, and were compact and globular. Embryos differentiated from such compact colonies and developed into plantlets on media with reduced auxin level. The cells of colonies growing in liquid medium were vacuolated, oval and loosely attached to each other. The results described above and others indicated that agarose embedding method is superior with respect to morphogenetic capacity of protoplast-derived colonies, although in liquid cultures first divisions occurs faster and plating efficiency is better in many cases (see Section 1). In *Abies alba*, for sustained divisions it is necessary to culture protoplasts in a modified KM medium for the first few days but it leads to the formation of fast growing, unorganized cell colonies. The development of proembryos could only be achieved when the KM medium was replaced by SH medium after 1-2 weeks of protoplast culture (Hatmann et al., 1992).

In many species, plantlets are readily obtained from embryos formed in protoplast cultures when they are transferred to a medium devoid of hormone or containing low levels of cytokinin and/or auxin. In some cases, GA stimulated plant regeneration from embryo (Lehminger-Mertens and Jacobnsen, 1989; Kunitake and Mii, 1990). It is interesting that in *Helianthus giganteus*, healthy plantlets were obtained only if clusters of embryos or embryos together with a small piece of callus tissue were transferred to hormone-free medium. Individual embryos usually failed to germinate (Krasnyanski et al., 1992).

5 GENTIC VARIATIONS IN PROTOPLAST-DERIVED PLANTS

The genetic characters of plants regenerated from protoplasts reflect the nature of the protoplast donor cells and the consequences of all the alterations and variations occurred during protoplast culture (Binding, 1985). Some of phenotypic changes (such as shoot vitrification, change of leaf shape, etc) may occur due to physiological abnormalities caused during protoplast isolatation and culture. However, such phenotypic abnormalities of regenerated plants can gradually disappear by manipulation culture conditions and medium compositions, or after transplantation. Frequently, morphologically normal plants may be obtained from the seeds of the regenerated plants.

5.1 Ploidy and chromosome number of protoplast-derived plants

The plant materials used for protoplast isolation may affect the ploidy and chromosome number of the plants regenerated from them. Generally, the plants regenerated from protoplasts of mesophyll tissue, shoot tips and embryogenic tissue maintain the original ploidy of the donor plants. However, various readily occur in the species with genetic instability in ploidy level or chromosome number, and those propagated vegetatively, or if long term suspension cultures are used as the source of protoplasts. For example, He et al. (1995) observed considerable variations in chromosome numbers of the protoplast-derived plants of *Actinidia deliciosa*. Most of the 29 plants checked were aneuploids, and their chromosome numbers ranged from 142 to 310. Only 6 plants showed the same ploidy as the parent plant (a pistillate plant; $2n=6x=174$). Grosser and Collins (1984) reported that where as the plants regenerated from mesophyll protoplasts of *Trifolium rubens* were diploid (2n=16), those from the protoplasts of suspension cultured cells were tetraploid (2n=32). Jones et al. (1989) observed that frequency of aueuploid plants from protoplasts of potato tuber tissue was much higher than that from mesophyll protoplasts. Even so, a lot of chromosomal variations occurred among the plants regenerated from mesophyll protoplasts of potato. Creisson and Karp (1985) checked 200 potato plants regenerated from mesophyll protoplasts and found that 57% of them were with normal chromosome number (2n=48), and 43% were aneuploids. Plants regenerated from protoplasts via somatic embryogenesis usually maintain the original karyotypic character of the donor plant. For example, Sun et al. (1989) reported that out of 11 plants regenerated from protoplasts isolated from haploid suspension cultured cells of a commercial supersweet maize cultivar, 10 plants were haploid and only one diploid. In *Asparagus officinalis*, all the plants regenerated from protoplasts via embryogenesis were diploid (Kunitake and Mii, 1990). However, it should be noted that during enzymatic degradation of cell walls, some of the adjacent protoplasts fuse together, forming multinucleate fusion bodies. Such a spontaneous fusion frequently occur when protoplasts are isolated from embryogenic tissues, and may cause an increase in the frequency of polyploid plants, observed in carrot (Dudits et al., 1976). The frequency of spontaneous fusion can be reduced by using a sequential method for protoplast isolation (instead of the treatment with a mixture of enzymes), or exposing the protoplast source materials to a high osmoticum solution for plasmolysis, and breaking the plasmodesmatal connections. However, it may be pointed out here that the frequency of polyploids is also high in some cases with low level of spontaneous protoplast fusion (see Gleba and Sytnik. 1984).

5.2 Variation for agronomic traits in protoplast-derived plants

Besides the karyotypic changes, a number of variations for morphological and agronomic traits have been observed in protoplast-derived plants. Shepard (1980) reported apparent variation for over 20 horticultural characters observed among 800 protoclones from potato mesophyll protoplasts, In the case of *Actinidia deliciosa,* described above, about 1/3 of the plants regenerated from protoplasts of female planting were male (He et al., 1995). She et al. (1990) observed changes in the agronomic characters, such as plant height, length of flag leaf and main panicle, number of effective panicles, grains per panicle, fertility and growth duration, changed in the progenies of protoplast-regenerated rice plants, although their chromosome number was maintained normal. Protoclonal variations were also reported in other plant species, e.g. lettuce (Engler and Grogan, 1984), and cabbage (Xiong et al., 1988).

In *Nicotiana sylvestris* (a diploid species), protoplast culture induced high level of genetic variations. About 50% of the diploid plants regenerated from a pure line T contained several mutations of recessive nuclear genes, (e.g. leaf colour and shape, germination and fertility) (Prat, 1983). When a second cycle of protoplast culture was carried out with the selfed progenies from a morphologically normal protoplast-regenerated plant, most of the regenerated plants were found to be mutated for either recessive, semi-dominant or cytoplasmic genes (e.g. specific mtDNA diletions leading to male sterility, see Li et al., 1988).

Regeneration of albino plantlets from protoplasts mainly occurred in Gramineae (see Table 2). It is consistent with high frequency of albino plants observed in another culture of some cereal crops and grasses. It might be explained that cereals and grasses are more sensitive to the culture procedures than the dicots, or addition of high level of 2,4-D may be responsible for such a genetic instability.

From the examples described above, it may be concluded that protoclonal variations are caused by different factors. One of the factors is the tissue used to prepare protoplast. Such variations are very readily observed in the plants which are routinely multiplied by vegetative means. For example, in a chimera of geranium, the plants regenerated from the protoplasts of albino tissue were albino, while the plants from green region of the leaf were normally green (Kameya, 1975). Another main factor for protoclonal variations could be protoplast manipulation itself. The studies from Carlberg et al. (1984) and Sree Ramuli et al. (1984) demonstrated that in potato the changes in DNA level and karyotype (aneuploidy, polyploidy and endopolyploidy) in the cells of protoplast-derived calli occurred during the early phase of protoplast culture. Protoclonal variations also vary with plant species and varieties. Great differences for genetic stability have been observed in protoplast cultures of different plant species, as in tissue and cell cultures. The variations produced from protoplast culture also provide another source of somavariations. Some of them may be useful for genetic studies and breeding work

6 REFERENCES

Akashi R & Adachi T (1992) Plant regeneration from suspension cultured-derived protoplasts of apomictic dallisgrass (*Paspalum dilatatum* Poir.). Plant Sci. 82: 219-225.

Al-Atabee JS, Mulligan BJ & Power JB (1990) Interspecific somatic hybrids of *Rudbeckia hirta* and *R.laciniata* (Compositae). Plant Cell Rep. 8: 517-520.

An LJ, Luo XM, Li XW, Li FX, He MY & Hao S (1991) Plant regeneration from mesophyll protoplasts of *Astragalus tenuis*. Acta Bot. Sinica 33: 38-42.

An LJ, Zhang JM, Li FX, Luo XM, He MY & Hao S (1993) Study on high-frequency plant regeneration from monocell suspension cultures of *Astragalus grubovii*. Acta Bot.Sinica, 35 (Suppl.) : 153-156.

Arcioni S, Davey MR, dos Santos AVP & Cocking EC (1982) Simatic embryogenesis in tissues from mesophyll and cell suspension protoplasts of *Medicago coerulea* and *M.glutinose*. Z. Pflanzenphysiol. 106: 105-110.

Arcioni S, Mariotti D & Pezzotti M (1985) *Hedysarum coronarium* L. *in vitro* conditions for plant regeneration from protoplasts and callus of various explants. J. Plant Physiol. 121: 141-148.

Arcioni S, Mariotti D & Pezzotti M (1985) Plant regeneration of *Medicago arborea* from protoplast culture and preliminary experiments of somatic hybridization with *Medicago sativa*. Genet.Agraria 39: 307.

Arcioni S, Mariotti D & Pezzotti M (1985a) Plant regeneration of *Medicago arborea* from protoplast culture and preliminary experiments of somatic hybridization with *Medicago sativa*. Genet. Agraria 39: 307.

Ahuja PS, Lu DY, Cocking EC & Davey MR (1983b) An assessment of the cultural capabilities of *Trifolium repens* L. (white clover) and *Onobrychis viciifolia* Scop. (sainfoin) mesophyll protoplasts. Plant Cell Rep. 2:269-272.

Asano Y & Sugiura K (1990)Plant regeneration from suspension culture-derived protoplasts of *Agrostic alba* L. (Redtop). Plant Sci. 72: 267-273.

Attree SM, Bekkaoui F, Dunstan DI & Fowke LC (1987) Regeneration of somatic embryos from protoplasts isolated from an embryogenic suspension culture of white spruce (*Picea glauca*). Plant Cell Rep. 6: 480-483.

Attree SM, Dunstan DI & Fowke LC (1989) Plantlet regeneration from embryogenic protoplasts of white spruce (*Picea glauca*). Biotechnology 7:1061-1062.

Bapat VA, Gill R & Rao PS (1985) Regeneration of somatic embryos and plantlets from stem callus protoplasts of sandalwood tree (*Santalum album* L.). Curr. Sci. 54:978-982.

Barakat MN & Cocking EC (1983) Plant regeneration from protoplast-derived tissues of *Linum usitatissimum* L. (flax). Plant Cell Rep. 2:314-317.

Bhadra SK, Hammatt N, Power JB & Davey MR (1994) A reproducible procedure for plant regeneration from seedling hypocotyl protoplasts of *Vigna sublobata* L. Plant Cell Rep. 14:174-179.

Binding H & Nehls R (1977) Regeneration of isolated protoplasts to plants in *Solanum dubcamara* L.. Z.Planzenphysiol. 85:279-280.

Binding H & Nehls R (1980) Protoplast regeneration to plants in *Senncio vulgaris* L.. Z.Pflanzenphysiol. 99:183-185.

Binding H (1974) Regeneration of haploid and diploid plants from protoplastts of *Petunia hybrida* L.. Z.Pflanzenphysiol 74: 327-356.

Binding H (1985) Regeneration of plants. In: Fowke LC & Constabel F (eds.) Plant Protoplasts (pp:21-27) CRC Press, Boca Raton.

Binding H, Bunning D, Gorschen E, Jorgensen J, Kollmann R, Krumbiegal-schroeren G, Ling HQ, Monzer J, Mordhorst G, Rudnick J, Sauer A, Witt D & Zuba M (1988) Uniparental, fusant, and chimeric plants regenerated from protoplasts after streak plating in agarose gels. Plant Cell Tissue Organ Cult. 12:133-135.

Binding H, Jain SM, Finger J, Mordhorst G, Nehls R & Gressel J (1982) Somatic hybridization of an atrazine resistant biotype of *Solanum nigrum* L. Clonal variation in morphology and in ztrazine sensitivity. Theor.Appl. Genet. 63:273-277.

Binding H, Nehls R & Kock R (1980) Nersuche zur protoplastenregeneration dikotyler pflanzen unterschiedlicher systematischer zugehorigkeit. Ber. Dtsch.Bot.Ges. 93:667-671.
Binding H, Nehls R, Kock R, Finger J & Mordhorst G (1981) Comparative studies on protoplast regeneration in herbaceous species of the dicotyledoneae. Z .Pflanzenphysiol 101:119-130.
Binding H, Rgensen J, Krumbiegel-Schroeran G, Finger J, Mordhorst G & Suchowiat G (1983) Culture of apical protoplasts from shoot cultures in the orders Fabales, Tosales, and Caryophyllales. In: Potrykus I et al. (eds.) Protoplasts 1983, Poster Proceedings (pp.90) Birkhauser, Basel.
Bokelmann GS & Roest S (1983) Plant regeneration from protoplasts of potato (*Solanum tuberosum* cv. Bintje). Z.Pflanzenphysiol. 109:259-263.
Burrus M, Chanace C, Alibert G & Bidney D (1991) Regeneration of fertile plants from protoplasts of sunflower (*Helianthus annuus* L.). Plant Cell Rep. 10:161-166
Caboche M (1980) Nutritional requirements of protoplast-derived, haploid tobacco cells grown at low cell densities in liquid medium. Planta 149:7-18.
Cai QG, Guo ZC, Qian YQ, Jiang RX & Zhao YL (1987) Plant regeneration from protoplasts of corn (*Zea mays* L.). Acta Bot.Sinica 29:453-459.
Carlberg I, Glimelius K & Eriksson T (1984) Nuclear DNA content during the initiation of callus formation from isolated protoplasts of *Solanum tuberosum*. Plant Sci Lett 35:225-230.
Chand PK, Ochatt SJ, Rech EL, Power JB & Davey MR (1988) Electroporation stimulates plant regeneration from protoplasts of the woody medicinal species *Solanum dulcamara* L.. J.Exp.Bot. 39:1267-1274.
Chen MH & Chen CC (1992) Somatic embryogenesis and plant regeneration from embryo and protoplast culture of *Carica*. Abstract Symp. on Botany, Inst. of Botany, Acad.Sinica, Taipei, pp:69-70.
Chen SC,Tam YY, Chang RL, Chen HR,Len TH & Chang WC (1990) *In vitro* flowering of ginseng protoplast-derived plantlets. Abstract plant Biotechnology Symp. (Taipei, Nov.19-29, 1990). p.33.
Chen WH, Davey MR, Power JB & Cocking EC (1988) Sugarcane protoplasts: factors affecting division and plant regeneration. Plant Cell Rep. 7:344-347.
Chen ZX, Li SJ, Yue JX, Jiao GL, Liu SX, She JM, Wu JY & Wang HB (1989) Plantlet regeneration from protoplasts isolated from an embryogenic suspension culture of cotton (*Gossypium hirsutum* L.) Acta Bot.Sinica 31:966-969.
Coulibaly MY & Demarly Y (1986) Regeneration of plantlets from protoplasts of rice, *Oryza sativa* L.. Z.Pflanzenzucht. 96:79-81.
Creissen GP & Karp A (1985) Karyotypic changes in potato plants regenerated from protoplasts. Plant Cell Tissue Organ Cult. 4:171-182.
Dalton SJ (1988a) Plant regeneration from cell suspension protoplasts of *Festuca arundinacea* Schreb.(tall fescue) and *Lolium perenne* L.(perennial ryegrass). J.Plant Physiol. 132:170-175.
Dalton SJ (1988b) Plant regeneration from cell suspension protoplasts of *Festuca arundinacea* Schreb.*Lolium perenne* L. and *L.multiflorum* Lam.. Plant Cell Tissue Organ Cult. 12:137-140.
Davey MR & Power JB (1980) Aspects of protoplast culture and plant regeneration. Plant Cell Tissue Organ Cult. 112:115-125.
Davey MR, Bush E & Power JB (1974) Cultural studies of a dividing legume leaf protoplast system. Plant Sci.Lett. 3:127-133.

Debeaujon I & Branchard M (1992) Induction of somatic embryogenesis and caulogenesis from cotyledon and leaf protoplast-derived coloniesof melon (*Cucumis melo* L.). Plant Cell Rep. 12:37-40.

Deng JJ & Xian JA (1989) Plantlet regeneration *Setaric italica* protoplasts. Plant Physiol.Commic. (Shanghai) (2):56-57.

Deng XX, Deng ZA & Wan SY (1989) Formation of embryogenic calli and regeneration of plants from protoplasts of *Citrus sinensis* (cv. Jincheng) and *Fortunella hindsii.* Chinese Bull. Agr. 3:13-15.

Deng XX, Zhang WC & Wan SY (1988) Studies of the isolation and plant regeneration of protoplast in *Citrus.* Acta Hortic. Sinica 15:99-102.

Dhir SK, Dhir S & Wildhom JM (1991) Plantlet regeneration from immature cotyledon protoplasts of soybean (*Glycine max* L.). Plant Cell Rep. 10:39-43.

Ding AP, Wang HP & Cao YP (1994) Protoplast culture and plant regeneration of *Malus pumila*. Acta Bot.Sinica 36:271-277.

Dong JJ & Xia ZA (1990) Protoplast culture and plant regeneration of foxtail amillet. Chinese Sci.Bull. 35:1560-1564.

Donn G (1978) Cell division and callus regeneration from leaf protoplasts of *Vicia narbonensis*. Z.Pflanzenphysiol. 86:65-75.

Dos Santos AVP, Davey MR & Cocking EC (1983) Cultural studies of protoplasts and leaf callus of *Trigonella corniculata* and *T.foenum-graecum* Z. Pflanzenphysiol. 109:227-234.

Dos Santos AVP, Outka DE,Cocking EC and Davey MR (1980) Organogenesis and somatic embryogenesis in tissues derived from leaf protoplasts and leaf explants of *Medicago sativa*. Z.Planzenphysiol. 99:261-270.

Dudits D, Kao KN, Constabel F & Gamborg, OL (1976) Embryogenesis and formation of tetraploid and hexaploid plants from carrot protoplasts. Can. J.Bot. 54:1063-1067.

Eigel L & Koop HU (1989) Nurse culture of individual cells: Regeneration of colonies from single protoplasts of *Nicotiana tabacum*, *Brassica napus* and *Hordeum vulgare*. J.Plant Physiol. 134:577-581.

Engler DE & Grogan RG (1984) Variation in lettuce plants regenerated from protoplasts. J.Hered 75:426-430.

Eriksson T & Johansson L. V Intl.Cogr.On Plant Tissue and Cell Culture -Abstracts, IAPTC,p.210.

Eriksson TR (1985) Protoplast isolation and culture. In: Fowke LC & Constabel F (eds.) Plant Protoplasts (pp.1-19) CRC Press.Inc.Boca Raton, Florida

Ferreira PCG, Hemerly AS, de Almeida Engler J, Van Montagu M, Engler G & Inze D (1994) Developmental expression of the *Arabidopsis* cyclin gene *cycl At.* Plant cell 6:1763-1774.

Fu YY, Jia SR & Lin Y (1985) Plant regeneration from mesophyll protoplast culture of cabbage (*Brassica oleracea* var. *capitata*). Theor.Appl.Genet. 71: 495-499.

Fujimura T, Sakurai M, Akagi H, Negishi T & Hirose A (1985) Regeneration of rice plants from protoplasts. Plant Tissue Cult. Lett. 2:74-75.

Galbraith DW & Shields BA (1982) The effects of inhibitors of cell wall synthesis on tobacco protoplast development. Physiol.Plant. 55:25-30.

Gamborg OL, Davis BP & Stahlhut RW (1983) Cell division and differentiation on protoplasts from cell cultures of *Glycine* species and leaf tissues of soybean.Plant Cell Rep. 2:213-215.

Ge Kouling, Wang Yunzhu, Yuan Xunmei, Huang Peiming, Yang Jinshui, Nie Zhipin, Testa D & Lee N (1989) Plantlet regeneration from protoplasts isolated from mesophyll cells of adsuke bean (*Phoseolus angularis*, Wight). Plant Sci. 63:209-216.

Gill R & Eapen S (1986) Plant regeneration from hypocotyl protoplasts of mothbean (*Vigna aconitifolia*). Curr. Sci. 55:100-102.

Gill R, Rashid A & Maheshwari SC (1979) Isolation of mesophyll protoplasts of *Nicotiana rustica* and their regeneration into plants flowering *in vitro*. Physiol.Plant 47:7-10.
Gilmour DM, Davey MR & Cocking EC (1987) Plant regeneration from cotyledon protoplasts of wild *Medicago* species. Plant Sci 48:107-112.
Gilmour DM, Davey MR, Cocking EC & Pental D (1987) Culture of low numbers of forage legume protoplasts in membrane chambers. J.Plant Physiol. 126:457-465.
Glimelius K (1984) High growth rates and regeneration capacity of hypocotyl protoplasts in some Brassicaceae. Physiol.Plant. 61:38-44.
Golds TJ, Babczinsky J & Mordhorst AP (1994) Protoplast preparation without centrifugation: plant regeneration of barley (*Hordeun vulgare* L.). Plant Cell Rep. 13:188-192.
Gosch G, Bajaj YPS & Reinert J (1975) Isolation, culture, and induction of embryogenesis in protoplasts from cell-suspensions of *Atropa belladonna*. Protoplasma 86:405-410.
Gresshoff PM (1980) *In vitro* culture of white clover: callus, suspension, protoplast culture, and plant regeneration. Bot.Gaz. 141:157-164.
Grosser JW & Collins GB (1984) Isolation and culture of *Trifolium rubens* protoplasts with whole plant regeneration. Plant Sci. Lett. 37:165-170.
Guo GQ, Xia GM, Li ZY & Chen HM (1990) Direct somatic embryogenesis and plant regeneration from protoplast-derived cells of wheat (*Triticum aestivum* L.). Sci.Sinica 9:970-974.
Gupta JN, Kaur R & Cheema GS (1994) Abstract 8th Intern.Congress of Plant Tissue and Cell Culture, p.30. Firenze, Italy.
Hammatt N, Knn HA, Davey MR, Nelson RS & Cocking EC (1987) Plant regeneration from cotyledon protoplasts of *Glycine canescens* and *C.clandestina*. Plant Sci. 48:129-135.
Harris R, Wright M, Byrne M, Varnum J, Brightwell B & Schubert K (1988) Callus formation and plantlet regeneration from protoplasts derived from suspension cultures of wheat (*Triticum aestivem* L.). Plant Cell Rep.7:337-340.
Hartmann S, Lang H & Reuther G (1992) Differentiation of somatic embryos from protoplasts isolated from embryogenic suspension cultures of *Abies alba* L.. Plant Cell Rep. 11:554-557.
He Z, Cai Q, Ke S, Qian Y & Xu L (1995) Cytogenetic sutdies on regenerated plants derived from protoplasts of *Actinidia deliciosa*. I. Variation of chromosome number of somatic cells. J.Wuhan Bot.Res. 13:97-101.
Heyser JW (1984) Callus and shoot regeneration from protoplasts of proso millet (*Panicum miliaceum* L.). Z.Pflanzenphysiol. 113:293-299.
Holm PB, Knudsen S, Mouritzen R, Negri D, Olsen FL & Roue C (1994) Regeneration of fertile barley plants from mechanically isolated protoplasts of the fertilized egg cell.Plant Cell 6:531-543.
Holm PB, Knudsen S, Mouritzen R, Negri D, Olsen FL & Roue C (1995) Fertile barley plants can be regenerated from mechanically isolated protoplasts of fertilized egg cells. In:Terzimn M et al.(eds). Current Issue in Plant Molecular and Cellular Biology, (pp. 213-218) Kluwer Academic Publishers, Dordrecht.
Horn ME, Conger BV & Harms CT (1988) Plant regeneration from protoplasts of embryogenic suspension cultures of orchardgrass (*Dactylis glomerata* L.) Plant Cell Rep. 7:371-374.
Jahne A, Lazzeri PA & Lorz H (1991) Regeneration of fertile plants from protoplasts derived from embryogenic cell suspensions of barley (*Hordeum vulgare* L.). Plant Cell Rep. 10:1-6.
Jia JF, Wu NF, Shi YJ & Zhang Q (1989) Plant regeneration from callus protoplasts in sainfoin (*Onobrychis viciaefolia* scop.). Acta Biol. Exp. Sinica, 22:353-355.

Jones H, Karp A & Jones MGK (1980) Isolation, culture and regeneration of plants from potato protoplasts. Plant Cell Rep. 8:307-310.
Kameya T (1975) Culture of protoplasts from chimeral plant tissue of nature. Japan.J.Genet. 50:417-420.
Kamo KK, Chang KI, Lynn ME & Hodges TK (1987) Embryogenic callus formation from maize protoplasts. Planta 172:245-251.
Kao KN & Michayluk MR (1975) Nutrient requirements for growth of *Vicia hajastana* cells and protoplasts at very low population density in liquid media. Planta 126:105-110.
Kao KN & Michayluls MR (1980) Plant regeneration from mesophyll protoplasts of alfalfa. Z.Pflanzenphysiol. 96:135-141.
Kirti PB (1988) Somatic embryogenesis in hypocotyl protoplast culture of rapeseed (*Brassica napus* L.). Plant Breeding 100:222-224.
Klein AS, Montezinos D & Delmer DP (1981) Cellulose and 1,3-glucan synthesis during the early stages of wall regeneration in soybean protoplasts. Planta 152:105-114.
Klimaszewska K (1989) Recovery of somatic embryos and plantlets from protoplast cultures of *Larix x eurolepis.* Plant Cell Rep. 8:440-444.
Kobayashi S, Uchimiya H & Ikeda I (1983) Plant regeneration from 'Trovita' orange protoplasts. Japan J.Breed. 33:119-122.
Kolenbach HW, Wanzel G & Hoffman F (1982) Regeneration of *Brassica napus* plantlets in cultures from isolated protoplasts of haploid stem embryos as campared with leaf protoplasts. Z.Pflanzenphysiol. 105:131-142.
Kouider M, Hauptamann R, Widholm JM, Skirvin RM & Korban SS (1984) Callus formation from *Malus* × *domestica* cv. "Jonathan" protoplasts. Plant Cell Rep. 3:142-145.
Kranz E & Lorz H In Vitro fertilization with isolated, single gametes results in zygotic embryogenesis and fertile maize plants. Plant Cell 5:739-746.
Krasnyanski S, Polgar Z, Nemeth G & Lupotto E (1992) Plant regeneration from callus and protoplast cultures of *Helianthus giganteus* L.. Plant Cell Rep. 11:7-10.
Kunitake H & Mii M (1990) Somatic embryogenesis and plant regeneration from protoplasts of asparagus (*Aspaaragus officinalis* L.). Plant Cell Rep. 8:706-710.
Kyozuka J, Hayashi Y & Shimamoto K (1987) High frequency plant regeneration from rice protoplasts by novel nurse culture methods. Mol.Gen.Genet. 206: 408-413.
Lehminger-Mertens R & Jacobsen HJ (1989) Plant regeneration from pea protoplasts via somatic embryogenesis. Plant Cell Rep. 8:379-382.
Li LC & Kohlenbach HW (1982) Somatic embryogeneis in quite a direct way in cultures of mesophyll protoplasts of *Brassica napus* L.. Plant Cell Rep. 1:209-211.
Li LC, Chen YM & Chen Y (1988) Studies on protoplast culture of rice (Oryza sativa L.) and plant regeneration from protoplast-derived calli.Acta Genet. Sinica 15:321-328.
Li SJ, Meng ZZ, Sheng FJ & Li DB (1992) Plant regeneration from mesophyll protoplasts from axenic of Chinese cabbage (*Brassica pekinensis* Rupr.) Chinese J.Biotech. 8:249-254.
Li WA & Wang YQ (1979) Observation on the induction of flower bud initiation from mesophyll cell protoplast culture of *Nicotiana rustica.* Acta Phytophysiol.Sinica 5:379-384.
Li WA (1987) Flower development from *Nicotiana rustica* mesophyll protoplasts. Sci.Sinica (Ser.B) 30:481-485.
Li XB & Xu ZH (1997) Protoplast culture of leguminous plants. In: Xu ZH & Wei ZM (eds.) Plant Protoplast Culture and Genetic Transformation. Chapter 11 (p. 105). Shanghai Sci. & Tech. Publ. Shanghai.
Li XB, Xu ZH, Wei ZM & Bai YY (1993) Somatic embryogenesis and plant regeneration from protoplasts of cowpea (*Vigna sinensis*). Acta Bot.Sinica, 35:632-636.

Li XQ, Chatrit P, Mathieu C, Vedel F, Paepe D, Remy R & Ambard-Bretteville F(1980) Regeneration of cytoplasmic male sterile protoclones of *Nicotiana sylvestris* with motochondrial variations. Curr.Genet. 13:261-266.

Li Z, Xia GM & Chen HM (1992) Somatic embryogenesis and plant regeneration from protoplasts isolated from embryogenic cell suspensions of wheat (*Triticum aestivem* L.). Plant Cell Tissue Organ Cult. 28:79-85.

Liao ZZ, Chen MZ, Liao QX, Yan QS & Zhang YQ (1994) Plant regeneration from protoplasts of sugarcane. Acta Bot.Sinica, 36:375-379.

Lillo C & Shahin EA (1986) Rapid regeneration of plants from hypocotyl protoplasts and root segments of cabbage. Hort.Sci. 21:315-317.

Ling HQ & Binding H (1987) Plant regeneration from protoplasts in *Linum*. Plant Breed. 98:312-317

Liu CM & Xu ZH (1988) Plant regeneration and transformation of mesophyll protoplasts of *Nicotiana glutinosa* L.. Acta.Biol.Exp.Sinica 21:273-283.

Loo SW & Xu ZH (1986) Recent advances in plant tissue and cell culture work in China. Advances in Science of China, Biology 1:295-354.

Lu CY, Vasil V & Vasil IK (1981) Isolation and culture of protoplasts of *Panicum maximum* Jacq. (guinea grass):somatic embryogenesis and plantlet formation. Z.Planzenphysiol. 104:311-318.

Lu DY, Davey MR & Cocking EC (1982) Somatic embryogenesis from mesophyll protoplasts of *Trigonella corniculata* (*Leguminosae*). Plant Cell Rep. 1:278-280.

Lu DY, Davey MR & Cocking EC (1983) A comparison of the cultural behaviour of protoplasts from leaves, cotyledons and roots of *Medicago sativa*. Plant Sci.Lett. 31:87-99.

Lu DY, Li FL, Chen YM & Chen Y (1987) Somatic embryogenesis and plant regeneration from protoplasts and explants of seedlings of crownvetch (*Forage legume*).Acta Biol.Exp.Sinica, 20:31-39.

Lu DY, Pental D & Cocking EC (1982) Plant regeneration from seedling cotyledon protoplasts.Z.Pflanzenphysiol. 107:59-63.

Luhrs R & Lorz H (1988) Initiation of morphogenic cell-suspension and protoplast cultures of barley (*Hordeum vulgare* L.). Planta 175:71-81.

Luo XM, Zhao GL, Xie XJ, Liu YZ, He MY & Hao S (1991) Plant regeneration from protoplast culture of *Astragalus huangheensis*. Acta Genet.Sinica, 18:239-243.

M et al. (eds.) Current Issues in Plant Molecular & Cellular Biology (pp:213-218) Kluwer Acad.Pulb., Dordrecht.

Maheshowari SC, Gill R, Maheshwari N & Gharyal PK (1986) Isolation and regeneration of protoplasts from higher plants. In:Reinert J & Binding H(eds.) Differentiation of Protoplasts and of Transformed Plant Cells (pp:3-36) Springer, Berlin.

Malaure RS, Davey MR & Power JB (1990) Plant regeneration from protoplasts of *Felicia* and *Brachycome*. Plant Cell Rep.9:109-112.

Mariotti D, Arcioni S & Pezzotti M (1984) Regeneration of *Medicago arborea* L. plants from tissue and protoplast cultures of different organ origin.Plant Sci.Lett. 37:149-156.

Meijer EGM & Steinbiss HH (1983) Plantlet regeneration from suspension and protoplast cultures of the tropical pasture legume *Stylosanthes guyanensis*(Aubl.). Sw.Ann.Bot. 52:305-310.

Merkle SA & Sommer HE (1987) Regeneration of *Liriodendron tulipifera*(family Magnoliaceae) from protoplast culture. Am.J.Bot. 74:1317-1321.

Meyer Y & Cooke R (1979) Time course of hormonal control of the first mitosis in tobacco mesophyll protoplasts cultivated *in vitro*. Planta 147:181-185.

Meyer Y & Herth W (1978) Chemical inhibition of cell wall formation and cytokinesis, but not of nuclear division in protoplasts of *Nicotiana tabacum* L. cultured *in vitro*. Planta 142:253-262.

Mii M & Ohashi H (1988) Plantlet regeneration from protoplasts of kiwifruit, Actinidia chinensis. Acta.Hort. 230:167-170.

Millam SR, Burns ATH, Hocking TJ & Cattell KJ (1988) Effects of a Ficollagarose system on early division and development of mesophyll protoplasts of winter oilseed rape (*Brassica napus* L.) Plant Cell Tissue Organ Cult. 12:285-290.

Nagata T & Takebe I (1971) Plating of isolated tobacco mesophyll protoplasts on agar medium. Planta 99:12-20.

Narasimhulu SB & Chopra VL (1988) Species specific shoot regeneration response of cotyledonary explants of Brassicas. Plant Cell Rep. 7:104-106.

Negrutiu I & Mousseau J (1980) Protoplasts culture from *in vitro* grown plants in *Nicotiana sylvestris* Spegg. and Comes. Z.Pflanzenphysiol. 100:373-379.

Newell CA & Luu HT (1985) Protoplast culture and plant regeneration in *Glycine canescens* F.J. Herm. Plant Cell Tissue Organ Cult. 4:145-149.

Nielsen KA, Larsen E & Knudsen E (1993) Regeneration of protoplast-derived green plants of kentucky blue grass (*Poa pratensis* L.). Plant Cell Rep. 12:537-540.

Nitsch JP & Ohyama K (1971) Obtention de plantes a partir de protoplasts haploides cultives *in vitro*. C.R.Acad.Sci.(Paris) 273:801-804.

Ochatt SJ & Caso OH (1986) Shoot regeneration from leaf mesophyll protoplasts of wild pear (*Pyrus communis* var. *Pyraster* L.). J.Plant Physiol. 122:243-249.

Ochatt SJ & Power JB (1988) Plant regeneration from mesophyll protoplasts of Williams' Bon Chretien (Syn.Bartiett) pear (*Pyrus communis* L.). Plant Cell Rep. 7:587-589.

Ochatt SJ (1990) Plant regeneration from root callus protoplasts of sourberry (*Prunus cerasus* L.). Plant Cell Rep. 9:268-271.

Ochatt SJ (1992) The development of protoplast-to-tree systems for *Prunus cerasifera* and *Prunus spinosa*. Plant Sci. 81:253-259.

Ochatt SJ, Chand PK, Rech EL, Davey MR & Power JB (1988) Electroporation-mediated improvement of plant regeneration from colt cherry *(Prunus avium ×Pseudocerasus*) protoplasts. Plant Sci. 54:165-169.

Ochatt SJ, Cocking EC & Power JB (1987) Isolation, culture and plant regeneration of colt cherry (*Prunus avium × pseudocerasus*) protoplasts. Plant Sci. 50:139-143.

Oka S & Ohyama K (1985) Plant regeneration from leaf mesophyll protoplasts of *Broussonetia kazinaki* Sieb. (paper mulberry). J.Plant Physiol. 119:455-460.

Oliviea MM & Pais MSS (1991) Plant regeneration from protoplasts of long-term callus cultures of *Actinidia deliciosa* var. *deliciosacv*. Hayward (kiwifruit). Plant Cell Rep. 9:643-646.

Park YG & Son SH (1992) In vitro shoot regeneration from leaf mesophyll protoplasts of hybrid poplar (*Populus nigra × P.maximowiczii*). Plant Cell Rep. 11:2-6.

Perales EH & Schieder O (1993) Plant regeneration from leaf protoplasts of apple. Plant Cell Tissue Organ Cult. 34:71-76.

Petersen WL, Sulc S & Armstrong CL (1992) Effect of nurse cultures on the production of macro-calli and fertile plants from maize embryogenic suspension culture protoplasts. Plant Cell Rep. 10:591-594.

Pillai V, Davey MR & Power JB (1990) Plant regeneration from mesophyll protoplasts of *Centaurea cyanus, Senecio × Hybridus* and *Callistephus chinensis*. Plant Cell Rep. 9:402-405.

Power JB & Berry SF (1979) Plant regeneration from protoplasts of *Browallia viscosa*. Z.Planzenphysiol. 94:469-471.

Pua EC (1990) Somatic embryogenesis and plant regeneration form hypocotyl protoplasts of *Brassica juncea* (L.) Czern & Coss. Plant Sci 68:231-238.
Puonti-Kaerlase J & Eriksson T (1988) Improved protoplast culture and regeneration of shoots in pea (*Pisum sativum* L.). Plant Cell Rep. 7:242-245.
Rao IVR, Mehta U & Mohan Ram HY (1982) Whole plant regeneration form cotyledonary 3.protoplasts of *Crotalaria juncea*. In: Fujiwara A (ed.) Plant Tissue Culture (pp.595-596) Japanese Association for Plant Tissue Culture, Tokyo.
Rao PS & Ozias-Akins P (1985) Plant regeneration through somatic embryogenesis in protoplast cultures of sandalwood (*Santalum album* L.). Protoplasma 124:80-86.
Rasheed JH, Al-Mallah MK, Cocking EC & Davey MR (1990) Root hair protoplasts of *Lotus corniculatus* L. (birdsfoot trefoil) express their totipotency. Plant Cell Rep. 8:65-569.
Ren YG, Jia JF & Zheng GC (1990) Plantlet regeneration from protoplasts of millet (*Setaria italica* L.). Chinese J.Bot. 2:103-108.
Rhodes CA, Lowe KS & Ruby KL (1988) Plant regeneration from protoplasts isolated from embryogenic maize cell cultures. Biotechnology 6:56-60.
Robertson D & Earle ED (1986) Plant regeneration from leaf protoplasts of *Brassica oleracea* var. *italica* cv. Green Comet broccoli. Plant Cell Rep. 5: 61-64.
Robertson D, Earle ED & Mutschler MA (1988) Increased totipotency of protoplasts from *Brassica oleracea* plants previously regenerated in tissue culture. Plant Cell Tissue Organ Cult. 1415-24.
Roest S & Gilissen LJW (1989) Plant regeneration from protoplasts: a literature review. Acta Bot.Nearl 38:1-23.
Russell JA & McCown BH (1986) Culture and regeneration of *Populus* leaf protoplasts isolated from non-seedling tissue. Plant Sci. 46:133-142.
Russell JA & McCown BH (1988) Recovery of plants from leaf protoplasts of hybrid-poplar and aspen clones. Plant Cell Rep. 7:59-62.
Saxena PK & Gill R (1987) Plant regeneration from mesophyll protoplasts of the tree legume *Pithecellobium dulce* Benth. Plant Sci. 53:257-262.
Schieder O (1980) Somatic hybrids between a herbaceous and two tree-*Datura* speies. Z.Pflanzenphysiol. 98:119-127.
Schopke C, Muller LE & Kohlenbach HW (1987) Somatic embryogenesis and regeneration of plantlets in protoplast culture from somatic embryos of coffee (*Coffea canephora* P.ex Fr). Plant Cell Tissue Organ Cult. 8:243-248.
Scowcroft WR, Davey MR & Power JB (1973) Crown gall protoplasts-isolation, culture and ultrastructure. Plant Sci. Lett. 1:451-456.
Shahin EA & Shepard JF (1980) Cassava mesophyll protoplasts: isolation, proliferation, and shoot formation. Plant Sci. Lett. 17:459-465.
She JM, Zhou HY, Lu WZ, Wu HM, Li XH & Sun YR (1990) Characters of the regenerated plants and their progenies (Rz) from rice protoplasts. Acta Genet.Sinica 17:438-442.
Shekhawat NS & Galston AW (1983 a) Mesophyll protoplasts of fenugreek (*Trigonella foenum-graecum*): isolation, culture and shoot regeneration. Plant Cell Rep. 2:119-121.
Shekhawat NS & Galston AW (1983 b) Isolation, culture and regeneration of moth bean *Vigna aconitifolia* leaf protoplasts. Plant Sci. Lett. 32:43-51.
Shepard JF (1980) Abscisic acid-enhanced shoot initiation in protoplast-derived calli of potato. Plant Sci. Lett. 18:327-333.
Shepard JF (1980) Mutant selection and plant regeneration from potato mesophyll protoplasts. In: Rubenstein I et al. (eds.) Genetic Improvement of Crops. (pp:185-219).
Sim GE, Loh CS & Goh CJ (1988) Direct somatic embryogenesis from protoplasts of *Citrus mitis* Blanco. Plant Cell Rep. 7:418-420.

Sinha RR & Das K (1986) Anther-derived callus of *Dolichos biflorus* L., its protoplast culture and their morphogenic potential. Curr. Sci. 55:447-452.
Sinha RR, Das K & Sen SK (1983) Embryoids from mesophyll protoplasts of *Vigna mungo* L. Hepper, a seed legume crop plant. In: Plant Cell Culture in Crop Improvement (pp.209-214) Plenum Press, New York.
Sree Ramulu K, Dijkhuis P, Roest S, Bokelmann GS & De Groot B (1984) Early occurrence of genetic instability in protoplast cultures of potato. Plant Sci. Lett. 36:79-86.
Srinivasan C & Vasil IK (1985) Callus formation and plantlet regeneration from sugarcane protoplasts isolated from embryogenic cell suspension cultures. Am. J. Bot. 72:833.
Sun JS, Lu TG & Sondahl MR (1989) Establishment and characterization of haploid suspension cells in supersweet maize. Acta Bot. Sinica 31:742-749.
Tegeder M, Schieder O & Pickardt T (1995) Breakthrough in the plant regeneration from protoplasts of *Vicia faba* and *V.narbonensiis*. In: Terzimn et al. (eds.) Current Issues in Plant Molecular and Cellular Biology(pp:75-80) Kluwer Academic Publishers, Dordrecht.
Terakawa T, Sato T & Koike M (1992) Plant regeneration from protoplasts isolated from embryogenic suspension cultures of creeping bentgrass (*Agrostis palustris* Huds.). Plant Cell Rep. 11:457-461.
Thompson JA, Abdullah R & Cocking EC (1986) Protoplast culture of rice (*Oryza sativa* L.) using media solidified with agarose. Plant Sci. 47:123-133.
Toriyama K, Hmata K & Sasaki T (1986) Haploid and diploid plant regeneration from protoplasts of anther callus in rice. Theor.Appl.Genet. 73:16-19.
Tsai CK (1988) Plant regeneration from leaf callus protoplasts of *Actinidia chinensis* Planch. var. *chinensis*. Plant Sci. 54:231-235.
Van Der Valk P & Zaal MACM (1988) Regeneration of plantlets from protoplasts of *Poa pratensis* L.(Kentucky bluegrass). In: Puite KJ et al. (eds.) Progress in Plant Protoplast Rcscarch (pp.59-60) Kluwcr Academic Publishers, Dordrecht.
Vardi A & Spiegel-Roy P (1982) Plant regeneration form *Citrus* protoplasts: variability in methodological requirements among cultivars and species. Theor.Appl.Genet. 62:171-176.
Vardi A, Hutchison DJ & Galun E (1986) A protoplast-to-tree system in *Microcitrus* based on protoplasts derived from a sustained embryogenic callus. Plant Cell Rep. 5:412-414.
Vardi A, Spiegel-Roy P & Galun E (1975) Citrus cell culture: isolation of protoplasts, plating densities, effect of mutagens and regeneration of embryos. Plant Sci. Lett. 4:231-236.
Vardi A, Spiegel-Roy P & Galun E (1983) Protoplast isolation, plant regeneration and somatic hybridization in different citrus species and microcitrus. In: Potrykus et al. (eds.) Protoplast 1983, Poster Proc. of 6th Intern.Protoplast Symp. Experientia(Suppl.) 45:284-285.
Vasil IK & Vasil V (1992) Advances in cereal protoplast research. Physiol.Plant. 85:279-283.
Vasil IK (1995) Cellualr and molecular genetic improvement of cereals. In: Terzi M et al. (eds.) Current Issues in Plant Molecular and Cellular Biology (pp.5-18) Kluwer Academic Publishers, Dordrecht.
Vasil V & Vasil IK (1980) Isolation and culture of cereal protoplasts. Part.2: Embryogenesis and plantlet formation from protoplasts of *Pennisetum americanum*. Theor.Appl.Genet. 56:97-99.
Vasil V & Vasil IK (1987) Formation of callus and somatic embryos from protoplasts of a commercial hybrid of maize (*Zea mays* L.). Theor.Appl. Genet. 73:793-798.
Vasil V, Redway F & Vasil IK (1990) Regeneration of plants from embryogenic suspension culture protoplasts of wheat (*Triticum aestivum* L.). Bio/Technology 8:429-434.
Vasil V, Wang DY & Vasil IK (1983) Plant regeneration from protoplasts of napier grass (*Pennisetum purpureum* Schum.). Z.Pflanzenphysiol. 111:233-239.

Wang D, Miller PD & Sondahl MR (1989) Plant regeneration from protoplasts of Indica type rice and CMS rice. Plant Cell Rep. 8:329-332.
Wang F, Shao XH & Chen RM (1992) Studies on protoplast culture and plant regeneration of *Dimocarpus longan*. Proc of the 1st Symp. of Chinese Soc. of Agricultural Biotechnology (pp.109-112).
Wang GY & Xia ZA (1987) Mature plant regeneration from rice protoplasts. Acta Biol.Exp.Sinica 20:253-257.
Wang GY, Xia ZA & Wang LF (1986) Regenerated plantlets from cultured mesophyll protoplasts of onion (*Allium cepa* L.). Acta Bio,.Exp.Sinica 19:409-413.
Wang SP, Xu ZH & Wei ZM (1991) Culture and regeneration of poplar mesophyll protoplasts. Sci.Sinica 34:587-592.
Wang Y, Huang MR, Chen DM, Xu N, Wei ZM & Xu ZH (1995a) Plants regenerated from mesophyll protoplasts of cottonwood new clone. Acta Bot.Sinica 37:379-385.
Wang Y, Huang MR, Wei ZM, Sun YR, Chen DM, Xu ZH, Zhang LM & Xu N (1995b) Regeneration of simon poplar (*Populus simonii*) from protoplast culture. Plant Cell Rep. 14:442-445.
Wei ZM & Xu ZH (1988) Plant regeneration from protoplasts of soybean (*Glycine max* L.). Plant Cell Rep. 7:348-351.
Wei ZM & Xu ZH (1989) Plant regeneration from *Sorghum* protoplast culture.Plant Physiol. Commic (Shanghai) 6:45-46.
Wei ZM & Xu ZH (1990) Factors affecting culture of hypocotyl protoplasts of cauliflower and plantlet regeneration. Acta Phytophysiol.Sinica 16:394-400.
Wei ZM & Xu ZH (1990) Protoplast culture and plant regeneration of soybean (*Glycine max* and G.soja). Acta Bot.Sinica 32:582-588.
Wei ZM & Xu ZH (1990) Regeneration of fertile plants from embryogenic suspension culture protoplast of *Sorghum vulgare*. Plant Cell Rep. 9:51-53.
Wei ZM, Xu ZH, Huang JQ & Huang MR (1994) Plants regenerated from mesophyll protoplasts of white mulberry. Cell Res. 4:183-189.
Wei ZM, Xu ZH, Xu N & Huang MR (1991) Mesophyll protoplast culture and plant regeneration of oriental plnetree (*Platanus orientalis*). Acta Bot.Sinica 33:813-818.
Wei ZM, Zhang Y & Xu ZH (1993) Plant regeneration iin peanut protoplast culture. Ann.Rep. of National Laboratory of Plant Mol.Genet.(1992). p.22-23.
White DWR (1983) Plant regeneration from mesophyll protoplast of white clover (*Trifolium repens* L.). In: Potrykus I et al. (eds) Protoplast 1983, Poster Proc. of 6th Intern.Protoplast Symp.(pp.60-61) Birkhauser Verlag,Basel.
Wilson VW, Haq N & Evans PK (1985) Protoplast isolation, culture and plant regeneration in the winged bean, *Psophocarpus tetragonolobus*(L.). Plant Sci.41:61-68.
Xia ZA (1992) Plant protoplast culture in China, Advaces in Science of China, Biology 3:77-86.
Xiong XS, Jia SR & Fu YY (1988) Protoclonal variation in cabbage (*Brassica oleracea* L. var. *capitata*). Acta Horticult.Sinica 15:120-124.
Xu XX & Xu ZH (1988) Plant regeneration from mesophyll protoplasts of *Orychophragmus violaceus*. Acta Phytophysiol.Sinica 14:170-174.
Xu ZH, Davey NR & Cocking EC (1981) Isolation and sustained division of *Phaseolus aureus* (mung bean) root protoplasts. Z.Pflanzenphysiol. 104:289-298.
Xu ZH & Wei ZM (1993) Protoplast culture of grain legumes. In: Soh WY et al.(eds.) Advances in Developmental Biology and Biotechnology of Higher Plants. (pp.197-209) Korean Soc. Plant Tissue Culture, Taejeon

Xu ZH & Xue HW (1997) The species of plants regenerated from protoplasts of higher plants. In:Xu ZH & Wei ZM (eds.) Plant Protoplast Culture and Genetic Transformation. Chapter 3. Shanghai Sci. & Tech. Publ. P246.

Xu ZH, Davey MR & Cocking EC (1982) Organogenesis from root protoplasts of the forage legumes *Medicago sativa* and *Trigonella foenum-graecum* Z. Pflanzenphysiol. 107:231-235.

Xu ZH, Davey MR & Cocking EC (1984) Root protoplast isolation and culture in higher plants. Sci.Sinica (11):1012-1018.

Xue HW, Wei ZM & Xu ZH (1994) Construction of high efficient regeneration system from hypocotyl protoplasts of *Brassica oleracea*. Acta.Biol.Exp.Sinica 27:259-269.

Yamada Y, Yang ZQ & Tang DT (1986) Plant regeneration from protoplast-derived callus of rice (*Oryza sativa* L.). Plant Cell Rep. 5:85-88.

Yamashita Y & Shimamoto K (1989) Regeneration of plants from cabbage (*Brassica oleracea* var. *capitata*) protoplasts. In: Bajajyps (ed) Biotechnology in Agriculture & Forestry, 8:193-205. Springer-Verlag, Berlin.

Yan CQ, Bian ZX, Zhang DS, Yang JS & Ge KL (1991) Plantlet regeneration from mesophyll protoplasts of *Canavallis ensiformis*. J.Fudan Univer. (Natural Sci.) 30:381-386.

Yan Q, Zhang X, Shi J & Li J (1990) Green plant regeneration from protoplasts of barley (*Hordeum vulgare* L.). Kexue Tongbao (20):1581-1583.

Yan Q, Zhang X, Teng S, Huang C, Yan Q & Wang J (1995) Establishment of a system of culture technique from rice protoplasts. Acta Agric.Sinica,No2. Special issue on protoplast culture of field crops (pp.20-26).

Yang JB, Wu JD, Wei ZM & Xu ZH (1992) Plant regeneration from protoplasts of Indica rece. Chinese J.Biotech.8:60-64.

Yang YM, He DD & Scott KJ (1993) Plant regeneration from protoplasts of durum wheat (*Triticum durum* Dest.cv.D6962). Plant Cell Rep. 12:320-323.

Yang ZN & Xu ZH (1997) Protoplast culture of Cruciferous plants. In: Xu ZH & Wei ZM (eds.) Plant Protoplast Culture and Genetic Transformation. Chapter 12.Shanghai Sci. & Tech. Publ. (in Chinese) 124-141.

Yang ZN, Xu ZH & Wei ZM (1994) Cauliflower inflorescence protoplast culture and plant regeneration. Plant Cell Tissue Organ Cult. 36:191-195.

Yang ZQ, Shikanai K, Mori K & Yamada Y (1989) Plant regeneration from cytoplasmic hybrids of rice(*Oryza sativa* L.). Theor.Appl.Genet. 77:305-310.

Yarrow SA & Barsby TL (1988) Recent advances in protoplast culture of horticultural crops: Vegetable crops. Sci.Horticult. 37:179-200.

Yasugi S (1989) Isolation and culture of orchid protoplasts. In: Bajaj YPS (ed.) Biotechnology in Agriculture & Forestry 8:235-253. Springer, Berlin.

Yin Y, Li S, Chen Y, Guo H, Tian W, Chen Y & Li L (1993) Fertile plants regenerated from suspension culture-derived protoplasts of an indica type rice (*Oryza sativa* L.). Plant Cell Tissue Organ Cult. 32:61-68.

Zakri AH (1984) Plant regeneration from protoplast-derived callus of winged bean. In: Laned W et al. (eds.) Efficiency in Plant Breeding (pp.363) PUDOS,Wageningen.

Zapata FJ & Sink KC (1981) Somatic embryogenesis from *Lycopersicon peruvianum* leaf mesophyll protoplasts. Theor.Appl.Genet. 59:265-268.

Zhang GF, Huang BL, Wei L, Luo XM, Zheng XF, He MY & Hao S (1993) Plant regeneration from protoplasts derived from suspension cultures in *Ceymus regemosus*. Acta Bot.Sinica, 35:422-428.

Zhang GF, Luo XM, Li FX & An LJ (1994) Plantlet regeneration of protoplasts derived from cell-suspension cultures of *Oxytropis leptophylla*. Acta Biol.Exp.Sinica, 27:117-121.

Zhang H, Tian HQ & Li QM (1994) Plant regeneration from cell and protoplast cultures of *Triticum aestivum-Haynaldia villosa* hybrid. Acta Bot.Sinica, 36:479-482.

Zhang YJ, Wu XJ, Cai QG, Zhou YL & Qian YQ (1995) Plantlet regeneration from leaf protoplasts in seedlings of *Actinidia eriantha* benth. Acta Bot.Sinica, 37:48-52.

Zhang ZQ, Zhong WJ, Tang KX, Zhou Y & Zhu MF (1994) Plant regeneration from allotriploid (*O.sativa* X *O.minuta*) Hybrid protoplast. Acta.Crops.Sinica. 20: 578-581.

Zhao GL & Luo XM (1990) Plant regeneration from protoplasts soybean (*Glyciine max* L.). Acta Bot.Sinica, 32:616-627.

Zhao GL (1990) Plant regeneration from saintoin (*Onobrychis viciaefolia*) protoplasts derived from cell suspensions of hypocotyl. Acta Bot.Sinica, 32:973-976.

Zhao YX & Yao DY (1993) Plant regeneration from protoplasts isolated from callus tissue of *Hibiscus syriacus* L. Chinese J.Bot. 3:29-33.

Zhao YX (1993) The isolation, culture and genetic manipulation of protoplast from salt and drought tolerant leguminous plants. Ph.D. Thesis, Coventry University, U.K.

Zhao YX, Harris PJC & Yao DY (1995) Plant regeneration from protoplasts isolated from cotyledons of *Sesbania bispinosa*. Plant Cell Tissue Organ Cult. 40:119-123.

Chapter 3

MORPHOGENESIS IN HAPLOID CELL CULTURES

Sant S. BHOJWANI[1], Woong-Young SOH[2] and Himani PANDE[1]
[1]Department of Botany, University of Delhi, Delhi 110007, India and [2]Department of Biological Sciences, Chonbuk National University, Chonju 561-756, Korea

1 INTRODUCTION

In higher plants, where the main plant body is diploid, the haploid phase is represented by short-lived, highly specialized cells of the male (pollen grains) and female (embryo sac) gametophytes. These gametophytes originate from a sporophytic cell through a meiotic division; one or more of the four products of meiosis develop into multicellular gametophytes (Bhojwani and Bhatnagar, 1999). The gametophytic differentiation involves the expression of a set of gametophyte-specific genes (Scott et al., 1991). The long felt need to produce large numbers of haploids for genetics and breeding research led to continued search for methods to achieve this goal. In 1964 it was shown for the first time that male apogamy can be induced in the anther cultures of a flowering plant (Guha and Maheshwari, 1964). Since then this morphogenic potential of male gametophytes (androgenesis; Fig. 1) has been demonstrated for over 134 species belonging to 25 families (Bhojwani and Razdan, 1996). Protocols have also been developed for androgenesis in isolated microspore cultures. Extensive research during the last 15 years has identified several factors which induce or promote androgenesis (Bhojwani et al., 1996) and has given some insight into the cellular and sub-cellular changes associated with the switch-over from gametophytic mode to sporophytic mode of development (Cordewener et al., 1996). Isolated microspore culture offers a unique opportunity to study the process of direct embryogenesis from single, haploid cells. These aspects are discussed in this article.

Figure 1. A cultured anther of *Brassica juncea* with a load of pollen embryos at different stages of development.

Haploid plants have also been obtained from the cells of the female gametophyte (in vitro parthenogenesis and apogamy) but due to technical difficulties this process of gynogenesis is not well understood.

2 ORIGIN OF ANDROGENIC HAPLOIDS

In the normal course of pollen development the uninucleate microspores undergo an asymmetric division, cutting a small generative cell and a large vegetative cell. The former is initially attached to the intine (inner wall of the pollen grain) but eventually it comes to lie freely in the cytoplasm of the vegetative cell. Whereas the generative cell divides further, forming sperms, the vegetative cell nucleus remains arrested in G_1 phase of the cell cycle (Zarsky et al., 1992). Thus, in situ development of pollen is programmed for terminal differentiation into highly specialized gametophytic cells. Manipulation at an early stage can alter this highly controlled developmental programme. Induction of androgenesis, thus, involves the inhibition of the existing gametophytic developmental pathway and acquisition of competence for sporophytic development via direct embryogenesis or through a callus phase.

2.1 Induction

For most species a suitable stage for the induction of androgenesis lies between just before or just after pollen mitosis. During this period the cells are noncommital in their developmental potential (Porter et al., 1984; Scott et al., 1991). After the first mitotic division the cytoplasm gets populated with gametophytic information and it gradually becomes irreversibly programmed to form male gametophyte. This stage may be reached within 24 hrs after pollen mitosis (Mascarenhas, 1971).

A variety of stresses applied during the labile developmental period of the pollen grain can mask the gametophytic programme and induce the expression of sporophyte-specific genes and, thus, induce the pollen to switchover from gametophytic mode to sporophytic mode of development. In some cases the stress of excising the anthers and placing them under culture conditions is adequate to bring about this shift (Sunderland, 1974; Nitsch and Nitsch, 1969). In others, treatments such as temperature shocks (high or low), high osmolarity, and starving the pollen grains of sugar or other nutrients are required to induce or promote the induction of androgenesis.

2.1.1 Molecular aspects

The induction of embryogenesis from microspores/pollen grains has to be accompanied by the initiation of a new programme of gene expression. This aspect of androgenesis has been investigated in some details only with *Nicotiana tabacum* and *Brassica napus*. The stage of pollen most susceptible to external stresses for shift from gametophytic development to sporophytic development is different in the two systems. In tobacco the optimum stage lies between late uninucleate to early binucleate stage, and the pollen embryos are formed by the vegetative cell of 2-celled pollen. In *B. napus*, on the other hand, the late uninucleate microspores are most suitable for androgenesis. In this case, microspores divide by a symmetrical division, and both the daughter cells contribute in the formation of embryo.

Aruga et al. (1985) reported that placing the anthers of *N. tabacum* in sugar-free medium for a few days immediately after culture suppressed gametogenesis and induced androgenesis. Sugar-starvation caused a loss of the ability of the generative cell nucleus to synthesize DNA even after transfer to sugar-containing medium. However, the vegetative cell nucleus acquired the unique potentiality to synthesize DNA even in the absence of sucrose (normally the vegetative nucleus remains arrested in the G1 phase of the cell cycle). Induction of androgenesis by sugar-starvation has also been observed in microspore cultures of *Hordeum vulgare* (Wei et al., 1986). In tobacco, glutamine-starvation (in the absence of sucrose) of the mid-bicellular pollen for the initial 24-48 h of culture also favoured androgenic development on glutamine-containing medium; direct culture of pollen on basal medium containing glutamine produced only mature pollen (Kyo and Harada, 1985, 1986).

During the starvation treatment a large fraction of pollen showed DNA replication in the vegetative nucleus. However, the presence of hydroxyurea in the starvation medium did not effect the formation of embryos after transfer to sugar-containing, hydroxyurea-free medium, suggesting that DNA replication during starvation is not essential for embryo formation (Zarsky et al., 1990,1992). Probably an event preceding S-phase is important for embryogenesis. No newly synthesized protein could be detected during starvation (Garrido et al., 1993; Kyo and Harada, 1990a) but two major mRNA appeared during this period which were stored in inactive form in the embryogenic grains (Garrido et al., 1993). Earlier, Harada and his co-workers had observed the appearance of four new phosphoproteins in the embryogenic grains which were located on the plasma membrane (Harada et al., 1988; Kyo and Harada, 1990a,b).

Zarsky et al. (1990) have suggested that the starvation of pollen grains before exposing them to full nutrient medium probably causes repression of the gametophytic cytoplasm and differentiation of vegetative cell, pushing it from G1 to S phase of cell cycle which is required for embryogenic pathway of pollen development.

Most of the information on the induction of pollen embryogenesis generated during the last 5 years is based on *B. napus* microspore cultures (Cordewener et al., 1996). In this system the late uninucleate microspores or early binucleate pollen cultured at 18℃ form only gametophytes. At 32℃ at least 40% of the microspores exhibit sporophytic division within the first two days of culture (Cordewener et al., 1994; Custer et al., 1994). About 90% of these microspores acquire irreversible commitment to embryogenesis within 8 h of high temperature treatment. Besides uninucleate microspores, some 2-celled pollen also form embryos after treatment at 32℃. Recently, Hause et al. (1996) achieved embryogenesis in the cultures of late 2-celled pollen of *B. napus* by exposing the pollen to 41℃ for 1-2 h before culture at 32℃. Two celled pollen form embryo by reactivation of the vegetative cell. Pechan et al. (1991) reported a decrease in the total protein content of rapeseed microspores under embryogenic (32℃) and non-embryogenic (18℃) condition. In contrast, Cordewener et al. (1994) reported a two fold increase in the overall rate of protein synthesis during the induction period. A comprehensive qualitative and quantitative computer analysis of the 2D protein pattern from these cultures revealed six polypeptides exclusively synthesized under embryogenic conditions, 18 polypeptides which were synthesized in the embryogenic pollen at a faster rate than in the non-embryogenic pollen and one polypeptide preferentially produced by non-embryogenic pollen. Two of the polypeptides of the 2nd category were identified as members of the 70 KDa heat shock proteins (HSP68, HSP70) (Cordewener et al., 1995). The HSP68 were co-distributed with DNA-containing organelles, presumably mitochondria. During normal gametogenesis the HSP70 are located in the nucleoplasm during the S-phase and predominantly in the cytoplasm during the remaining part of the cell cycle. During the induction of pollen embryogenesis the HSP70 occurred mainly in the nucleoplasm of the cell which formed embryo, suggesting a strong correlation among the induction of embryogenesis, HSP70 synthesis and nuclear localization.

As in rapeseed, induction of pollen embryogenesis in *Hyoscyamus niger* occurs within a few hours of culture under inductive conditions (Raghavan, 1979). Autoradiographic studies revealed that an intense period of protein synthesis is associated with the deflection of the gametophytic programme to the embryogenic pathway. Radioisotope incorporation has established that transcription of rRNA and mRNA in the generative cell (in this plant pollen embryo develops by the sole activity of the generative cell) is an important prerequisite for embryogenic divisions; isotope incorporation being negligible during pollen development.

2.1.2 Cytological aspects

The cytology of pollen embryogenesis has been studied chiefly in *N. tabacum* and *B. napus*. In *N. tabacum*, where androgenic embryos arise from the vegetative cell, for some time after pollen mitosis all the pollen grains develop similarly: increase in size, achieve higher stainability with acetocarmine and accumulate starch grains in the vegetative cell. Whereas most of the grains continue to gain further in these features up to maturity, a small proportion of the pollen (ca 0.7%), called "S-grains" or "P-grains" or simply "androgenic grains", do not increase much in size, their cytoplasm remains faintly staining, and their nuclei clearer. The occurrence of structurally two types of grains in an anther is described as "pollen dimorphism". Normally the embryogenic grains would be non-functional male gametophytes but given appropriate conditions they form embryos (Horner and Street, 1978; Horner and Mott, 1979; Heberle-Bors, 1985). Pollen dimorphism has also been observed in the cereals, wheat (Zhou, 1980), rice (Cho and Zapata, 1990) and barley (Dale, 1975). In rice the androgenic microspores (ca 10%) are larger and more vacuolated than the non-emryogenic microspores (Fig. 2). The frequency of embryogenic grains can be increased above the level of its natural occurrence by dedifferentiation of the gametophytically programmed grains under inductive conditions (Bhojwani et al., 1973; Dunwell, 1978; Heberle-Bors, 1983). During dedifferentiation lysosome-like multivesicular bodies appear which may be involved in the breakdown of the cytoplasm. Towards the end of the inductive period, in terms of cell organelles, only a few mitochondria and structurally simplified plastids are left in the vegetative cell (Dunwell, 1978). The ribosomes are completely washed out. Following the first sporophytic division a fresh population of ribosomes and other organelles appear.

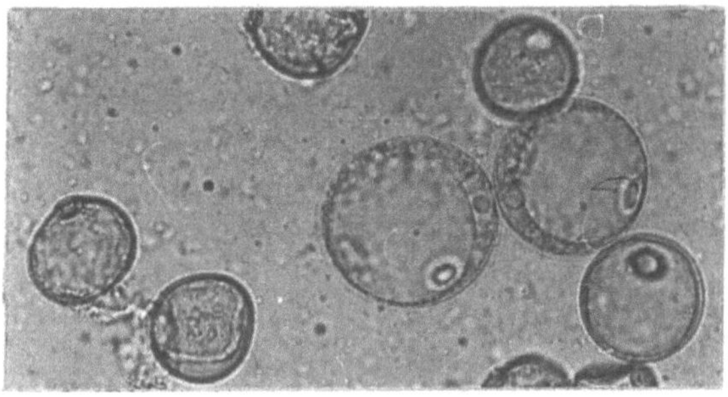

Figure 2. Pollen grains from an anther of *Oryza sativa* showing pollen dimorphism.

In *B. napus*, where the late uninucleate stage of microspore development is most labile to the induction of embryogenesis, the first cytological changes associated with the switch to sporophytic development is the loss of vacuole, the movement of the asymmetrically placed nucleus to the central position due to apparent loss of microtubule cytoskeleton investing the nucleus, and the appearance of starch containing plastids and globular domain within the cytoplasm (Zaki and Dickinson, 1990). A 12 h pulse treatment of pollen grains with 25 mg l^{-1} of colchicine, an antimicrotubule drug, before the first mitosis significantly increased the number of microspores undergoing symmetrical division and the production of embryos in isolated microspore cultures (Zaki and Dickinson, 1991).

Before a cell of the male gametophyte undergoes the first androgenic division it synthesizes a thick somatic cell wall around the plasma membrane giving it a sporophytic characteristic (Sangwan-Norreel, 1978; Rashid et al., 1982; Zaki and Dickinson, 1990). The first division is clearly a somatic type division in which a normal cell plate is formed which is traversed by plasmodesmata and contains no callose (Zaki and Dickinson, 1990).

2.2 Early cell divisions

Based on the few initial divisions in the microspores/pollen four modes of in vitro androgenesis have been identified (Bhojwani and Razdan, 1996). (i) As commonly observed in *B. napus* (Zaki and Dickinson, 1992), the microspores divide by an equal division, and the two identical daughter cells contribute to the sporophyte development; distinct vegetative and generative cells are not formed. (ii) The uninucleate microspores divide by a normal unequal division, and the sporophytes arise through further divisions in the vegetative cell. This mode of development is commonly encountered in *N. tabacum, H. vulgare Triticum aestivum, Triticale* and *C. annum*. (iii) In *H. niger* the pollen embryos are predominantly formed from the generative cell alone; the vegetative cell either does not divide at all or does so only to a limited extent. (iv) As in pathway ii, vegetative and generative cells are formed but both the cells divide further and participate in the development of the sporophyte (e.g., *D. innoxia*). A system may show more than one the four modes of androgenesis.

2.3 Later development

Irrespective of the early pattern of divisions, the responsive pollen grains finally become multicellular and burst open to release the tissue which has an irregular outline. In several cases (e.g., *Atropa, Brassica, Datura, Hyoscyamus, Nicotiana*) this cellular mass gradually assumes the form of a globular embryo and undergoes the normal stages of post-globular embryogeny (heart-shaped, torpedo-shaped, and cotyledonary stage). By this method a pollen grain forms only one plant. However, in several species where androgenic sporophytes have been obtained through anther culture (*Arabidopsis, Asparagus, Triticale*) the multicellular mass liberated from the bursting pollen grain undergoes further proliferation and forms a callus which may later differentiate plants on the same medium or on a modified medium. In some plants, such as *Oryza sativa*, by varying the composition of the medium androgenic plants can be obtained via embryo formation (Guha et al., 1970) or through callusing (Niizeki and Oono, 1968). This is also possible in *H. niger* (Raghavan, 1978).

2.4 Plant regeneration from pollen embryos

Regeneration of plants from pollen callus or pollen embryos may occur on the original medium or it may require transfer to a different medium. The pollen embryos exhibit considerable similarity with zygotic embryos in their morphology and certain biochemical features (Holbrook et al., 1990).

However, often the pollen embryos do not germinate normally. In *D. innoxia* pollen embryos frequently produce secondary embryos on hypocotyl surface, and all such embryos which produce secondary embryos are haploid and the others non-haploids (Lenee et al., 1987). To raise full plants from pollen embryos it is necessary to excise a cluster of the secondary embryos along with a part of the parent embryo and plant them on fresh medium. They do not germinate if left on the pollen embryo or removed individually (Sangwan-Norreel, 1983).

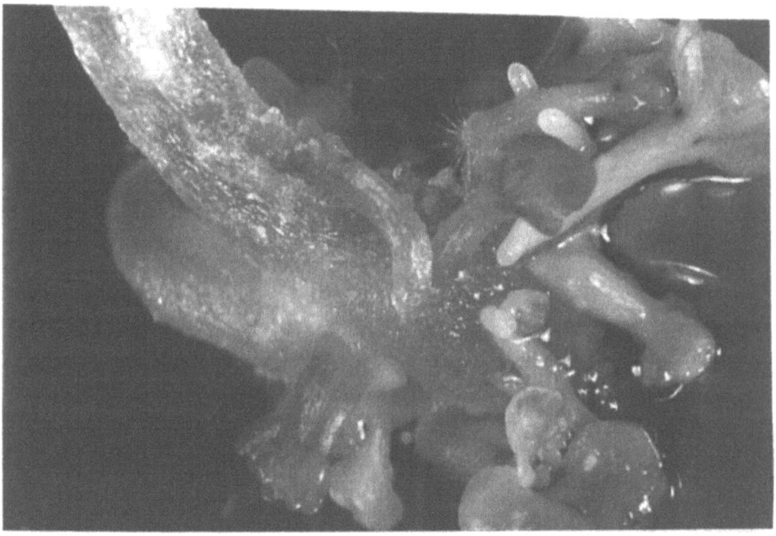

Figure 3. A pollen embryo of *B. juncea*, showing secondary embryogenesis from the radicular end of the hypocotyl.

In most of the *Brassica* species pollen embryos exhibit very poor germination (10-30%). On the germination medium the hypocotyl elongates, the cotyledons turn green, a primary root develops but the plumule rarely produces a shoot. The recalcitrant embryos regenerate plants through adventitious shoot bud differentiation or secondary embryogenesis (Fig. 3) from the epidermal cells of the hypocotyl (Loh and Ingram, 1982; Chuong and Beversdorf, 1985; Sharma and Bhojwani, 1989). ABA and cold treatment (at 4℃ for 6-12 h) and partial desiccation of the embryos of *B. napus* promoted their germination (Kott and Beversdorf, 1990). The best method for the germination of *Brassica* pollen embryos is to transfer the dicot embryos to B_5 + 0.1 mg l^{-1} GA_3 and expose them to 5℃ for 10 days before transferring to 25℃, in light. With this protocol up to 90% germination of pollen embryos has been achieved in *B. napus* (Kott and Beversdorff, 1990). Swanson et al. (1987) reported that shaking of globular and heart-shaped embryos obtained from isolated microspores of *B. napus* dramatically improved the speed and synchrony of embryo development as well as the quality of the embryos produced. ABA and cold treatment also enhanced normal germination of pollen embryos of *B. juncea* (Agarwal and Bhojwani, 1993).

9 FACTORS AFFECTING ANDROGENESIS

In vitro androgenesis is highly dependent on the genetic potential and physiological state of the donor plants and the developmental stage of the male gametophyte. External factors, such as culture medium and a range of chemical and physical treatments play a critical role in the induction of androgenesis.

9.1 Genetic Potential

Although androgenesis has been reported in more than 200 species, this approach to produce large numbers of pollen plants is limited to only a few crop plants. Considerable genotypic variation for androgenic response has been observed with several systems (Wenzel, 1977; Jacobsen and Sopory, 1978; Bhojwani and Sharma, 1991; Duijs et al., 1992; Bhojwani et al., 1996). The observed intraspecific variation is often so great that while some lines of a species exhibit good androgenesis, others are extremely poor performers or completely non-responsive. Duijs et al. (1992) tested 64 accessions of *B. oleracea* of which 86% were responsive. The responding accessions exhibited large genotypic differences for the yield of pollen embryos and the frequency and mode of plant regeneration from pollen embryos. In general, japonica rice cultivars are more responsive than indica rice cultivars (Cho and Zapata, 1990). Similarly, among crop brassicas, *B. napus* is most responsive and *B. juncea* the least.

Although it may be possible to overcome the genotypic limitation by developing genotype-specific protocols for androgenesis through an extensive manipulation of culture conditions, a more realistic approach to circumvent this problem would be to transfer the androgenic trait from highly androgenic lines to recalcitrant lines by breeding methods. This approach has been successfully tried with crops such as potato (Jacobsen and Sopory, 1978), barley (Foroughi-Wehr et al., 1982) and maize (Petolino et al., 1988; Barloy et al., 1989).

In recent years considerable effort has been devoted to identify genes associated with androgenesis in cereals, particularly maize (Cowen et al., 1992; Wan et al., 1992; Devaux and Zivy, 1994) and wheat (Agache et al., 1989). Based on RFLP analysis of the products of a number of crosses between a highly androgenic and a non-androgenic line of maize, Cowen et al. (1992) concluded that two major genes, which are epistatic, and two minor recessive genes are involved in the androgenic response of this plant. According to these authors, one of the major genes is located in the proximal region of the long arm of chromosome 3 near the indeterminate gametophyte gene (ig1) and the other one in the centromeric region of chromosome 9. The two minor genes are mapped on chromosomes 1 and 10. Similarly, two major and two minor genes are suspected to be associated with androgenesis in barley (Devaux and Zivy, 1994).

In *Melandrium album*, which shows a chromosomal basis of sex determination, only pollen with X chromosome are competent to form pollen plants and are, thus, phenotypically and cytologically female (Wu et al., 1990). In tetraploid *Melandrium* even a single Y chromosome is able to suppress the effect of three X chromosomes.

9.2 Physiological Status of the Donor Plant

The environmental conditions and the age of the donor plants, which influence their physiology, significantly affect the androgenic response in anther and microspore cultures. Generally, the first flush of buds show better response than those borne later (Sunderland, 1971; Olesen et al., 1988; Sato et al., 1989), thereby limiting the number of buds that can be used for culture. It may be possible to extend the responsive period, by removing the unused buds so that fruit formation is prevented and the plants remain physiologically young (Dunwell, 1985; Sato et al., 1989). Contrary to this, Takahata et al. (1991) and Burnett et al. (1992) observed that microspores isolated from older, sickly looking plants of *B. napus* and *B. rapa*, respectively, produced a higher number of embryos than those from younger, healthy plants. Similarly, late sown plants of *B. juncea* yielded more androgenic anthers than the plants sown at normal time (Agarwal and Bhojwani, 1993).

Exposure of donor plants to nutrient stress (Sunderland, 1978) or water stress (Wilson et al., 1978) is known to promote androgenesis. Efforts have been made to increase the initial frequency of androgenic grains, *in situ*, by treating them with feminizing agents. Application of Naphthaleneacetic acid (NAA) or Alar 85 to tobacco plants shifted the sex balance towards femaleness and increased the frequency of androgenic grains as well as dead pollen grains (Heberle-Bors, 1983). Ethrel treatment of plants induced additional mitosis resulting in the formation of supernumerary nuclei in wheat microspores (Bennett and Hughes, 1972). Subsequently, it was shown that ethrel-treated plants of rice (Wang et al., 1974) and *B. juncea* (Agarwal and Bhojwani, 1993) provided androgenically more productive anthers.

Application of fernidazon-potassium, a gametocidal agent, to donor plants not only enhanced the yield of androgenic haploids and homozygous diploids, but it also accelerated pollen embryo development by 8 days (Picard et al., 1987).

9.3 Stage of Pollen Development

Competence of microspores to respond to various external treatments depends on the stage of their development and the phase of the cell cycle. Using anthers of different developmental stages, Porter et al. (1984) had shown that the potentialities of microspores at the uninucleate stage are not fixed, as the sporophyte-specific gene products are eliminated from the cytoplasm before meiosis and the gametophyte-specific genes are generally transcribed after pollen mitosis. It is only 24 hours after pollen mitosis that pollen become irrevocably determined in the gametophytic pathway. Stress (temperature shock, high osmolarity, starvation) applied during the non-committal phase can thwart the gametophytic programme and induce expression of sporophyte-specific genes. However, the most vulnerable stage for androgenic development may differ with the plant. Anthers of *D. innoxia, N. tabacum* and *Paeonia hybrida* give best response when the microspores are just before, at or just after pollen mitosis. Early bicellular stage is absolutely necessary for *N. knightiana* and best for *Atropa belladonna* and *N. sylvestris*. Late uninucleate stage is most suitable for androgenesis in rice (Raghavan, 1990) and for most crop *Brassica* sp. (Leelavathi et al., 1984; Dunwell et al., 1985; Sharma and Bhojwani, 1985).

Kott et al. (1988) concluded that in *B. napus* older binucleate microspores release certain hydrophilic, heat stable toxins that suppress embryo formation by younger, embryogenic pollen and also induce abnormalities in the developing pollen embryos. However, according to Duijs et al. (1992) and Takahata et al. (1993) maximum yield of pollen embryos in *B.napus* and *B. oleracea*, respectively, occurred when the pollen population contained a large proportion of 2-celled pollen grains and a mean nucleus number of 1.68 per pollen (i.e. 50-80%). Recently, it has been clearly shown that even 2-celled pollen of *B. napus* are capable of forming embryos (Hause et al., 1996).

In *D. innoxia* (Engvild et al., 1972) and *Petunia* sp. (Engvild, 1973; Raquin and Pilet, 1972) the stage of microspore development at culture affected the ploidy level of the pollen plants. Whereas uninucleate microspores produced mainly haploids, the binucleate pollen formed plants of higher ploidy.

9.4 External Factors

9.4.1 Culture medium

Normally, only two mitotic divisions occur in a microspore but androgenesis involves repeated divisions. A variety of treatments are known to induce additional divisions in the pollen grains. In field-grown plants of wheat sprayed with ethrel (2-chloroethylphosphonic acid), Bennett and Hughes (1972) recorded multicellular grains. Incorporation of this compound in the nutrient medium was later shown to enhance the androgenic response in anther cultures of tobacco (Bajaj et al., 1977).

That sucrose is essential for androgenesis was first demonstrated by Nitsch (1969) for tobacco and later by Sunderland (1974) for *D. innoxia*. Therefore, sucrose is included in all the anther culture media and is generally used at a concentration of 2-4%. However, in wheat 6% sucrose was found to promote pollen callusing and inhibit the proliferation of somatic tissues (Ouyang et al., 1973). Similarly, for potato 6% sucrose proved distinctly superior to 2% or 4% sucrose in terms of the number of anthers forming pollen embryos (Sopory et al., 1978). All *Brassica* species require 12-13% sucrose for androgenesis in anther and pollen cultures.

Last and Brettell (1990) reported that most of the cultures of wheat showed higher androgenic responses when sucrose was substituted by maltose in the medium. Maltose also improved the androgenic response in *O. sativa* (Xie et al., 1995).

Pretreatment of barley anthers with high concentrations of mannitol brought about a linear gain in the number of dividing microspores, an enhancement of the green to total plant ratio and improvement of response in cultivars otherwise showing low androgenic ability (Cistue et al., 1994). The beneficial effect of mannitol may be due to its effect on sugar uptake with a large concentration of glucose building up inside the anther.

Although androgenic development of pollen grains in *N. tabacum* and *D. innoxia* can be induced on agar plates containing only sucrose, on such a simple medium the development proceeds only up to the globular stage. For further development of the embryos mineral salts are required. Possibly, the nutrients and growth factors necessary for the induction and early development of the androgenic embryos are supplied by the anther wall or pollen itself.

Most species exhibit androgenesis on a complete nutrient medium (mineral salts, vitamins and sucrose), with or without growth substances (see Bhojwani and Razdan, 1996). However, the available literature does not allow recommendation of an anther culture medium of general applicability.

The requirements may vary with the genotype and, probably, the age of the anther and the conditions under which the donor plants are grown.

In China considerable work has been done to develop media that would favour the formation of green haploid plants in anther cultures of cereals at a high frequency. Low inorganic nitrogen, particularly ammonium, in the medium is reported to promote androgenesis (Clapham, 1973, Chu et al., 1975) and the yield of green plants (Olesen et al., 1988) in some cereals. Ten times reduction in NH_4NO_3 concentration in LS medium substantially enhanced the overall androgenic response in *Lolium perenne* and *L. multiflorum* (Bante et al., 1990). Even KNO_3 in the medium was inhibitory for embryogenesis. Halving the concentration of KNO_3 in MS basal medium remarkably enhanced pollen embryogenesis in *Hevea brasiliensis* (Chen, 1990). A potato medium, with the 6 major salts at reduced concentration , iron (no minor salts), thiamine and 10% potato extracts, besides growth regulators and sucrose, gave considerably higher numbers of green plants in anther cultures of wheat as compared to N6 medium (Chuang et al., 1978). The media used for isolated pollen culture are generally very low in overall salt concentration (Bhojwani and Razdan, 1996).

In the solanaceous plants, where androgenesis occurs via direct pollen embryo formation, the presence of a growth adjuvent is generally not required, and full haploid plants are formed on basal media. In some cases even vitamins are not essential. However, most *Brassica* species require an auxin and a cytokinin for direct pollen embryogenesis. For the majority of non-solanaceous species known to exhibit androgenesis *via* a callus phase it is essential to fortify the medium with growth regulators, complex nutrient mixtures (yeast extract, casein hydrolysate, potato extract, coconut milk), either alone or in different combinations.

Sometimes it may be possible to change the mode of androgenic development by a judicious change of growth adjuvents in the medium (Raghavan, 1978). For example, in *T. aestivum* callusing of pollen occurs if the medium contains 2,4-D and lactalbumin hydrolysate, but in the presence of coconut milk pollen grains directly develop into embryos (Ouyang et al., 1973, *cited in* Clapham, 1977).

9.4.2 Culture density

The culture density is a critical factor in isolated pollen culture, as also in single cell and protoplast culture. Based on a detailed study, Huang et al. (1990) reported that in pollen cultures of *B. napus* the minimum density required for direct embryogenesis is 3000 pollen ml^{-1} of the culture medium but highest embryo yield was obtained at 10,000-40,000 pollen ml^{-1}. This high plating density is crucial only for the initial couple of days. Dilution from 30,000-40,000 to 1,000 pollen ml^{-1} after two days of culture did not reduce the embryogenic frequency. The media conditioned by growing pollen at high densities (30,000-40,000 ml^{-1}) for two days also stimulated embryogenesis in pollen cultures at low plating densities (3,000-10,000 pollen ml^{-1}).

Arnison et al. (1990) reported the effect of culture density in anther cultures of *B. oleracea*. The frequency of pollen embryogenesis was enhanced if the anther culture density was increased from 3 anther/4 ml to 12-24 anthers ml^{-1} of the medium. However, this observation is contradictory to the findings of Cardy (1986), according to which in *B. napus* the response was better at low density (2 anthers ml^{-1}).

9.4.3 Colchicine treatment

In *Brassica* species symmetrical division of microspores is considered a key factor in embryo formation (Fan et al., 1988; Zaki and Dickinson, 1990,1995; Hause et al., 1993). Promotion of equal division in microspores of tobacco (Nitsch, 1977) and tulip (Tanaka and Ito, 1981) by colchcine treatment has been known for quite some time. However, it was only recently that Zaki and Dickinson (1991, 1995) demonstrated that in anther and microspore cultures of *B. napus* the number of symmetrically dividing microspores and the number of embryogenic microspores increased significantly after colchicine treatment. Subsequently, promotional effect of colchicine on androgenesis has also been observed in *T. aestivum* (Szakacs and Barnabas, 1995), *O. sativa* (Alemanno and Guiderdoni, 1994) and *Zea mays* (Barnabas et al., 1991). Colchicine probably disrupts the microtubular cytoskeleton, which is responsible for positioning the nucleus on one side to maintain asymmetric division. Consequently, the nucleus moves to a central position followed by equal division of the microspore (Zaki and Dickinson, 1990). Similar observations for *Triticum* were reported by Szakacs and Barnabas (1995).

Both, the concentration and the duration of colchicine treatment are important for promoting androgenesis. According to Zaki and Dickinson (1991), treatment with 25 mg l^{-1} of colchicine for 12 h was optimum for two cultivars of *B. napus*. It caused 3-4 fold increase in androgenic response. For the same species, Iqbal et al. (1994) found 100 mg l^{-1} colchicine treatment for 24 h to give best results. Although colchicine promotes androgenesis significantly, very few of the embryos attain full development (Zaki and Dickinson, 1995).

9.4.4 Starvation

Initial starvation of developing microspores of some important nutrients has been shown to suppress gametophytic differentiation and induce sporophytic development. Cultivation of tobacco anthers on sugar-free medium for a few days before transfer to sugar-containing medium suppressed gametogenesis and induced androgenesis (Aruga et al., 1985). Maintaining the anthers in sugar-free medium for 6 days gave maximum androgenic response upon transfer to complete medium. Induction of androgenesis by sugar-starvation has also been observed in microspore cultures of *H. vulgare* (Wei et al., 1986). Similarly, whereas direct culture of tobacco pollen, at the mid binucleate stage, on basal medium containing glutamine favoured gametophytic development leading to the formation of mature pollen, glutamine starvation of the pollen for the initial 24-48 h favoured androgenic development on glutamine containing medium (Kyo and Harada, 1985, 1986).

9.4.5 Temperature shock

In many species the incubation of anther/pollen cultures at low temperatures (4℃-13℃) for varying periods before shifting them to 25℃ (culture room conditions) enhanced the androgenic response. The optimum duration of treatment may vary with the temperature, stage of pollen development, and the genotype. Cold treatment seems to delay the rate of degeneration of the anther wall and, thereby, enhances the proportion of the surviving microspores.

In some other plants (*Brassica sp., Capsicum, Avena*) an initial high temperature shock has proved beneficial. The efficiency of pollen embryogenesis in anther cultures of *B. campestris* treated at 35℃ for 24 h was 20 times higher than the untreated control (Hamaoka et al., 1991). For most *Brassica* species a high temperature treatment (30-35℃) for the initial 1-4 days of culture is essential to induce androgenesis. In the cultures of late uninucleate microspores of *B. napus* the minimum duration of high temperature (32℃) required to induce embryogenesis is 2 h, and the maximum response (70% grains forming embryos) is achieved after 4 days of heat shock (Pechan et al., 1991). Almost 90% of the potential androgenic pollen acquired the competence within 8 h of incubation at the elevated temperature (Pechan et al., 1991). Exposure to 41℃ for one or 2 h followed by incubation at 32℃ for 4 days induced androgenesis even in late binucleate pollen grains (Hause et al., 1996).

9.4.6 Centrifugation

In *D. innoxia* the centrifugation of anthers at 40 g for 5 min after cold treatment of buds at 3℃ for 48 h improved the percentage of androgenic anthers (Sangwan-Norreel, 1977). The promotion was especially striking (60%) in the cultures of anthers at the early binucleate stage; in the cultures at the late uninucleate stage the enhancement over the control was only 7%. Similarly, in the cultures of pollen grains from cold-treated buds centrifugation at 120 g for 15 min not only increased the number of pollen embryos but also brought about rapid and synchronous development of embryos.

One of the reasons for substantially higher androgenic response in pollen cultures than in anthers cultures of *B. napus* could be centrifugation, which is required to obtain clean pollen preparation for culture. To test this hypothesis, Aslam et al. (1990) compared the androgenic response of anther cultures from untreated buds and those from buds centrifuged at 400 g for 12 min or at 280 g for 5-10 min. All centrifugation treatments markedly improved the embryo yield over the control. With the optimum treatment (400 g for 8 min) the mean number of embryos per anther was 9.5 as compared to 0.2 in the untreated control.

9.4.7 γ-Irradiation

Judicious application of γ-irradiation to anthers before culture has been reported to promote pollen callusing and pollen embryogenesis in *Nicotiana* and *Datura* (Sangwan and Sangwan, 1986), wheat (Wang and Yu, 1984; Yin et al., 1988b), rice (Yin et al., 1988a) and *B. napus* (MacDonald et al., 1988). In wheat, γ-irradiation at 1, 3 and 5 Gys quadrupled the yield of pollen embryos and enhanced the frequency of regeneration of green plants and the production of doubled haploids (Ling et al., 1991). Irradiation induced pollen embryogenesis in otherwise non-amenable genotypes of wheat.

Low doses (10 Gys) of irradiation greatly enhanced anther culture efficiency (number of responding anthers) in two cultivars of *B. napus* sp. *oleifera* (MacDonald et al., 1988). In the cv Ariana, irradiation of young buds (2.6-3 mm) doubled or tripled the frequency of responsive anthers and almost quadrupled the number of embryos per 1000 plated anthers. Even old buds (3.1-3.5 mm), which normally do not exhibit androgenesis, became responsive after γ-irradiation. For the other cultivar (Primor) γ-irradiation was detrimental when applied to young buds but proved promotory in the case of older buds. γ-irradiation also caused a decline of pollen embryogenesis in isolated pollen cultures. The anther wall is implicated in the promotory effect of this factor as irradiation of intact anthers produced better results than irradiation of isolated microspores.

Irradiation is known to inactivate nuclei (Zelcer et al., 1978) and alter the levels of auxins and cytokinins in the tissues (Degani and Pickholtz, 1982). These actions may be involved in the promotion of androgenesis by low irradiation in some systems.

10 CONCLUDING REMARKS

Regeneration of plants from microspores is of both practical and theoretical significance. The practical applications of this technique are discussed in chapter 12 of this book. In many plants the normal gametophytic development of microspores can be suppressed and these haploid single cells induced to follow an altogether different morphogenic pathway to produce sporophytes. Surprisingly, this dramatic change in the destiny of a cell can often be brought about by a simple treatment, such as a brief temperature shock. Microspore culture of *B. napus* has emerged as a model system to study the induction of androgenesis and the biochemical and cellular changes associated with the induction of embryogenesis. Some of the advantages offered by this system are: (1) embryogenesis occurs in freshly isolated microspores rather than cells from suspension cultures, as in the case of somatic embryogenesis; (2) embryos develop directly from the microspores, without a callus phase, within 14 days of culture initiation; (3) a single stress treatment (32℃ for 8 h) can induce and sustain the embryogenic process; (4) up to 70% of the microspores in a given culture can undergo embryogenesis, and (5) induction of embryogenesis is dependent on the stage of microspore, thus allowing the study of a cell cycle regulated morphogenic process.

The recent progress in identifying the genes involved in androgenesis may prove rewarding to achieve androgenesis in hitherto recalcitrant species.

11 REFERENCES

*Cardy BJ (1986) Production of anther-derived doubled haploids for breeding oil-seed rape (*Brassica napus* L.). Ph.D. Thesis, Univ. Guelph, Ontario, Canada.

Agache S, Bachelier B, de Buyser J, Henry Y & Snape J (1989) Genetic analysis of anther culture response in wheat using aneuploid, chromosome substitution and translocation lines. Theor. Appl. Genet. 77: 7-11.

Agarwal PK & Bhojwani SS (1993) Enhanced microspore embryogenesis and plant regeneration in anther cultures of *Brassica juncea* cv. PR45. Euphytica 70: 191-196.

Alemanno L & Guiderdoni E (1994) Increased doubled haploid plant regeneration from rice (*Oryza sativa* L.) anthers cultured on colchicine-supplemented media. Plant Cell Rep. 13: 432-436.

Arnison PG, Donaldson P, Ho LCC & Keller WA (1990) The influence of various physical parameters on anther culture of broccoli (*Brassica oleracea var. italica*). Plant Cell Tissue Organ Cult. 20: 147-155.

Aruga K, Nakajima T & Yamamoto K (1985) Embryogenic induction of pollen grains of *Nicotiana tabacum* L. Jpn J. Breed. 35: 50-58.

Aslam FN, MacDonald MV, Loudon PT & Ingram DS (1990) Rapid cycling *Brassica* species: Inbreeding and selection of *Brassica napus* for anther culture ability and an assessment of its potential for microspore culture. Ann. Bot. 66: 331-339.

Bajaj YPS, Reinert J, & Heberle E (1977) Factors enhancing in vitro production of haploid plants in anthers and isolated microspores. In: Gautheret, R..J.(ed.) La Culture des Tissue et des Cellules des Vegetaux. pp. 47-58. Masson, Paris.

Bante I, Sonke T, Tandler RF, van der Bruel AMR, & Meyer EM (1990) Anther culture of *Lolium perenne* and *L. multiflorum*. In: Sangwan, R.S., and Sangwan, B.S. (eds) The Impact of Biotechnology in Agriculture. pp. 105-127. Kluwer, Dordrecht.

Barloy D, Dennis L & Beckert M (1989) Comparison of the aptitude for anther culture in some androgenetic doubled haploid maize lines. Maydica 34: 303-308.

Barnabas B, Pfahler PL & Kovacs G (1991) Direct effect of colchicine on the microspore embryogenesis to produce dihaploid plants in wheat (*Triticum aestivum* L.). Theor. Appl. Genet. 81: 675-678.

Bennett MD & Hughes WG (1972) Additional mitosis in wheat pollen induced by ethrel. Nature 240: 566-568.

Bhojwani SS, Agarwal PK & Satpathy MR (1996) Androgenesis in anther and pollen culture of *Brassicas*: A Review. In: Islam, A.S. (ed.) Plant Tissue Culture. pp. 169-202. Oxford & IBH, India.

Bhojwani SS & Bhatnagar SP (1999) The Embryology of Angiosperms. Vikas Publishing House, New Delhi

Bhojwani SS, Dunwell JM & Sunderland N (1973) Nucleic acid and protein contents of embryogenic pollen. J. Exp. Bot. 24: 863-871.

Bhojwani SS & Razdan MK (1996) Plant Tissue Culture: Theory and Practice, A Revised Edition. Elsevier, Amsterdam.

Bhojwani SS & Sharma KK (1991) Anther and pollen culture for haploid production. In: Mandal, A.K., et al. (ed.) Advances in Plant Breeding, Vol. 2. pp. 65-91. CBS Publ. and Distributors, New Delhi.

Burnett L, Yarrow S & Huang B (1992) Embryogenesis and plant regeneration from isolated microspores of *Brassica rapa* L. ssp. *oleifera*. Plant Cell Rep. 11: 215-218.

Chen Z (1990) Haploid induction in perennial crops. In: Chen, Z. et al. (ed.) Handbook of Plant Cell Culture. Vol. 6. Perennial Crops. pp. 62-75. McGraw-Hill, New York.

Cho MS & Zapata FJ (1990) Plant regeneration from isolated microspore of indica rice. Plant Cell Physiol. 31: 881-885.

Chu CC, Wang CC, Sun CS, Hsu C, Yin KC, Chu CY & Bi FY (1975) Establishment of an efficient medium for anther culture of rice through comparative experiments on the nitrogen sources. Sci. Sin. 18: 559-668.

Chuang C, Ouyang T, Chia H, Chou S & Ching C (1978) A set of potato medium for wheat anther culture. In: Proceedings of Symposium on Plant Tissue Culture. pp. 51-56. Science Press, Peking,

Chuong PV & Beversdorf WD (1985) High frequency embryogenesis through isolated microspore culture in *Brassica napus* and *B. carinata* Braun. Plant. Sci. 39: 219-226.

Cistue L, Ramos A, Castillo AM & Romagosa I (1994) Production of large number of doubled haploid plants from barley anthers pretreated with high concentrations of mannitol. Plant Cell Rep. 13: 709-712.

Clapham D (1973) Haploid *Hordeum* plants from anthers in vitro. Z. Pflanzenzucht. 69: 142-155.

Clapham D (1977) Haploid induction in cereals. In: Reinert, J., and Bajaj, Y.P.S. (eds), Applied and Fundamental Aspects of Plant Cell, Tissue and Organ Culture. pp. 279-298. Springer, Berlin.

Cordewener JHG, Busink R, Traas JA, Custer JBM, Dons JJM & Van Lookeren Campagne MM (1994) Induction of microspore embryogenesis in *Brassica napus* is accompanied by specific changes in protein synthesis. Planta 195: 50-56.

Cordewener JHG, Custers JBM, Dons HJM & van Lookeren MM (1996) Molecular and biochemical events during the induction of microspore embryogenesis. In: Jain, S.M. et al. (eds), In Vitro Haploid Production in Higher Plants. Vol. 1. pp. 111-124 Kluwer, Dordrecht.

Cordewener JHG, Hause G, Gorgen E, Busink R, Hause B, Dons HJM, Van Lammeren AAM, Van Lookeren Campagne MM & Pechan P (1995) Changes in synthesis and localization of the 70 kDa class of heat shock proteins accompany the induction of embryogenesis in *Brassica napus* microspores. Planta 196: 747-755.

Cowen NM, Johnson CD, Armstrong K, Miller M, Woosley A, Pescitelli S, Skokut M, Belmar S & Petolino JF (1992) Mapping genes conditioning in vitro androgenesis in maize using RFLP analysis. Theor. Appl. Genet. 84: 720-724.

Custers JBM, Cordewener JHG, Nollen Y, Dons HJM & Van Lookeren Campagne MM (1994) Temperature controls both gametophyte and sporophyte development in microspore cultures of *Brassica napus*. Plant Cell Rep. 13: 267-271.

Dale PJ (1975) Pollen dimorphism and anther culture in barley. Planta 127: 213-220.

Degani N & Pickholtz D (1982) The effect of kinetin on the growth of gamma irradiated soybean culture. In: Fujiwara, A. (ed.) Proceedings of 5th International Congress of Plant Tissue and Cell Culture. pp. 459-460. Jpn. Assoc. Plant Tissue Cult., Tokyo.

Devaux P & Zivy M (1994) Protein markers for anther culturability in barley. Theor. Appl. Genet. 88: 701-706.

Duijs JG, Voorrips RE, Visser DI & Custers JBM (1992) Microspore culture is successful in most crop types of *Brassica oleracea* L. Euphytica 60: 45-55.

Dunwell JM (1978) Division and differentiation in cultured pollen. In: Thorpe, T. (ed.) Frontiers of Plant Tissue Culture 1978. pp. 103-112. Univ. Calgary Press, Canada.

Dunwell JM (1985) Embryogenesis from pollen in vitro. In: Zaitlin, P. et al. (eds), Biotechnology in Plant Science. pp. 49-76. Academic Press, Orlando.

Dunwell JM, Cornish M & De Courcel AGL (1985) Influence of genotype, plant growth temperature & anther incubation temperature on microspore embryo production in *Brassica napus* ssp. *oleifera*. J. Exp. Bot. 36: 679-689.

Engvild KC (1973) Triploid petunias from anther cultures. Hereditas 74: 144-147.

Engvild KC, Linde-Laursen Ib & Lundquist A (1972) Anther cultures of *Datura innoxia*. Flower bud stage and embryo level of ploidy. Hereditas 72: 331-332.

Fan Z, Armstrong KC & Keller WA (1988) Development of microspores in vivo and in vitro in *Brassica napus* L. Protoplasma 147: 191-199.

Foroughi-Wehr B, Friedt W & Wenzel, G. (1982) On the genetic improvement of androgenetic haploid formation in *Hordeum vulgare* L. Theor. Appl. Genet. 62: 233-239.

Garrido D, Eller N, Heberle-Bors E & Vicente O (1993) *De novo* transcription of specific messenger RNA's during the induction of tobacco pollen embryogenesis. Sex. Plant. Reprod. 6: 40-45.

Guha S, Iyer RD, Gupta N & Swaminathan MS (1970) Totipotency of gametic cells and the production of haploids in rice. Curr. Sci. 39: 174-176.

Hamaoka Y, Fujita Y & Iwai S (1991) Effects of temperature on the mode of pollen development in anther culture of *Brassica campestris*. Physiol. Plant. 82: 67-72.

Harada H, Kyo M & Imamura, J. (1988) The induction of embryogenesis in *Nicotiana* immature pollen culture. In: Bock, G., and March, J. (eds), Applications of Plant Cell and Tissue Culture. pp. 59-74. Wiley, Chichester.

Hause B, Hause G, Pechan P & Van Lammeren A (1993) Cytoskeletal changes and induction of embryogenesis in microspore and pollen cultures of *Brassica napus*. Cell. Biol. Inter. 17: 153-168.

Hause G, Cenklova V, Cordewener JGH, Van Lookeren Campagne MM & Binarova P (1996) Induction of embryogenesis in late bicellular pollen of *Brassica napus* by high temperature treatment. In: Plant Embryogenesis Workshop (Abstr.). p. 69. Univ Hamburg, Germany.

Heberle-Bors E (1983) Induction of embryogenic grains in situ and subsequent in vitro pollen embryogenesis in *Nicotiana tabacum* by treatments of the pollen donor plants with feminizing agents. Physiol. Plant. 59: 67-72.

Heberle-Bors E (1985) In vitro haploid formation from pollen: a critical review. Theor. Appl. Genet. 71: 361-374.

Holbrook LA, Scowcroft WR, Taylor DC, Pomeroy M., Wilen RW & Moloney MM (1990) Microspore-derived embryos: A tool for studies in regulation of gene expression. In: Nijkamp, H.J.J. et al. (eds) Progress in Plant Cellular and Molecular Biology. pp. 402-406. Kluwer, Dordrecht.

Horner M & Mott RL (1979) The frequency of embryogenic pollen grains is not increased by in vitro anther culture in *Nicotiana tabacum* L. Planta 147: 156-158.

Horner M & Street HE (1978) Pollen dimorphism-origin & significance in pollen plant formation by anther culture. Ann. Bot. 42: 763-771.

Huang B, Bird S, Kemble R, Simmonds D, Keller W & Miki B (1990) Effects of culture density, conditioned medium and feeder cultures on microspore embryogenesis in *Brassica napus* L. cv *Topas*. Plant Cell Rep. 8: 594-597.

Iqbal MCM, Mollers C & Robbelen G (1994) Increased embryogenesis after colchicine treatment of microspore cultures of *Brassica napus* L. J. Plant Physiol. 143: 222-226.

Jacobsen E & Sopory SK (1978) The influence and possible recombination of genotypes on the production of microspore embryoids in anther cultures of *Solanum tuberosum* & dihaploid hybrids. Theor. Appl. Genet. 52: 119-123.

Kott LS, Polsoni L, Ellis B & Beversdorf WD (1988) Autotoxicity in isolated microspore cultures of *Brassica napus*. Can. J. Bot. 66: 1665-1670.

Kott LS & Beversdorf WD (1990) Enhanced plant regeneration from microspore-derived embryos of *Brassica napus* by chilling, partial desiccation and age selection. Plant Cell Tissue Organ Cult. 23: 187-192.

Kyo M & Harada H (1985) Studies on conditions for cell division and embryogenesis in isolated pollen culture of *Nicotiana rustica*. Plant Physiol. 79: 90-94.

Kyo M. & Harada H (1986) Control of the developmental pathway of tobacco pollen in vitro. Planta 168: 427-432.

Kyo M & Harada H (1990) Specific phosphoproteins in the initial period of tobacco pollen embryogenesis. Planta 182: 58-63.

Last DI & Brettell IS (1990) Embryo yield in wheat anther culture is influenced by the choice of sugar in the culture medium. Plant Cell Rep. 9: 14-16.

Leelavathi S, Reddy VS & Sen SK (1984) Somatic cell genetic studies in *Brassica* species . I. High frequency production of haploid plants in *Brassica alba* (L.) H.F. & T. Plant Cell Rep. 3: 102-105.

Lenee P, Sangwan-Norreel BS & Sangwan RS (1987) Nuclear DNA contents and ploidy in somatic embryos derived androgenic plants of *Datura innoxia* Mill. J. Plant Physiol. 130: 37-48.

Ling DX, Luckett DJ & Darvey NL (1991) Low-dose gamma irradiation promotes wheat anther culture response. Aust. J. Bot. 39: 467-474.

Loh CS & Ingram, D.S. (1982) Production of haploid plants from anther cultures and secondary embryoids of winter oilseed rape, *Brassica napus* ssp. *oleifera*. New Phytol. 91: 507-516.

Lowen NM, Johnson CD, Armstrong K, Miller M, Woosley A, Pescitelli S, Skokut M, Belmar S & Petolino JF (1992) Mapping genes conditioning in vitro androgenesis in maize using RFLP analysis. Theor. Appl. Genet. 84: 720-724.

MacDonald MV, Hadwiger MA, Aslam FN & Ingram DS (1988) The enhancement of anther culture efficiency in *Brassica napus* ssp. *oleifera* Metzg. (Sinsk.) using low doses of gamma irradiation. New Phytol. 110: 101-107.

Mascarenhas JP (1971) RNA and protein synthesis during pollen development and tube growth. In: Heslop-Harrison, J. (ed.), Pollen - Development and Physiology. pp. 201-222. Butterworth, London.

Niiizeki H & Oono K (1968) Induction of haploid rice plants from anther culture. Proc Jpn Acad 44: 554-557.

Nitsch C (1977) Culture of isolated microspores. In: Reinert, J., and Bajaj, Y.P.S. (eds), Fundamental and Applied Aspects of Plant Cell Tissue Organ Culture. pp. 268-278. Springer, Berlin.

Nitsch JP (1969) Experimental androgenesis in *Nicotiana*. Phytomorphology 19: 389-404.

Nitsch JP & Nitsch C (1969) Haploid plants from pollen grains. Science, NY. 163: 85-87.

Olesen A, Anderson SB & Due IK (1988) Anther culture response in perennial ryegrass (*Lolium perenne* L.). Plant Breed. 101: 60-65.

Ouyang T, Hu H, Chuang C & Tseng C (1973) Induction of pollen plants from anthers of *Triticum aestivum* L. cultured *in vitro*. Sci. Sin. 16: 79-95.

Pechan PM, Bartels D, Brawn DCW & Schell J (1991) Messenger-RNA and protein changes associated with induction of *Brassica* microspore embryogenesis. Planta 184: 161-165.

Petolino JF, Jones AM & Thompson SA (1988) Selection for increased anther culture response in maize. Theor. Appl. Genet. 76: 157-159.

Picard E, Hours C, Gregoire S, Phan TH & Meunier JP (1987) Significant improvement of androgenetic haploid and doubled haploid induction from wheat plants treated with a chemical hybridization agent. Theor. Appl. Genet. 74: 289-297.

Porter EC, Parry D, Bird J & Dickinson HG (1984) Nucleic acid metabolism in the nucleus and cytoplasm of angiosperm meiocytes. In: Evans, C., and Dickinson, H.G. (eds) Controlling events in meiosis. pp. 363-369. Company of Biologists, Cambridge.

Raghavan V (1978) Origin and development of pollen embryoids and pollen calluses in cultured anther segments of *Hyoscyamus niger* (henbane). Am. J. Bot. 65: 984-1002.

Raghavan V (1979) Embryogenic determination and ribonucleic acid synthesis in pollen grains of *Hyoscyamus niger* (henbane). Am. J. Bot. 66: 36-39.

Raghavan V (1990) Gene expression during anther and pollen development transformation in rice. In: Sangwan, R.S., and Sangwan-Norreel, B.S. (eds) The Impact of Biotechnology in Agriculture. pp. 85-98. Kluwer, Dordrecht.

Raquin C & Pilet V (1972) Production de plantules a partir d'antheres de petunias cultivees in vitro. CR Acad. Sci. Paris 274: 1019-1022.

Rashid A, Siddiqui AW & Reinert J (1982) Subcellular aspects of origin and structure of pollen embryos of *Nicotiana*. Protoplasma 113: 202-208.

Sangwan RS & Sangwan BS (1986) Effects of gamma radiation on somatic embryogenesis and androgenesis in various in vitro cultured plant tissues. In: Nuclear Techniques and In Vitro Cultured Plant Tissues. pp. 181-185. International Atomic Energy Agency, Vienna.

Sangwan-Norreel BS (1977) Androgenic stimulating factors in the anther and isolated pollen grain culture of *Datura innoxia* Mill. J. Exp. Bot. 28: 843-852.

Sangwan-Norreel BS (1978) Cytochemical and ultrastructural peculiarities of embryogenic pollen grains and of young androgenic embryos in *Datura innoxia* Mill. Can. J. Bot. 56: 805-817.

Sangwan-Norreel BS (1983) Male gametophyte nuclear DNA content evolution during androgenic induction in *Datura innoxia* Mill. Z. Pflanzenphysiol. 111: 47-54.

Sato T, Nishio T & Hirai M (1989) Varietal differences in embryogenic ability in anther culture of Chinese cabbage (*Brassica campestris* ssp. *pekinensis*. Jap. J. Breed. 39: 149-157.

Scott R, Dagless E, Hodge R, Wyatt P, Soufleri I & Draper J (1991) Patterns of gene expression in developing anthers of *Brassica napus*. Plant Mol. Biol. 17: 195-209.

Sharma KK & Bhojwani SS (1985) Microspore embryogenesis in anther cultures of two Indian cultivars of *Brassica juncea* (L.) Czern. Plant Cell Tissue Organ Cult. 4: 235-239.

Sharma KK & Bhojwani SS (1989) Histological and histochemical investigations of pollen embryos of *Brassica juncea* (L.) Czern Biol. Plant 31: 276-279.

Sopory SK, Jacobsen E & Wenzel G (1978) Production of monohaploid embryoids and plantlets in cultured anthers of *Solanum tuberosum*. Plant Sci. Lett. 12: 47-54.

Sunderland N (1971) Anther culture: a progress report. Sci. Prog. (Oxford) 59: 527-549.

Sunderland N (1974) Anther culture as a means of haploid induction. In: Kasha, K.J. (ed.) Haploids in Higher Plants-Advances and Potential. pp. 91-122. Univ. Guelph, Canada.

Sunderland N (1978) Strategies in the improvement of yields in anther culture. In: Proc. Symp. Plant Tissue Culture. pp. 65-86. Sci. Press, Peking.

Swanson EB, Coumans MP, Wa SC, Barsby TL & Beversdorf WD (1987) Efficient isolation of microspores and the production of microspore derived embryos from *Brassica napus*. Plant Cell Rep. 6: 94-97.

Szakacs E & Barnabas B (1995) The effect of colchicine treatment on microspore division and microspore-derived embryo differentiation in wheat (*Triticum aestivum* L.). Euphytica 83: 209-213.

Takahata Y, Brown DCW & Keller WA (1991) Effect of donor plant age and inflorescence age on microspore culture of *Brassica napus* L. Euphytica 58: 51-55.

Tanaka I & Ito M (1981) Control of division in explanted microspores of *Tulipa gesneriana*. Protoplasma 108: 329-340.

Wan Y, Rocheford TR & Widholm JM (1992) RFLP analysis to identify putative chromosomal regions involved in the anther culture response and callus formation of maize. Theor. Appl. Genet. 85: 360-365.

Wang BK & Yu YQ (1984) A study of the effect of radiation on anther culture in wheat. Appl. Atom. Energy. Agric. 3: 28-33.

Wang CC, Sun CS & Chu ZC (1974) On the conditions for the induction of rice pollen plantlets and certain factors affecting the frequency of induction. Acta Bot. Sin. 16: 43-54.

Wei ZM, Kyo M & Harada H (1986) Callus formation and plant regeneration through direct culture of isolated pollen of *Hordeum vulgare* cv. *Sabarlis*. Theor. Appl. Genet. 72: 252-255.

Wilson HM, Foroughi-Wehr B & Mix G (1978) Haploids in *Hordeum vulgare* through anther culture: potential as starting material for genetic manipulation in vitro. In: Sanchez-Monge, E., and Garcia-Olmedo, F. (eds) Interspecific Hybridization in Plant Breeding. pp. 243-252. Univ. Politecnica, Madrid.

Wu Y, Ye D, Installe P, Jacobs M & Negrutiu I (1990) Sex determination in the dioecious *Melandrium*: The in vitro culture approach. In: Nijkamp, H.J.J. et al. (eds) Progress in Plant Cellular and Molecular Biology. pp. 239-243. Kluwer, Dordrecht.

Xie J, Gao M, Cai Q, Cheng X, Shen Y & Liang Z (1995) Improved isolated microspore culture efficiency on medium with maltose and optimized growth regulator combination in japonica rice (*Oryza sativa*). Plant Cell Tissue Organ Cult. 42: 245-250.

Yin DC, Wei QJ, Yu QC & Wang L (1988b) Effects of gamma radiation on wheat pollen development in anther culture. In: Genetic Manipulation in Crops. pp. 38-39. Cassell Tycooly, U.K..

Yin DC, Wei QJ, Yu QC & Wang L (1988a) Effects of gamma radiation on anther culture of rice. In: Genetic Manipulation in Crops. pp. 37-38. Cassell Tycooly, U.K..

Zaki MA & Dickinson HG (1992) Microspore derived embryogenesis. In: Cresti, M., and Tiezzi, A. (eds) Sexual Plant Reproduction. pp. 18-29. Springer, Berlin.

Zaki MA & Dickinson HG (1995) Modification of cell development in vitro: The effect of colchicine on anther and isolated microspore culture in *Brassica napus*. Plant Cell Tissue Organ Cult. 40: 225-70.

Zaki MAM & Dickinson HG (1990) Structural changes during the first divisions of embryos resulting from anther and free microspore culture in *Brassica napus*. Protoplasma 156: 149-162.

Zaki MAM & Dickinson HG (1991) Microspore-derived embryos in *Brassica*: The significance of division symmetry in pollen mitosis I to embryogenic development. Sex. Plant Reprod. 4: 48-55.

Zarsky V, Garrido D, Rihova L, Tupy J, Vincente O & Heberle-Bors E (1992) Derepression of the cell cycle by starvation is involved in the induction of tobacco pollen embryogenesis. Sexual Plant Reprod. 5: 189-194.

Zarsky V, Rihova L & Tupy J (1990) Biochemical and cytological changes in young tobacco pollen during in vitro starvation in relation to pollen embryogenesis. In: Nijkamp, H.J.J. et al. (eds) Progress in Plant Cellular and Molecular Biology. pp. 228-233. Kluwer, Dordrecht.

Zelcer A, Aviv D & Galun E (1978) Interspecific transfer of cytoplasmic male sterility by fusion between protoplasts of normal *Nicotiana sylvestris* & X-ray protoplasts of male sterile *N. tabacum*. Z. Pflanzenphysiol. 90: 397-407.

Zhou JY (1980) Pollen dimorphism and its relation to the formation of pollen embryos in anther culture of wheat (*Triticum aestivum*). Acta Bot. Sin. 22: 117-121.

∗ not seen original paper

Chapter 4

LIGHT AND ELECTRON MICROSCOPIC STUDIES OF SOMATIC EMBRYOGENESIS IN SPRUCE

Larry C. FOWKE[1], Steve ATTREE[2] and Pa la BINAROVA[3]
[1]Department of Biology, University of Saskatchewan, Saskatoon, SK, Canada S7N 5E2, [2]Pacific Regeneration Techonlogies, 455 Gorge Road E, Victoria, B.C., Canada, V8T2W1, And [3]De Montfort University, Norman Borlaug Centre for Plant Science, Institute for Experimental Botany, Academy of the Czech Republic, Sokolovska 6, 77200, Olomouc, Republic of Czech.

1 INTRODUCTION

Since 1985 when somatic embryogenesis was first reported for spruce and larch (Hakman et al.,1985; Chalupa,1985; Nagmani and Bonga, 1985), this morphogenic phenomenon has been observed in a wide variety of conifers in laboratories around the world. Conifers, which have been traditionally considered recalcitrant for plant regeneration in tissue cultures are now readily amenable for establishment as embryogenic cultures. The availability of techniques to grow the embryos in culture will provide a key method for clonal propagation of these forest trees and also provide a useful tool for fundamental research related to the cell biology, physiology, biochemistry and molecular biology of conifer embryo development. A number of recent reviews have described the process of somatic embryogenesis in conifers, their conversion into artificial seeds and the usefulness of this technique for micropropagation in the forest industry (e.g. Tautorus et al., 1991; Attree et al., 1991a; Attree and Fowke, 1993; Gupta et al., 1993; Roberts et al., 1993; Misra, 1994). This chapter focuses on somatic embryogenesis in spruce with an emphasis on the morphology of somatic embryos and embryogenic protoplasts as revealed by light and electron microscopy.

2 SPRUCE SOMATIC EMBRYOGENESIS

An overview of the process of somatic embryogenesis in spruce is presented in Figure 1. A more complete discussion of the steps depicted in Figure 1 is presented by Attree et al. (1991a) and Attree and Fowke (1993). Somatic embryos are induced by culturing zygotic embryos or seedling explants on an agar medium containing an auxin and a cytokinin, in the dark, for 4-6 weeks (Fig. 1a). With a number of spruces it is possible to induce somatic embryogenesis (Fig. 1b) from both immature and mature zygotic embryos; however, with some conifers it has proved difficult or impossible to achieve it using mature zygotic embryos. The origin of somatic embryos in both Norway and white spruce has been traced to single cells of the hypocotyl regions of cultured zygotic embryos (Nagmani et al., 1987). Embryogenic cultures generally appear shiny and translucent and contain numerous somatic embryos which proliferate by cleavage polyembryogenesis. With many cell lines this embryogenic material can be transferred to liquid medium to establish suspension cultures (Fig. 1c). Such cultures contain somatic embryos which proliferate but remain immature, free vacuolated single cells and small cell clusters. Embryogenic suspension cultures of spruce grow rapidly and require sub-culture approximately once a week. Methods are now available for the cryopreservation of conifer embryogenic cultures from both solid and liquid cultures (e.g. Gupta et al., 1987; Kartha et al., 1988; Klimaszewska et al., 1992; Cyr et al., 1994). In our laboratory, one white spruce line frozen in liquid nitrogen from suspension cultures has been stored for twelve years with no apparent detrimental effect on its viability. In fact, new embryogenic cultures are routinely re-established from this culture on an annual basis. Large quantities of viable protoplasts (Fig. 1d) are readily obtained from embryogenic suspension cultures of white spruce. These protoplasts are capable of sustained cell division to regenerate directly to immature somatic embryos; with one white spruce line plantlets can be obtained (Attree et al., 1989). When immature embryos in suspension culture are transferred to medium containing abscisic acid (ABA), they undergo further development to form mature cotyledonary embryos (Fig. 1e). Development can often be enhanced by including an osmoticum, such as polyethylene glycol (PEG), in the medium. Under appropriate conditions these embryos can be desiccated (Fig. 1f) to low moisture levels, stored for prolonged periods and then imbibed and germinated (Fig. 1g) at high frequencies to form plantlets (see Attree and Fowke, 1993). The following discussion highlights the structure of immature and maturing spruce somatic embryos as well as protoplasts isolated from embryogenic suspension cultures of spruce.

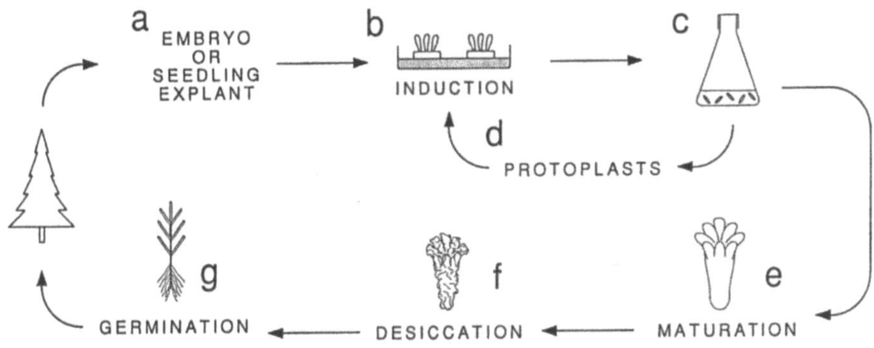

Figure 1. Diagram showing stages in the spruce somatic embryogenesis system under study in the Department of Biology, University of Saskatchewan. a. Zygotic embryos or pieces of young seedlings used as initial explants. b. Induction of somatic embryos after 4-6 weeks of culture in the presence of an auxin and a cytokinin, in dark, c. Suspension culture containing immature somatc embryos proliferating by cleavage polyembryogenesis. d. Protoplasts isolated from immature somatic embryos are capable of regenerating to somatic embryos. e. Maturation of somatic embryos following treatment with ABA + PEG. f. Desiccation of somatic embryos to low moisture levels. g. Germination of dry embryos following imbibition.

3 IMMATURE SOMATIC EMBRYOS

Immature spruce somatic embryos derived from either agar or suspension cultures are similar in morphology to immature seed embryos and are characterized by a meristematic embryonal region composed of tightly packed cells, subtended by a suspensor composed of elongate highly vacuolated cells (Hakman et al., 1987; Joy et al., 1991 and Figs. 2-3a). Embryo growth occurs primarily by active cell division in the embryonal region and cell expansion of the suspensor cells. Electron microscopic studies of the embryonal region reveal cells with an ultrastructure typical of actively dividing plant cells. These cells have a high nucleus to cytoplasm ratio, numerous ribosomes, small vacuoles, frequent plasmodesmata on internal cell walls and relatively undifferentiated leucoplasts (Fig. 3b, 3c). Leucoplasts are characteristic of a variety of immature somatic and zygotic embryos; green non-embryogenic callus contains mature chloroplasts (Feirer, 1988; Hakman and von Arnold, 1988). It is difficult to analyze the frequency and distribution of microtubules and actin microfilaments by electron microscopy, and most information about the cytoskeleton has been

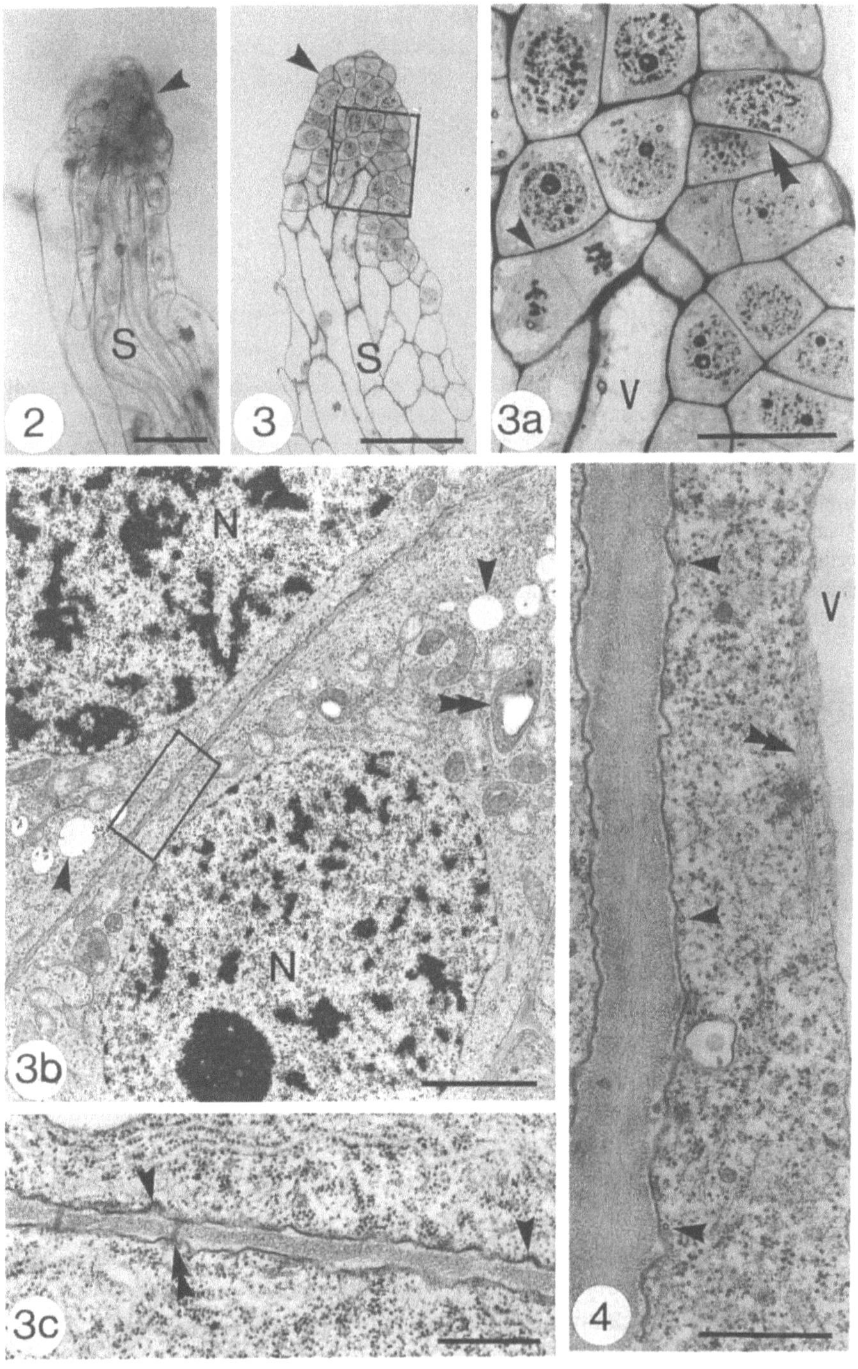

Figure 2-4. Light and electron micrographs of spruce somatic embryos.

Fig. 2. Immature white spruce somatic embryo consisting of a meristematic embryonal region (arrow) and suspensor (S) of large vacuolated cells. Bar = 100 μm. Fig. 3. Section of a white spruce immature somatic embryo showing embryonal region (arrow) and suspensor (S). Area outlined in back shown enlarged in Fig. 3a. Bar = 100 μm. Fig. 3a Enlargement of the area outlined in Fig. 3, showing a cell at telophase (single arrow) and upper region of a suspensor cell with large vacuole (V). The cell wall region designated by the double arrow is depicted in more detail in Fig. 3b. Bar = 300 μm. Fig. 3b. Electron micrograph of area identified by double arrow in Fig. 3a. Note the large nuclei (N), tiny vacuoles (single arrows) and leucoplasts (double arrow). The area outlined in black is enlarged in Fig. 3c. Bar = 3 μm. Fig. 3c. Enlargement of area outlined in black in Fig. 3b, showing coated pits (single arrows) and plasmodesmata (double arrow). Note the high density of ribosomes. Bar = 0.5 μm. Fig. 4. Electron micrograph showing suspensor cell wall and peripheral layer of cytoplasm. Note the transversely oriented cortical microtubules (single arrows) and part of a bundle of actin microfilaments (double arrow). V = vacuole. Bar = 0.5 μm.

provided by fluorescence light microscopy (see below). Organelles involved in endocytosis, such as coated pits, coated vesicles, the partially coated reticulum and multivesicular bodies, are also present in meristematic cells of the embryonal region (Hakman et al., 1987; Galway et al., 1993).

In contrast to the small highly cytoplasmic embryonal cells, the suspensor cells of immature somatic embryos are enormous and contain a central vacuole bounded by a thin layer of cortical cytoplasm (Fig. 3, 3a). Ultrastructural analysis reveals transversely oriented cortical microtubules and longitudinally oriented actin cables (Fig. 4) with plasmodesmata primarily localized in the end walls (Hakman et al., 1987). These cells also contain leucoplasts similar to those observed in the embryonal region. In liquid cultures, some of the most basipetal cells appear to be degenerating (Hakman et al., 1987) and basipetal suspensor cells are often sloughed into the medium.

Fluorescence light microscopy using fluorescent labelled antibodies against microtubules and fluorescent phalloidin which reacts specifically with filamentous actin have been very helpful in assessing the 3-dimensional distribution of cytoskeletal elements in intact immature spruce somatic embryos. Labelling with anti-tubulin clearly shows that individual cells in the embryonal region exhibit oriented cortical microtubule arrays but the cells themselves seem to be randomly arranged (Fowke et al., 1990 and Fig. 5). Both transition cells which link the embryonal cells to the suspensor and suspensor cells are characterized by transversely oriented microtubules; however, these microtubules often appear to be poorly preserved When the embryonal region is stained with rhodamine-phalloidin an extensive network of fine actin microfilaments is revealed in the cortical cytoplasm (Fig. 6). Actin filaments also extend deep within the cytoplasm and form a cage around individual nuclei. This distribution of actin is very sensitive to preparative methods and actin microfilaments disappear when embryos are

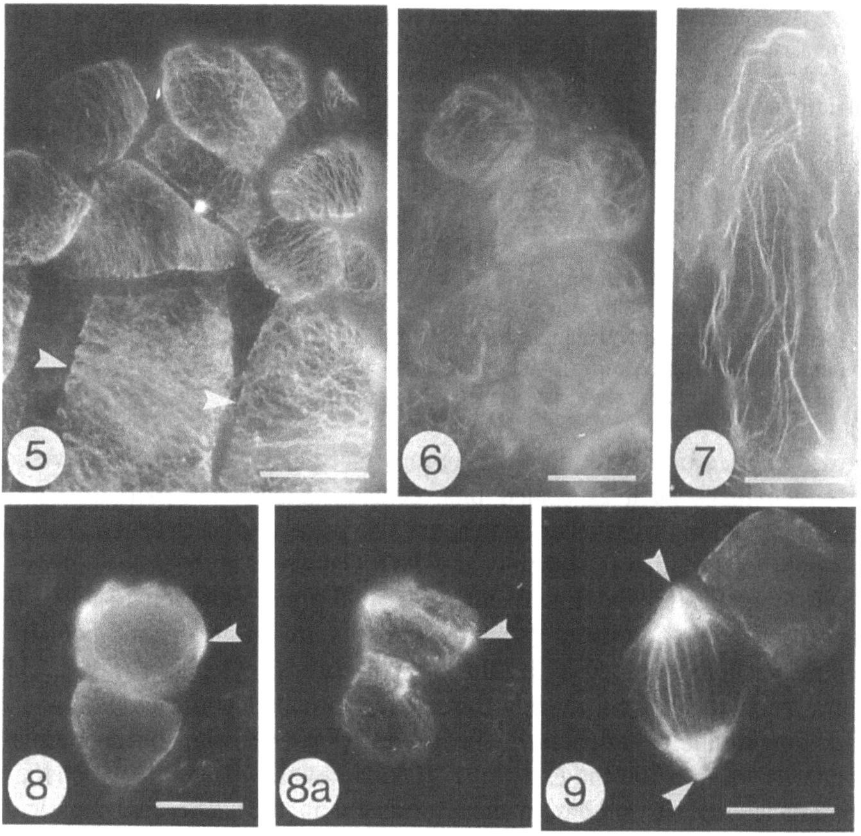

Figure 5-9. Fluorescence micrographs of spruce somatic embryos

Fig. 5. Fluorescence micrograph of embryonal region of immature white spruce somatic embryo stained with anti-tubulin to show distribution of microtubules. Note the transverse orientation of microtubules in the elongate cells (arrows) forming the transition with the suspensor cells below. Bar = 30 μm. Fig. 6. Fluorescence micrograph of embryonal region of immature white spruce somatic embryo stained with rhodamine-phalloidin to show the network of fine actin microfilaments. Bar = 20 μm. Fig. 7. Fluorescence micrograph of suspensor cell of immature white spruce somatic embryo stained with rhodamine-phalloidin to show the longitudinal actin bundles. Bar = 20 μm. Fig. 8. Fluorescence micrograph of dividing white spruce embryonal cell stained with anti-tubulin to show PPB (arrow) in median optical section. Bar = 20 μm. Fig. 8a. Surface view of same cell as in Fig. 8 to show PPB (arrow). Fig. 9. Fluorescence micrograph of white spruce embryonal cell at anaphase stained with anti-tubulin to show pointed spindle poles (arrows). Bar = 20 μm. Figures 8-9 kindly provided by Dr. H. Wang.

fixed with aldehydes (Binarova et al., 1996). Rhodamine-phalloidin staining of suspensor cells, with or without aldehyde fixation, confirms the presence of the longitudinally oriented actin cables observed by electron microscopy (Hakman et al., 1987 and Fig. 7). Rapidly dividing embryonal cells have been useful for immunofluorescence studies of microtubules in conifer cells during cell division. It has been possible to rapidly examine the various microtubule arrays during mitosis. White spruce cells display preprophase bands (PPBs) of microtubules (Fig. 8, 8a) typical of angiosperms and rather unusual pointed spindles at prophase and anaphase (Fig. 9). Microtubule arrays at other stages of mitosis appear similar to those in most dividing angiosperm cells.

4 MATURING SOMATIC EMBRYOS

When immature conifer somatic embryos are grown for 4-8 weeks on medium containing sufficient ABA, cleavage polyembryogenesis is inhibited and the embryos develop to a stage similar to that of zygotic embryos in mature seed (reviewed by Attree and Fowke, 1993). The addition of high molecular weight PEG to the medium dramatically improves the synchrony of development, the quality of mature embryos, particularly in terms of storage reserves, and prevents precocious germination (Attree et al., 1991b, 1992). Polyethylene glycol (Mol. Wt. 1000) is unable to cross the cell wall resulting in a water stress similar to that produced by drought conditions (Attree and Fowke, 1993).

The scanning electron micrographs in Figures 10-13 illustrate the developmental changes in surface morphology which occur during the maturation of white spruce somatic embryos in the presence of ABA and

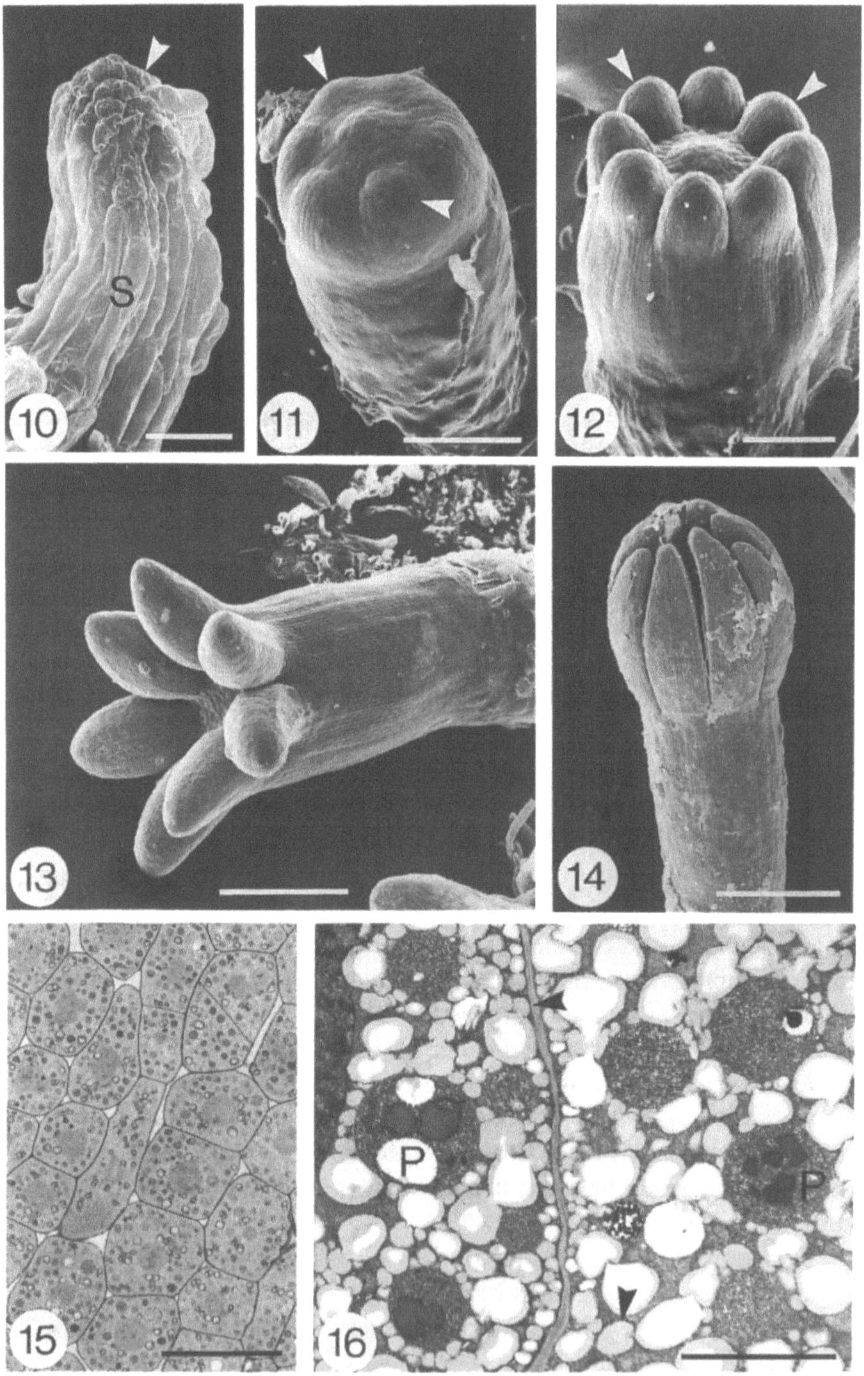

Figure 10-16. Light and scanning electron micrographs of spruce somatic embryos

Fig. 10. Scanning electron micrograph of immature white spruce somatic embryo, showing embryonal region (arrow) and suspensor (S). Bar = 100 μm. Fig. 11. Scanning electron micrograph of a spruce somatic embryo at an early stage of maturation, showing cotyledons (arrows) appearing on the apical dome. Bar = 200 μm. Fig. 12. Scanning electron micrograph of a maturing white spruce somatic embryo with developing cotyledons (arrows) surrounding a domed apex. Bar = 200 μm. Fig. 13. Scanning electron micrograph of a mature white spruce somatic embryo with cotyledons flaring outwards. Bar = 500 μm. Fig. 14. Scanning electron micrograph of a white spruce zygotic embryo, showing compact arrangement of cotyledons. Bar = 400 μm. Fig. 15. Light micrograph showing high density of storage reserves in cortical cells of a white spruce somatic embryo matured with ABA + PEG. Bar = 50 μm. Fig. 16. Electron micrograph showing storage reserves in cells of a white spruce somatic embryo matured with ABA + PEG. Note the numbers of lipid bodies (arrows) and the well developed protein bodies (P). Bar = 5 μm.

PEG (see also Fowke et al., 1994). The embryos first enlarge both in length and width and cotyledons appear as small bulges surrounding the apical region (Fig. 11 cf. Fig.10). The cotyledons increase in size and a pronounced domed apex is visible in the center of the whorl of developing cotyledons (Fig. 12). By 6-8 weeks embryo maturation is complete and the fully developed cotyledons appear turgid (Fig. 13). The cotyledons often flare outwards but with newer methods the cotyledons tend to remain tightly packed together as observed in zygotic embryos of mature seed illustrated in Fig. 14. It is interesting to note that somatic embryos of conifers tend to closely resemble their zygotic counterparts while somatic embryos of angiosperms usually differ markedly from their respective zygotic embryos (e.g. Xu and Bewley, 1992; Brisibie et al., 1993).

Maturation of immature spruce somatic embryos in the presence of ABA results in an increase of storage reserves (see Attree and Fowke, 1993; Misra, 1994). With low osmoticum, embryos show an increase in lipids, starch and protein but with white spruce protein bodies often do not appear to fill with proteins. Biochemical analysis indicates that these embryos do not contain the full complement of storage proteins (Joy et al., 1991; Misra et al., 1993). In contrast, the inclusion of PEG in the medium has been reported to limit the production of starch and result in higher levels of storage lipids and proteins (Attree et al., 1992; Misra et al., 1993 and Figs. 15, 16). Kong and Yeung (1992) also noted a decrease in starch and concomitant increase in protein when white spruce embryos cultured with low osmoticum were exposed to partial drying. Moisture stress, therefore, appears to result in an accumulation of protein at the expense of starch. With Norway spruce (Hakman et al., 1990) and Sitka spruce (Flinn et al., 1991), however, somatic embryos cultured in ABA and low osmoticum medium contained a normal complement of storage proteins. Ultrastructural observations showed well developed protein bodies in the Norway spruce embryos (Hakman and von Arnold, 1988), however, the use of increased osmoticum with these conifers would probably yield still higher levels of storage reserves by preventing precocious germination during later maturation stages. In white spruce the storage reserves tend to increase most dramatically in the cotyledons, in the region adjacent to the shoot apex and in the cortical cells of the hypocotyl and root. White spruce somatic embryos matured with ABA and PEG have the full complement of storage polypeptides, and Northern blot analysis indicates that the levels and patterns of certain storage protein m-RNA transcripts is comparable to those observed in zygotic embryos (Misra et al., 1993; Leal et al., 1995). The mechanism of protein deposition in spruce somatic embryos has not been investigated. However, in megagametophytes of white spruce transport of storage proteins is apparently mediated by the Golgi (Hakman, 1993).

Mature somatic embryos acquire the ability to tolerate some degree of moisture loss by either partial drying (Roberts et al., 1990) or slow drying using a series of chambers with decreasing relative humidities (Attree et al., 1991b). An exciting consequence of the improvement of white spruce somatic embryo maturation due to the addition of PEG to culture medium is the enhancement of desiccation tolerance allowing rapid drying of embryos to very low moisture levels (Attree et al., 1995). This enhanced desiccation tolerance appears to be due in part to markedly increased storage reserves and inhibition of precocious germination and results in embryos with high germination vigour. Desiccated embryos appear opaque and waxy yellow in colour. Scanning electron microscopy reveals that desiccated embryos are characterized by severely wrinkled surface cells due to the drying process (Fowke et al., 1994). Despite such changes to the surface cells, desiccated embryos can be stored frozen for at least one year, imbibed and germinated at high frequencies to yield plantlets with good vigour (Attree et al., 1995).

5 PROTOPLASTS FROM IMMATURE SPRUCE SOMATIC EMBRYOS

Immature conifer somatic embryos provide an excellent experimental material for the isolation of large quantities of viable protoplasts (reviewed by Attree et al., 1991a; Tautorus et al., 1991; Bekkaoui et al. 1994). Non-embryogenic conifer protoplasts are typically recalcitrant to culture and it is difficult to achieve sustained cell division. The advantage of using embryogenic tissue as a source of protoplasts is that, with some species, the protoplasts divide rapidly and regeneration to embryos and subsequently to plantlets can be achieved (spruce - Attree et al., 1989; larch - Klimaszewska, 1989).

Figure 17 illustrates the range of protoplasts released from immature spruce somatic embryos and their fate during short term culture. Protoplasts from somatic embryos fall into two categories, (i) embryogenic protoplasts derived from the embryonal region which can regenerate into new embryos and (ii) non-embryogenic protoplasts from the vacuolated cells of the suspensor which are not capable of regenerating into embryos. Two types of embryogenic protoplasts can also be distinguished, uninucleate and multinucleate. The multinucleate protoplasts result from spontaneous fusion of the closely packed embryonal cells during enzyme treatment. Since non-embryogenic protoplasts fail to show sustained division in culture, the following discussion will focus on the embryogenic protoplasts with only brief mention of the non-embryogenic protoplasts.

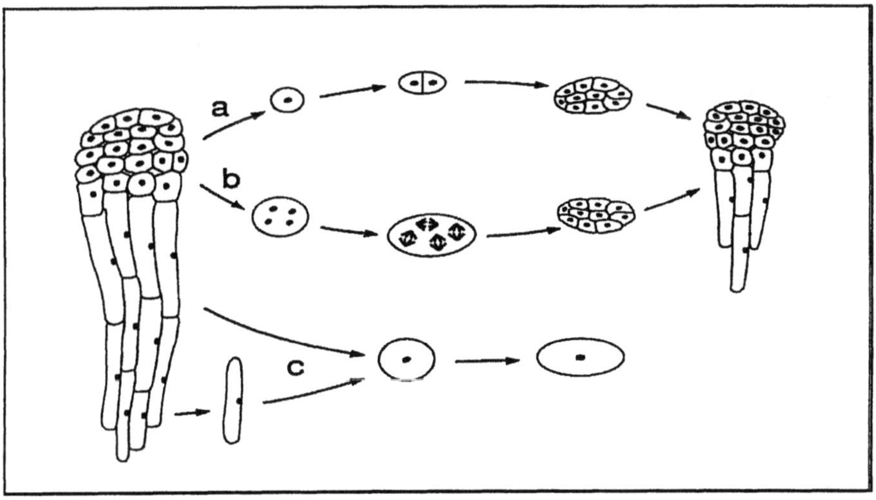

Figure. 17. Diagram illustrating the types of protoplasts produced from immature spruce somatic embryos and the fate of the protoplasts in culture. a. Uninucleate protoplasts divide to form cell clusters, and within 8-12 days, develop to immature embryos. b. Multinucleate protoplasts, resulting from spontaneous fusion of embryonal cells, exhibit synchronous division of their nuclei and form cell clusters which develop into immature embryos. Protoplasts from vacuolated suspensor cells and free floating cells may divide but do not form embryos.

5.1 Uninucleate protoplasts

The freshly isolated embryogenic protoplasts of white spruce derived from the embryonal region typically contain many small vacuoles (Fig. 18). Uninucleate protoplasts are characterized by a randomly oriented population of cortical microtubules and actin microfilaments (Fowke et al., 1990 and Figs. 19, 20). Differences in the distribution of cortical microtubules in embryogenic and non-embryogenic spruce protoplasts have not been detected. With larch, however, non-embryogenic protoplasts apparently

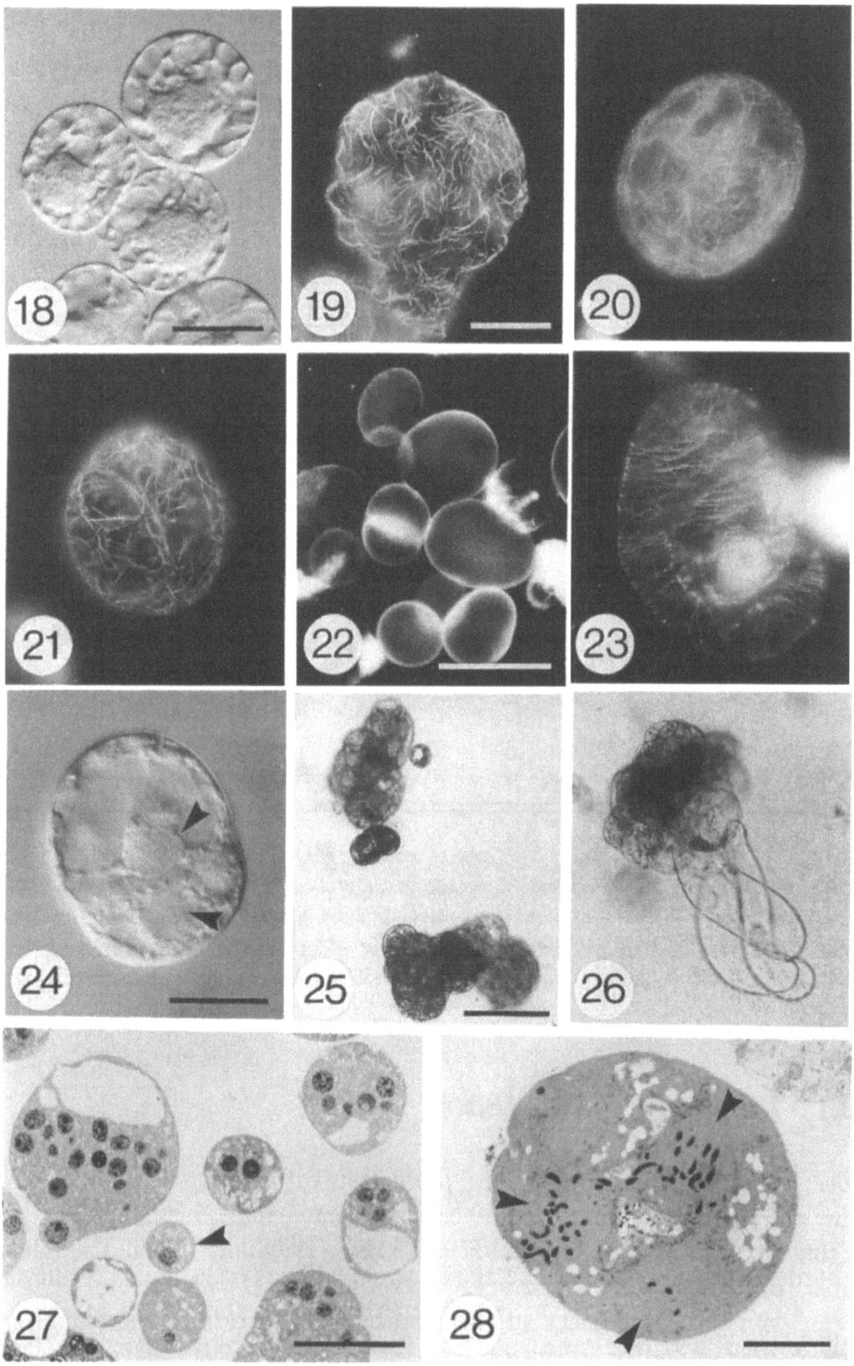

Figure 18-28. Micrographs of spruce protoplasts and derivatives

Fig. 18. Freshly isolated uninucleate protoplasts from immature white spruce somatic embryos. Bar = 30 µm. Fig. 19. Fluorescence micrograph of freshly isolated white spruce protoplast derived from embryonal cells stained with anti-tubulin to show random array of cortical microtubules. Bar = 20 µm. Fig. 20. Fluorescence micrograph of freshly isolated white spruce protoplast derived from embryonal cell, stained with rhodamine-phalloidin to show network of thin actin microfilaments typical of embryogenic protoplasts. Same magnification as Fig. 19. Fig. 21. Fluorescence micrograph of freshly isolated white spruce protoplast derived from highly vacuolated cell, stained with rhodamine-phalloidin to show thick actin cables typical of non-embryogenic protoplasts. Same magnification as Fig. 19. Fig. 22. White spruce protoplasts 24h after culture, showing cell walls stained with Calcofluor white. Note that most of the protoplasts are no longer spherical. Bar = 60 µm. Fig. 23. Fluorescence micrograph of white spruce uninucleate protoplast cultured for 48 h, to show transverse orientation of microtubules in elongating protoplast. Same magnification as Fig. 19. Fig. 24. Recently divided white spruce protoplast containing two nuclei (arrows). Bar = 30 µm. Fig. 25. Cell colonies formed from embryogenic white spruce protoplasts after 10 days of culture. (Reprinted with permission from Attree et al. 1989). Bar = 100 µm. Fig. 26. Immature white spruce somatic embryo formed from embryogenic protoplast after 12 days of culture. (Reprinted with permission from Attree et al. 1989). Same magnification as Fig. 25. Fig. 27. Sectioned white spruce uninucleate (arrow) and multinucleate protoplasts. Bar = 70 µm. Fig. 28. Sectioned multinucleate white spruce protoplast 48 h after culture, showing synchronously dividing nuclei (arrows). Bar = 20 µm.

retain a random orientation of microtubules for a longer period in culture than do embryogenic protoplasts (Staxen et al., 1994). A distinct difference is observed between embryogenic and non-embryogenic spruce protoplasts in terms of actin distribution. The embryogenic protoplasts contain a fine network of actin similar to their cells of origin (Fig. 20 cf. Fig. 6), while non-embryogenic protoplasts derived from the vacuolated transition and suspensor cells tend to be characterized by thick actin cables (Binarova 1996. and Fig. 21).

It has been suggested that the potential for organized plant growth might be correlated with a high PPB index, which is the ratio of the number of PPBs to the number of phragmoplasts in a dividing population (Gorst et al., 1986). Results from a study of PPBs in conifers (Tautorus et al., 1992) are consistent with this proposal since embryogenic black spruce cells have a significantly higher PPB index than non-embryogenic jack pine cells. This correlation, however, does not always hold for protoplast systems as demonstrated by Wang et al. (1989) who found a very high PPB index for non-morphogenic protoplasts of soybean. Work from Von Arnold's laboratory in Sweden suggests that it may be possible to distinguish embryogenic from non-embryogenic spruce cultures by the proteins they secrete into the culture medium (Egertsdotter et al., 1993).

When embryogenic white spruce protoplasts are cultured in agarose, they form a new cell wall and assume an oval shape within the first 24-48 hours (Attree et al., 1987, 1989 and Fig. 22). The cortical microtubules assume a transverse orientation during this period (Fowke et al., 1990 and Fig. 23) which is consistent with their suggested role in directing the deposition of cellulose microfibrils during cell wall formation in elongating cells (e.g. Giddings and Staehelin, 1991; Williamson, 1991). The fine network of actin microfilaments remains unchanged during this early culture period. Both embryogenic and non-embryogenic protoplasts may divide after 24 hours (Fig. 24), but only the embryogenic protoplasts continue to divide to form cell clusters (Fig. 25). Research with angiosperm somatic embryos suggests that embryo initiation and development require the formation of a cluster of highly cytoplasmic cells connected by plasmodesmata (Emons, 1994). By 8-12 days following a gradual decrease in osmoticum, suspensor cells emerge from the clusters yielding clearly recognizable immature somatic embryos (Fig. 26).

5.2 Multinucleate protoplasts

Plant protoplast isolation by enzyme digestion of cell walls often results in spontaneous fusion of adjacent cells because of plasmodesmatal connections between source cells (see Fowke and Gamborg, 1980). Cell walls of the tightly packed embryonal cells are richly endowed with plasmodesmata and thus it is not surprising that multinucleate protoplasts are formed during enzyme treatment. The multinucleate protoplasts contain numbers of nuclei ranging from two to more than twenty (Fig. 27). The distribution of cortical microtubules in freshly isolated multinucleate protoplasts resembles that of uninucleates in that they are first randomly distributed but become transversely oriented during culture (Fowke et al., 1990). These protoplasts also exhibit an extensive cortical network of fine actin microfilaments and cages which enclose the nuclei similar to those observed in the embryonal cells. In general, protoplasts with a small number of nuclei (<10), will form a new cell wall, change shape and undergo synchronous nuclear division (Fig. 28). The nuclei remain separate during cell division suggesting that perhaps the actin cage surrounding nuclei functions to prevent contact between adjacent nuclei and subsequent fusion of nuclei. Nuclear fusion is commonly

Fig. 29. Electron micrograph showing part of a freshly isolated white spruce protoplast. Note the leucoplasts (single arrows) and tannin deposits (double arrow) on the tonoplast of the vacuole (V). Bar = 2 μm. Fig. 30. Electron micrograph of a multinucleate protoplast showing 3 nuclei (N). Note the small vacuoles (single arrows) and leucoplasts (double arrow). Bar = 5 μm. Inset - Enlargement of the area outlined in black, showing remnants of cell wall material. Note the high density of ribosomes in the cytoplasm. Bar = 0.5 μm. Fig. 31. Electron micrograph showing surface region of white spruce protoplast. A remnant of a plasmodesma (double arrow) is visible. Note coated pit (small arrow) and numerous microtubules (large arrow) adjacent to the plasma membrane. Bar = 0.5 μm.

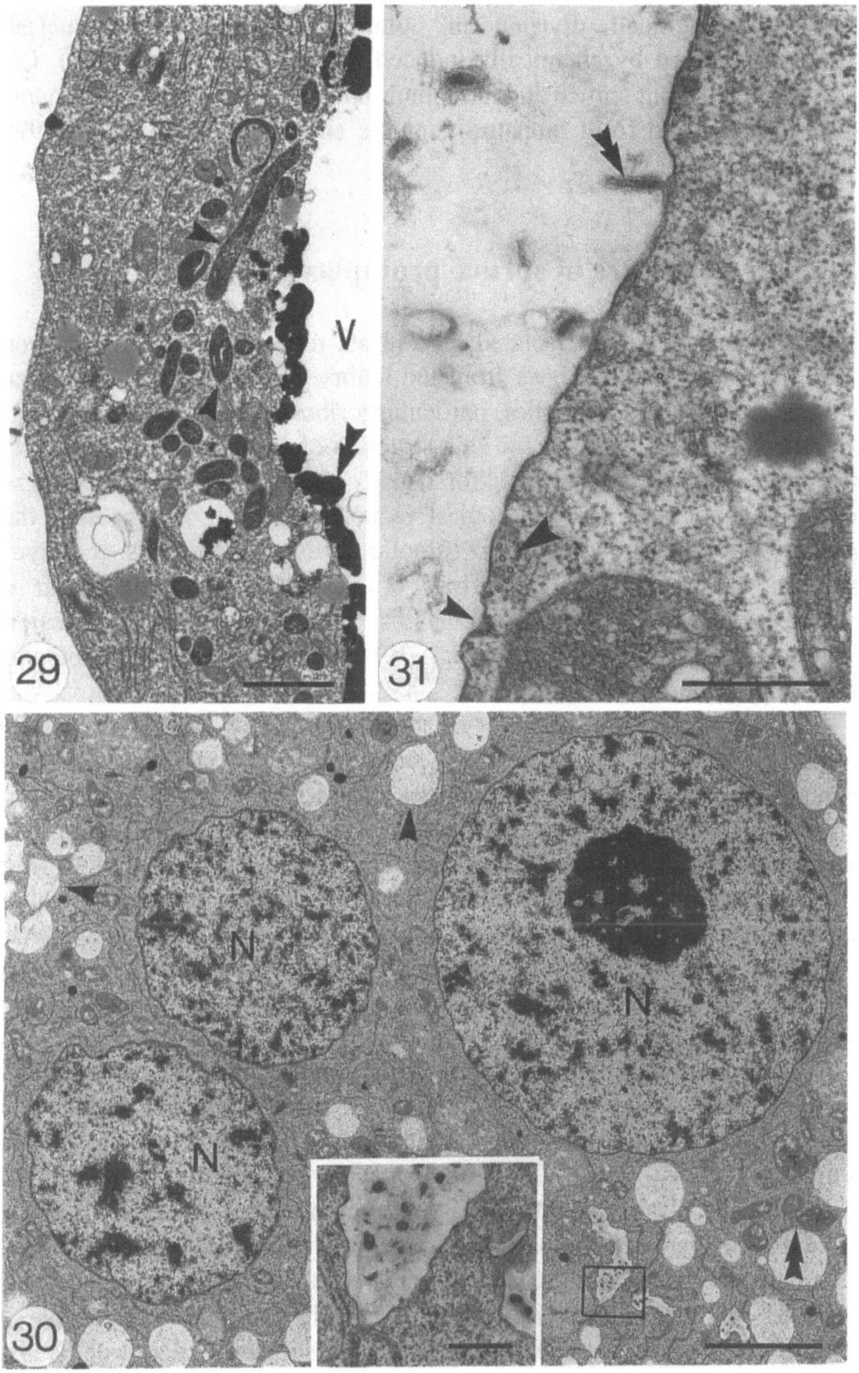

Figure 29-31. Electron Micrographs of spruce protoplasts.

observed during cell division in cultured angiosperm multinucleate protoplasts derived by chemically induced fusion (see Fowke, 1989). Cell clusters formed from spruce and larch multinucleate protoplasts can generate suspensor cells and form immature somatic embryos (Fowke et al., 1990; Von Aderkas, 1992).

5.3 Ultrastructure of spruce protoplasts

The cytoplasm of freshly isolated protoplasts reflects the cell source from which they are derived. Those from the embryonal cells, for example, are richly populated with organelles, particularly ribosomes (Figs. 29-31). One difference observed between the fine structure of embryonal cells and their protoplasts is the presence of electron dense deposits, believed to be tannins, along the tonoplast of many protoplast vacuoles (Fig. 29). It is possible that these are deposited in response to osmotic stress or the effects of enzyme digestion of the cell wall (Galway et al., 1993). The cytoplasm of multinucleate protoplasts also contains irregularly shaped inclusions containing cell wall remnants (Fig. 30) trapped during the protoplast isolation procedure.

A closer look at the structure of the cortical cytoplasm of protoplasts reveals a number of interesting features. First, as expected from immunofluorescence observations, numerous microtubules are seen adjacent to the plasma membrane (Fig. 31 cf. Fig. 19). In a few instances, remnants of plasmodesmata are observed projecting from the plasma membrane (Fig. 31). The retention of plasmodesmata remnants following protoplast isolation has been reported previously (Attree and Sheffield, 1985) and it suggests the presence of rather rigid plasmodesmata in the embryonal region of immature embryos. The plasma membrane is also characterized by many clathrin coated pits (Fig. 31).

A study of white spruce protoplasts employing both conventional chemical fixation and freeze fixation/freeze substitution has revealed an active endocytotic pathway originating with coated pits on the plasma membrane (Galway et al., 1993). Cationized ferritin supplied to protoplasts attaches to the plasma membrane, is taken into coated vesicles via coated pits and is delivered to the partially coated reticulum, multivesicular bodies and finally to the vacuole. Coated vesicle mediated endocytosis has been described in other plants (Hawes et al., 1991) and may play a role in membrane recycling during active cell wall secretion by Golgi or perhaps in detoxifying cells exposed to fungal toxins or elicitors (Fowke et al., 1991).

In summary, this chapter has described structural studies of somatic embryos of spruce with a particular emphasis on white spruce (*Picea glauca*). During the past ten years, somatic embryogenesis has been adapted to spruce and reliable methods are now available for embryo induction, establishment of stable embryogenic suspension cultures, cryopreservation of embryogenic lines, protoplast isolation and regeneration as well as embryo maturation, desiccation and germination to produce plantlets. The light and electron microscope data indicate that immature somatic embryos show many similarities to their zygotic counterparts. Furthermore, somatic embryos matured using recent methods involving ABA and osmoticum are strikingly similar to mature seed embryos in terms of morphology as well as their complement of storage reserves. Somatic embryogenesis thus offers a powerful experimental system to examine aspects of embryo development in conifers. Also, structural studies of immature somatic embryos and protoplasts derived from them, including an analysis of the cytoskeleton by fluorescence methods, have revealed some differences between embryogenic and non-embryogenic cultures. Further research with this system should help to define embryogenic potential more clearly by providing further markers of embryogenic potential.

6 ACKNOWLEDGEMENTS

Special thanks to Dennis Dyck, Evelyn Peters and Pat Clay for assistance with manuscript preparation. We gratefully acknowledge the financial support from the Natural Sciences and Engineering Research Council of Canada and the Province of Saskatchewan Environmental Technology Development Program.

7 REFERENCES

Attree M, Bekkaoui F, Dunstan DI & Fowke LC (1987) Regeneration of somatic embryos from protoplasts isolated from an embryogenic suspension culture of white spruce (*Picea glauca*). Plant Cell Rep. 6: 480-483.

Attree SM, Dunstan DI & Fowke LC (1989) Plantlet regeneration from embryogenic protoplasts of white spruce (*Picea glauca*). Bio/Technology 7: 1060-1062.

Attree SM, Dunstan DI & Fowke LC (1991a) White spruce [*Picea glauca* (Moench) Voss] and Black spruce [*Picea mariana* (Mill) B.S.P.]. In: Bajaj YPS (ed) Biotechnology in Agriculture and Forestry. Vol.16. pp.423-445. Springer, Berlin.

Attree SM & Fowke LC (1993) Embryogeny of gymnosperms: advances in synthetic seed technology of conifers. Plant Cell Tissue Organ Cult. 35: 1-35.

Attree SM, Moore D, Sawhney VK & Fowke LC (1991b) Enhanced maturation and desiccation tolerance of white spruce [*Picea glauca* (Moench) Voss] somatic embryos: effects of a non-plasmolysing water stress and abscisic acid. Ann. Bot. 68: 519-525.

Attree SM, Pomeroy MK & Fowke LC (1992) Manipulation of conditions for the culture of somatic embryos of white spruce for improved triacylglycerol biosynthesis and desiccation tolerance. Planta 187: 395-404.

Attree SM, Pomeroy MK & Fowke LC (1995) Development of white spruce (*Picea glauca* (Moench.) Voss) somatic embryos during culture with abscisic acid and osmoticum, and their tolerance to drying and frozen storage. J. Exp. Bot. 46 (285): 433-439.

Attree SM & Sheffield E (1985) Plasmolysis of *Pteridium* protoplasts: A study using light and scanning-electron microscopy. Planta. 165: 151-157.

Bekkaoui, F, Tautorus TE & Dunstan DI (1994) Gymnosperm protoplasts. In: Jain S, Gupta P & Newton R (eds) Somatic Embryogenesis in Woody Plants. Vol. 1. pp. 167-191. Kluwer Academic Publishers, The Netherlands.

Binarova P, Cihalikova C, Dolezel J, Gilmer S & Fowke LC (1996) Actin distribution in somatic embryos and embryogenic protoplasts of white spruce (*Picea glauca*). In Vitro Cell. Dev. Biol. 32p : 59-65.

Brisibie EA, Nishioka D, Miyake H, Taniguchi T & Malda E (1993) Developmental electron microscopy and histochemistry of somatic embryo differentiation in sugarcane. Plant Sci. 89: 85-92.

Chalupa V (1985) Somatic embryogenesis and plantlet regeneration from cultured immature and mature embryos of *Picea abies* (L.) Karst. Comm. Inst. For. 14: 57-63.

Cyr DR, Lazaroff WR, Grimes SMA, Quan G, Bethune TD, Dunstan DI & Roberts DR (1994) Cryopreservation of interior spruce (*Picea glauca engelmanni* complex) embryogenic cultures. Plant Cell Rep. 13: 574-577.

Egertsdotter U, Mo LH & von Arnold S (1993) Extracellular proteins in embryogenic suspension cultures of Norway spruce (*Picea abies*). Physiol. Plant. 88: 315-321.

Emons AMC (1994) Somatic embryogenesis: cell biological aspects. Acta Bot. Neerl. 43: 1-14.

Feirer RP (1988) Chloroplast ultrastructure and gene expression in embryogenic conifer callus. In: Cheliak,WM & Yapa AC (eds) Molecular Genetics of Forest Trees. Proc. 2nd IUFRO Working Party on Molec. Genet. pp. 89-94. Petawawa Nat. For. Inst.

Flinn BS, Roberts DR & Taylor IEP (1991) Evaluation of somatic embryos of interior spruce. Characterization and developmental regulation of storage proteins. Physiol. Plant. 82:624-632.

Fowke LC (1989) Ultrastructural studies of plant protoplast fusion. In: Bajaj YPS (ed) Biotechnology in Agriculture and Forestry. Vol.8. pp. 289-303. Springer, Berlin.

Fowke LC, Attree SM, Wang H & Dunstan DI (1990) Microtubule organization and cell division in embryogenic protoplast cultures of white spruce (*Picea glauca*). Protoplasma 158: 86-94.

Fowke LC, Attree SM & Rennie PJ (1994) Scanning electron microscopy of hydrated and desiccated mature somatic embryos and zygotic embryos of white spruce (*Picea glauca* [Moench] Voss.) . Plant Cell Rep. 13: 612-618.

Fowke LC, Tanchak MA & Galway ME (1991) Ultrastructural cytology of the endocytotic pathway in plants. In: Hawes CR et al. (eds) Endocytosis, Exocytosis and Vesicle Traffic in Plants. pp. 15-40. Cambridge Univ. Press, Cambridge.

Fowke LC & Gamborg OL (1980) Applications of protoplasts to the study of plant cells. Int. Rev. Cyt. 68: 9-51.

Galway ME, Rennie PJ & Fowke LC (1993) Ultrastructure of the endocytotic pathway in glutaraldehyde-fixed and high-pressure frozen/freeze-substituted protoplasts of white spruce (*Picea glauca*). J. Cell Sci. 106: 847-858.

Giddings TH Jr & Staehelin LA (1991) Microtubule-mediated control of microfibril deposition: a re-examination of the hypothesis. In: Lloyd CW (ed) The Cytoskeletal Basis of Plant Growth and Form. pp. 85-99.

Gorst J, Wernicke W & Gunning BES (1986) Is the preprophase band of microtubules a marker for organization in suspension cultures? Protoplasma 134: 130-140.

Gupta PK, Durzan DJ & Finkle BJ (1987) Somatic polyembryogenesis in embryogenic cell masses of *Picea abies* (Norway spruce) and *Pinus taeda* (loblolly pine) after thawing from liquid nitrogen. Can. J. For. Res. 17: 1130-1134.

Gupta PK, Pullman G, Timmis R, Kreitinger M, Carlson WC, Grob J & Welty E (1993) Forestry in the 21st Century: The biotechnology of somatic embryogenesis. Bio/Technology 11: 454-459.

Hakman I (1993) Embryology in Norway spruce (*Picea abies*). Immunochemical studies on transport of a seed storage protein. Physiol. Plant. 88: 427-433.

Hakman I, Fowke LC, von Arnold S & Eriksson T (1985) The development of somatic embryos in tissue cultures initiated from immature embryos of *Picea abies* (Norway spruce). Plant Sci. 38: 53-59.

Hakman I, Rennie P & Fowke L (1987) A light and electron microscope study of *Picea glauca* (white spruce) somatic embryos. Protoplasma 140: 100-109.

Hakman I, Stabel P, Engstrom P & Eriksson T (1990) Storage protein accumulation during zygotic and somatic embryo development in *Picea abies* (Norway spruce). Physiol. Plant. 80: 441-445.

Hakman I & von Arnold S (1988) Somatic embryogenesis and plant regeneration from suspension cultures of *Picea glauca* (white spruce). Physiol. Plant. 72: 579-587.

Hawes CR, Coleman JOD & Evans DE (eds) (1991) Endocytosis, Exocytosis and Vesicle Traffic in Plants. Cambridge Univ. Press, Cambridge.

Joy RW, Yeung EC, Kong L & Thorpe TE (1991) Development of white spruce somatic embryos: I. Storage product deposition. In Vitro Cell. Dev. Biol. 27P: 32-41.

Kartha KK, Fowke LC, Leung NL, Caswell KL & Hakman I (1988) Induction of somatic embryos and plantlets from cryopreserved cell cultures of white spruce (*Picea glauca*). J. Plant Physiol. 132: 529-539.

Klimaszewska K (1989) Recovery of somatic embryos and plantlets from protoplast cultures of *Larix* × *eurolepis*. Plant Cell Rep. 8: 440-444.

Klimaszewska K, Ward C & Cheliak WM (1992) Cryopreservation and plant regeneration from embryogenic cultures of larch (*Larix* × *eurolepis*) and black spruce (*Picea mariana*). J. Exp. Bot. 43: 73-79.

Kong L & Yeung EC (1992) Development of white spruce somatic embryos: II Continual shoot meristem development during germination. In Vitro Cell. Dev. Biol. 28P: 125-131.

Leal I, Misra S, Attree SM & Fowke LC (1995) Effect of abscisic acid, osmoticum and desiccation on 11S storage protein gene expression in somatic embryos of white spruce. Plant Sci. 106: 121-128.

Misra S (1994) Conifer zygotic embryogenesis, somatic embryogenesis, and seed germination: Biochemical and molecular advances. Seed Sci. Res. 4: 357-384.

Misra S, Attree SM, Leal I & Fowke LC (1993) Effect of abscisic acid, osmoticum, and desiccation on synthesis of storage proteins during the development of white spruce somatic embryos. Ann. Bot. 71: 11-22.

Nagmani R, Becwar MR & Wann SR (1987) Single-cell origin and development of somatic embryos in *Picea abies* (L.) Karst. (Norway spruce) and *P. glauca* (Moench) Voss (white spruce). Plant Cell Rep. 6: 157-159.

Nagmani R & Bonga JM (1985) Embryogenesis in subcultured callus of *Larix decidua*. Can. J. For. Res. 15: 1088-1091.

Roberts DR, Sutton BCS & Flinn BS (1990) Synchronous and high frequency germination of interior spruce somatic embryos following partial drying at high relative humidity. Can. J. Bot. 68: 1086-1090.

Roberts DR, Webster FB, Flinn BS, Lazaroff WR & Cyr DR (1993) Somatic embryogenesis of spruce. In: Redenbaugh K (ed) Synseeds: Applications of Synthetic Seeds to Crop Improvement. pp. 427-450. CRC Press, Boca Raton.

Staxen I, Klimaszewska K & Bornman CH (1994) Microtubular organization in protoplasts and cells of somatic embryo-regenerating and non-regenerating cultures of *Larix*. Physiol. Plant. 91: 680-686.

Tautorus TE, Fowke LC & Dunstan DI (1991) Somatic embryogenesis in conifers. Can. J. Bot. 69: 1873-1899.

Tautorus TE, Wang H, Fowke LC & Dunstan DI (1992) Microtubule pattern and the occurrence of pre-prophase bands in embryogenic cultures of black spruce (*Picea mariana* Mill.) and non-embryogenic cultures of jack pine (*Pinus banksiana* Lamb.). Plant Cell Rep. 11: 419-423.

Von Aderkas P (1992) Embryogenesis from protoplasts of haploid European larch. Can. J. For. Res. 22: 397-402.

Wang H, Cutler AJ & Fowke LC (1989) High frequencies of preprophase bands in soybean protoplast cultures. J. Cell Sci. 92: 575-580.

Williamson RE (1991) Orientation of cortical microtubules in interphase plant cells. Int. Rev. Cyt. 129: 135-206.

Xu N & Bewley D (1992) Contrasting pattern of somatic and zygotic embryo development in alfalfa (*Medicago sativa* L.) as revealed by scanning electron microscopy. Plant Cell Rep. 11: 279-284.

Chapter 5

PHYSIOLOGICAL AND MORPHOLOGICAL ASPECTS OF SOMATIC EMBRYOGENESIS

Kojn NOMURA[1] and Atsushi KOMAMINE
1Institute of Agriculture and Forestry, University of Tsukuba, Tsukuba, Ibaraki 305, and The Research Institute of Evolutionary Biology, 2-4-28, Kamiyoga, Setagaya-ku, Tokyo 158-0098, Japan

1 INTRODUCTION

Somatic embryogenesis is an ideal system for investigate the entire process of differentiation of plants, as well as of the mechanisms of expression of totipotency in plant cells, having major advantages as compared to zygotic embryogenesis. For example, (a) the process of embryogenesis is easily monitored, (b) the environment of the embryo can be controlled, and (c) large numbers of embryos can easily be obtained. Haberlandt's goal in his initial attempt to establish a plant tissue culture system was to provide evidence for totipotency of plant cells, and the mechanism of somatic embryogenesis remains one of the most fundamental problems in plant physiology.

Recent studies on mechanisms of somatic embryogenesis have mainly involved biochemical and molecular approaches. However, morphological approaches using recently improved techniques, such as immunochemical methods, can also provide important information about changes *in situ* during embryogenesis.

In this chapter, we first describe the establishment of high frequency and synchronized development of somatic embryos in carrot suspension cultures, and then we review the events that occur *in situ* during somatic embryogenesis in this system. The importance of the establishment of polarity at the early stage of somatic embryogenesis is emphasized.

2 DEVELOPMENT OF SOMATIC EMBRYOS

2.1 A system for high frequency and synchronized development of somatic embryos from embryogenic cell clusters

When carrot cells are cultured in the presence of 2,4-dichlorophenoxyacetic acid (2,4-D) and then transferred to a medium without an auxin, certain clusters of cells develop into somatic embryos. There are many types of cells in a suspension culture, and somatic embryogenesis is not synchronous if some types of cells are transferred to auxin-free medium. Large numbers of somatic embryos at specific stages are clearly necessary for biochemical and molecular biological studies. An experimental system in which somatic embryogenesis occurs synchronously and at a high frequency was established by Fujimura and Komamine (1979 a). In the system, certain embryogenic cell clusters that were similar in terms of size and density were collected from suspension cultures by sieving through nylon screens with 47- and 31- μm openings and density gradient centrifugation in 18% Ficoll. Almost all of the collected cell clusters developed synchronously to somatic embryos when they were transferred to an auxin-free medium. These embryogenic cell clusters seemed similar to the proembryogenic masses described by Halperin (1966). Although it is difficult to identify the origin of these cell clusters in suspension cultures, Halperin and Jensen (1967) suggested that fragmentation of large clusters (or clumps) might possibly induce formation of such embryogenic cell clusters or proembryogenic masses. Each large "parent" cell cluster had two histological regions: an inner region of highly vacuolated, nondividing cells, and a peripheral region of small meristematic cells. Fragmentation of such a parent cluster released smaller cell clusters which exhibited morphological polarity. The daughter cell clusters had large, vacuolated cells at one end and smaller, meristematic cells at the other. The structural differences among the cells in a cell cluster resembled those between the dividing part and the suspensor-like structure of the zygotic embryo.

Development of somatic embryos from cell clusters in auxin-free medium is associated with some asymmetry from the very beginning. Serial observations of embryogenesis by Fujimura and Komamine (1980) revealed that formation of a globular embryo occurred from a single site on each embryogenic cell cluster. When cell clusters were transferred to medium without auxin, cell division occurred throughout the cell clusters at a relatively low rate for the first 3 days of culture. The shape of the cell cluster was irregular at this stage. Thereafter, between 3 and 4 days after transfer, very rapid cell division occurred at a specific site on each cell cluster. The volume of individual cells at this site decreased considerably because of rapid division of cells. After this rapid division phase, this part of the cluster became spherical in shape and the cluster developed into an early globular embryo. By contrast, no cell division was subsequently observed in parts of the cluster other than the actively dividing region, and the non-dividing cells formed a suspensor-like structure. At this stage, somatic embryos were composed of three distinguishable regions: an actively dividing region, which developed into shoot on further culture, a region adjacent to the first region, which developed into roots, and a third region, within the non-dividing areas which did not grow any further. Thus, each region was already destined to differentiate to a specific organ at this stage.

Fujimura and Komamine (1979) and Nomura and Komamine (1985) succeeded in establishing systems in which high-frequency synchronized embryogenesis occurred, and these systems have provided various types of physiological information. Heterogeneous populations of cells from carrot suspension cultures were fractionated by sieving through nylon screens and subsequent density gradient centrifugation. In this way, small, round and cytoplasm-rich competent single cells (State 0), and embryogenic cell clusters (State 1), which consisted of fewer than ten cytoplasm-rich cells were isolated. Single cells at State 0 could divide to form embryogenic cell clusters at State 1 in the presence of auxin (2,4-D at 5×10^{-8} M was optimal). Embryogenic cell clusters at State 1 differentiated to globular, heart-shaped and torpedo-shaped embryos synchronously and at high frequency when they were transferred to auxin-free medium that contained zeatin at 10^{-7} M. The frequency of embryogenesis from single cells was between 85 and 90%. When single cells at State 0 were cultured in the absence of auxin, they did not divide but, rather, elongated. Elongated cells could not divide or differentiate even when cultured in the presence of auxin and, therefore, may appeared to have lost totipotency. Thus, the development of cells at State 0 to cell clusters at State 1 and embryos can be regarded as the process whereby totipotency is expressed, while the conversion of cells at State 0 to elongated cells in the absence of auxin corresponds to loss of totipotency. The polarity of the synthesis of DNA and RNA and of levels of mRNA, free Ca^{2+} ions and, probably, auxin appears at a late stage during conversion of cells at State 0 to cell clusters at Sate 1. Polarity might play an important role during development of embryos from single cells. Therefore, it seems that auxin is required to induce embryogenesis from competent single cells, while auxin is inhibitory during the development of embryos from embryogenic cell clusters. We can assume, then, that determination of embryogenesis has already occurred in cell clusters at State 1, because the fate of each cell has already been determined. This assumption was confirmed by a study of the distribution of cells labeled with a fluorescent dye that had been introduced by microinjection (Nomura and Komamine, unpublished data).

At the early stage of the process, from cell clusters at State 1 to embryos, three phases can be recognized. During Phase 1, the first three days after transfer to auxin-free medium, division of undifferentiated cells occurs slowly. During Phase 2, between 3 and 4 days, very rapid division of cells occurs, giving rise to globular embryos that subsequently develop into heart-shaped and torpedo-shaped embryos. Active turnover of RNA and protein occurs during Phase 1 and synthesis of new species of mRNA and protein occurs in Phases 1-2. During Phase 2, the active synthesis of DNA occurs which is due, at least partially, to reduction in the size of the replicon. High template activity and changes in relative levels of histones have also been observed in Phases 1-2. The cell-doubling time in Phase 2 was found to be only 6.3 h, while doubling times in Phases 1 and 3 were 58 and 20 h, respectively.

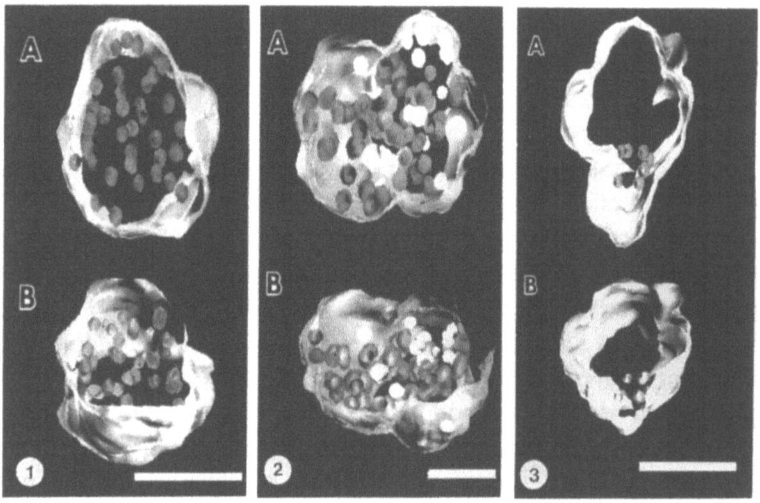

Figure 1-3. Embryogenic cell clusters (State 1) were cultured for 48 h in an auxin-free medium and puls-labelled with [^{3}H]-thymidine for 24 h. Fig. 1: Cell clusters were fixed immediately after puls labelling. Serial sections and autoradiograms from them were prepared. Data of serial sections and autoradiograms were input into a computer and the original cell clusters were reconstructed three-dimensionally. Brown balls indicate heavily labelled nuclei, indicating that DNA synthesis, that is, cell division occurs randomly. Fig. 2: Cell clusters of Fig. 1 were treated in the same way as in Fig. 1, but chased with cold thymidine in an auxin-free medium for 24 h and fixed. White balls indicated weakly labelled nuclei, indicating that cell division occurs in this area, resulting in dilution of radioactivity. Fig. 3: Cell clusters in Fig. 1 were chased in an auxin-free medium for 48 h and differentiated to proglobular-globular embryos. No labelled nuclei appeared in the "head" part of embryos A: vertical view of cell clusters. B: 50 degrees oblique view from vertical axis.

Several molecular markers associated with the totipotent nature of cells have been isolated. Four polypeptides, a protein (21D7) and two mRNAs have been specifically detected during the induction and development phases of somatic embryogenesis from single cells. When single cells are cultured in the absence of auxin, they form elongated cells with a concomitant loss of totipotency, and these markers are no longer detectable. These results are summarized in Figures 1 and 2.

2.2 Polarized synthesis of DNA and cell division in cell clusters

As described in the first section, observations by electron and light microscopy revealed that embryogenic cell clusters, or pro-embryogenic masses that can develop into embryos, had morphological polarity in auxin-free media. These embryogenic cell clusters included regions of cytoplasm-rich cells and other regions that contained vacuolated cells (Halperin and Jensen, 1967). It has been reported that globular embryos are initiated by the rapid localized growth of cells or from one cell in an embryogenic cell cluster (Fujimura and Komamine, 1980; Haccius, 1978; DosSantos et al. 1983). Undifferentiated cell clusters that have not yet acquired the competence to form embryos in auxin-free medium have no structural polarity. However, when embryogenic cell clusters are transferred to auxin-free medium, the polarity of DNA synthesis becomes apparent. It can be analyzed by three-dimensional computer associated reconstruction of the clusters from serial sections of cells that have been puls-chase labelled with [^{3}H]-thymidine. Embryogenic cell clusters, collected by the procedure described in the previous section, are labelled with [^{3}H]-thymidine for 24 h in auxin-free medium. Some cell clusters are fixed immediately after labelling, but others are incubated in auxin-free medium without [^{3}H]-thymidine for 24 h, 48 h and more (the chase). Autoradiograms are prepared from serial sections of each cell cluster. Three-dimensional reconstruction yields results such as those shown in Figure 3, confirming that DNA synthesis occurs randomly during the first 3 days of culture in the absence of auxin (Phase 1), while polarized synthesis of DNA occurs during days 3-4 of culture (Phase 2). In the globular embryos, evidence can be detected of high-level DNA-synthetic activity in procambial and proepidermal cells, while no DNA-synthetic activity is detected in the suspensor like structures.

The occurrence of the polarized synthesis of DNA in embryogenic cell clusters was observed on the third and fourth days after transfer to auxin-free medium (Tsukahara and Komamine, 1996). The cells that were actively synthesizing DNA were separated from cells that were not synthesizing

DNA by maceration of cell clusters into individual protoplasts and centrifugation in a Percoll density gradient. Three polypeptides were found that were specifically expressed in cells that were actively synthesizing DNA and these polypeptides might be candidates for markers of the polarity of DNA synthesis and for the active division of cells that are specific to embryogenesis.

The polarity of embryogenic cell clusters might play an important role in the expression of totipotency during somatic embryogenesis. Thus, the timing of events associated with the expression of totipotency in embryogenesis should be revealed by investigations of the development of the polarity of cell clusters. However, details of the formation of such embryogenic cell clusters are obscure since many workers have used the cell clusters themselves as the starting material for embryogenesis. These embryogenic clusters or pro-embryogenic masses develop into embryos at high frequency when they are transferred to an auxin-free medium (Fujimura and Komamine, 1979a), an indication that the determination of embryogenesis has already occurred in these embryogenic clusters. The system described above, in which single cells develop into embryos at high frequency, is useful for studies on the initial stages of somatic embryogenesis from single cells.

2.3 Gradients of endogenous auxin

Somatic embryogenesis is induced by the transfer of cell clusters from a medium that contains auxin to an auxin-free medium after fractionation of the cell clusters by sieving and density gradient centrifugation. Since the rapid division and differentiation of the cells occur during embryogenesis in auxin-free medium, endogenous auxins in the cells might be expected to play an important role in embryogenesis as auxin is considered to be essential for cell division and differentiation. Fujimura and Komamine (1979b) examined the effects of exogenous auxins and antiauxins, and measured levels of endogenous auxins. Indoleacetic acid (IAA) inhibited embryogenesis when added to the medium at concentrations higher than 10^{-7}M. Addition of 2,4,6-trichlorophenoxyacetic acid and p-chlorophenoxyiso-butyric acid, which are known as antiauxins, also inhibited embryogenesis. Results of experiments in which an auxin or an antiauxin was applied at different stages of embryogenesis indicated that the early stage of embryogenesis was sensitive to both auxin and antiauxin. The presence of endogenous auxins in the cells at every stage of the embryogenesis was demonstrated by the so-called *Avena* curvature test. Fujimura and Komamine (1979b) found evidence that the polarized localization of auxins in embryogenic cell clusters is essential for embryogenesis; exogenously supplied auxins might eliminate the polarity of endogenous auxins with resultant inhibition of embryogenesis. Furthermore, antiauxins would inhibit the actions of endogenous auxin, thereby inhibiting embryogenesis.

2.4 Localization of calcium ion

Calcium ions play an important role in the regulation of cellular events. Calcium ions mediate signal transduction in many cases. The role of Ca^{2+} ions and calmodulin in carrot somatic embryogenesis was examined by Overvoorde and Grimes (1994). Formation of embryos was not affected until the concentration of Ca^{2+} ions was below 200 μM and beyond this threshold the rate of embryo formation decreased with decreasing concentrations of Ca^{2+} ions. Treatment of developing embryos with Ca^{2+}-channel blockers or a Ca^{2+} ionophore inhibited embryo formation. The results of Overoorde and Grimes indicated that exogenous Ca^{2+} ions or the maintenance of a gradient of Ca^{2+} ions was required for embryogenesis. These researchers also found that calmodulin-Ca^{2+} complexes were localized in regions that contained the developing meristem of both the cotyledon tips and rhizoid regions while the calmodulin protein appeared to be more uniformly distributed. Levels of mRNA for calmodulin increased slightly when cell clumps were induced to form embryos.

The distribution of cytosolic Ca^{2+} ions was traced by use of the fluorescent indicator fluo 3, antimonate precipitation and proton-induced X-ray emission analysis. Embryogenesis was found to coincide with an increase in the level of free cytosolic Ca^{2+} ion. The highest level of Ca^{2+} ions was found in the protoderm of embryos from the late globular to the torpedo stage. The gradient of Ca^{2+} ions was observed along the longitudinal axis of the embryo. The nucleus gave the most conspicuous signals.

Signal transduction systems mediated by calmodulin have been investigated by Nomura et al. (unpublished data). Levels of some calmodulin-binding proteins were found to decrease before the formation of globular embryos. The proteins appeared when developing embryos were transferred to a medium that contained 2,4-D. These results suggested that some signal transduction system(s) might play a role in embryogenesis and the proliferation of undifferentiated somatic cells.

2.5 Ionic currents and electrical polarity

Endogenous electrical currents across somatic embryos of wild carrot were first observed by Brawley et al. (1984). Development of a vibrating probe allowed them to measure currents around small cell clusters and embryos. In fully developed globular embryos, the efflux of ionic current was observed in the region near the suspensor, with an influx at the apical pole. Current also entered the exposed surfaces of early globular embryos that were developing from parts of large clusters of cells. In addition to the morphological asymmetry, this electrical current is the first detectable evidence of polarity in developing somatic embryos. A localized current can be observed at both ends of the embryo at subsequent stages of embryogenesis. At later stages, an inward current is found at the cotyledon and an outward current is found at the root in heart- and torpedo-shaped embryos and in plantlets. This current is reversibly inhibited by exogenous indole-3-acetic acid at 3 x 10^{-6} M.

Electrical polarity was also recognized in cell clusters during apparently unorganized proliferation in the presence of 2,4-D (Gorst et al. 1987). This electrical polarity is similar to that found in developing somatic embryos in auxin-free medium. This observation suggested that embryogenesis was suppressed in pro-embryogenic masses that were proliferating in the presence of 2,4-D. Gorst et al. (1987) confirmed this suppression of embryogenesis by scanning electron microscopy. From their observations, they concluded that the potential for embryogenesis exists even in the presence of 2,4-D, but exogenous auxins inhibit its expression.

The inward current at the cotyledon is composed largely of K^+ ions and the outward current at the radicle is mainly the result of the active extrusion of protons (Rathore et al., 1988). In the heart-shaped embryos, an inward current of 1.2 + 0.1 μAcm^{-2} was detected at the cotyledon, and an outward current of 1.0 + 0.1 μAcm^{-2} was found at the radicle at pH 5.5. When the pH was raised to 5.75, the currents increased by 0.2 to 0.3 μAcm^{-2}. The sites of entry and exit of the current were more acidic than the rest of the medium. Removal of K^+ ions from the medium reversibly reduced the currents to about 25% of their original value at both the cotyledon and radicle; removal of Cl^- ions decreased the currents slightly; and removal of Ca^{2+} ions resulted in a rapid doubling of currents. Addition of N, N'-dicyclohexylcarbodiimide, an inhibitor of plasma-membrane ATPase, or tetraethyl ammonium chloride, a K^+-channel blocking agent, substantially reduced the overall currents, and their removal resulted in partial recovery of the currents. These observations suggest that such currents are due mainly to K^+ and H^+ ions.

Changes in electrical and ionic currents might lead to alterations in the pattern of development. Application of a low-voltage electrical field enhances the development of somatic embryos from protoplasts of alfalfa (*Medicago sativa*) (Dijak et al., 1986). An electrical field might alter the distribution of proteins or channels on the cellular membrane.

3 FORMATION OF EMBRYOGENIC CELL CLUSTERS FROM SINGLE CELL

3.1 Embryogenesis from single cells

Several experimental systems in which single cells develop to embryos have been reported. Nomura and Komamine (1985) reported a system for high-frequency embryogenesis from single cells. In this system, selected small, round and cytoplasm-rich single cells develop to embryogenic cell clusters when they are cultured in a medium containing auxin. These embryogenic cell clusters develop into embryos at high frequency when transferred to an auxin-free medium. However, single cells are not able to form embryogenic cell clusters nor do they develop to form embryogenic cell clusters or embryos when they are cultured directly in auxin-free media. Using this system, Nomura and Komamine (1986) investigated polarity at the early stages of embryogenesis from single cells by serial observations and they examined the polarized synthesis of DNA during the initial phase of embryogenesis from single cells.

From microscopic observations of embryogenesis from single carrot cells, the earliest event in embryogenic differentiation of a single cell has been proposed to be unequal division (Backs-Hüsemann and Reinert, 1970; Nomura and Komamine, 1986; Guzzo et al., 1994). Serial observations were reported by Nomura and Komamine (1986). Spherical single cells with a diameter of 10-12 μm (Fig. 4A) were prepared by sieving through nylon screens and subsequent density gradient centrifugation in Percoll. The single cells failed to develop to somatic embryos in a medium without auxins. For expression of embryogenic competence, they were cultured in a medium that contained 10^{-8} M 2,4-D and 10^{-6} M zeatin. The first divisions of singles cell were observed on the second day of culture in this medium (Fig. 4B). This first division was morphologically unequal in most single cells, giving rise to two distinct daughter cells. One was smaller than the other, which contained most of the original vacuole in most cases. This observation coincided with the serial observations reported by Backs-Hüsemann and Reinert (1970) and Guzzo et al. (1994). In the latter cases, embryogenesis was initiated from elongated single cells, but the first cell division also yielded a small, cytoplasm-rich daughter cell and a large, vacuolated daughter cell. Subsequent preferential proliferation of the smaller daughter cells was observed, while the larger cells did not divide further and became vacuolated. The subsequent divisions of the smaller cells gave rise to embryogenic cell clusters on the seventh day (Fig. 4C and D). Such cell clusters were able to develop to somatic embryos in the absence of auxins.

Figure 4E shows embryos on the eighth day after transfer to a medium without 2,4-D. It appeared that the formation of embryogenic cell clusters was initiated from the unequal first division and proceeded by growth of cell clusters from only one of the daughter cells that was present at the two-cell stage. Autoradiography was then performed to monitor changes in the uptake of [^{3}H]-thymidine into nuclear DNA during this phase.

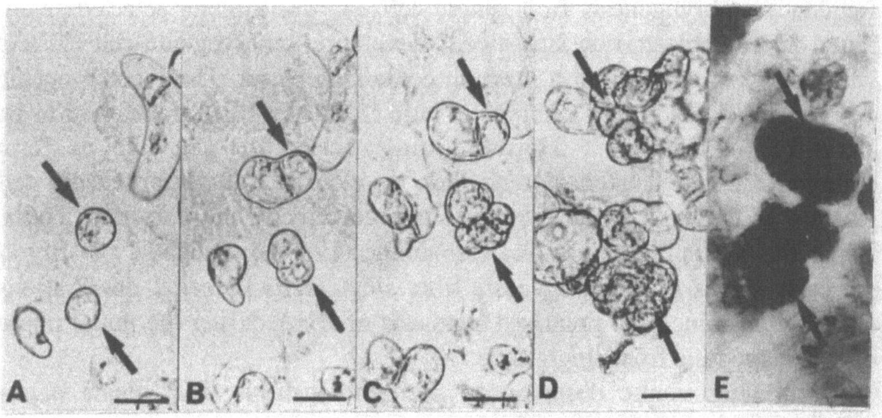

Figure 4. Serial observations of the formation of embryogenic cell clusters from single cells (arrow marked). A: Single cells. B: First division of a single cell on the second day of culture in a medium that contained $5x10^{-7}$ M 2,4-D, 10^{-6} M zeatin and 0.2 M mannitol. C: The fifth day of culture. D: An embryogenic cell cluster formed on the seventh day. E: An embryo formed on the eighth day after transfer of the cell cluster (shown in D) to embryo-inducing medium. Arrows indicate serially observed cells and cell clusters during the differentiation of embryogenic cell clusters (A to D) or embryos at the early globular stage. Bars indicate 20μm.

Figure 5 shows the autoradiographs of single cells and embryogenic cell clusters formed from single cells. Nuclei and vacuoles are recognizable in single cells and cell clusters are stained with toluidine blue. There was a two-day lag before the first division of single cells. Incorporation of [^{3}H]-thymidine into nuclei was observed in single cells during this lag stage (Fig. 5A). When two-cell clusters were examined by autoradiography, both nuclei were labelled in some cases. In other cases, [^{3}H]-thymidine was incorporated into only one nucleus. Subsequent division of only one of the two daughter cells occurred, and proliferation of the other cell was seldom observed. Polarity of DNA-synthetic activity was clear after the second division. When cell clusters had reached the three- or four-cell-stage, incorporation of [^{3}H]-thymidine was observed in only one cell or in two adjacent cells, indicating that localization of DNA synthesis was already cell-specific in cell clusters at this stage. Incorporation of [^{3}H]-thymidine was also examined in serial sections of competent embryogenic cell clusters (Fig. 5). The autoradiographs indicated that there was just one site of active DNA synthesis in each embryogenic cell cluster.

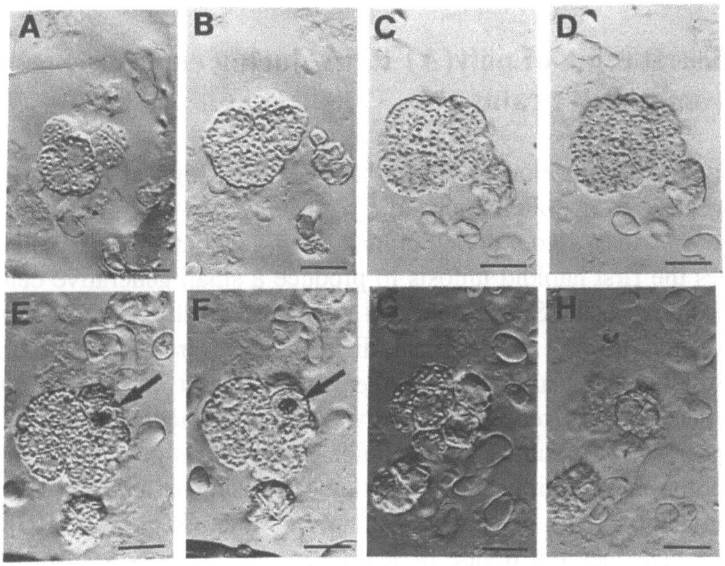

Figure 5. Autoradiographs of serial sections of an embryogenic cell cluster formed from a single cell on the fifth day in medium that contained $5x10^{-8}$ M 2,4-D and 10^{-6} M zeatin. The arrow indicates cells in which nuclei incorporated ^{3}H-thymidine. A to E, G and H: Autoradiograms of sections with focus on sections. F: Same as E but with focus on silver grains. Arrow indicates cells that had incorporated [^{3}H]-thymidine. Photographed with Nomarski optics. Bars represent 20μm.

The polarized incorporation of the radio label was observed after the cell clusters had been transferred to an auxin-free medium for induction of the development of somatic embryos. However, it was negated by the prolonged culture of the cell clusters in the presence of 2,4-D, which inhibited embryogenesis. From these observations, it seems that the polarity of DNA-synthetic activity is strongly correlated with somatic embryogenesis at its early stages.

Backs-Hüsemann and Reinert (1970) reported that an unequal first division is necessary if single cells are to develop into embryos, and this conclusion was supported by similar observations reported by Nomura and Komamine (1986) and Guzzo et al. (1994). A two-celled cluster is not able to accomplish somatic embryogenesis in an auxin-free medium. By contrast, cell clusters after the four-cell stage can develop to somatic embryos. The polarized synthesis of DNA was observed in embryogenic cell clusters after the four-cell stage. This polarity was induced in cell clusters in media that contained auxin, and it was maintained after transfer to an auxin-free medium for induction of embryonic development. When the cell clusters were cultured in the presence of 2,4-D, the polarity was eliminated. Thus, the polarity was strongly correlated with embryogenesis.

3.2 Distribution of poly(A)$^+$RNA during embryogenesis from pollen grains

The polarized localization of RNAs has been investigated in detail during embryogenesis from pollen grains of *Hyoscyamus niger*. A single uninuclear microspore matures into a pollen grain. During its ontogeny, a pollen grain undergoes the first haploid mitosis to produce a small generative cell and a large vegetative cell. Subsequent cell divisions are strictly programmed in pollen grains, leading to terminal differentiation into pollen tubes and gametes. However, culture of anthers *in vitro* at a certain stage of development induces repeated cell divisions in a fraction of the enclosed pollen grains. The resultant multicellular pollen grains give rise to embryos under appropriate conditions. A large number of such embryos are produced by the division of the generative cell in *H. niger* (Raghavan, 1978). The vegetative cell does not divide or it divides only to form a suspensor-like structure on the organogenetic part of the pollen embryo that originates from the generative cell. The determination for embryogenic cell division seems to occur at the first mitosis in cultured pollen grains. When the first haploid mitosis in culture results in pollen grains with two nearly identical nuclei, those in which both nuclei synthesize RNA become embryogenic. In binucleate pollen grains in which the incorporation of [^{3}H]-uridine occurs exclusively in the vegetative nucleus, the cells gradually became starch-filled and nonembryogenic. The extent of involvement of the vegetative nucleus in the formation of embryos from pollen was studied by Raghavan (1979), who noted differences between the incorporation of [^{3}H]-uridine in the generative and vegetative nuclei during embryogenesis. To confirm that such embryogenic determination is a result of gene expression, Raghavan (1981) monitored the distribution of poly(A)$^+$RNA in pollen grains during normal gametophytic development and embryogenic development by hybridization on tissue sections with [^{3}H]-polyuridylic acid as the probe. No binding of the label to pollen grains occurred during the uninucleate phase of their development. Although binding sites for [^{3}H]-polyuridylic acid were observed in the generative and vegetative cells of maturing pollen grains, such sites disappeared almost completely from mature grains that were ready to germinate. During germination of pollen, formation of poly(A)$^+$RNA occurred transiently and was due to the activity of the generative nucleus;

the vegetative nucleus and the sperm cells failed to interact with the [^{3}H]-polyuridylic acid probe. In cultured segments of anthers, moderate amounts of poly(A)$^+$RNA were detected in uninucleate, nonvacuolated, embryogenic pollen grains. The accumulation of poly(A)$^+$RNA in these grains was sensitive to actinomycin D, suggesting that the poly(A)$^+$RNA might represent newly transcribed mRNA. After the first haploid mitosis in the embryogenically determined pollen grains, only those grains in which the generative nucleus alone, or together with the vegetative nucleus, accumulated poly(A)$^+$RNA in the surrounding cytoplasm were found to divide in the embryogenic pathway. These results suggest that embryogenic development in the uninucleate pollen grains of cultured segments of anthers of *H. niger* might be due to activation of transcription of an informational type of RNA. Subsequent divisions in the potentially embryogenic, binucleate pollen grains appear to be mediated by the continued synthesis of mRNA either in the generative nucleus or in both the generative and the vegetative nuclei. These observations also suggested that the axis of the pollen embryo is formed during the initial division of a pollen grain.

Embryogenic cultures of carrot have been studied in a similar manner. Methacrylate sections of cells and of cell clusters from carrot suspension culture were probed with [^{3}H]-polyuridylic acid to identify sites of poly(A)$^+$RNA by autoradiography. Polarized localization of transcription activity was observed in cell clusters that were competent with respect to embryogenesis until the globular stage. However, such polarized distribution of transcription was no longer apparent in cell clusters after prolonged culture in the presence of 2,4-D (Nomura and Komamine, 1986). Polarization of an embryogenic cell cluster seemed to be associated with localized transcription.

4 DEVELOPMENT OF *FUCUS* ZYGOTES AS A MODEL FOR THE ESTABLISHMENT OF POLARITY DURING EMBRYOGENESIS

Establishment of polarity during early development has been well studied in *Fucus* zygotes. It is a brown alga, and the pattern of division and development of its spherical zygote is quite similar to that during early embryogenesis in *Arabidopsis*. A zygote of *Fucus* is apolar but a polar axis becomes fixed before the first division, which occurs 10 h after fertilization. Polarity of the developmental axis of the zygote is induced and fixed by environmental stimuli, such as light. The emergence and fixation of the polarity in *Fucus* embryos can be separated into several phases: signal perception, axis formation, axis fixation, emergence of polarity, and the unequal first division that leads to formation of a rhizoid.

The first phase involves perception of a signal and transduction of the signal. While there are no reports of light-receptor molecules in *Fucus* membranes, there must exist some kind of receptor, as well as some signal transduction system in the cytoplasm.

The second phase is formation of the polar axis. The polar axis of the zygote is labile for 10 h after fertilization. If a zygote is incubated in uniform light or in darkness after an oriented pulse of light, a rhizoid is formed 14 to 18 h after fertilization on the shaded side of the zygote. However, if a second pulse of light is given to a different side of the zygote before 10 h have passed, the rhizoid appears on the side opposite the second stimulus. This result indicates that there is a 'memory' system in the *Fucus* zygote that can be stimulated and overcome a light pulse during the initial 10 h period. Little is known about the molecular basis of this memory system, but it seems likely that the gradient of light results in the asymmetric distribution of proteins on the plasma membrane of the zygote. There have been extensive investigations of the passage of electrical current through the zygote during axis formation (Robinson and Jaffe, 1975; Nuccitelli and Jaffe, 1976; Jaffe, 1990). The current flows into the presumptive rhizoid pole and out of the thallus pole.

Formation of a polar axis requires K^+ ions and intact actin microfilaments. It has been proposed that actin microfilaments become linked to membrane proteins on the shaded side of the zygote, and this link between the filaments and membrane proteins becomes stabilized at a later time. The mechanism of formation and stabilization of the axis is still unclear, and the roles of ion channels and membrane proteins need to be investigated.

The third phase is fixation of the axis, which occurs 10 to 12 h after fertilization. Once the axis is fixed, a second pulse of light to a different site on the zygote no longer affects the site of rhizoid formation. Axis fixation is not affected by colchicine, cycloheximide or high osmoticum (Quatrano, 1973). However, cytochalasin B (Quatorano, 1973) or removal of cell wall (Kropf et al., 1988) allows reorientation of the axis by a second pulse of light. These observations suggest that microfilaments and a cell wall are required for stabilization of a previously formed axis (Quatrano, 1990). The localized accumulation of F-actin was observed at the presumptive site of rhizoid formation in a zygote with a fixed polar axis (Kropf et al., 1989). Even though the localization of actin filaments is the first detectable evidence of polarization, the mechanism by which this localization occurs is not yet known.

The cell wall is also essential for fixation of the polar axis. Kropf et al. (1988) proposed that linkage of the cytoskeleton and cell-wall fibrils at the site of the future rhizoid, through integral membrane proteins, might occur during fixation of the axis. Cell walls at the rhizoid pole are different from those at the thallus pole in terms of both structure and in molecular composition (Brawley and Quatrano, 1979). In the elongating tip of the rhizoid, both a sulfated fucan polysaccharide and a vitronectin-like protein accumulate (Quatrano et al. 1979; Wagner et al., 1992). The roles of these molecules in fixation of the polar axis should be examined.

An intracellular current can be detected in a tip-growing zygote. A transcellular current flows inward at the rhizoid tip and outward on the thallus side. A similar current can also be detected in carrot somatic embryos. There is much evidence for the importance of a gradient of Ca^{2+} ions in the development of zygote polarity. The localization of unique proteins in the plasma membrane and of Ca^{2+} ions in the cytoplasm at the rhizoid tip appears to be essential for maintaining tip growth, in addition to the transcellular ionic current (Goodner and Quatrano, 1993).

The signals that induce the polarity of somatic embryos are not yet understood, and there may be many different mechanisms by which polarity is established. However, the *Fucus* system can serve as a basic model in our efforts to understand reception of the inductive signal, the initial transduction of the signal and the early events in the establishment of polar axis. Striking changes in microtubules (Halperin and Jensen, 1967) and changes in molecules associated with signal transduction systems (Nomura et al. unpublished data) have been detected. Thus, the Fucus model should be a focus of research into the earliest events of somatic embryogenesis in single plant cells.

5 REFERENCES

Backs-Hüsemann D & Reinert J (1970) Embryobildung durch isolierte Einzelzellen aus Gewebekulturen von *Daucus carota*. Protoplasma . 70: 49-60.

Brawley SH, Wetherell DF & Robinson KR (1984) Electrical polarity in embryos of wild carrot precedes cotyledon differentiation. Proc. Natl. Acad. Sci. USA 81: 6064-6067.

Brawley SH & Quatrano RS (1979) Sulfation of fucoidin in *Fucus distichus* embryos. 4. Autoradiographic investigation of fucoidin sulration secretion during differentiation and the effect of cytochalasin treatment. Dev. Bio. 73: 193-205.

Dijak M, Smith DL, Wilson TJ & Brown D (1986) Stimulation of direct embryogenesis from mesophyll protoplasts of *Nechiego sativa.* Plant Cell Rep. 5:468-470

DosSantos AVP, Cutter EG & Davey MR (1983) Origin and development of somatic embryos in *Medicago sativa* L. (alfalfa). Protoplasma. 117: 107-115.

Fujimura T & Komamine A (1979a) Synchronization of somatic embryogenesis in a carrot cell suspension culture. Plant Physiol. 64: 162-164.

Fujimura T & Komamine A (1979b) Involvement of endogenous auxin in somatic embryogenesis in a carrot cell suspension culture. Z. Pflanzenphysiol. 95: 13-19.

Fujimura T & Komamine A (1980) The serial observation of embryogenesis in a carrot cell suspension culture. New Phytol. 86: 213- 218.

Goodner B & Quatrano RS (1993) *Fucus* embryogenesis: a model to study the establishment of polarity. Plant Cell 5: 1471-1481.

Gorst J, Overall RL & Wernicke W (1987) Ionic currents traversing cell clusters from carrot suspension cultures reveal perpetuation of morphogenetic potential as distinct from induction of embryogenesis. Cell Diff. 21: 101-109.

Guzzo F, Barbara B, Mariani P, LoSchiavo F & Terzi M (1994) Studies on the origin of totipotent cells in explants of *Daucus carota* L. J. Exp. Bot. 45: 1427-1432.

Haccius B (1978) Question of unicellular origin of non-zygotic embryos in callus culture. Phytomorphology 78: 74-81.

Halperin W (1966) Alternative morphogenetic events in cell suspensions. Am. J. Bot. 53: 443-453.

Halperin W & Jensen WA (1967) Ultrastructural changes during growth and embryogenesis in carrot cell suspension cultures. J. Ultrastruc. Res. 18: 428-443.

Jaffe LF (1990) Calcium ion currents and gradients in fucoid eggs. In: Leonard, P.T. and Hepler, P.K. (eds). Calcium in Plant Growth and Development. pp. 120-126. Amer. Soc. Plant Physiol. Rockville.

Kropf DL, Berge SK & Quatrano RS (1980) Actin localization during *Fucus* embryogenesis. Plant Cell 1: 191-200.

Kropf DL, Kloereg B & Quatrano RS (1988) Cell wall is required for fixation of the embryonic axis of *Fucus* zygotes. Science 239: 187-190.

Nomura K & Komamine A (1985) Identification and isolation of single cells that produce somatic embryos at a high frequency in a carrot suspension culture. Plant Physiol. 79: 988-991.

Nomura K & Komamine A (1986) In situ hybridization on tissue sections. Plant Tissue Cult. Lett. 3: 92-94.

Nucciteilli R & Jaffe LF (1976) The ionic components of the current pulses generated by developing fucoid eggs. Dev. Biol. 49: 518-531.

Overoorde PJ & Grimes HD (1994) The role of calcium and calmodulin in carrot somatic embryogenesis. Plant Cell Physiol. 35: 135-144.

Quatrano RS (1973) Separation of processes associated with differentiation of two-celled *Fucus* zygotes. Dev. Biol. 30: 209-213.

Quatrano RS (1990) Polar axis fixation and cytoplasmic localization in *Fucus*. In: Mahowald, A. (ed). Genetics of Pattern Formation and Growth Control. pp. 31-46. Alan R. Liss. New York.

Quatrano RS, Brawley SH & Hogsett WE (1979) The control of polar deposition of a sulfated polysaccharide in *Fucus* zygotes. In : Subtelny, I.R.K.S. (ed). pp. 77-96. Determinants of Spatial Organization. Academic Press: New York.

Raghavan V (1978) Origin and development of pollen embryoides and pollen calluses in cultured anther segments of *Hyoscyamus niger* (henbane). Am. J. Bot. 65: 984-1002.

Raghavan V (1979) An autoradiographic study of RNA synthesis during pollen embryogenesis in *Hyoscyamus nigar* (henbane). Am. J. Bot. 66: 784-795.

Raghavan V (1981) Distribution of poly(A)-containing RNA during normal pollen development and during induced pollen embryogenesis in *Hyoscyamus niger*. J. Cell Biol. 89: 593-606.

Rathore KS, Hodges TK & Robinson KR (1988) Ionic basis of currents in somatic embryos of *Daucus carota*. Planta 175: 280-289.

Reinert J (1958) Untersuchungen uber die Morphogenese an Gewebekulturen. Berichte der Deutschen Botanischen Gesellschaft. 71: 15.

Robinson KR & Jaffe LF (1975) Polarizing fucoid eggs drive a calcium current through themselves. Science. 187: 70-72.

Steward FC, Mapes MO & Mears K (1958) Growth and organized development of cultured cells. II. Organization in cultures grown from freely suspended cells. Am. J. Bot. 45: 705-708.

Wagner VT, Brian L & Quatrano RS (1992) Role of a vitronectin-like molecule in embryo adhesion of the brown alga Fucus. Proc. Natl. Acad. Sci. USA 89: 3644-3648.

Chapter 6

DEVELOPMENTAL AND STRUCTURAL ASPECTS OF ROOT ORGANOGENESIS

Woong-Young SOH[1], Sant S. BHOJWANI[2] and Sukchan LEE[3]
[1]Department of Biological Sciences, Chonbuk National University, Chonju 561-756, Korea, [2] Department of Botany, University of Delhi, Delhi 110007, India, and [3] Department of Genetic Engineering, SungKyunKwan University, Suwon 440-746, Korea

1 INTRODUCTION

The study on adventitious root formation is not only to throw light on the mechanism of plant organogenesis but also to provide information to improve the techniques for clonal multiplication of plants by conventional methods as well as in tissue cultures. In tissue cultures various explants derived from structurally, developmentally and physiologically diverse plant tissues can be used for *de novo* root organogenic experiments. Another advantage of this experimental system (tissue culture) is that the bulk of the cells which are structurally not involved in root organogenesis can be eliminated by selecting an explant comprising of the exact region where roots are formed. For example, the explants for root organogenesis can be minimized to tiny slices (Van der Krieken et al., 1993) or microcallus (Kim et al., 1995) and examined to determine the exact region (Gutmann et al., 1996; Soh et al., 1998). However, in recent years the field of adventitious root organogenesis in tissue culture systems has been overshadowed by studies on shoot organogenesis (Moncousin, 1991), even though in conventional propagation the induction of root organogenesis was more actively studied than shoot organogenesis.

The structural analysis of root primordium initiation is a prerequisite to understand the process of adventitious root organogenesis. Adventitious roots, in general, arise endogenously from dedifferentiated parenchyma cells around vascular tissues by auxin action. During dedifferentiation many cells from pith to cortex regain the potential for cell division. However, only definite cells, such as meristematized phloem parenchyma cells, can differentiate into root primordia. Other dividing cells form groups of transversely dividing cells but do not take part in the root differentiation of primordium (Mitsuhashi-Kato, 1978a; Harbage et al., 1993). Furthermore, cell divisions also occur in nonrooting explants treated with auxin (Geneve et al., 1988). Thus, cell divisions in parenchyma do not necessarily lead to root primordium formation; only localized and specifically oriented cell divisions lead to root primordium formation. To understand the process of root initial initiation the cells leading to root primordia must be distinguished from the other dividing or divided cells which are not associated with root initiation. Developmental and structural aspects of the newly differentiated root primordia from induced cell divisions may allow correlating biochemical and physiological changes with anatomical events associated with root initiation.

During root primordium development, vascular differentiation occurs in parental tissue and connection is established between the newly differentiated vascular tissues and those in parental tissue. On the practical side, it has been shown that incomplete vascular connection at root-shoot interface may result in poor success during the *in vitro* establishment of regenerated plantlets (Thorpe, 1984). Although roots can arise from callus tissue *in vitro*, poor vascular connections between root and shoot are reported since vascular tissues in callus are haphazard in their arrangement and orientation (Roberts et al., 1988). Thus, a basic structure of organized tissues cannot exist in callus tissue and this phenomenon is related to a mulfunction of vascular tissue. For successful establishment of regenerated plantlets, their adventitious roots should be structurally and functionally normal. The development of such healthy and functional roots needs to be established by anatomical studies.

This chapter describes adventitious root formation in the cultures of various organs, explants and tissues, with emphasis on developmental and anatomical aspects. Physiological aspects of adventitious root formation has been previously reviewed by Thorpe (1980), Van Staden and Harty (1988), Gaspar and Coumans (1987).

2 ROOT PRIMORDIUM DIFFERENTIATION

2.1 Site of root primordium initials

The terminology used to identify the anatomical stages in root primordium development must be precise and universally acceptable for clear understanding of descriptions of root formation. The process of adventitious root formation can be divided into stages (Moncousin, 1991): (1) cell structure modification, (2) cell division and meristematic activity initiation (root initials), (3) root meristem (root primordium) differentiation, and root growth and emergence (Favre 1973 a. b. c.; Vieitez et al., 1981). However, there is a lack of agreement on the number and nature of the stages and also in the terminology used. The definitions of 'root initials' and 'root primordium' during root organogenesis were summarized by Lovell and White (1986). The early events during root organogenesis are characterized by the appearance of cells with a centrally located enlarged nucleus, prominent nucleoli and dense cytoplasm. The cells are capable of division and may give rise to initials of organized root tips if the primordium has differentiated *in situ* (Blazich and Heuser, 1979a; Patel et al., 1986; Apter et al., 1993a; Harbage et al., 1993). The term 'root initials' can be used for these cells but only in those cases where the cells do, in fact, ultimately become the initials which give rise to root primordium. At this early stage, the initials are an unorganized cell mass and are not necessarily 'determined' because a root might not be produced from the initials (Mitsuhashi-Kato, 1978a). At a somewhat later stage some differentiation may occur by the meristematic acitvity of these initials. At this point they are 'determined' and the term 'primordium' can be used for this stage (Stangler, 1956; Girouard 1967a; Blazich and Heuser, 1979a; Hicks, 1987; Harbage et al., 1993).

Adventitious roots mostly originate endogenously from different sites in extreme case, around the vascular tissues (Attfield and Evans, 1991) and from the pith, as in *Linum usitatissimum* and *Portulaca oleracea* (Crooks, 1933; Fahn, 1990). But in some instances the origin has been exogenous, from epidermal cells, as in *Cardamine pratensis*, *Crassula multicava* and *Rorippa austriaca* (Mc Veigh, 1938; Fahn, 1990) and *Nicotiana tabacum* (Tran Thanh Van, 1973). The sites and modes of origin of adventitious root primordia in stem cuttings were listed by Lovell and White (1986, Tables 1-3). Salicaceous members and other woody plants almost exclusively have 'preformed root primordia' (Haissig, 1970), and the primordia are mainly initiated in the cambial region (*Acer pseudoplatanus*), ray parenchyma (*Populus simoni*), ray tissue and leaf or bud gaps (*Lonicera japonica, Salix discolar, S. viminalis* and *Taxus baccuta* (van der Lek, 1925; Sandison,

1934, Fink, 1982; Fjell, 1985 a,b). On the other hand, such preformed primordia may be locally present in the stems of some herbaceous plants. Root primordia may also be initiated, occasionally, in the interfascicular parenchyma cells near the phloem in the basipetal 5 mm of sunflower hypocotyls (Fabijan et al., 1981), or from vascular cambium below the apical 6th nodes and internode of *Ipomoea batatus* . Or, they may be present in the first or second internode, but not higher up, as in pea seedlings (Schmidt, 1956/ 57). Therefore when sampling for synchronized rooting it is necessary to confirm whether such primordia are already present in explants by microscopic observations of explants cleared in chromic acid.

Root formation in conifers has been difficult because the shoots do not usually possess preformed root initials, unlike easy-to-root species of most woody plants. Thus, the adventitious roots of conifers can be called wound induced roots (Eriksson, 1984; Patel et al., 1986; Mott and Amerson, 1981). The primordium of wound induced adventitious roots is initiated from cells in interior cortex in *Cercis canadensis*, phloem parenchyma and the cambial zone in 'Gala' apples, the interfascicular region in *Agathis australis*, ray tissue in *Ficus pumila*, bud and leaf gaps in *Malus pumila,* lenticels in *Tamarix aphylla*, pericycles in *Abies procera*, and callus in *Abies firma* (Wilcox, 1955; Satoo, 1956; Ginzburg, 1967; Doud and Carlson, 1977; Davies et al., 1982; White and Lovell, 1984a; Geneve and Kester, 1990; Harbage et al., 1993). Successful development of adventitious roots depends on the cambium participating in callus formation in the cuttings of apple rootstock (Mackenzie et al., 1986). In shoot explants of black and white spruce, on SH medium containing IBA, root primordia differentiated from a meristemoidal mass formed from cells surrounding tracheidal nests which differentiated in the vicinity of vascular tissues at the base of the shoot (Patel et al., 1986). Similarly, adventitious roots developed from tracheidal nests in hypocotyl cuttings of *Pinus sylvestris* (Grönroos and von Arnold, 1985; Patel et al., 1988).

Adventitious roots may be initiated in different tissues according to the culture conditions of the parent organs. In shoot explants of 'Gala' apple cultured on medium containing IBA and sucrose, root primordia differentiated from meristemoids formed through meristematic activity of some phloem parenchyma cells (Harbage et al., 1993) but the preformed root primordia were associated with leaf gaps in field grown *Malus domestica* (Swingle, 1927) and layered induced roots on *M. Pumila* (Doud and Carlson, 1977). Although initiation of cell division by the rooting treatment in 'Gala' apple occurred in the phloem parenchyma, cortices, cambial regions, xylem, and pith, only the phloem parenchyma cells formed root primordia, as also observed in micropropagated shoots of other woody species, such as apple (KSC-3), cherry, camellia and chest nut (Vieitez and Vieitez, 1983; Samartin et al., 1986; Hicks 1987; Ranjit et al., 1988). In English ivy, root primordium from petiole initiated in epithelial cells of ducts adjacent to vascular bundles and inner cortical cells (Geneve et al., 1988).

Adventitious root primordia in some herbaceous plants may be formed by the following tissues: stem pericycle in *Coleus*, tomato, *Phaseolus vulgaris* and *Zea mays*; ray parenchyma between pericycle and cambium in *Tropaeolum majus* and *Lonicera japonica*; interfascicular parenchyma in *Azukia angularis*; interfascicular cambium and pericycle in *Dianthus caryophyllus*, *Pelargonium* x *hortorum* and *Portulaca oleracea*; interfascicular cambium, pericycle and phloem in *Begonia*; and tissues of leaf margins and petioles in *Begonia* and *Kalanchoe* (Hayward, 1938; Boureau, 1954; Ginzburg, 1967; Girouard 1967a, b; Byrne et al., 1975; Cline and Neely, 1983; Lovell and White, 1986; Fahn, 1990). In the cultures of *Solanum melongena* in vitro adventitious roots originated from callus formed on the parent organ (Matsuoka and Hinata, 1979).

In the hypocotyl cuttings of *Phaseolus aureus* and *P. vulgaris* adventitious roots developed in four rows parallel to and between the four pairs of vascular bundles, in contrast to their irregular development in petioles and epicotyl cuttings where the distribution pattern of xylem bundles are also irregular (Blazich and Heuser, 1979b; Friedman et al., 1979). Similar results were observed in *Pisum sativum* and *Vigna unguiculata* (Fig. 1).

In cotyledon and leaf explant cultures root primordia developed close to vascular strands (Ladeinde and Soh, 1991; Sharma et al., 1991; Gurel and Wren, 1995; Gutman et al., 1996). The distance from the explant base to the site of adventitious roots initiation differs among the species examined. In *Phaseolus vulgaris* the formation of adventitious roots occurred sequentially from the cut surface upward to nearly the distal end of the hypocotyl cuttings (Friedman et al., 1979). However, adventitious roots were formed about 1 mm above the cut surface of epicotyl cuttings in *Azukia angularis* (Mitsuhashi-Kato et al, 1978a), in the basal 1 mm of microcuttings in cultivars of apple, Gala and KSC-3 regardless of treatment or water incubation (Hicks,1987; Harbage et al., 1993), and at 0.2-1.0 mm from the cut surface of hypocotyl cuttings in *Pinus sylvestris* (Grönroos and Arnold, 1985). In walnut cotyledon fragments adventitious root formation occurred at the initially wounded end of elongated petioles (Gutmann et al., 1996), and in many other plants at the proximal cut edge of the cotyledon (Tepper and Mante, 1990).

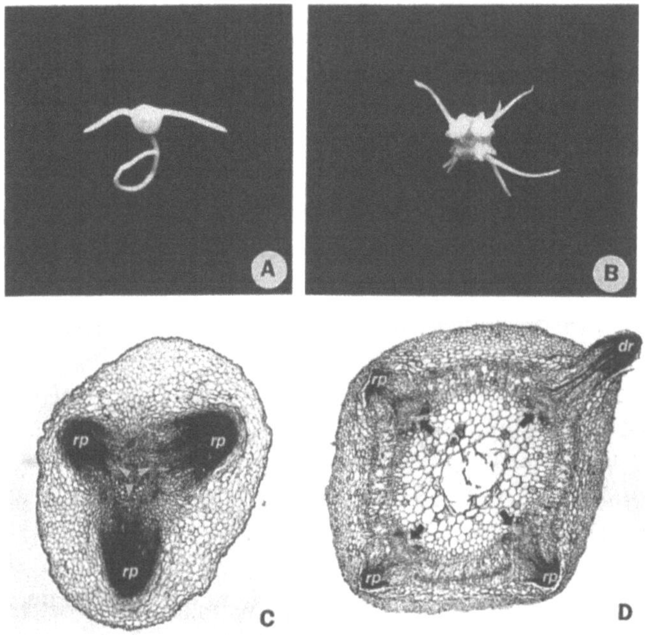

Figure 1. Hypocotyl segments with developing roots of *Pisum sativum* (A) and *Vigna unguiculata* (B) and transverse sections of the hypocotyls with root primordia (C, D). Adventitious roots were formed in 3 or 4 rows parallel to the main vascular bundles of the hypocotyls (A, B). Transverse sections of the hypocotyls with root primordia(rp) or developing root (dr) of *Pisum sativum* (C) and *Vigna unguiculata* (D) show root primordia developed from the outer phloem parenchyma cells of the 3 or 4 main vascular bundles (arrows) in the hypocotyls.

2.2 Root primordium initiation

Information on the time course of changes in the cell division plane in the initiation of root initials in tissue culture systems is sparse. Because root formation from seedling cuttings has been examined mainly by transverse sectioning (Blazich and Heuser, 1979a; White and Lovell, 1984b), root primordium differentiation from the first cell division in cuttings has not yet been fully elucidated. The anatomical study on the differentiation of root initial and primordium formation in water-treated epicotyl cuttings of *Azukia angularis* was made by transverse and longitudinal sections by Mitsuhashi-Kato et al. (1978a, b). At the basal region of the cuttings there are six large and six small alternately arranged vascular bundles. Root primordia appeared in the region between the large and the small bundles. Root initials were initiated only by the transverse division of pericycle cells, first observable at 10 hr after preparing the cuttings. All the daughter cells produced by the first transverse division underwent the second transverse division. Half of the daughter cells from the second transverse division underwent the first longitudinal division at 18 hr after preparing the cuttings. Cells in this newly formed meristematic cell mass continued division without an increase in gross volume of the cell mass up to about 30 hr, and then the cell mass began to enlarge and protrude as a root. On the basis of the analysis of the plane of cell division the process of a root primordium formation was divided into five stages; (A) cell production by the 1st transverse division; (B) cell production by the 2nd transverse division before the 1st longitudinal division; (C_1) cells formed by the 1st or 2nd longitudinal division; (C_2) a cell mass formed by further divisions without an increase in the volume ; (D) an increase of cell mass volume by the expansion of cells, and beginning of protrusion of the cell mass (A-C_2: root initials, D: root primordium). The initiation and stages in the differentiation of root primordia occurred somewhat synchronously like that of *in vitro* root formation in *Trachelospermum asiaticum* (Apter et al., 1993a). The initiation of each stage occurred at highly regular intervals in the epicotyl cuttings of *A. angularis* (Mitsuhashi-Kato, 1978a), showing a unique characteristic in the initiation of root primordium. It was clarified from anatomical studies that wound stimulus (cutting) triggered the first cell division at specific sites in the basal region of the *A. angularis* epicotyls after a lag period of 8 hr, which is incredibly prompt when considering that first cell division in the cuttings of *Phaseolus aureus* occurred 20-24 hr after preparing the cuttings (Blazich and Heuser, 1979a). The 2nd and 3rd cell divisions took place about 4 hr after the preceding divisions. The fact that the root initials in stage C_2 appeared about 4 hr after the appearance of those in stage C_1 suggests that the 4th division also occurs 4 hr after the 3rd division. It took about 8 hr from the appearance of root initials in stage C_2 to that of the initials in stage D. The number of initial clusters proceeding to stages C_1, C_2, and D was limited, and only a very few entered stage D by hr 36; these eventually became visible roots. These results seem to be related to the fact that about half of the daughter cells produced by the transverse second division underwent longitudinal division, and the remaining half did not divide further. And groups of cells which consisted only of transversely divided cells were still observed in some interfascicular regions. Therefore, probably the cell mass at stage D was a determined state although from the description of Mitsuhashi-Kato (1978a) it is not clear whether the mass had an organized structure.

Indole-3-acetic acid (IAA) treatment during the first 24 hr shortened the lag period for the 1st transverse division and hastened the formation of stage A initials (Mitsuhashi-Kato, 1978b). The treatment also increased the number of initials entering stage C_1 from stage B, and all the initials at the 12 possible sites proceeded to stage C_1 by hr 24. Moreover, in the presence of IAA during the second 24 hr, initials in stages C_1 and C_2 developed into roots. In its absence many initials in stages C_1 and C_2 did not develop into roots, but primordia in stage D did. Therefore, IAA not only hastened the transverse 1st division of cells but also their entrance into the longitudinal division stage from the transverse division stage in the rooting zone, although auxin did not alter the timing of root initiation in *Phaseolus aureus* cuttings (Blazich and Heuser, 1979a). Gibberellic acid inhibited both transverse and longitudinal divisions and significantly suppressed the progression from stage A to stage B, thus decreasing the number of primordia developing into roots.

The effect of auxins on root primordium elongation is different from its effect on the initiation of root initials. In *Choisya ternata* shoot explants (Nicholas et al., 1986) a maximum of 8 roots per shoot were ultimately observed during the experiment after a 7-day-culture on rooting medium with 0.1 mgl^{-1} indole butyric acid (IBA). Shoots left on the rooting medium for more than 7 days produced fewer roots, with shoots planted after 20-24 days on rooting medium averaging only 3 roots per explant. These results indicate that prolonged exposure to this auxin is deleterious to root initiation. Observations by fluorescence microscopy on the process of root initiation and elongation confirmed that shoots left on the rooting medium for more than 7 days had only a limited number of expanding root initials; the majority of them remained unelongated. Welander(1983) described excessive callusing and poor root elongation in apple rootstock M.26 after prolonged exposure of shoots to rooting medium, and it is almost similar to the result in *Bupleurum falcatum* and *Lactuca sativum* (Kim et al., 1995; Kang et al., 1996). Thus, it is clear from these observations that auxins in rooting medium favours the appearance of root initials but inhibits root primordium elongation.

2.3 Molecular markers and genetics of root initiation

Several biochemical markers of rooting, such as polyamines and peroxidases, have been proposed by several research groups, but the supporting evidence is often diametrically opposed (Haissig, 1986; Jarvis, 1986; Davis et al., 1988; Wilson and van Staden 1990). It is difficult to measure events which are localized in a small number of cells, and to isolate the key factors from an extremely complex set of biochemical interactions. Exogenous putrescine, but not other polyamines, promoted root formation from shoots grown on auxin-free medium (Hausman et al., 1994). Furthermore, application of polyamine biosynthesis inhibitors blocked both polyamine accumulation and rhizogenesis (Altamura et al., 1991b). Polyamines have been involved in root formation in tissue culture systems (Biondi et al., 1990; Hausman et al., 1995a), showing an increase during root primordium formation but a decrease upon completion of root formation (Torrigiani et al., 1989). Experiments with different species *in vitro* linked an increase in putrescine levels with an increase in mitotic activity during primordium development after the induction phase (Biondi et al., 1990; Geneve and Kester, 1991). On the other hand, studies of free and conjugated polyamines during root formation have produced contradictory results (Altamura et al., 1991b).

Peroxidases could be another potential biochemical marker to study root formation, as several studies have shown a positive correlation between peroxidase activity and root formation (Molnar and LaCroix, 1972; Quorin et al., 1974). A positive correlation occurred between the activity of peroxidase in stems of *Malus domestica* shoots at the time of transfer to rooting medium and the number of roots 21 days after the transfer (de Klerk et al., 1990). An increase in endogenous peroxidase activity was detected preceding rooting (Hausman et al., 1995b). However, plant peroxidases are found in multiple isozyme forms and also participate in several important responses and metabolism so that the function of peroxidase in root formation is not clear. Pressey (1990) suggested that peroxidase activity was related to auxin oxidation and it possibly induces hairy root initiation by reducing IAA levels, which could induce physiological changes in nearby cells showing hairy root formation. The specific role of peroxidases in rooting is, therefore, difficult to ascertain because plant peroxidases participate in the various processes of metabolism (Hand, 1994).

Although most efforts have been focused on the function of established meristems, some information has appeared on the *de novo* induction of meristematic activity. The protein kinase product of *cdc2* gene found in several plant species including *Arabidopsis* (Martinez et al., 1992), is a key component in the cell cycle for the G_1 to S-phase transition and entry into mitosis. Expression of *cdc2* gene is identified in all the organs tested but the histochemical staining of fused reporter genes between promotor of *cdc2* gene and GUS gene is specifically localized at the root meristem, in the initial stages of the activation of a new meristem and at the sites of lateral root development. This data suggest that *cdc2* expression is an early event in the root meristem redifferentiation. Expression of *cdc2* is also strong in root pericycle and stelar parenchyma cells, which are arrested in the G_2 stage of mitosis (Martinez et al., 1992). The authors suggested that *cdc2* expression may be involved in determining competence for cell multiplication, but additional signals may be required for cell division to proceed. Genes like *cdc2* can be good candidates for further study, both as markers of the competent state and for their function in the root induction process, because molecular studies of rooting have only recently begun and there is no enough evidence linking a biochemical or molecular marker to competence to root (Hand, 1994).

The accomplishment of hairy root initiation and development in plants is a complex and progressive process that involves positional indications, environmental stimuli, and internal genetic programs (Masucci and Schiefelbein, 1996).

Hairy root development involves a series of complex events, and the molecular mechanisms controlling hairy root development are not well understood in plant systems even though the physiological and anatomical studies have been reported. Apical meristems exert the greatest effect on organogenesis in plant development relative to other meristematic tissues, and are, therefore, important aspect in studying plant development (Scheres et al., 1996). However, the general biology of hairy root apical meristem formation is still poorly understood in plant science, and will require a concerted effort in the scientific society to break down the developmental events involved in the formation of this meristem or meristem-like cells.

Development of hairy root is clearly a very complex multicellular event, and a mixture of cellular, tissue and organ systems will be required to identify and characterize the many genetic and biochemical observations of this developmental process. It is clear that such a major change in the developmental destinations of cells involves complex changes at the biochemical level and the level of gene expression (Hand, 1994). There is concrete evidence that auxins play an important role in the formation of hairy roots, but they are not the sole determinant. Many other environmental and endogenous factors are involved in the rooting process.

In the recent reports about gene expression for hairy root initiation and formation, six loci are identified to influence the formation of root hair and/or root hairless cells in the *Arabidopsis* root: transparent testa glabra (TTG, Galway et al., 1994), glabra 2 (GL2, Masucci et al., 1996), constitutive triple response 1 (CTR1, Dolan et al., 1994), auxin resistant 2 (AXR2, Wilson et al., 1990), root hair defective 6 (RHD6, Masucci et al., 1996) and lateral root primordia 1 (LRP1, Smith and Fedoroff, 1995). Among them, two (CTR1 and AXR2) mutants involved in root development and expression of these genes were each correlated with plant growth regulator (ethylene and/or auxin) effects on root epidermis development where hairy roots are initiated from. CTR1 was isolated from mutant which constitutively exhibits the triple response in the presence of inhibitors of ethylene biosynthesis and binding (Dolan et al., 1994). This CTR1 has been cloned (Kieber et al. 1993). The highest similarity obtained in searches of the protein databases with CTR1 is to Raf family of protein kinases (Dolan et al., 1994). Therefore, the CTR1 gene product is suggested to negatively regulate the ethylene signal transduction pathway, and recessive *ctr1* mutations cause root hairs to form on epidermal cells which are normally hairless. Auxin mutant ARX2 was originally discovered from EMS-mutagenized seedlings screened for auxin resistance. The dominant mutations of the *axr2* gene present insensitivity to high concentrations of auxin, ethylene, and abscisic acid, and they reduce hairy root formation (Wilson et al., 1990). This mutant indicates a role for auxin in hairy root development. But this AXR2 seems to be gain-of-function mutation so that it is hard to conclude the function of wild-type gene.

Biochemical analysis represented that ethylene plays an important role in root hair formation by using ethylene synthesis inhibitor (aminoethoxyvinylglycine, AVG) or inhibitor of ethylene action (Ag or trans-cyclo-octene). These inhibitors cause a reduction in the frequency of hairy root (Masucci and Schiefelbein, 1996). These observations suggest that ethylene and auxin may act as positive regulators of hairy root cell differentiation in *Arabidopsis*. Musucci and Schiefelbein (1996) reported that the *Arabidopsis* root produces a position-dependent pattern of hair-bearing and hairless cell types during epidermis development. They showed that the hairless cell promoting genes TTG and GL2 are likely to act early to negatively regulate the ethylene and auxin pathways by showing the epistasis test results and reporter gene studies (Galway et al., 1994; Masucci et al., 1996). With this observation, they proposed new model in which the patterning of root epidermal cells in *Arabidopsis* can be regulated by the cell position-dependent action of the TTG/GL2 pathway, and the ethylene and auxin hormone pathways act to promote hairy root outgrowth at a relatively late stage of differentiation. Another gene LRP1 is cloned and the LRP1 expression is activated during the early stages of root primordium development, and is turned off prior to the emergence of lateral roots from the parent root (Smith and Fedoroff, 1995).

These genetic analysis and molecular biological studies about hairy root development provide a foundation for further investigations of the pathways that control epidermal cell differentiation and formation. For example, many aspects of the suggested model for hairy root differentiation and formation may be tested by using available gene probes, such as the genes listed above. The continued study of hairy root development should provide new sagacity into the mechanisms by which hormones influence plant cell differentiation and the regulations of genes involved in different stages of hairy root formation.

3 ORGAN CULTURE SYSTEM

3.1 Shoot explant culture

Induction of adventitious root formation allows clonal propagation of many plant species (Hartmann et al., 1990). Shoots developed from explants in culture spontaneously form adventitious roots in some cases but the majority of them have to be transferred to root induction medium in order to produce complete plants. Changes during induction and post-induction stages of adventitious root formation were anatomically examined in shoot explants of 'Gala' apple by Harbage et al. (1993). The shoot explants on induction medium containing 1.5 μM IBA plus 44 mM sucrose formed 11.9 meristemoids in the basal portions of shoot explants after 4 days of treatment but only a few meristermoids were formed on medium containing either IBA or sucrose. Time-course analysis of the induction phase indicated that some phloem parenchyma cells became densely cytoplasmic, having nuclei with enlarged nucleoli, within 1 day, as in *Phaseolus aureus* cuttings (Blazich and Heuser, 1979a); meristematic activity in the phloem was spread over 2 days; continued division of phloem parenchyma cells advanced into the cortex by day 3, and identifiable root primordia appeared by day 4. The number and location of root primordia remained unchanged when root induction treatment was extended to 8 days. Time-course analysis of the post-induction phase revealed that root formation shifted from initiation of primordia to development of organized root tissue systems within 1 day; primordia with a conical shape and several cell layers at the distal end formed by day 2 to 3 ; roots with organized tissue systems emerged from the stem by day 4; and numerous roots emerged by day 6. Organized roots with vascular systems continuous with that of the stem could be observed by the time they emerged, as in other plants (*Trachelospermum asiaticum*; Apter et al., 1993a).

The time required for adventitious root initiation in micropropagated woody plants varies with the species but shows some lag in comparison to herbaceous species (Blazich and Heuser, 1979b), even though determination of the earliest signs of meristematic activity has sometimes been unclear. Meristematic activity in response to inductive treatments was observed within 8 days in *Camellia japonica* L. (Samartin et al., 1986), 4 days in *Rosa* x *hybrida* (Bressan et al., 1982), and 2 days in *Prunus avium* L. x *P. pseudocerasus* Lindl. 'Colt' (Ranjit et al., 1988). In apple root-stock (*M. pumila*, 'KSC-3'), meristematic activity was seen within 1.5 to 3 days and meristemoids were formed within 5 days of treatment (Hicks, 1987). In vitro developed shoot explants of apple, 'Gala' and KSC-3 rarely showed differentiation of sclerenchymatous cells in phloem; most cells in the tissue were parenchymatous. Root induction in micropropagated explants is generally easier than those from field-grown plants, perhaps because of the lack of sclerenchymatous cell differentiation in the phloem (McCown, 1988). And shoot explants of many cultivars of apple, which proved difficult-to-root *in vitro* initially gained a progressive improvement in root formation with increasing number of subcultures under continuous illumination and at relatively high temperatures (Sriskandarajah et al., 1982). Anatomical studies of difficult-to-root, field-grown stem cuttings of *M. domestica* showed varying levels of sclerenchymatous cell differentiation in primary phloem in different cultivars. The amount of differentiation correlated well with the difficulty of root induction of the cultivar (Beakbane, 1961, 1969), although the distribution of fibre strands in phloem of *Hedera helix* did not impede adventitious root development (Girouard, 1967 a,b).

The structural characteristics of root organogenesis were confirmed with chestnut microcuttings by Vieitez et al.(1981). The chestnut shoot microcuttings in transverse sections showed a discontinuous ring of sclerenchyma surrounding the primary phloem which was composed of fibers separated into groups by regions of parenchymatous tissue. Analysis of the transverse sections from the point of cutting preparation to the time of root emergence revealed a sequence of four distinguishable structures: meristemoid, root primordium, root primordium with its own vascular system, and the adventitious root. Prominent changes in cellular structure occurred as early as 2 days after cutting; certain cells in the phloem parenchyma and vascular rays were dividing and showed dedifferentiation giving rise to cells of smaller size with prominent nucleus and dense cytoplasm. Four days later, the number of dividing cells increased and meristematic cell masses were formed in the phloem region near the cambium; this increase of cells undergoing division could be identified as meristemoids on the 6th day, which were composed of smaller isodiametric and nonvacuolated cells. Typical root primordia appeared with organized structure on the 8th day. The primordia at this stage were generally located at the level of the sclerenchyma ring, but protruded from the sclerenchyma between days 9 and 10. Between days 11 and 12, most of the primordia developed their own vascular systems. Similar evolution, but out of phase, was noted with *Picea* (Patel et al. 1988) and *Pinus* microcuttings (Mott and Amerson 1981).

In shoot explant culture of *Castanea dentata*, Xing et al. (1997) developed a three-step medium sequence for improvement of root formation rate and reduction of shoot-tip necrosis. First, individual shoots or clumps of shoots were cultivated on shoot elongation medium which consisted of modified Woody Plant Medium, with 50 mgl^{-1} polyvinylpyrrolidone (MW 40,000), and 0.89 μM N^6-benzyladenine (BA) for 4-8 weeks until shoots were 2-3 cm long. Microcuttings were then excised, vertically split at the base to approximately 2 mm through the pith, dipped in 5 or 10 mM indolebutyric acid for 1 min, and cultivated on half-strength Murashige and Skoog (MS, Murashige and Skoog, 1962) basal medium plus 0.2 g l^{-1} charcoal for 2 weeks. During this time, roots were induced and became visible. Finally, the microcuttings were transferred back to shoot-elongation medium and cultivated for 3 weeks, allowing growth of both roots and shoots. Using this protocol with 3 genotypes derived from one mature tree and two 1-yr-old seedlings, 57 - 73% rooting was obtained with less than 23% shoot-tip necrosis, as against earlier reports of 33 -55% root formation with shoot-tip necrosis in this species (Read and Fellman 1985 ; Maynard et al., 1993). Thus, plant regeneration from shoot explants of *Castanea dentata* was improved by a three-step medium culture.

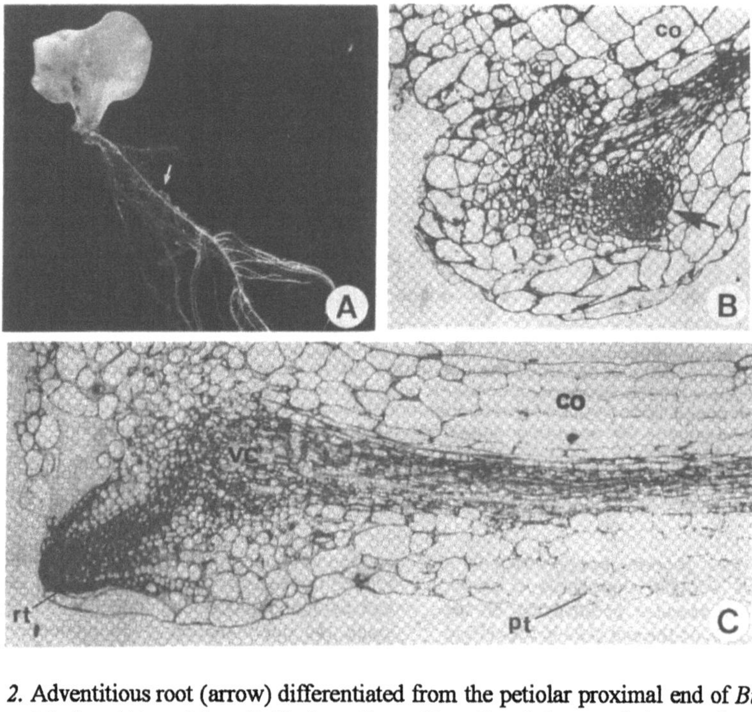

Figure 2. Adventitious root (arrow) differentiated from the petiolar proximal end of *Brassica juncea* cotyledon (A) and longitudinal sections (B , C) of the cotyledon explants with developing root or root primordia. The cotyledon explant cultured on MS medium for 4 days shows the initiation of a root primordium(arrow) close to the vascular supply (B). In cotyledon with petiole (pt) cultured on MS medium for 5 days (C), a well developed root (rt) is just emerging from the cortex (co), showing vascular connection (vc)between petiole and developed root.

3.2 Cotyledon explant culture

Excised cotyledon culture systems are potentially useful for basic studies on organogenesis as well as for more applied work such as genetic transformation. In cotyledon cultures of *Brassica juncea* adventitious roots invariably differentiated at the cut ends of petioles which were in contact with MS basal medium (Sharma et al., 1991; Fig. 2). In the absence of the petiole, differentiation from the lamina was rare, and this was also confirmed using mature cotyledon culture of 16 species by Tepper and Mante (1990) and using ginseng cotyledon culture by Choi and Soh (1995). In an experiment of *B. juncea* involving lamina removal at different time intervals, it was shown that the presence of the lamina even for a day enhanced the rhizogenic response . For maximum rhizogenesis the presence of lamina was required for 7 days. The requirement of the lamina for root formation from the cotyledon petiole on basal MS medium supports the idea that the lamina factor may be an auxin-like substance.

A new plant model for studying adventitious root development in detail and with greater rapidity has been established using *in vitro* culture of cotyledon fragments of walnut without growth regulators (Jay-Allemand et al., 1991). The initial structure of the cotyledon fragments is quite typical for a storage tissue; the cells contain plenty of proteins and lipids as energy sources for thc gcrminating embryo but without starch, and the vascular tissue for the transport of mobilized reserves is not yet fully developed (Gutmann et al., 1996). Within 24 to 72 hr of culture the most striking morphological event is the rapid differentiation of the vascular bundles which is observed at the very onset of petiole development. The further development of the cotyledon fragments is characterized by metabolic activation of storage proteins, final differentiation of the vascular tissue, limited elongation of the cotyledon petioles and accumulation of phenolic compounds in cortical and vascular tissues. After 3 days of culture, the cotyledon fragment was greatly elongated and a morphogenetic zone was formed which surrounded each of the vascular bundles initially wounded by removal of the embryonic axis. Growing roots, appearing at the initially wounded end of the cotyledon petiole were visible after 4 days of culture, as in *Brassica juncea* (Sharma et al., 1991). The morphogenetic zone around the vascular bundles gave rise to this rapid root organogenesis, in continuity with a corresponding vascular bundle of the cotyledon petiole. Possibly, after detachment of the embryonic axis only those cells of the cotyledon fragments that were immediately adjacent to the rooting pole had the capability to develop morphogenetic zones that finally resulted ir the formation of adventitious roots.

Since each vascular bundle in the cotyledon fragments gaves rise to a single root, these bundles are supposed to play a major role in this model of organogenesis, as in leaf disk culture of cowpea (Ladeinde and Soh, 1991), possibly as a kind of *cellular organizational matrix*, causing controlled cell division after wounding. The vascular system might participate in a polar transport of hormonal signals (Goldsmith, 1977; Sachs, 1981; Stinemetz, 1995) and sugars, thereby promoting intense cell divisions at the tip of the petiole. A short auxin treatment to the cotyledon fragments suppressed the formation of large roots and induced numerous tiny rootlets dispersed all over the surface of the fragments. Thus, the formation of typical roots from the wounded end of the petiole was disturbed with the reduced elongation of the cotyledon petiole. This phenomenon suggests that the dispersed tiny root formation all over the surface of cotyledon fragment was caused by the apolar distribution of endogenous auxin by treated exogenous auxin. Therefore, the hormone-free plant model of walnut cotyledon merits further investigation for the possible involvement of endogenous auxin in adventitious root development (Blakesley et al., 1991).

Cotyledon slice cultures of *Corylus avellana* L. is another simple experimental system to study root formation as they give a high yield of differentiating root primordia on half-strength basal medium with auxin and cytokinin (Gonzalez et al., 1991). In the system, typical parenchyma cells and some preformed vascular strands were histologically detected by day 8 of culture (pre-initiation stage). Small nodules of calli were formed on the cut surface of the cotyledons and cell divisions occurred periclinally and anticlinally on the proximal surface of the cotyledon slices after 12 days of culture. At this stage meristmatic centers had started to differentiate (initiation stage). The development of root primordia occurred mostly from around vascular strands on the proximal surface as in ginseng cotyledon cultures (Choi and Soh, 1997). Well developed roots appeared through the nodular callus after 15 days (manifestation stage). The number of roots was higher in the cotyledons of germinated seeds than those of ungerminated seeds, and darkness favoured adventitious root formation, not only during the manifestation phase but also increased rooting potential which in turn shortened the induction period. Adventitious root formation was more actively induced in 5 to 7 mm thick cotyledonary portions cultured on half-strength basal medium with 50 μM IBA and 5 μM kinetin than on medium with only one of these growth regulators. Thus, an adequate relationship of both regulators-stimulated root formation was established, as in tissue cultures of *Helianthus tuberosus* and *Antirrhinum majus* (Gautheret, 1961; Sangwan and Harada, 1975). The cotyledons of *Corylus avellana* on hormone-free medium formed roots poorly yet differently from those of walnut (Gutmann et al., 1996).

3.3 Sliced explants

Research on root formation has been conducted with relatively large tissue system; whole seedlings of maize (Ludwig-Muller and Epstein, 1991), whole cuttings of mung bean hypocotyls (Wiesman et al., 1988), and 3 to 4 cm long cuttings in pea (Nordstrom et al., 1991). However, only a very small proportion of the cells in these systems is structurally involved in rooting; in 'Gala' apple the initiation of root primordium occurred in the basal 1 mm of microcuttings regardless of treatment (Harbage et al., 1993). A test system for root regeneration was developed with apple root-stock M. 'Jork' by Van der Krieken et al. (1993). In this system thin stem slices (ca. 0.5 mm thick; fresh weight ca. 1 mg) were incubated in inverted petri dishes with the apical side attached to auxin-containing medium, *via* nylon mesh. Although a stem slice had a fresh weight of only 1 mg, still 8 roots per slice could be induced by incubation of the slices, for 3 days, in a medium with 3.2 μM of IBA. The roots regenerated synchronously between 7 and 10 days with no intervening callus between the slices and roots. In contrast, root formation on whole shoots was more asynchronous and slower (between 10 and 21 days) and with a callus interphase (Van der Krieken et al., 1992). The main advantage of this test system over the systems generally used is that the metabolic events related to root regeneration can be studied with very little interference from the bulk of the cells which are not structurally involved in thc regeneration process. Another advantage of the system is that no compounds originating from other parts of the plant can interfere with auxin-induced root formation. Furthermore, the position of the slice in the stem did not effect rooting, which enabled experiments with larger number of slices.

4 TISSUE CULTURE SYSTEM

4.1 Thin cell layers

Thin cell layers stripped from the outer layers of tobacco floral branches are able to undergo root, vegetative shoot and flower organogenesis (Tran Thanh Van, 1973). The organogenic potential of thin cell layers from various organs of a number of species have justified that the type of organogenesis is dependent upon the species, the tissue used, and the hormonal condition of the medium (Mohnen, 1994). It is important that organogenesis in thin cell layers is truly *de novo* organ formation because the starting explant contains no meristems or centers of cell division.

Thin cell layers have been used as an experimental system to study adventitious root differentiation (Tran Thanh Van, 1981; Mohnen et al., 1990; Altamura and Capitani; 1992) which could be obtained on thin cell layers of tobacco on rooting medium containing 7 μM IBA and 0.2 μM kinetin in an unbuffered medium with an initial pH of 5.8 (Tran Thanh Van and Gendy, 1993). In general, roots arise from the subepidermal layer or from layers slightly deeper within the tissue (Altamura et al., 1991a; Tran Thanh Van and Gendy, 1993). In experiments on the competence and determination of tobacco thin cell layers for root formation (Mohnen et al., 1990), the layers were cultured for zero to 16 days on root induction medium with IBA and kinetin and subsequently transferred to basal medium for a total culture period of 25 days. From the results it was suggested that thin cell layers require three days of exposure to root induction medium to become determined for root formation. In subsequent experiments thin cell layers were first precultured on basal medium for three days before being transferred to root induction medium for zero to 16 days (Mohnen, 1994). Thin cell layers precultured on basal medium formed roots after only two days of culture on root induction medium. Thus, the three days of culture required for thin cell layers to become determined for root formation can be subdivided into one day on root induction medium for development of competence and two days on root induction medium for determination for root formation.

It was shown by Tran Thanh Van and Gendy (1993) that root differentiation in the thin cell layer is correlated with a peak of putrescine synthesis at day 7 after culture. Analysis of tobacco thin cell layer explants revealed that both free and bound putrescine and spermidine increased at the time of root meristemoid emergence but that polyamines decreased when root formation was completed (Torrigiani et al., 1989). It is, therefore, linked to polyamine synthesis in tobacco thin cell layers that determination for root formation occurs well before root meristems are formed (Warnick, 1992). The thin cell layers as a tissue culture system are particularly useful for organogenesis studies since organogenesis can occur directly without intervening callus formation (Mohnen, 1994), and thin cell layers have a variety of morphogenic potential.

4.2 Callus cultures

4.2.1 Culture on solid medium

For meristemoid initiation alfalfa callus must be transferred to a hormone-free medium after culture on an inductive medium with 2,4-

dichlorophenoxyacetic acid (2,4-D) and kinetin. The callus does not exhibit organogenic meristemoid initiation if the callus is continuously cultured on a hormone-containing medium (Walker et al., 1979). This is different from the work of Skoog and Miller (1957) in which culture of tobacco callus on medium with relatively high auxin to cytokinin concentrations led to root formation. If alfalfa callus is subsequently transferred to a hormone-free regeneration medium, or if the callus is subcultured on medium containing 1-naphthaleneacetic acid (NAA) and kinetin (non-inductive medium), root formation could not occur. However, after the culture of the callus on a root-inducing medium containing low level of 2,4-D and high level of kinetin (inductive medium) if it is subsequently transferred to a hormone-free regeneration medium, roots develop from the callus. Maximum root formation occurred in alfalfa callus that had been cultured for four days on the root-inducing medium. It is concluded that 2,4-D is an inducer of root formation in alfalfa callus cultures. Since maximum root formation occurred within four days, it is clear that the alfalfa callus cultures were determined for root formation within four days of exposure to the root-inducing medium.

In callus cultures of *Bupleurum falcatum* also adventitious roots were formed from the callus which was transferred to hormone-free medium after culture on medium with 2,4-D for 5 days (Bae et al., 1994; Kim et al., 1995). Callus of *Aralia elata* was induced on medium with BA and 2,4-D but root formation did not occur on the same medium. When the callus was subcultured on hormone-free medium adventitious roots differentiated from the callus (Yoshizawa et al., 1994). Almost similar result was obtained from immature leaf explant cultures of *Sorghum bicolor* (Wernicke et al., 1982). Therefore, adventitious root primordia were induced from cultures on medium containing 2,4-D but root development occurred on medium without 2,4-D or with decreased 2,4-D level.

Yellowish compact callus, induced from cowpea hypocotyls on MS medium containing 0.2 mgl^{-1} (0.93 μM) kinetin and 0.4 mgl^{-1} (1.81 μM) 2,4-D, actively proliferated on the same medium but did not show any organogenic activity macroscopically as well as microscopically (Soh et al., 1998). On medium with BA and NAA, the yellowish compact callus first changed to pale green compact callus and then many green spots appeared on its surface under light culture. These spots gradually became white nodular structures. Adventitious root formation from the nodular structures occurred not only on the same medium, but also on medium with either auxin or cytokinin, but not both (Table 1). Yellowish compact callus on medium with auxin alone was transformed to yellowish friable callus, which did not develop adventitious roots. The yellowish friable callus could gain rhizogenic activity only after morphological modification to pale green

Table 1. The average numbers of roots formed from calluses of various types on media with auxin, cytokinin or their combinations

Callus types	Growth regulators									
	IAA+ Kin	IBA+ BA	NAA +BA	NAA +Kin	NAA	IBA	IAA	BA	Kin	Zea
Yc	2.3	11.3	32.3	7.7	0.3	0.0	0.0	4.0	6.3	4.7
Pgs	18.3	13.8	23.7	17.8	15.5	12.5	12.8	7.5	10.0	6.9
Pgw	2.3	6.5	11.8	4.1	0.0	0.0	0.0	2.4	1.3	0.3
Ycw	0.0	4.0	12.3	2.0	0.0	0.0	0.0	2.0	1.7	1.7

Yc: yellowish compact callus induced from *Vigna unguiculata* hypocotyl explant. Pgs: pale green compact callus with green spots derived from yellowish compact callus. Pgw: pale green compact callus with white spots derived from yellowish compact callus in dark. Ycw: yellowish compact callus with white spots derived from yellowish compact callus in dark. Kin: kinetin, Zea: zeatin. Maximum standard error of each value was less than 10%.

compact callus on medium with auxins and cytokinins. The modified callus did not form adventitious roots on medium with auxins but only with cytokinins. Therefore, it is suggested that cytokinins have stimulating effects on root formation from callus that previously did not show rhizogenic activity on medium with auxins alone. The phenomena that root organogenesis did not occur directly from the initial callus but could occur only after the appearance of nodular structures were also observed in callus of *Glycine clandestina* (Sharma and Kothari, 1993). In addition, the rhizogenic potential of cowpea callus was discriminated from that of leaf explants (Ladeinde and Soh, 1991), which formed adventitious roots directly on medium with auxin alone.

4.2.2 Suspension culture

The experimentally controllable regeneration of roots from cell clumps in liquid cultures of pea was reported by Torrey and Shigemura (1957). Partial organization of root meristem-like structures had occurred in friable clumps of tissue. These organized structures were shown to be capable of developing into roots in auxin-free medium. Steward et al. (1958) described the origin of roots from cell masses in domestic carrot cell suspension cultures in White's medium containing coconut milk. In such cultures, the root primordia developed from cells around tracheary elements in the cell clumps. When these minute, root-bearing cell masses were transferred to semi-solid media, whole carrot plantlets eventually appeared. In microcallus suspension cultures of *Bupleurum falcatum*, tracheary elements differentiated within the callus and cell divisions occurred around the elements (Bae et al., 1994; Kim et al., 1995), followed by root primordium initiated from such cell divisions.

In the embryogenic cell cultures of carrot, in liquid medium, root development was also observed (Halperin, 1966). When carrot petiole explant cultures were started on media lacking reduced nitrogen, the callus usually consisted of a mass of large parenchymatous cells and few multicellular units. This may indicate that in the absence of reduced nitrogen the cells tend to expand and separate after division. The multicelluar clumps which are eventually formed, when such a suspension is sieved and inoculated into low-auxin media, are made up of completely disorganized, loosely cohering cells. After a considerable production of parenchymatous or tracheid-like cells in such clumps, a relatively limited number of internal cells underwent the process of dedifferentiation and a meristematic area appeared. These cells gave rise to a root primordium, although it could not be established whether or not primordia arose in the cambium-like regions associated with tracheary elements. In every case examined the root primordia appeared to arise endogenously. If carrot petiole explant cultures were started on media containing only nitrate nitrogen, the cell suspensions obtained from such explants formed adventitious roots. Conversely, cell suspensions derived from explants initially cultured on a medium containing ammonium nitrogen invariably produced embryos. The cytological events involved in rhizogenesis contrast strongly with those described for embryogenesis (Halperin, 1966), as in the cotyledon cultures of ginseng (Choi and Soh, 1997).

In callus cultures derived from leaf tissue of *Populus spp.* many hundreds of small (1-3 mm in diameter) microcalli could be generated by plating of the cells from suspension cultures, and such calli had the capacity to regenerate roots by moving the web containing the callus to a medium supplemented with 0.1 mg l^{-1} IBA (McCown, 1988). The root primordia often were as large as the original microcalli source, although the minimum size of calli for the induction of root formation was unclear in that observation. Alfalfa callus below 150 μm in diameter did not form roots but they regained the ability to form roots if the calli were allowed to grow to a larger size (Walker et al., 1979). In carrot cell suspensions disorganized cell clumps, 500-1,000 μm in diameter, developed. The cell clumps lacked any trace of root primordia as determined by examination of serial microtome sections but gave rise to roots after several weeks of growth (Halperin, 1966). Adventitious root formation also occurred on microcalli of *Bupleurum falcatum* within one month (Kim et al., 1995; Fig.3). The highest frequency of root production appeared in calli, 900-1,000 μm in diameter (Fig. 4). The root formation rate of calli larger than 1,000 μm in diameter was similar to that of 1,000 μm; calli below 900 μm showed a prominently lower frequency of rooting, and calli smaller than 200μm in diameter rarely formed roots. Thus, it is apparent that a minimum size is required for callus to become competent for induction of root formation.

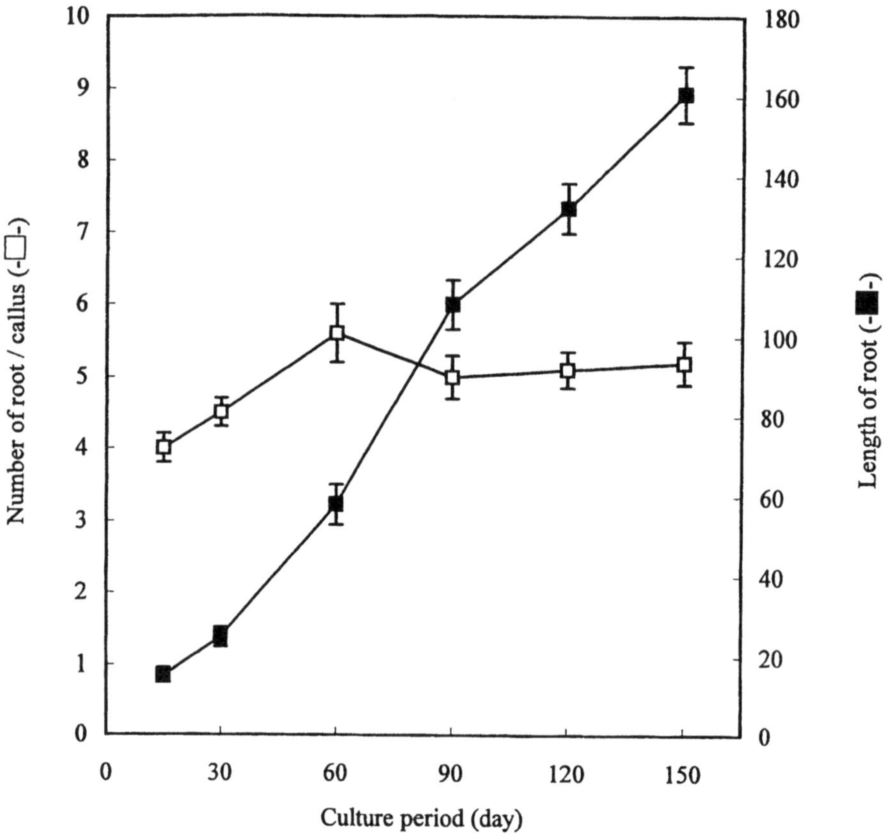

Figure 3. The effect of culture period on the number and length of adventitious root formed from *Bupleurum falcatum* microcallus. The adventitious roots were cultured in MS basal liquid medium after pretreatment in MS medium containing 0.1 mg l^{-1} 2,4-D for 5 days.

Microcallus has some advantages as a research tool because it is a relatively small uniform cell mass and a simple unorganized system without the complexities of correlative interactions between tissues and organ systems. Such microcalli have recently become the centre of research interest as a useful tissue system to study adventitious root formation (McCown, 1988; Kim et al., 1995). First, experiments such as the effect of specific exogenously applied compounds on adventitious root primordium initiation and development can be done near or at the cellular level. Such

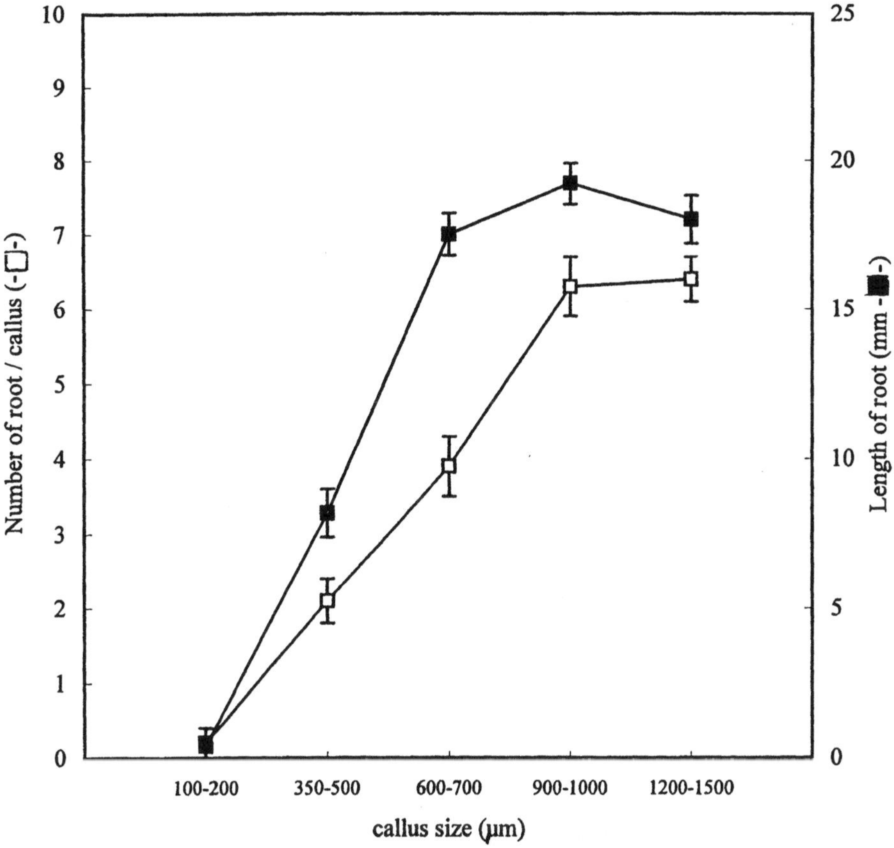

Figure 4. Adventitious root formation from *Bupleurum falcatum* microcallus of different sizes. The callus was cultured in MS basal liquid medium for 4 weeks after pretreatment in 0.1 mg l^{-1} 2,4-D for 5 days.

microcalli are reproducible and lightly responsive because of their standardized biological history and amenability to infusion of agents on both a long- term or pulse dosage basis.

Thus, the microcalli may be particularly useful in studies of the progenitor cells of root initials (Nougarede and Rondet, 1983) and on the inductive phase of rooting. Second, microcalli experiments may be appropriate in determining the exact cellular effects of some environmental variables such as the control of water stress implicated in the inductive phase of rooting by addition of osmotic components (Haissig, 1986). Third, the root formation from microcalli is applicable to root cultures for the production of secondary metabolite (Kim et al., 1995); the production of root biomass can be established from mass culture of microcalli in some medicinal plants, such as *Bupleurum falcatum* (Kim et al., 1995).

5 VASCULAR DIFFERENTIATION IN ROOT PRIMORDIUM

Although the cytodifferentiation leading to vascular differentiation during adventitious root formation in tissue cultures is almost similar to that of intact plant growth (Gonzalez et al., 1993), the vascular differentiation at the root-to-shoot interface needs to be examined in association with the function of vascular tissue.

During root primordium development in the parental organ (tissue), vascular differentiation occurs and then the differentiating vascular tissues are connected with those of the shoot system. In the cuttings of *Agathis australis* the first indication of root primordium initial development is when the end of the induced vascular strand is in the mid cortex and the cells immediately ahead of it divide but do not differentiate (White and Lovell, 1984a). The cells abutting the ends of the strand remain undifferentiated during early root primordium development but eventually form the final vascular connection. After the beginning of organized cell arrangement at the apex of the developing root primordium three rows of periclinally divided cells become the vascular and columella initials. Differentiation of the cells remaining undifferentiated takes place at about the time of root emergence. These cells form vascular connection between the primordium and the parental organ. Therefore, it is very important to understand the normal vascular connection and function of roots formed in tissue cultures; how these undifferentiated cells at the root-to-shoot interface are differentiated into vascular tissue. The sequence of events in the later stages is similar for most species but may differ in detail depending on the location of the root primordium. The sequence in *Agathis australis* differs from that in *Griselinia littoralis* in terms of whether cortical cells are incorporated into the primordium or not (White and Lovell, 1984b).

In *Trachelospermum asiaticum*, well defined root primordia were present by day 5 of stem cutting cultures (Apter et al., 1993a). Vascular differentiation in early root primordium and occasional root penetration through the periderm occurred by day 7. Early xylem cells with helical wall thickenings and retaining their nuclei were still evident at the root primordium base and were connected to the secondary xylem of the shoot system. Continued elongation of the emerging root and further xylem development within the adventitious root and at the root-shoot interface occurred up to days 8 to 10. Developmentally and anatomically, no difference existed between tissue culture explants and conventional macropropagation cuttings. While tissue culture explants exhibited 1 or 2 days lag in root primordia initiation, by day 9 these differences were no longer evident. There was no evidence that xylem elements were discontinuous between the elongating primordia and the base of the primordia adjacent to the cambium after 8 to 10 days. Also, in other plants, such as apple, vascular discontinuity between adventitious roots and parental shoots was not detected (Hicks, 1987). Furthermore, since root-shoot vascular connections were confirmed by ^{32}P uptake experiments (Apter et al., 1993b), it is concluded that the vascular tissues are functionally normal. The vascular connections are established at the beginning of vascular differentiation since early vascular differentiation occurs at the root primordium base.

In *Pseudotsuga menziesii* shoot cuttings tracheidal nests surrounded by elongated meristematic cells were formed at the cutting base in peat-perlite (Mohammed et al., 1989). Actively dividing meristematic cells at the periphery of the tracheidal nests became nodular structures and then formed root primordia. As the root primordia elongated, their vascular systems developed simultaneously, originating from a tracheidal nest. Vascular tissues of each root were connected to the central vascular system of the shoot through a nest. Thus, a vascular connection between two organs was established at an early stage of vascular differentiation. On agar medium, tracheidal nests do not contain the enveloping meristematic cells that were present in peat-perlite, and cell divisions were disorganized, producing a callus. Agar may impede gas exchange and, thus, inhibit vascular differentiation in *Pseudotsuga menzisii*. In other gymnosperms, such as *Pinus sylvestris* and black and white spruce, it was also confirmed that the vascular connection between root primordium and the parental organ passes through the tracheidal nest (Gronroos and von Arnold, 1985; Patel et al., 1986).

On the other hand, it is known that incomplete vascular differentiation at the root-to-shoot interface restricts acropetal water transfer, and becomes a cause for poor success during plantlet acclimatization (Grout and Aston, 1977; Ziv, 1986). In regenerated plantlets of *Brassica oleracea* var. botrytis cv. 'Armado Tardo' the basal part of the plantlets formed a limited amount of callus, and it is in this region that abnormal xylem regeneration occurs (Grout and Aston, 1977). Roots that develop spontaneously arise in this callus mass, and have poor connections with the main vascular system. Three weeks after transplanting, the development of greater vascular continuity between the roots and shoots was reflected by increased rates of water transfer. When roots arose from callus tissue in *Salpiglossis sinuata* and some conifers, plantlets failed to survive transplantation (Hughes et al., 1973; Thorpe, 1984), probably because of poor vascular connections. To aid vascular connections between shoots and roots, it is important that the roots are initiated directly from the shoots, at the shoot bases, and particularly not from callus. Therefore, any unorganized tissue should be removed from the shoot bases during their subculture onto a rooting medium. Similarly, vascular connections between petioles and stems of microcultured shoots were smaller and more poorly structured than those of seedlings and greenhouse samples of Asian white birch (Smith et al., 1986). Furthermore, there is an opinion that adventitious roots formed in tissue cultures are nonfunctional and, therefore, must be replaced by new roots when the plants are to be acclimated (Davis et al., 1977; Debergh and Maene, 1981; Maene and Debergh, 1983; Read and Fellman, 1985; Pierik, 1987). These contradictory results on vascular connections seem to be the most striking differences between the species examined. Therefore, the differentiation of vascular tissue at the root-to-shoot interface must be paid attention for the fullest understanding of normal vascular connections and functions of tissue cultured roots.

6 ROOT EMERGENCE AND GROWTH

Morphological appearance of roots of micropropagated plants are roughly similar to those of macropropagated plants but under closer scrutiny morphological differences may exist between roots cultured under the two propagative conditions. Roots of white spruce produced in agar medium were reported to be stubby, thick and unbranched (Toivonen, 1985). This was possibly caused by declining photosynthetic rates in aerial parts because roots do not provide efficient sinks for photosynthate. In radiata pine, plantlets with roots that elongated in agar medium withered and turned black when transferred to soil (Horgan and Aitken, 1981). Agar may have physically restricted root elongation, and the roots became thicker by radial expansion of the cells (Salisbury and Ross, 1985). Therefore, root growth *in vitro* is influenced mainly by the substrate and medium. In *Acacia*, roots produced in agar medium lacked both root hairs and a fully developed root system, while roots produced in liquid medium were normal (Skolmen and Mapes, 1978). In *Trachelospermum asiaticum* root hairs produced under tissue culture conditions were one-third to one-half as long as those occurring under macropropagative conditions (Apter et al 1993a). Root hairs of macropropagated plantlets were long, thin, almost wiry, and interwoven as a mat but those of tissue cultured plantlets were shorter, thicker, and straight. The distances from the root tips to root-hair zones and root-hair density did not seem to differ significantly between the two propagation systems. The presence of shorter root hairs on tissue culture-formed roots could have significant implications for acclimatization of tissue culture-produced plant material. Root hairs play an important role in the rhizosphere, effectively increasing root surface area; similarly, water uptake is made more effective by the presence of root hairs (Kramer, 1983). The shortening or absence of root hairs on tissue culture-formed roots could render those roots less functional to the plantlet, increasing the stress imposed during and after acclimatization. Roots grown in agar medium frequently have few or no root hairs, and those present can die after transplanting, causing the loss of the plantlets or at least, cessation of their growth (Preece and Sutter, 1991). McKeand and Allen (1984) compared, in loblolly pine (*Pinus taeda* L.), nutrient uptake (N and P) by tissue culture-formed and seedling roots of identical age after a 4 week of acclimatization. Tissue culture-formed roots showed a thick and unbranched morphology and were inefficient in nutrient uptake.

The alterations in root hair development may frequently take (a) the form of changes in the direction of growth, (b) a response that requires the perception of an external signal, (c) transduction of the signal, (d) alteration in gene regulation and/or protein activity, and (e) modification of programs of cell division, expansion, and differentiation (Aeschbacher et al., 1994). Among the many factors and signals, the plant hormones such as auxin and ethylene are known to influence many physiological and developmental processes during root development, including some aspects of cell dedifferentiation and differentiation.

Genetic analysis of root hair development is expedited by the fact that root hairs are dispensable, and plants lacking root hairs are viable and able to grow normally (Scheifelbein and Somerville, 1990). The loci that have been identified by mutations about root hair development may be subdivided into two groups: (1) those that affect the emergance of root hairs and (2) those that are responsible for root hair growth (Baskin, et al., 1992; Scheifelbein and Somerville, 1990).

One *Arabidopsis thaliana* locus that plays a role in the localized swelling of the epidermal cells is RHD1 (Scheifelbein and Somerville, 1990). The *rhd1* mutants form an abnormally large swelling or protrusion on the epidermal cell surface during the initial phase of hair formation (Scheifelbein and Somerville, 1990). Other loci involved in root tip growth, including RHD2, RHD3, RHD4 and TIP1, have been described (Aeschbacher et al., 1994). These genes may encode products that affect known tip-growth factors, such as ingredient of the actin cytoskeleton or Ca^{2+} fluxes (Heath, 1990; Scheifelbein, et al., 1992). According to the recent review paper about these gene actions in root development (Aeschbacher, 1994), double mutant analyses show that the RHD2 gene product is required before the RHD3 or RHD4 products, and the RHD3 and RHD4 products probably act in separate pathways.

Several *A. thaliana* loci are probably required for normal root hair differentiation, although these were not originally identified for the studying hairy root development. Some anxin-resistant mutants fall into this category. The *dwf* and *axr2* mutants show abnormal phenotypes in root hair formation (Maher and Martindale, 1980; Wilson, et al., 1990). The cytokinin-resistant mutants (*ckr1*) also produce shorter root hairs than wild-type (Su and Howell, 1992), and the *hy3* mutants, which have mutations in the phytochrome B gene, produce longer root hairs than normal when they are grown under light (Reed, et al., 1993). Although the abnormal root hairs in these mutants need to be analyzed and characterized in very detail, these studies may examplify the utility of root-hair cells for analyzing the control of plant cell differentiation by diverse regulatory pathways.

Auxins, which stimulate the production of ethylene causing inhibition of root elongation (Salisbury and Ross, 1985), have usually been omitted from the medium for root elongation (Cheng and Voqui, 1977; Kaul, 1987; Mott and Amerson, 1981; Patel et al., 1986; Poissonnier et al, 1980; Harbage et al., 1993). Root primordia formed by microcallus of *Bupleurum falcatum* in medium containing 2,4-D did not grow out from the microcallus during culture in the same medium but did so after transfer to medium without 2,4-D (Bae et al., 1994). On the other hand adventitious roots originated from callus of *Robina hispida* grew in liquid half-strength MS medium with 1.0 mg l^{-1} IAA (Deguchi et al., 1994). In *Castanea stativa* excised shoots which were preconditioned for root initiation in a medium containing NAA (40 μM), for 7 days and then transferred to a fresh medium with IBA (0.8 μM) enhibited root development (Rodriguez, 1982). The newly produced shoots of *Digitalis thapsi* developed normally and rooted directly after 4 weeks of shoot tip culture in a single medium containing both auxins and cytokinins (Herrera et al., 1990). The concentration of the basal medium and sucrose is often reduced for root elongation (Patel et al., 1986; Rumary and Thorpe, 1984; Von Arnold, 1982). In English ivy, sucrose is required for the outgrowth of root primordia but not for the initiation of primordia (Geneve et al., 1988). Rose shoots grown in media containing high sucrose concentrations (146-236 mM) produced more and longer roots than those grown in media containing a low sucrose concentration (0-87 mM). This respose to sucrose was related to the metabolism of sucrose rather than to its osmotic properties, since the use of mannitol as an osmotic substitute did not replicate this effect on root formation (Hyndman et al., 1982). On the other hand, there is contradictory result that the osmotic effects of non- or poorly-metabolized carbohydrates provided an important trigger for root initiation in minicrown cultures of asparagus (Conner and Falloon, 1993). Thus relationship between carbohydrates as an osmotic substitute and root initiation is far from clear.

7 REFERENCES

Aeschbacher RA, Schiefelbein JW & Benfey PN (1994) The genetic and molecular basis of root development. Ann. Rev. Plant Physiol. Plant Mol. Biol. 45: 25-45.

Altamura MM, Capitani F, Serafini-Fracassini D, Torrigiani P & Falasca G (1991a) Root histogenesis from tobacco thin cell layers. Protoplasma 161: 31-42 .

Altamura MM, Torrigiani P, Capitani F, Scaramagli S & Bagni N (1991b) *De novo* root formation in tobacco thin layers is affected by inhibition of polyamine biosynthesis J. Exp. Bot. 42: 1575-1582.

Altamura MM & Capitani F (1992) The role of hormones on morphogenesis of thin layer explants from normal and transgenic tobacco plants. Physiol. Plant. 84: 555-560.

Apter RC, Davies FTJr & McWilliams EL (1993a) *In vitro* and *ex vitro* adventitious root formation in Asian jasmine (*Trachelospermum asiaticum*). I. Comparative morphology. J. Amer. Soc. Hort. Sci. 118: 902-905.

Apter RC, Davies FTJr & McWilliams EL (1993b) *In vitro* and *ex vitro* adventitious root formation in Asian jasmine (*Trachelospermum asiaticum*). II. Physiolosical comparisons. J. Amer. Soc. Hort. Sci. 118: 906-909.

Attfield EM & Evans PK (1991) Stages in the initiation of root and shoot organogenesis in cultured leaf explants of *Nicotiana tabacum* cv. Xanthi nc. J. Exp. Bot. 42: 59-63.

Bae HH, Cho DY, Kim SG, Soh W-Y & Seong RS (1994) Effects of 2,4-dichlorophenoxyacetic acid on adventitious root formation from callus of *Bupleurum falcatum* L. and its histological observation. Korean J. Plant Tiss. Cult. 21: 41-46.

Baskin Tl, Betzner AS, Hoggart R, Cork A & Williamson RE (1992) Root morphology mutants in *Arabidopsis thaliana*. Aust. J. Plant Physiol. 19: 427-437.

Beakbane AB (1961) Structure of the plant stem in relation to adventitious rooting. Nature, London, 192: 954-955.

Beakbane AB (1969) Relationships between structure and adventitious rooting. Comb. Proc. Intern. Plant Prop. Soc. 19: 192-201.

Biondi S, Daiz T, Iglesias I, Gamberini G & Bagni N (1990) Polyamines and ethylene in relation to adventitious root formation in *Prunus avium* shoot cultures. Physiol. Plant. 78: 474-483.

Blakesley D, Weston GD & Hall JF (1991) The role of endogenous auxin in root initiation. Part 1: Evidence from studies on auxin application and analysis of endogenous levels. Plant Growth Regul. 10: 341-353.

Blazich FA & Heuser CW (1979b) The mung bean rooting bioassay: a re-examination. J. Amer. Soc. Hort. Sci. 104: 117-120.

Blazich FA & Heuser CW (1979a) A histological study of adventitious root initiation in *Mung bean* cuttings. J. Amer. Soc. Hort. Sci. 104:63-67.

Boureau E (1954) *Anatomie vegetale*, Vol. I. Presses Universitaires de France, Paris.

Bressan OG, Kim YJ, Hyndman SE, Jasegawa PM & Bressan RA (1982) Factors affecting in vitro propagation of rose. J. Amer. Soc. Hort. Sci. 107: 979-990.

Byrne JM, Collins KA, Cashau PF & Aung LH (1975) Adventitious root development from the seedling hypocotyl of *Lycopersicon esculentum*. Amer. J. Bot. 62: 731-737.

Cheng TY & Voqui TH (1977) Regeneration of Douglas fir plantlets through tissue culture. Science 198: 306-307.

Choi YE & Soh W-Y (1995) Effects of growth regulators on somatic embryogenesis from ginseng zygotic embryos. Korean J. Plant Tiss. Cult. 22: 157-163.

Choi YE & Soh W-Y (1997) Effect of ammonium ion on morphogenesis from cultured cotyledon explants of *Panax ginseng*. J. Plant Biol. 40: 21-26.

Cline MN & Neely D (1983) The histology and histochemistry of the wound-healing process in *Geranium* cuttings. J. Amer. Soc. Hort. Sci. 108: 496-502.

Conner AJ & Falloon PG (1993) Osmotic versus nutritional effects when rooting in vitro asparagus minicrowns on high sucrose media. Plant Sci. 89: 101-106.

Crooks DM (1933) Histological and regenerative studies on the flax seedling. Bot. Gaz. 95: 209-239.

Davies FTJr, Lazarte JE & Joiner JN (1982) Initiation and development of roots in juvenile and mature leaf bud cuttings of *Ficus pumila* L. Amer. J. Bot. 69:804-811.

Davies FTJr & Hartmann HT (1988) The physiological basis of adventitious root formation. Acta. Hortic. 227: 113-120.

Davis M J, Baker R & Hanan JJ (1977) Clonal multiplication of carnation by micropr - opagation. J. Amer. Soc. Hort. Sci. 102: 48-53.

De Klerk G-J, Smulders R & Benschop M (1990) Basic peroxidases and rooting in microcuttings of *Malus*. Acta. Hort. 280: 29-36.

Debergh DL & Maene LJ (1981) A scheme for commercial propagation of ornamental plants by tissue culture. Scientia Hort. 14: 335-345.

Deguchi M, Gyokusen K & Saito A (1994) Propagation of plantlets by the culture of multiple shoots and roots induced from callus originating in the cambium of *Robinia hispida* L. J. Jap. For. Soc. 76: 346-354.

Dolan L, Duckett C, Grierson C, Linstead P, Schneider K, Lawson E, Dean C, Poethig S & Roberts K (1994) Clonal relations and patterning in the root epidermis of *Arabidopsis*. Development 120: 2465-2474.

Doud SL & Carlson RF (1977) Effects of etiolation, stem anatomy and starch resources on root initiation of layered *Malus* clones. J. Amer. Soc. Hort. Sci. 102: 487-491.

Eriksson T (1984) Progress and problems in tissue cultures of Norway spruce, *Picea abies*. Proc. Intern. Symp. In Vitro Prop. Forest Tree Species, Bologna, pp. 104-107.

Fabijan D, Taylor JS & Reid DM (1981) Adventitious rooting in hypocotyls of sunflower (*Helianthus annuus*) seedlings. II. Action of gibberellins, cytokinins, auxins and ethylene. Physiol. Plant. 53: 589-597.

Fahn A (1990) Plant Anatomy (4th ed), Pergamon Press, Oxford, England.

Favre JM (1973a) Effets corrélatifs de facteurs internes et externes sur la rhizogenèse de la Vigne cultivée in vitro. Thèse d'Etat. Cent. Sci. D'Orsay, Univ. Paris-Sud. pp. 89.

Favre JM (1973b) Effets corrélatifs de facteurs internes et externes sur la rhizogenèse d'un clone de vigne (*Vitis riparia* × *Vitis rupestris*) cultivée in vitro. Rev. Gén. Bot. 80: 279-361.

Favre JM (1973c) Divers aspects du rô le du bourgeon et des noeuds sur la rhizogenèse de la vigne cultivée in vitro. Rev. Cytol. Biol. Végét. Paris 37: 393-406.

Fink S (1982) Adventitious root primordia- the cause of abnormally broad xylem rays in hard and softwoods. Intern. Asso. Wood Anat. 3: 31-38.

Fjell I (1985a) Preformation of root primordia in shoots and root morphogenesis in *Salix viminalis*. Nord. J. Bot. 5. 357-376.

Fjell I (1985b) Morphogenesis of root cap in adventitious roots of *Salix viminalis*. Nord. J. Bot. 5: 555-573.

Friedman R, Altman A & Zamski E (1979) Adventitious root formation in bean hypocotyl cuttings in relation to IAA translocation and hypocotyl anatomy. J. Exp. Bot. 30: 769-777.

Galway M, Masucci J, Lloyd A, Walbot V, Davis R & Schiefelbein J (1994) The TTG gene is required to specify epidermal cell fate and cell patterning in the *Arabidopsis* root. Dev. Biol. 166: 740-754.

Gaspar Th & Coumans M (1987) Root formation. In: Bonga, J.M., and Durzan, D.J.(eds). Cell and Tissue Culture in Forestry. pp. 202-217. Martinus Nijhoff Pub., Dordrecht, Netherlands.

Gautheret RJ (1961) Action conjugueede l'acde gibberellique de la cinetine et de l'acide indole-acetique sue les tissus cultives "in vitro" , particulierement sur ceux de Topinambur. C.R. Acad. Sci. Paris 253: 1381-1385.

Geneve RL, Hackett WP & Swanson BT (1988) Adventitious root initiation in de-bladed petioles from the juvenile and mature phases of English ivy. J. Amer. Soc. Hort. Sci. 113: 630-635.

Geneve RL & Kester ST (1990) The initiation of somatic embryos and adventitious roots from developing zygotic embryo explants of *Cercis canadensis* L. cultured *in vitro*. Plant Cell, Tiss. Org. Cult. 22: 71-76.

Geneve RL & Kester ST(1991) Polyamines and adventitious root formation in the juvenile and mature phase of English ivy. J. Exp. Bot. 42: 71-75.

Ginzburg C (1967) Organization of the adventitious root apex in *Tamarix aphylla*. Amer. J. Bot. 54: 4-8

Girouard RM (1967a) Initiation and development of adventitious roots in stem cuttings of *Hedera helix*. Anatomical studies of the juvenile growth phase. Can. J. Bot. 45: 1877-1882

Girouard RM (1967b) Initiation and development of adventitious roots in stem cuttings of *Hedera helix*. Anatomical studies of the mature growth phase. Can. J. Bot. 45:1883-1886.

Goldsmith MHM (1977) The polar auxin transport. Ann. Rev. Plant Physiol. 28: 439-478.

Gonzalez A, Sanchez Tames R & Rodriguez R (1991) Ethylene in relation to protein, peroxidase and oxidase activities during rooting in hazelnut cotyledons. Physiol. Plant. 83: 611-620.

Gonzalez A, Sanchez Tames R, Rodriguez R (1993) Adventitious root differentiation in hazelnut (*Corylus avellana* L.) cotyledons. Phyton 54: 119-126.

Gronroos R & von Arnold S (1985) Initiation and development of wound tissue and roots on hypocotyl cuttings of *Pinus sylvestris in vitro*. Physiol. Plant. 64: 393-401.

Grout BWW & Aston MJ (1977) Tranplanting of cauliflower plants regenerated from meristem culture. I. Water loss and water transfer related to changes in leaf wax and to xylem regeneration. Hort. Res. 17: 1-7.

Gurel E & Wren MJ (1995) *In vitro* development from leaf explants of sugar beet (*Beta vulgaris* L.) rhizogenesis and the effect of sequential exposure to auxin and cytokinin. Ann. Bot 75: 31-38.

Gutmann N, Charpentier JP, Doumas P & Jay-Allemand C (1996) Histological investigation of walnut cotyledon fragments for a better understanding of *in vitro* adventitious root initiation. Plant Cell Rep. 15: 345-349.

Haissig BE (1970) Preformed adventitious root initiation in brittle willow grown in a controlled environment. Can. J. Bot. 48: 2309-2312.

Haissig BE (1986) Metabolic processes in adventitious rooting. In: Jockson, M.B. (ed.) New Root Formation in Plant and Cuttings. pp. 141-189. Martinus Nijhoff Pub., Dordrecht, Netherlands.

Halperin W (1966) Alternative morphogenetic events in cell suspensions. Amer. J. Bot. 53: 143-153.

Hand P (1994) Biochemical and molecular makers of cellular competance for adventitious rooting. In: Davis, T.D., and Haissing, B.E. (eds). Biology of Adventitious Root Formation. pp. 111-122. Plenum Press, New York.

Harbage JF, Stimart DP & Evert RF (1993) Anatomy of adventitious root formation in microcuttings of *Malus domestica* Borkh. 'Gala'. J. Amer. Soc. Hort. Sci. 118: 680-688.

Hartmann JT, Kester DE & Davies FTJr (1990) Plant Propagaion: Principles and Practices (5th ed). Prentice-Hall, Englewood Cliffs.
Hausman JF, Kevers C & Gaspar T (1994) Involvement of putrescine in the inductive rooting phase of poplar shoots raised *in vitro*. Physiol. Plant. 92: 201-206.
Hausman JF, Kevers C & Gaspar T (1995a) Auxin-polyamine interaction in the control of the rooting inductive phase of poplar shoots *in vitro*. Plant Sci. 110: 63-71.
Hausman JF, Kevers C & Gaspar T (1995b) Putrescine control of peroxidase activity in the inductive phase of rooting in poplar shoot *in vitro*, and adversary effect of spermidine. J. Plant Physiol. 146: 681-685.
Hayward HE (1938) The Structure of Economic Plants. MacMillan, New York.
Heath IB (1990) Tip Growth in Plant and Fungal Cells. Academic Press. San Diego.
Herrera MT, Cacho M, Corchete MP & Fernandez-Tarrago OJ (1990) One step shoot tip multiplication and rooting of *Digitalis thapsi* L. Plant Cell Tiss. Org. Cult. 22: 179-182.
Hicks GS (1987) Adventitious rooting of apple microcuttings in vitro: An anatomical study. Can. J. Bot. 65: 1913-1920.
Horgan K & Aitken J (1981) Reliable plantlet formation from embryos and seedling shoot tips of radiata pine. Physiol. Plant. 53: 170-175.
Hughes H, Lam S & Janik J (1973) *In vitro* culture of *Salpiglosis sinuata* L. Hort. Sci. 8: 335-336.
Hyndman SE, Hasegawa PM & Bressan RA (1982) The role of sucrose and nitrogen in adventitious root formation on cultured rose shoots. Plant Cell Tiss. Org. Cult. 1: 229-238.
Jarvis BC (1986) Endogenous control of adventitious rooting in non-woody cuttings, In: Jackson, M.B. (ed.) New Root Formation in Plants and Cuttings. Martinus Nijhoff, Dordrecht.
Jay-Allemand C, De Pons V, Doumas P, Capelli P, Sossountzov S & Cornu D (1991) *In vitro* root development from walnut cotyledons: a new model to study the rhizogenesis processes in woody plants. C. R. Acad. Sci. Paris. 312: 369-375.
Kang MK, Cho DY & Soh W-Y (1996) Effects of auxins on adventitious root formation on cotyledon-derived microcalli in lettuce (*Lactuca sativa* L.). Korean J. Plant Tiss. Cult. 23: 135-139.
Kaul K (1987) Plant regeneration from cotyledon-hypocotyl explants of *Pinus strobus* L. Plant Cell Rep. 6: 5-7.
Kieber JJ, Rothenberg M, Romon G, Feldman KA & Ecker JR (1993) CTR1, a negative regulator of the ethylene response pathway in *Arabidopsis*, encodes a member of the Raf family of protein kinases. Cell 72: 427-441.
Kim SK, Cho DY & Soh W-Y (1995) Saikosaponin content in adventitious root formed from callus of *Bupleurum falcatum* Korean J. Plant Tiss. Cult. 22: 29-33.
Kramer PJ (1983) Water Relations of Plants. Academic Press. New York.
Ladeinde TAO & Soh W-Y (1991) Effects of different growth regulators on organogenesis and total fresh weight gain in cultured leaf tissue of cowpea (*Vigna unguiculata*(L.) Walp.) Phytomorphology. 41: 199-207.
Lloyd G & McCown G (1980) Commercially feasible micropropagation of Mountain Laurel, *Kalmia latifolia*, by use of shoot tip culture. Comb. Proc. Intern. Plant Prop. Soc. 30: 421-427.
Lovell PH & White J (1986) Anatomical changes during adventitious root formation. In: Jackson, M.B.(ed) New Root Formation in Plants and Cuttings. pp. 111-140. Martinus Nijhoff Publishers, Dordrecht, Netherands.
Ludwig-Muller J & Epstein E (1991) Occurrence and *in vivo* biosynthesis of indole-3-butyric acid in corn (*Zea mays* L.) Plant Physiol. 97: 765-770.

Mackenzie KAD, Howard BH & Harrison-Murray RS (1986) The anatomical relationship between cambial regeneration and root initiation in wounded winter cuttings of the apple rootstock M. 26. Ann. Bot. 58: 649-661.

Maene LM & Debergh PC (1983) Rooting of tissue cultured plants under *in vivo* conditions. Acta Hort. 131: 201-208.

Maher EP & Martindale SJB (1980) Mutants of *Arabidopsis thaliana* with altered responses to auxins and gravity. Biochem. Genet. 18: 1041-1053.

Martinez MC, Jörgensen J-E, Lawton MA, Lamb CJ & Doerner PW (1992) Spatial pattern of *cdc2* expression in relation to meristem activity and cell proliferation during plant development. Proc. Natl. Acad. Sci. USA. 89: 7360-7364.

Masucci JD & Schiefelbein JW (1996) Hormones act downstream of TTG and GL2 to promote root outgrowth during epidermis development in the *Arabidopsis* root. Plant Cell 8: 1505-1517.

Masucci JD, Rerie WG, Foreman DR, Zhang M, Galway ME, Marks MD & Schiefelbein JW (1996) The homeobox gene GLABRA2 is required for position-dependent cell differentiation in the root epidermis of *Arabidopsis thaliana*. Development 122: 1253-1260.

Matsuoka H & Hinata K (1979) NAA-induced organogenesis and embryogenesis in hypocotyl callus of *Solanum melongena* L. J. Exp. Bot. 30: 363-370.

Maynard CA, Satchwell M & Rieckermann H (1993) Micropropagation of American chestnut (*Castanea dentata* (Marsh.) Borkh.); rooting and acclimatization. In: Mohn, C. A., (ed). Proc. Sec. North. For. Genet. Assoc. Conf. Roseville, MN 29-30: 161-170.

McCown BH (1988) Adventitious rooting of tissue cultured plants. In: Davis, T. D., Hassing, B. E., and Sankhla, N. (eds). Adventitious Root Formation in Cuttings. pp. 289-302. Dioscorides Press, Portland, U.S.A.

McKeand SE & Allen HL (1984) Nutritional and root development factors affecting growth of tissue culture plantlets of loblolly pine. Physiol. Plant. 61 : 523-528.

McVeigh I (1938) Regeneration in *Crassula muticava*. Amer. J. Bot. 25: 7-11.

Mitsuhashi-Kato M, Shibaoka H & Shimokoriyama M (1978a) Anatomical and physiological aspects of developmental processes of adventitious root formation in *Azukia* cuttings. Plant Cell Physiol. 19: 867-874.

Mitsuhashi-Kato M, Shibaoka H & Shimokoriyama M (1978b) The nature of the dual effect of auxin on root formation in *Azukia* cutting. Plant Cell Physiol. 19: 1535-1542.

Mohammed GH, Patel KR & Vidaver WE (1989) The control of adventitious root production in tissue-cultured Douglas-fir. Can. J. For. Res. 19: 1322-1329.

Mohnen D (1994) Novel experimental system for determining cellular competence and determination. In: Darvis, T.D., and Haissig, B.C. (eds) Biology of Adventitious Root Formation. pp. 87-98. Plenum Press, New York.

Mohnen D, Eberhard S, Marfa V, Doubrava N, Toubart P, Gollin DJ, Gruber TA, Nuri W, Albersheim P & Davill A (1990) The control of root, vegetative shoot and flower morphogenesis in tobacco thin cell-layer explants (TCLs). Development 108: 191-201.

Molnar JM & LaCroix LJ (1972) Studies of the rooting of cuttings of *Hydrangea macrophylla*: enzyme changes. Can. J. Bot. 50: 315-322.

Moncousin C (1991) Rooting of *in vitro* cuttings. In: Bajaj, Y.P.S. (ed.). Biotechnology in Agriculture and Forestry vol. 17. High-Tech and Micropropagation I. pp. 222-261. Springer-Verlag, Berlin.

Mott RL & Amerson HV (1981) A tissue culture process for the clonal production of loblolly pine plantlets. North Carolina Agr. Res. Sr. Tech. Bull. 271, 14p.

Murashige T & Skoog F (1962) A revised medium for rapid growth and bioassays with tobacco tissue. Physiol. Plant. 15: 473-497.

Nicholas JR, Gates PJ & Grierson D (1986) The use of fluorescence microscopy to monitor root development in micropropagated explants. J. Hort. Sci. 61: 417-421.

Nordstrom AC, Alvarado JF & Eliasson L (1991) Effect of endogenous indole-3-acetic acid on internal levels of the respective auxin and their conjugation with aspartic acid during adventitious root formation in pea cuttings. Plant Physiol. 96: 856-861.

Nougare'de A & Rondet P (1983) Bases cytophysiologiques de l'induction rhizogen en responese a un traitement auzinuque dans l'epicotyle du pois nain. Ann. Sce. Nat. Bot. Paris. 13e: 121-149.

Patel KR, Rumary C & Thorpe TA (1986) Plantlet formation in black and white spruce. III. Histological analysis of in vitro root formation and the root-shoot union. N. Z. J. For. Sci. 16: 289-296.

Pierik RLM (1987) In Vitro Culture of Higher Plants. Martinus Nijhoff, Dordrecht, Netherlands.

Poissonnier M, Franclet A, Dumant MJ & Gautry JY (1980) Enracinement de tigelles in vitro de *Sequoia sempervirens*. Annales AFOCEL, 1980, pp 231-254.

Preece JE & Sutter EG (1991) Acclimation of micropropagated plants to the greenhouse and field, In: Debergh P.C., and Zimmerman, R.H. (eds). Micropropagation, technology and application. pp. 71-93. Kluwer Academic Publishers, Dordrecht, Netherlands.

Pressey R (1990) Anions activate the oxidation of indoleacetic acid by peroxidases from tomato and other sources. Plant Physiol. 93: 798-804.

Quoirin M, Boxus P & Gaspar T (1974) Root initiation and isoperoxidases of stem tip cuttings from mature *Prunus* Plant Physiol. Vég. 12: 165-171.

Ranjit M, Kester DE & Polito VS (1988) Micropropagation of cherry rootstocks: III. Correlations between anatomical and physiological parameters and root initiation. J. Amer. Soc. Hort. Sci. 113: 155-159.

Read PE & Fellman CD (1985) Accelerating acclimation of in vitro propagated woody ornamentals. Acta Hort. 166: 15-20.

Reed JW, Nagpal P, Poole DS, Furuya M & Chory J (1993) Mutations in the gene for the red/far-red light receptor phytochrome B alter cell elongation and physiological responses throughout *Arabidopsis* development. Plant Cell 5: 147-57.

Robert ML, Herrera JL, Contreras F & Cooper KN (1988) *In vitro* propagation of *Agave fourcroydes* Lem (Henequen). Plant Cell Tiss. Org. Cult. 8: 37-48.

Rodriguez R (1982) Multiple shoot-bud formation and plantlet regeneration on *Castanea sativa* Mill. seeds in culture. Plant Cell Rep. 1: 161-164.

Rumary C & Thorpe TA (1984) Plantlet formation in black and white spruce. I. *In vitro* techniques. Can. J. For Res. 14: 10-16.

Sachs T (1981) The control of patterned differentiation of vascular tissues. Adv. Bot. Res. 9: 151-262.

Salisbury FB & Ross CW (eds) (1985) Plant Physiology, 3rd ed, Wadsworth, Belmont

Samartin A, Vieitez AM & Vieitez E (1986) Rooting of tissue cultured camellias. J. Hort. Sci. 61: 113-120.

Sandison S (1934) The rooting of cuttings of *Lonicera japonica*. A preliminary account. New Phytol. 33: 211-217.

Sangwan RS & Harada H (1975) Chemical regulation of callus growth, organogenesis and plant regeneration in *Antirrhinum majus* tissue and cell cultures. J. Exp. Bot. 26: 868-882.

Satoo S (1956) Anatomical studies on the rooting of cuttings in coniferous species. Bull. Tokyo Univ. For. 51: 109-158.

Scheres Ben, McKhann H & Berg C (1996) Roots redefined: Anatomical and genetic anaylsis of root development. Plant Physiol. 111:959-964.

Schiefelbein JW, Shipley A & Rowse P (1992) Calcium influx at the tip of growing root hair cells of *Arabidopsis thaliana*. Planta 187: 455-59.

Schiefelbein JW & Somerville (1990) Genetic control of root hair development in *Arabidopsis thaliana*. Plant Cell 2: 235-243.

Schmidt E (1956/57) Anatomische Untersuchung uber das Vorkommen von Wurzelanlagen in verscheidenen Internodien von *Pisum sativum*. Flora 144: 151-153.

Sharma KK, Bhojwani SS & Thorpe TA (1991) The role of cotyledonary tissue in the differentiation of shoots and roots from cotyledon explants of *Brassica juncea* (L.) Czern. Plant Cell Tiss. Org. Cult. 24: 55-59.

Sharma VK & Kothari SL (1993) High frequency plant regeneration in tissue cultures of *Glycine clomdestina* - a wild relative of soybean. Phytomorphology 43: 29-34.

Skolmen RG & Mapes MO (1978) Aftercare procedures required for field survival of tissue culture propagated *Acacia Koa*. Comb. Proc. Intern. Plant Prop. Soc. 28: 156-164.

Skoog F & Miller CO (1957) Chemical regulation of growth and organ formation in plant tissues cultured *in vitro*. Symp. Soc. Exp. Biol. 11:118-131.

Smith DL & Fedoroff (1995) LRP1, a gene expression in lateral and adventitious root primordia of *Arabidopsis*. Plant Cell 7: 735-745.

Smith MAL, Palta JP & McCown BH (1986) Comparative anatomy and physiology of microcultured, seedling and greenhouse-grown Asian white birch. J. Amer. Soc. Hort. Sci. 111: 437-442.

Soh W-Y, Choi PS & Cho DY (1998) Effects of cytokinin on root formation in callus cultures of *Vigna unguiculata* (L.) Walp. In Vitro Cell. Dev. Biol. p, 34: 189-195.

Sriskandarajah S, Mullins MG & Nair Y (1982) Induction of adventitious rooting *in vitro* in difficult-to-propagate cultivars of apple. Plant Sci. Lett. 24: 1-9.

Stangler BB (1956) Origin and development of adventitious root in stem cuttings of chrysanthemum, carnation, and rose. Cornell Univ. Agri. Station Mem. 342: 1-4.

Steward FC, Mapes M & Mears K (1958) Growth and organized development of cultured cells. II. Organization in cultures grown from freely suspended cells. Amer. J. Bot. 45: 705-708.

Stinemetz CL (1995) Transport of ^{3}H-IAA label in gravistimulated primary roots of maize. Plant Growth Regul. 16: 83-92.

Su W & Howell SH (1992) A single genetic locus, *Ckr1*, defines *Arabidopsis* mutants in which root is resistant to low concentrations of cytokinin. Plant Physiol. 99: 1569-1574.

Swingle CF (1927) Burr knot fromation in relation to the vascular system of the apple stem. J. Agri. Res. 34: 533-544.

Tepper HB & Mante S (1990) The mature dicot cotyledon as an organogenic structure. Phytomorph. 40: 163-158.

Thorpe TA (1980) Organogensis *in vitro:* Sturctural, physiological, and biochemical aspects. Intern. Rev. Cytol. supplement 11A. 71-111.

Thorpe TA (1984) Clonal propagation of conifers. Proc. Intern. Symp. In Vitro Prop. Forest Tree Species, Bologna, 1984, pp. 35-50.

Toivonen PMA (1985) The development of a photophysiological assessment system for white spruce (*Picea glauca* (Moench) Voss) seedlings and micropropagated plantlets. Ph.D. Thesis, Simon Fraser University, Burnaby.

Torrey JG & Shigemura Y (1957) Growth and controlled morphogenesis in pea root callus tissue grown in liquid media. Amer. J. Bot. 44: 334-344.

Torrigiani P, Altamura MM, Capitani F, Serafini-Fracassini D & Bagni N (1989) *De nove* root formation in thin cell layers of tobacco: changes in free and bound polyamines. Physiol. Plant. 77: 294-310.
Tran Thanh Van K (1973) *In vitro* and *de nove* flower, bud, root and callus differentiation from excised epidermal tissue. Nature 246: 44-45.
Tran Thanh Van K (1981) Control of morphogenesis in *in vitro* cultures. Ann. Rev. Plant Physiol. 32: 291-311.
Tran Thanh Van K & Gendy CA (1993) Relation between some cytological, biochemical molecular markers and plant morphogenesis. In : Roubelakis-Angelakis, K.A., and Tran Thanh Van, K. (eds). Morphogenesis in Plants. pp. 39-54. Plenum Press, New York.
Van Staden J & Harty AR (1988) Cytokinins and adventitious root formation. In: Davis, T., Haissig, B.E., Sankhla, N. (ed). Adventitious Root Formation in Cuttings. pp. 185-201. Dioscorides Press, Portland, Oregon.
Van der Krieken WM, Breteler H, Marcel H, Visser M & Jordi W (1992) Effect of light and riboflavin on indolebutyric acid-induced root formation on apple *in vitro*. Physiol. Plant. 85: 589-594.
Van der Krieken WM, Breteler H, Marcel H, Visser M & Dimitra M (1993) The role of conversion of IBA into IAA on root regeneration in apple: introduction of a test system. Plant Cell Rep. 12: 203-206.
Van der Lek HAA (1925) Root development in woody cuttings. Mededelingen Land-bouwhogeschool Wageningen. 28: 211-230.
Vieitez AM, Vieitez L & Ballester A (1981) *In vitro* chestnut regeneration anatomical and chemical changes during the rooting process. Colloq. Int. sur la culture in vitro des assences forestières. IUFRO, Fountainebleau, pp. 149-152 .
Vieitez AM & Vieitez ML (1983) Secuencia de cambios anatomicos durante la rizogenesis in vitro del castano. Phyton 43: 185-191.
Von Arnolds S (1982) Factors influencing formation, development and rooting of adventitious shoots from embryos of *Picea abies* (l.) Karst. Plant Sci. Lett. 27: 275-287.
Walker KA, Wendeln ML & Jaworski EG (1979) Organogenesis in callus tissue of *Medicago sativa*. The temporal separation of induction processes from differentitation processes. Plant Sci. Lett. 16: 23-30.
Warnick DA (1992) Developmental Biology of Rhizogenesis *in vitro* in *Convolvulus arvensis*. M. S. thesis, San Jose State Univ., Dept. Biol. Sci., San Jose.
Welander M (1983) *In vitro* rooting of the apple rootstock M26 in adult and juvenile growth phases and acclimatization of the plantlets. Physiol. Plant. 58: 231-238.
Wernicke W, Potrykus I & Thomas E (1982) Morphogenesis from cultured leaf tissue of *Sorghum bicolor* - The morphogenetic pathways. Protoplasma 111: 53-62.
White J & Lovell PH (1984a) Anatomical changes which occur in cuttings of *Agathis australis* (D.Don) Lindl 1. Wounding responses. Ann. Bot. 54:621-632 .
White J & Lovell PH (1984b) Anatomical changes which occur in cuttings of *Agathis australis* (D.Don) Lindl 2. The initiation of root primordia and early root development. Ann. Bot. 54: 633-646.
Wiesman Z, Riov J & Epstein E (1988) Comparison of movement and metabolism of indole-3-acetic acid and indole-3-butyric acid in mung bean cuttings. Physiol. Plant. 74: 556-560.
Wilcox H (1955) Regeneration of injured root systems in noble fir. Bot. Gaz. 116:221-234.
Wilson AK, Pickett FB, Turner JC & Estelle MA (1990) A dominant mutation in *Arabidopsis* confers resistance to auxin, ethylene and abscissic acid. Mol.Gen Genet. 222: 377-433.
Wilson PJ & Van Staden J (1990) Rhizocauline, rooting co-factors and the concept of promoters and inhibitors of adventitious rooting - a review. Ann. Bot. 66: 479-490.

Xing Z, Satchell MF, Powell WA & Maynard CA (1997) Micropropagation of American chestnut: increasing rooting rate and preventing shoot-tip necrosis. In Vitro Cell. Dev. Biol. p. 33: 43-48.

Yoshizawa N, Shimizu H, Wakita Y, Yokota S & Idei T (1994) Formation of adventitious roots from callus cultures of taranoki (*Aralia elata* Seen.). Bull. Utsunomiya Univ. For. 30: 19-26.

Ziv M (1986) *In vitro* hardening and acclimatization of tissue culture plants, In: Withers, L.A. and Alderson, P.G. (eds). Plant Tissue Culture and its Agricultural Applications. pp. 187-196. Butterworths, London.

Chapter 7

SHOOT MORPHOGENESIS: STRUCTURE, PHYSIOLOGY, BIOCHEMISTRY AND MOLECULAR BIOLOGY

Richard W. JOY IV[1] and Trevor A. THORPE[2]
[1]Plant Biotechnology Institute, National Research Council of Canada, Saskatoon, SK S7N 0W9, Canada and [2] Plant Physiology Research Group, Department of Biological Sciences, University of Calgary, Calgary, AB T2N 1N4, Canada

1 INTRODUCTION

The process of organogenesis is considered to be complex, involving multiple internal and external factors. Thus, the predictability of organ formation and causal factors surrounding these events are difficult to dissect, especially *in planta*. Elucidation of the temporal and spatial patterns of primordium formation have been pursued using a number of techniques such as microsurgery, inhibitors, application of exogenous phytohormones to different areas of the plant at physiological and at pharmacological levels, mutant analysis (e.g., *Arabidopsis*), plant tissue culture, and molecular biology (Thorpe, 1980; 1993; Lyndon, 1990; 1994; Steeves and Sussex, 1989; Pyke, 1994). The study of *de novo* shoot organogenesis in vitro, which was first reproducibly demonstrated in tobacco by White (1939), has led to much of the understanding we have about organ formation. Organogenesis consists of several stages, including dedifferentiation of the target tissue and subsequent initiation of the various developmental stages which culminate in the production of a fully developed shoot.

This chapter will discuss the process of *de novo* shoot organogenesis, emphasizing the two morphogenic systems which have been extensively investigated in our laboratory, namely, the tobacco and the *Pinus radiata* shoot-forming systems. These represent not only two widely divergent vascular plant groups, an angiosperm and a gymnosperm, but also illustrate indirect and direct organogenesis. The chapter will touch on the anatomical, physiological, biochemical and molecular aspects of shoot organogenesis. In addition, the roles of transcription factors, signal transduction, competency and the cell cycle in *de novo* organ formation will be addressed

2 SHOOT MORPHOGENESIS

In general, the initiation of organized development in plants involves an interaction between the inoculum, the medium and the physical environment. For shoot formation, the manipulation of the process has been discussed in detail elsewhere, due to the importance of organogenesis in clonal propagation (see Murashige, 1974; Thorpe, 1980; 1994; Evans et al., 1981; Tran Thanh Van and Trinh, 1990). Examples of tissue manipulation for shoot formation are presented below for tobacco callus and excised cotyledons of radiata pine.

Tobacco (*Nicotiana tabacum* L. cv W38) callus is generally initiated from stem pith-cambium sections and maintained on MS salts (Murashige and Skoog, 1962) supplemented with White's organics (White, 1943), inositol, IAA (10 μM), kinetin (2 μM) and Difco casamino acids and sucrose (Thorpe and Murashige, 1970). For shoot-forming (SF) tissues, callus is cultured as above but without the casamino acids; with adenine sulfate (0.98 mM), L-tyrosine (0.55 mM), and $NaH_2PO4{\cdot}H_2O$ (1.2 mM) added to the medium; and kinetin at 10 μM. Cultures are usually maintained in darkness at 27°C, but some tissues have been grown in the light with a 16-hour photoperiod and a photon fluence rate of ca. 80 $\mu mol{\cdot}m^{-2}s^{-1}$. Comparative studies are carried out between shoot-forming tobacco callus and two control tissues, grown as non-shoot-forming and repressed shoot-forming (SF medium with 50μM gibberellic acid added) counterparts (Thorpe and Murashige, 1970).

The cotyledon explant system of *Pinus radiata* D. Don (radiata or Monterey pine) consists of cotyledons excised from aseptic, dark-germinated seed approximately one day after radicle emergence (Aitken et al., 1981). Radicle length at excision varies between 0.5 and 2.0 cm, and the cotyledons between 3 and 5 mm. Each radiata pine embryos has seven to ten cotyledons that respond uniformly in culture (Yeung et al., 1981; Villalobos et al., 1985). Excised cotyledons are cultured aseptically on modified Schenk and Hildebrandt medium (Reilly & Washer, 1977) containing 3% sucrose and 25 μM N^6-benzyladenine (BA) in the light (16 hr, ca. 80 $\mu mol{\cdot}m^{-}2{\cdot}s^{-1}$) at 27 ± 1 °C, for 21 days. At the end of 21 days in culture, meristematic tissue is formed along the entire length of the cotyledon in contact with the medium.

For shoot development, cotyledons are transferred to BA-free medium with 2% sucrose. During the initial 21 days in culture, cotyledons cultured in the absence of BA have served as controls. These cotyledons do not form meristematic tissue, but elongate rapidly until about day 10. Also, in a few studies old cotyledons (i.e. cotyledons excised 5 days post-germination) cultured in the presence of BA served as an additional control, as these do not form meristematic tissue.

2.1 Histology of bud formation

Studies with tobacco callus have provided a general picture of some of the events associated with primordium formation (Thorpe, 1982). The induction of organogenesis leads to distinct structural patterns in the lower half of the tissue in contact with the medium. The process is not synchronous, but by day 8 in culture, random cell division activity in organogenic tissue produces visible zones of high mitotic activity. These regions produce meristematic centers or meristemoids containing densely plasmatic cells by days 8-14 of culture (Thorpe and Murashige, 1970; Ross et al., 1973; Maeda and Thorpe, 1979). Meristemoids consist of spherical masses of small isodiametric meristem-like cells with dense cytoplasm and a high nucleo-cytoplasmic ratio. These cells are microvacuolated, they contain a higher content of all organelles, and the nuclei contain more nucleolar material (Ross et al., 1973; Asbell, 1977). Meristemoids are initially apolar but rapidly show directional and divisional activity to form unipolar primordia (Thorpe, 1980), which become visible starting at day 10 in culture. Primordia emerge from the base of the callus as early as day 12 (Ross et al., 1973; Thorpe, 1979).

In radiata pine at the time of excision, numerous mitotic figures could be seen throughout the cotyledon, and all cell divisions were anticlinal (Yeung et al., 1981; Patel and Thorpe, 1984). Large and prominent nuclei were present, the cytoplasm stained densely, and reserve substances were abundant, with numerous starch grains and protein bodies. In non-shoot-forming (NSF) controls, mitotic activity stopped by day 2 in culture. During the first few days in culture, storage products of NSF cotyledons were much reduced. Protein bodies disappeared by day 2, and plastids enlarged with a concomitant reduction of starch. Chloroplasts were fully developed by day 3. Stomatal complexes began to differentiate at day 1 and the stomata were fully developed by day 5 (Douglas et al., 1982; Villalobos et al., 1982; 1985). Small vacuoles, probably derived from protein bodies, gradually fused, resulting in large vacuoles. Intercellular air spaces gradually appeared in the mesophyll along with substantial vacuolation which caused the chloroplasts and cytoplasm to be confined to the periphery of the cells (Villalobos et al., 1982).

Under shoot-forming (SF) conditions, the pattern of cell divisions leading to organized structures in the subepidermal regions of the cotyledonary face in contact with the medium became apparent very early in culture (Villalobos et al., 1982; 1985). Initial random cell division became concentrated by day 3 in the epidermis and subepidermal parenchyma cells in contact with the medium. By day 5, organized 6-8-celled structures, promeristemoids, could be detected. These structures originated from a single subepidermal cell that divided periclinally, then peri- and anticlinally, after the third day in culture (Villalobos et al., 1985). Cells within each promeristemoid were tightly packed, with little or no intercellular space. Flinn et al. (1988) also found 'cell clusters,' which were similar to promeristemoids, in *Pinus strobus* cotyledons. The presence of cell clusters in cotyledons directly correlated with shoot formation. Promeristemoids have also been found in other species, including *Torenia* (Chlyah, 1974), *Picea abies* (Bornman, 1983), *Pinus brutia* (Abdullah and Grace, 1987), and *Pinus eldarica* (Sen et al., 1994).

Ultrastructurally, after two days in culture in the presence of BA (SF conditions), the subepidermal cells of the cotyledons still had a dense cytoplasm with prominent nuclei and lipoidal storage products. Vacuoles started to appear in the cytoplasm. At day 3, just prior to cell division, some of the vacuoles had coalesced to form larger ones. Chloroplasts were prominent and randomly distributed throughout the cytoplasm; starch granules appeared in their stroma as storage lipids were broken down. As the subepidermal cells divided to form promeristemoids, the starch content decreased. Plasmodesmata were easily discerned in walls between daughter cells; these cell walls were thinner than the parent cell wall. In contrast, plasmodesmata were absent between different organized clusters (Villalobos et al., 1982; 1985).

After day 10 of culture, cotyledons had a nodular appearance because of the increase in size of the meristemoids under the epidermis, and by day 21, leaf primordia were evident. The epidermis of the explants did not rupture but became the protoderm of the shoot primordium. There was evidence of cell division in the epidermis, indicating that in radiata pine cotyledons it could adjust to the size increases and changes in shape of the explant (Villalobos et al., 1982; 1985). In other systems, in contrast, e.g., bud-forming cotyledons of Douglas fir, developing shoot primordia rupture the explant epidermis (Kirby and Shalk, 1982).

Changes in ultrastructure and reserve substances in shoot-forming and non-shoot-forming cotyledons of *Pinus strobus* paralleled those found in radiata pine: Flinn et al. (1989) found that rapid lipid reserve depletion was delayed in SF tissues, which also had biphasic starch levels (increasing initially, then decreasing during meristemoid/primordia development).

Vacuolation and nuclear size increased in culture, but were distinctly different between the SF and NSF treatments

There are common features of the developmental sequence for shoot primordium formation in conifers and angiosperms, and the sequence leading to shoot formation is similar in various explants, including mature embryos, cotyledons, and epicotyls (Hicks, 1980, 1994; Thorpe, 1980; Thorpe and Patel, 1986). This sequence includes the formation of first, meristemoids (also referred to as meristematic bud centers or meristematic tissue), then bud primordia, and finally, adventitious shoots with well-organized apical domes and needle primordia. In radiata pine and other species, an important structural feature is the promeristemoid, which gives rise to the meristemoids (Villalobos et al., 1985). The observations made on the origin of the promeristemoid support the idea that organized development *in vitro* begins with changes in a single cell which becomes activated (Thorpe, 1980; Thorpe and Biondi, 1981) to become "mitosis determined" (in the sense of Stebbins, 1965).

2.2 Physiological aspects of organized development

The biggest question with regards to the process of *de novo* organogenesis is what makes a particular cell or a group of cells deviate from the normal pattern of development, and, in response to organ-inducing treatment, dedifferentiate and start to divide and form a new developmental pattern. Steward et al. (1958, 1964) proposed that in order for a cell to respond, it must become physiologically or physically isolated from the surrounding tissue. However, it has also been suggested that it is the interaction of gradients of inducing factors including nutrients and plant growth substances that allow cells to express their developmental potential by producing the correct conditions for morphogenesis within the tissues (Skoog and Tsui, 1948; Skoog and Miller, 1957). The apparent need for polarity in tissues in order for organogenesis to occur (Sinnott, 1960) seems to implicate concentration gradients, rather than the isolation of cells.

There is some evidence that isolation may be necessary or beneficial to organogenesis. The formation of thick cell walls around meristematic cells which become embryogenic has led to the inference that physiological isolation preceded the initiation of organized development in somatic embryogenic cultures (Button et al., 1974; Kohlenbach, 1977). The absence or presence of plasmodesmata is a histological indicator of cell isolation. As mentioned in the previous section, while plasmodesmata are present between cells within promeristemoids, none connect different promeristemoids in *Pinus radiata* (Villalobos et al. 1985). There is other evidence, though, that plasmodesmata are necessary for organogenesis: Artificial reaggregates of

Eucalyptus camaldulensis cells into pseudotissue (without plasmodesmata but with close cell contact) did not form trachearv elements as did the normal callus (Sussex and Clutter, 1967). Also, stem segments of *Torenia fournieri* in which the epidermis had been separated from the cortex did not produce the original pattern of epidermal meristem formation until callus, and presumably plasmodesmata connections, had been formed (Chlyah, 1974). Doubt has also been cast on this necessity of physical isolation by ultrastructural studies of cotton (*in vivo*) (Jensen, 1963) and carrot (*in vitro*) (Halperin and Wetherall, 1964; Street and Withers, 1974) embryogenesis. However, other than these types of data, there is no real evidence for or against the necessity for cell isolation in the differentiation process.

There is, on the other hand, some good evidence for the involvement of gradients of physiologically active substances in organogenesis. Explants in which organogenesis is induced require precise balances of nutrients and growth factors in their culture media. Simple sugars, one component of nutrient media, have been shown to enter the tissue through simple diffusion through the apoplast and possibly through hydrophilic pores (Opekarova and Kotyk, 1973). Most other factors supplied in the medium probably also enter the tissue along concentration, diffusion and/or physiological gradients. Factors coming from the medium must interact with endogenous factors in the tissue.

This type of interaction is evident in the fact that the earliest histological events leading to shoot initiation in tobacco callus, as well as the development of the shoot primordia, occur at discrete distances from the surface of the tissue, always in the lower half in contact with the culture medium (Ross et al., 1973; Maeda and Thorpe, 1979). The formation of shoot primordia in radiata pine cotyledons only in the subepidermal layer on the side in contact with the medium (Villalobos et al., 1985; Yeung et al., 1981) also supports this idea. Furthermore, when shoot-forming tobacco callus was inverted at various times during the culture period to test the existence of interacting gradients, shoots were produced on the lower half, both halves, or the upper half of the callus, depending on when the tissue was inverted (Ross and Thorpe, 1973). The importance of gradients in determining the points of organ primordium initiation has also been supported by studies of xylem formation in lilac callus (Wetmore and Rier, 1963) and bud formation in *Begonia rex* leaf fragments and in *Torenia fournieri* stem segments (Chlyah et al., 1975).

In addition, the timing of the shoot-forming process can be modulated by transfer of tobacco callus between permissive (SF medium) and non-permissive conditions (NSF medium) (Hammersley-Straw and Thorpe, 1987) and by the use of osmotic inhibition by high levels of sucrose (Hammersley-Straw and Thorpe, 1988). The size of the explant may also be critical, as the number of cells indirectly determines the content of endogenous phytohormones and other factors (Bonnett and Torrey, 1965; Okazawa et al., 1967). These findings also support the concept of organogenesis determination by gradients of influencing metabolites.

The requirement of polarity in cells and tissues for the process of differentiation is supported by the observation that in intact plants, as well as in cultured carrot cells (Backs-Husemann and Reinert, 1970), an asymmetrical cell division initiates separate differentiation patterns of daughter cells. There is also good evidence for the existence and influence of polar gradients of metabolites in explants, which cause differential organ generation: In stem or root cuttings, shoot buds are formed at the physiological apices and root buds at the physiological bases of the segments, regardless of the orientation in which the explants are placed during the process. Further evidence is found in *Torenia fournieri* stem segments, which have a specific distribution of cell division centers, increasing in frequency toward the base of the stem and between the vascular bundles (Chlyah et al., 1975). When organs are formed *de novo* from explants or callus, a polarity must also be imposed upon the initially apolar meristemoid by some mechanism which is as yet unknown, after which polarized growth and development to produce shoot or root primordia occurs.

2.3 Phytohormones

Phytohormones and other growth-active substances are included in the culture medium and are important in initiating and regulating organized development. For these to be effective in inducing organized development, there must be a critical balance and/or concentration within the tissue, and at specific loci. The balance required follows the basic ideas of Skoog and Miller (1957) that high cytokinin/auxin ratios promote shoot formation and low cytokinin/auxin ratios promote root formation.

Many plant species, like tobacco, require both exogenous auxin and cytokinin in suitable balances in order for shoot formation to occur (Chandler and Thorpe, 1986). BA or kinetin (0.05-46 μM) are the cytokinins used for 75% of shoot-forming tissues, and the auxins IAA and NAA (0.06-27 μM) are also commonly used (Evans et al., 1981). In some cases, a mixture of two auxins or two cytokinins is more effective in promoting shoot formation than a single one (Evans et al., 1981). In radiata pine (Aitken et al., 1981), as is generally found in conifers (Thorpe et al., 1991), exogenous cytokinin alone is sufficient for shoot induction. That the final endogenous concentrations, and not the exogenous ones, are the important factor in determining the organogenic outcome is demonstrated in the lack of necessity for exogenous auxin in the conifers, the lower cytokinin requirement for shoot formation in many graminaceous species (Evans et al., 1981), and by the requirement of a high exogenous auxin/cytokinin ratio (opposite to the high cytokinin/auxin ratio usually required) for shoot formation in some tissues such as alfalfa callus (Walker et al., 1978). This view is supported by the finding that transformation of cells with modified T-DNA coding for auxin or cytokinin allows for appropriate organogenesis (Schell et al., 1982)

The timing of application of exogenous phytohormones is also important. In some systems, one combination of phytohormones is necessary for the induction of organogenetic tissue, but the tissue must be transferred to medium with a different phytohormonal balance in order for organs to develop (Evans et al., 1981). In addition, in radiata pine, BA must be present during the first three days of culture for any shoot formation to take place, but a few shoots are ultimately formed by exposure to BA for three and seven days. About 50% of cotyledons form shoots with 14 days exposure to BA. Light also seems to be necessary to the shoot-forming process: exposure to light could be delayed at least until day 10, but after 21 days in darkness no shoots formed upon transfer to light (Biondi and Thorpe, 1982; Villalobos et al., 1984a; Villalobos et al., 1985). These data suggest that cytokinin is directly involved in the induction of shoot initials and that both light and cytokinin are required for the development of meristematic tissue and subsequent shoot formation.

Light is not the only factor other than auxins and cytokinins which can affect organogenic systems. In some cases, cytokinin-like (e.g. substituted purines, pyrimidines and ureas) or auxin-like substances can substitute for these phytohormones in inducing shoot formation (Thorpe, 1980; Chandler and Thorpe, 1986). Other metabolites including some nucleotide bases, amino acids, phenolic acids and anti-auxins stimulate organogenesis in some species (see Thorpe, 1980; Chandler and Thorpe, 1986). These substances may act by altering the endogenous levels of phytohormones. Other plant growth regulators, for example, gibberellins and abscisic acid, may also be implicated in organogenesis, but their effects vary in different species (Thorpe, 1980), as do the effects of substances such as polyamines (Bagni and Biondi, 1987; Torrigiani et al., 1987; 1989), oligosaccharines (Tran Thanh Van et al., 1985) and the primary cell wall (Fry, 1990).

In radiata pine, the addition of phytohormones or growth regulators other than cytokinin to the medium during the first 21 days in culture tended to reduce cytokinin-induced meristematic tissue formation and to promote callus production (Biondi and Thorpe, 1982). The effect was more pronounced with increasing concentration (in the range of 10^{-8} to 10^{-4} M) of the test compound in the medium. The auxins naphthalenacetic acid and

2,4-dichlorophenoxyacetic acid (but not indolebutyric acid), abscisic acid, and cAMP all caused callus formation. Transfer of cotyledons to BA-free medium after day 21 led to additional callus proliferation, reducing the number of shoots ultimately formed. The substances tested that were least inhibitory to shoot formation were the growth retardants CCC and AMO-1618, the aromatic amino acids, phenylalanine (Phe) and tyrosine (Tyr), and gibberellic acid (GA_3) (Biondi and Thorpe, 1982). GA_3 caused some reduction in meristematic tissue formation, when the explants were exposed to it between days 0-4 and 0-7, but not if they were exposed for longer periods or after day 4.

Exogenous GA_3 has also been shown to be inhibitory, especially during shoot initiation, in the tobacco shoot-forming system (Murashige, 1961; Thorpe and Meier, 1973). GA_3 causes inhibition of shoot formation, as well as decreased starch synthesis. However, tobacco callus contains gibberellin (GA)-like substances and can metabolize exogenous GAs (Lance et al., 1976a,b). Furthermore, the levels and the spectrum of endogenous GA-like substances change during shoot formation. Thus, GAs appear to be involved in normal tissue growth and differentiation. Tobacco callus synthesizes enough for the organogenetic process; inhibition of organ formation by exogenous GAs results from the supraoptimum levels imposed on the cells. These phenomena suggest that a stimulation of organogenesis in vitro by GAs may be indicative of low endogenous GA levels (Negrutiu et al., 1978, Thorpe, 1978).

In *Digitalis obscura* cultures, morphogenesis, whether promoted by auxins and/or cytokinins, was variably influenced by GA_3: organogenesis was inhibited but embryo development was promoted (Segura and Perez-Bermudez, 1993). This does not mean that the endogenously produced phytohormones play no role in bud induction; rather, their levels may vary over time. This hypothesis is supported by preliminary data with radiata pine which indicate that the contents of endogenous indoleacetic acid and abscisic acid change during bud induction in excised cotyledons (Macey, Reid and Thorpe, unpubl). Collectively, these studies indicate that exogenous growth regulators influence endogenous concentrations, which, by virtue of their function, control morphogenic directions.

The importance of ethylene in *de novo* organized development in both tobacco and radiata pine has been shown. In shoot-forming tobacco callus, less endogenous ethylene was produced than in non-shoot-forming tissue cultured in the light (16 h photoperiod) or the dark, although light also had an inhibitory effect on ethylene production by the tissues. In shoot-forming tissue more ethylene was produced early in culture than later; this neogenic biosynthesis was likely due, in part, to wounding which is necessary to initiate the culture. Furthermore, increased endogenous (by addition of ACC or Etherel) or exogenous ethylene added early in culture (days 0-5) inhibited organogenesis, but later (days 5-10) exogenous ethylene or increased endogenous ethylene production speeded up primordium formation (Huxter et al., 1981). Thus, it is conceivable that if the endogenous ethylene production could be reduced early in culture and stimulated during meristemoid and primordium formation, these structures would be produced earlier and proceed faster and more synchronously to shoot development than normally occurs. The changes associated with ethylene production and action are apparently independent of auxin action, as judged by changes in specific cathodic isoperoxidases (Thorpe and Gaspar, 1978, Thorpe et al., 1978).

In radiata pine, the role of ethylene and its interaction with CO_2 in bud induction in excised cotyledons was examined (Kumar et al., 1987). Shoot-forming cotyledons were found to produce considerable amounts of C_2H_4 and CO_2. The highest number of buds per explant was obtained when the flasks had accumulated about 5 to 8 $\mu l\ l^{-1}$ of C_2H_4 and about 10% CO_2 in the headspace during the first 15 days in culture. CO_2 and C_2H_4 continued to accumulate in bud-forming cultures between days 15 and 21, after the key events leading to bud primordia had occurred, but the removal of these gases at that time had no effect on morphogenesis (Kumar et al., 1987). When either one of the gases was absorbed from the atmosphere, differentiation was adversely affected, and if both gases were eliminated from the flasks, differentiation was completely inhibited, showing that they influenced morphogenesis even after they were released from the tissue into the culture vessel. These studies showed that the effects of C_2H_4 and CO_2 may be synergistic and necessary in order for the cytokinin (BA) in the medium to bring about the switch in morphogenesis from the normal maturation of the cotyledons to shoot bud differentiation.

Since the first ten days in culture are a period of intense, localized cell division leading to the formation of meristematic domes (Villalobos et al. 1985, Yeung et al. 1981), the stimulatory action of CO_2 and C_2H_4 on morphogenesis could be mediated through modification of the cell division process. This is further supported by an earlier study which indicated that C_2H_4 can bring about partial synchrony in cell suspension cultures (Constabel et al., 1977).

If excessive amounts of these two gases were allowed to accumulate within the flasks after the first 15 days in culture, the shoot buds dedifferentiated, the cotyledons turned brown at the cut ends, and the buds became unable to elongate and form normal shoots. Browning of the cut ends of the cotyledons in serum-capped flasks kept under a continuous flow of $C_2H_4 + O_2 + N_2$ indicated that this reversal of the organogenetic process may have been a result of the high concentration of C_2H_4 to which the cotyledons were exposed (Kumar et al., 1987).

From the known interaction of CO_2 and C_2H_4, three possible roles of CO_2 can be proposed First, in the early stages it might be acting primarily to enhance the biosynthesis of C_2H_4; second, at later stages (beyond day 10 or 15) it might be important in antagonizing the action of C_2H_4; and third, CO_2 might have an independent role in metabolism, which is essential for changing the morphogenic response of the tissues to phytohormones. It would, for example, increase the nonphotosynthetic carbon fixation, and there is evidence indicating that the activity of phosphoenolpyruvate carboxylase is high during the process of differentiation both in radiata pine cotyledons (Kumar et al., 1988) and in shoot-forming tobacco callus (Plumb-Dhindsa et al., 1979).

There is now strong evidence that ethylene plays a key role in *de novo* shoot formation (Chi et al., 1991; Pua and Chi, 1993). Using the Indian mustard (*Brassica juncea*) regeneration system, Pua and Lee (1995) produced transgenic plants containing antisense ACC oxidase (ACC oxidase, the enzyme which catalyzes ACC to ethylene is considered the key regulatory enzyme in ethylene biogenesis). These transgenic plants with lower C_2H_4 production were found to have a substantial increase in shoot regenerability which was reversed upon treatment with 2-chloro-ethylphosphonic acid.

Designating a specific role for ethylene in growth and organized development is difficult since this phytohormone is gaseous and diffuses easily through the tissue. Furthermore, because endogenous ethylene is measured after it has passed through the cells, and could affect them prior to being released, there is no direct method of separating cause and effect relationships. Although most of the ethylene released is a by-product of general metabolism, the initial endogenous amount produced could act as a stimulant to growth and differentiation (Huxter et al., 1979). In addition, it may be argued that in the study by Pua and Lee (1995), the down regulation of C_2H_4 biosynthesis may have caused an upregulation of another pathway (i.e. polyamine biosynthesis, since the pathways are connected) whose products could be the causal agents to increase morphogenic capacity. Notwithstanding these difficulties, additional investigations into the regulatory roles of the ethylene biosynthetic pathway must be undertaken to further refine our understanding of this pathway during *de novo* organogenesis.

2.4 Metabolism and organogenesis

Histo- and cyto-chemical approaches, as well as regular biochemical techniques, have been used to glean some data on the metabolic aspects of differentiation. Most of these studies are fragmentary, since a thematic approach to examining basic aspects of differentiation generally has not been undertaken. Nevertheless, direct and indirect evidence has shown that DNA, RNA, and protein synthesis occur during, and are probably necessary for, organ formation (see Thorpe, 1980; Thorpe and Biondi, 1981). For example, in tobacco callus more intense histochemical staining for RNA and protein, but not DNA, was found in shoot-forming tissue than in non-shoot-forming tissue (Thorpe and Murashige, 1970).

In radiata pine, precursor incorporation of [^{3}H]-uridine into RNA, [^{3}H]-leucine into protein, and [^{3}H]-thymidine into DNA in the epidermal and subepidermal cells from the cotyledonary face that was in contact with the medium during the first five days in culture has been examined (Villalobos et al., 1984b). From the time of excision (day 0) to day 2 of culture, incorporation of the precursors was randomly distributed, whether the cotyledons were cultured in the presence (SF) or absence (NSF) of BA. By days 3 to 5 in SF cotyledons, labeling became concentrated in the epidermal and subepidermal cells, while NSF cotyledons showed very little incorporation. Cytoplasmic (plastidic) and nuclear incorporation of [^{3}H]-thymidine were both evident, but after day 2, labelled nuclei were only found in SF cotyledons.

Changes in the rates of macromolecular synthesis preceded the histological localization of labeling patterns and the subsequent elongation/maturation of the NSF cotyledons and the differentiation of the SF tissues. The DNA synthetic rates decreased over the first 24 hours of culture in both treatments, followed by a secondary rise in synthetic rate by day 3. The rates of RNA and protein synthesis during the first 24 hr in culture increased in NSF but decreased in SF cotyledons (Villalobos et al., 1984b). After 24 hr the rate also decreased in the NSF cotyledons, so that by day 3 both BA-free and BA-treated cotyledons exhibited the same rate of RNA and protein synthesis.

The methods used above do not allow a determination of the type of RNA that was being synthesized, or of how much of the RNA was being processed and translated. Although more RNA and protein were being synthesized in the elongating cotyledons, qualitative differences in the type of RNA and protein being produced could take place in bud-forming cells. Some evidence is available to indicate that specific proteins are produced during *de novo* shoot formation (Syono, 1965; Hasegawa et al., 1979;

Yasuda et al., 1980; Stabel et al., 1990; Campbell et al., 1992; Thompson and Thorpe, 1997).

A histochemical analysis of the process of *de novo* bud formation in the excised cotyledons of radiata pine showed uniform localization of activity of various enzymes in the original explants and in NSF cotyledons; the intensity of staining in this tissue decreased over time (Patel and Thorpe, 1984). In the cotyledons cultured in the presence of BA, lipase activity was confined to the shoot-forming layers during the initial stages of shoot formation but during the later stages was detected in cells underlying the meristematic tissue. Increased staining intensities for acid phosphatases, adenosine triphosphatase, succinate dehydrogenase, and peroxidase were also found in shoot-forming regions of the cotyledons (Patel and Thorpe, 1984). The above findings suggest that the newly synthesized proteins in the shoot-forming radiata pine cotyledons could be enzymatic, related at least in part to energy metabolism (Villalobos et al., 1984b).

2.5 Carbon metabolism and utilization

The accumulation of starch in shoot-forming tobacco callus is a prominent feature of the organogenic process. The peak of starch accumulation occurred just prior to the formation of meristemoids (Thorpe and Murashige, 1968; Thorpe and Murashige, 1970; Thorpe and Meier, 1972). Activities of enzymes involved in starch metabolism (starch synthetase, Q-enzyme, R-enzyme, phosphorylase, α-amylase, and maltase) indicated that the starch accumulation resulted from increased synthetic activity, while the reduction during meristemoid and primordium formation involved enhanced rates of degradation (Thorpe and Meier, 1974). Also, both isozymes of α-amylase were enhanced in SF tissue (Leung and Thorpe, 1985).

Even though starch content was correlated to enzyme activities, starch which was synthesized during the different stages of organogenesis was not utilized immediately in *de novo* shoot formation. Hence, turnover rates were fastest during meristemoid and primordia formation (Thorpe et al., 1986). A continuous supply of free sugars from the medium was also required for shoot formation (Thorpe, 1974). While there was no difference in the uptake of ^{14}C-sucrose into the dark-grown shoot-forming and non-shoot-forming tobacco callus (Thorpe and Brown, unpubl.), there was a steady and linear incorporation of ^{14}C into starch in the former (Thorpe et al., 1986).

The addition of ascorbic acid to the cultures increased their starch content and enhanced organogenesis in tobacco callus (both young and old) (Joy et al., 1988). This indicates a probable need for a reductive environment for optimal shoot formation. Although there are quite a number of reports that either a reductive or oxidative environment is prevalent in proliferative and developing tissues, no real consensus can be drawn from the literature at present (cf. Allen, 1995; Dey and Kar, 1995 and references therein). It may be that organogenesis and embryogenesis have different antioxidant requirements (whether this be concentration dependent or reductive vs. oxidative environments), since work with carrot somatic embryos showed that ascorbic acid inhibited development (Earnshaw and Johnson, 1987).

Interestingly, it has recently been found that ascorbic acid affects the onset of cell proliferation in pea roots by enhancing the G_0 - G_1 transition at germination (Citterio et al., 1994). In addition, Gilissen et al. (1994) have shown that cell competence for regeneration and division is dependent on cellular location, ploidy and cell cycle phase in tobacco explants. Therefore, it is probable that in tobacco, the addition of ascorbic acid affects the ability of quiescent cells to proceed through the cell cycle (perhaps faster) which in turn produces a larger population of morphogenically competent cells.

Increased rates of respiration found in shoot-forming tobacco tissues indicate that the process probably has a high energy requirement (Thorpe and Meier, 1972; Ross and Thorpe, 1973). The mobilization of stored starch during organ formation may serve as a readily available source of energy for the organogenetic process. Shoot-forming tobacco callus had higher activities of enzymes of the Embden-Meyerhof glycolytic and the pentose phosphate pathways, and higher ^{14}C-glucose oxidation rates than NSF callus (Thorpe and Laishley, 1973). Shoot-forming tissues also had higher levels of total adenosine phosphates and NAD^+ and a lower energy charge (Brown and Thorpe, 1980a).

Studies using isolated mitochondria showed no differences in the capacity of mitochondrial enzymes between isolated shoot-forming and non-shoot-forming mitochondria, however, there was a trend toward higher and more efficient respiration in the former (Brown and Thorpe, 1982). A decrease in the relative level of the alternate cyanide-insensitive pathway and an increase in respiration via the normal cyanide-sensitive cytochrome pathway, which produces more ATP per molecule oxidized, was found during meristemoid and primordium formation in tobacco. In radiata pine, however, the rates of respiration, as well as the use of the alternate pathway, peaked at day 5 in the shoot-forming tissue (Winkler et al., 1994). SF tissue had a greater capacity for this pathway than did NSF tissue, even at stages in culture when relatively low activity was observed. In NSF tissue the activity of the alternate pathway was low early in culture and no activity was found after day 3. These findings suggest that the more efficient cytochrome pathway is fully engaged during meristemoid formation and that additional ATP and carbon skeletons are obtained by the less efficient alternate pathway.

The enhancement of the pentose phosphate pathway (PPP) may be used in the production of reducing power (NADPH) for reductive biosynthesis during shoot formation in tobacco callus. This role was supported by $^{14}CO_2$ incorporation into acid-stable compounds and heightened activities of enzymes involved in malate metabolism in dark-grown SF cultures (Plumb-Dhindsa et al., 1979). Thus, malate metabolism seems to be important in the production of NADPH *via* the $NADP^-$linked malate enzyme. Measurement of the NADPH and $NADP^+$ pools indicated a faster decline, more complete utilization of NADPH and a greater build up of $NADP^+$ levels in SF tissue, than in NSF tissue (Brown and Thorpe, 1980a). This variation in the NADPH/NADP ratios could be involved in *in vivo* regulation of the PPP. While the ratio dropped to < 0.05 in shoot-forming callus, it remained relatively constant (0.75-1.75) in non-shoot-forming tissue. Glucose-6-P dehydrogenase (G6PD) is competitively inhibited by NADPH and is completely inhibited by a ratio of > 2.0 (Thompson and Thorpe, 1980). Since the capacity of G6PD doubles during shoot formation to become 2 to 3 times that of 6-phosphogluconate dehydrogenase (which requires a ratio of >5 to be fully inhibited), it appears that regulation of the pathway could occur by coarse control of the amount of G6PD present and by fine control due to variation in the NADPH/NADP ratios.

Sucrose also fulfills an osmotic requirement in tobacco callus, since one-third of the medium sucrose could be replaced by mannitol (Brown et al., 1979). This sugar alcohol is considered non-metabolizable in most plant species and, therefore, could act in a colligative fashion intracellularly . Tobacco cells in suspension absorb less than 5% of the fed mannitol in 14 hours of culture. Of the mannitol taken up, 50% was metabolized: ca. 4% was released as CO_2, ca. 30% was converted into an unidentified low molecular weight sugar or other neutral compounds, and ca. 15% was converted into basic, acidic and insoluble compounds (Thompson et al., 1986). Thus, most of the osmotic effect of mannitol apparently takes place from the apoplast; i.e., from mannitol in the cell wall, intercellular spaces, etc. Much of the sucrose placed in the medium also remains outside of the cells, and whether or not the osmotic agent is outside the plasmalemma or inside the tonoplast, its osmotic effect on organelles in the cytoplasm is likely to be the same.

Shoot-forming tobacco callus grown in a medium with the same water potential as non-shoot-forming tobacco callus maintained greater (i.e., more negative) water and osmotic potentials, as well as greater (more positive) pressure potentials than non-shoot-forming tissue (Brown and Thorpe, 1980b). These differences were observed by day 2 in culture (prior to any visible histological changes in the tissue), peaked at day 6 (at the time of the first visible changes leading to organized development) and were maintained throughout the culture period. They did not result from differences in the uptake of sucrose from the medium. In addition, fresh weight/dry weight ratios and water content of the tissues were of the same order (Brown and Thorpe, 1980b).

One possible function of the increased osmotic potential of SF tissue is the enhancement of the activity of the mitochondria, perhaps due to changes in membrane properties (Brown and Thorpe, 1980b). When isolated mitochondria were tested at differing water potentials, their capacity to utilize substrate, the coupling of oxidative phosphorylation, and the efficiency of ATP production all increased when the osmotic potential changed from that found in growing tissue to that in the SF tobacco (Brown and Thorpe, 1980b; Thorpe, 1983). Therefore, the early osmotic adjustment in SF tissue could be a mechanism to increase the metabolic capacity of mitochondria to produce the energy required for primordium formation.

Greater water, osmotic, and pressure potentials in shoot-forming tobacco tissue may be maintained, in part, by (a) the accumulation of malate early in culture, (b) the accumulation of free sugars from the medium throughout the culture period, and (c) the degradation products of starch at the time of meristemoid and primordium formation (Thorpe, 1974; Thorpe and Meier, 1974; Plumb-Dhindsa et al., 1979). Further, shoot-forming tobacco callus has been found to have high levels of threonine/serine and proline (Hardy and Thorpe, 1990). Hence, morphogenic systems (as judged by tobacco) may use a variety of metabolites colligatively, with different metabolites contributing osmotically at different stages of the process. The importance of osmotic adjustment to growth and organized development remains to be determined. Biophysical events, however, play as important a role as do biochemical ones in the process.

Excised cotyledons of radiata pine are packed with reserve material, and as indicated previously, these disappear during *de novo* organogenesis. Lipid and free sugars were depleted approximately ten- and six-fold, respectively, and there was also a steady decline in free amino-N levels (Biondi and Thorpe, 1982). However, the protein-N pool remained relatively large during the 21-day culture period. Since lipids were the major polymeric storage substance, and since they declined most rapidly during organogenesis, their fate was followed in greater detail (Douglas et al., 1982).

Fatty acid and sterol analysis indicated that there were both quantitative and qualitative changes in the different classes of lipids. Neutral lipids (triglycerides and steryl/wax esters) decreased and polar lipids (predominately C16:0, C18:1, and C18:2 fatty acids) increased during bud induction. One fate of the fatty acid, glycerol, and sterol components was to produce new membranes, as the most abundant fatty acids found are the major components of phospholipids in many plant species. The C18:3 fatty acid, a glycolipid which is particularly abundant in chloroplast lamellae, increased markedly. This increase probably reflected plastid biogenesis, which was also shown ultrastructurally (Douglas et al., 1982; Villalobos et al., 1982).

The amounts of triglyceride fatty acids used by the cotyledons were far in excess of those needed for the observed increases in polar lipids. This suggests that the excised cotyledons, like germinating seeds, rely heavily upon the stored lipid reserves for energy production, particularly during the first three days in culture when respiration rates are highest (Biondi and Thorpe, 1982). Sterols are structural and functional components of plant cell membranes, and the specific changes observed in the spectrum of free 4-desmethylsterols suggested that the excised cotyledons behaved similar to those of germinating seeds (Douglas et al., 1982). There was a lower content of chlorophylls and carotenoids in the cultured cotyledons than in developing seedlings, probably due to the high level of sucrose (3% w/v) in the medium (Dalton and Street, 1976). Nevertheless, it appears that changes in lipids (quantity and type) and pigments are similar to those that occur in developing seedlings. This would suggest that the observed changes are not directly involved in the initiation of organized development in vitro, but play only an indirect role through energy production, membrane proliferation, etc. (Douglas et al., 1982).

All the above findings are consistent with the hypothesis that the initiation of organized development involves a shift in metabolism, which precedes and is coincident with the process (Thorpe, 1980; Thorpe and Biondi, 1981). To gain greater insight into the nature of this metabolic shift, an in-depth analysis of primary metabolism in the cultured cotyledons of radiata pine, as well as in tobacco callus, during shoot initiation was undertaken. Initial studies involved feeding the tissues with [^{14}C]-glucose, [^{14}C]-acetate, or [^{14}C]-bicarbonate on different days in culture (Obata-Sasamoto et al., 1984; Thorpe and Beaudoin-Eagan, 1984).

When the cotyledons or the callus were fed [^{14}C]-glucose and [^{14}C]-acetate, ^{14}CO2 was produced (no CO_2 measurement was made for [^{14}C]-bicarbonate feeding). Label from these precursors was also incorporated into ethanol-soluble (consisting of lipids, amino and organic acids, and sugars) and, to a lesser extent, ethanol-insoluble fractions. In general, there was a tendency towards a high rate of incorporation of label in elongating cotyledons during the period of rapid elongation (day 3). Also, there was a high rate of incorporation of label from [^{14}C]-glucose or [^{14}C]-acetate in shoot-forming cotyledons at days 10 and 21 (during meristematic tissue and shoot primordium formation) (Obata-Sasamoto et al., 1984). Tobacco also showed enhanced incorporation of label into various metabolites during meristemoid and primordium initiation and development (Thorpe and Beaudoin-Eagan, 1984). In contrast, feeding either the cotyledons or callus with [^{14}C]-bicarbonate led to higher incorporation into the NSF tissues.

Further studies were carried out in which [^{14}C]-glucose was supplied to the cotyledons at different days in culture for 3 hr, followed by a chase of 3 hr with [^{12}C]-glucose (Bender et al., 1987a,b). The most highly labelled metabolites were malate, citrate, glutamate (Glu), glutamine (Gln), and alanine (Ala). At day 0 the labeling of Gln was much higher in the BA-treated cotyledons. This suggests a positive influence of cytokinin on nitrogen incorporation at the time of excision. No labelled citrate was detected at day 0, but by day 3, large amounts were detected (Bender et al., 1987a). Additional studies have indicated that there is a high rate of turnover in citrate at day 1 and that the pool size is very small (Joy et al., 1994). At days 10 and 21, general metabolic patterns were qualitatively the same in SF and NSF cotyledons (Bender et al. 1987b), but metabolism leading to respiration and amino acid synthesis was strongly enhanced in SF cultures. More label was also incorporated into protein in SF than in NSF cotyledons at days 0 and 3.

The radioactivity present in the lipid fraction at days 0 and 3 indicated the synthesis of metabolically stable lipids, since the radioactivity of this fraction increased during the chase period. In contrast, later in culture there was considerable turnover in the lipid fraction, indicating that lipid synthesis for both structural and nonstructural components was taking place (Bender et al., 1987a, b).

The effects of excision, light and cytokinin on carbon metabolism have also been determined during shoot induction in radiata pine cotyledons. Labeling of examined metabolites (TCA metabolites and amino acids) was at least 50% greater in excised cotyledons than in intact seedlings. Although this could mask morphogenetically-related changes at the beginning of culture, light and cytokinin were found to cause an increase in metabolism even in the absence of wounding (Joy and Thorpe, 1990).

Although the findings above are not dramatic, and are in many respects predictable, it is clear that at day 0, during the first 3 hours of excision, BA had an effect on the cotyledonary metabolism (Bender et al., 1987a,b; Joy and Thorpe, 1990; Joy et al., 1994) and, as indicated earlier, major differences in the rate of RNA and protein synthesis were evident by 24 hr (Villalobos et al. 1984b). Thus, not only is BA necessary during the first 24 hours of culture for subsequent bud formation, it seems that the cotyledons must receive the cytokinin signal from the start of culture.

2.6 Nitrogen assimilation and amino acid metabolism

In tobacco, shoot-forming tissue maintained higher levels of total-N, protein-N, and nitrite-N, but similar levels of amino-N, in comparison to non-shoot-forming tissue (Hardy and Thorpe, 1990). Nitrate reductase activity was higher in shoot-forming tissue, but the activity of nitrite reductase was similar in both shoot-forming and non-shoot-forming tissues. In shoot-forming cultures of buds of white spruce, NH_4^+ from the medium was assimilated faster than NO_3^-, and no NO_3^- was taken up in the absence of NH_4^+ (Thorpe et al., 1989). As well, both NH_4^+ and NO_3^- were required for induction of nitrate and nitrite reductase.

Amino acid analyses of three stages of organogenesis in tobacco callus showed that pool sizes changed during culture, and that, in general, shoot-forming tissue had higher levels of amino acids, particularly proline (Pro) and threonine (Thr)/serine (Ser) (Thorpe, 1983). In radiata pine at the time of excision, glutamate (Glu), glutamine (Glu), arginine (Arg), Pro, asparagine (Asn), alanine (Ala), valine (Val) and γ-aminobutyric acid (GABA) were the most abundant in the free amino acid pool (Winkler and Thorpe, unpublished). Lysine (Lys), Arg, Gln and Pro all decreased, while GABA increased, during meristemoid and primordium formation. As well, Asn was higher in shoot-forming than in non-shoot-forming cotyledons throughout the culture period. These changes are thus indicative of differential synthesis and/or utilization of free amino-acids. In cultured white spruce buds, ^{15}N from inorganic N was found in Glu, Pro, Ala, Gln, Arg, Ornithine and GABA. Label was first incorporated into the amide group of Gln and then into the α-NH_4 pool. Based on the pattern of incorporation, it was concluded that the glutamine synthetase/glutamate synthase pathway was operative during culture of the buds (Thorpe et al., 1989).

Glu and Gln were the principle amphoteric metabolites into which label was incorporated when radiata pine cotyledons were fed [^{14}C]-acetate (Joy et al., 1994). The percentage label incorporated into Glu remained similar throughout the culture period for both SF and NSF tissues, although its specific activity was highest during promeristemoid formation. The percentage of label incorporated into Gln was highest during shoot bud initiation (day 10) but in NSF cultures, a continual increase in labeling and specific activity were found. The activity of the nitrogen assimilation enzymes glutamate synthase and glutamine synthetase increased from day 0 to day 21 in both SF and NSF tissues. Glutamate dehydrogenase activity was similar for both treatments up to day 10 but at day 21, activity was higher in NSF tissues than in SF tissues.

As mentioned previously, the pentose phosphate and glycolytic pathways both had enhanced enzyme capacities during shoot formation in tobacco callus. These pathways provide erythrose-4-phosphate and phosphoenol pyruvate, respectively, for aromatic amino acid biosynthesis via the shikimate pathway. Two to ten-fold higher levels of activity of DAHP synthetase (the coupling enzyme), shikimate kinase (rate limiting step in pathway), chorismate mutase and anthranilate synthetase (branch point enzymes) were observed in SF tissues, compared to NSF callus (Beaudoin-Eagan and Thorpe, 1983). Differences were evident by day 6 in culture and reached a maximum between days 12 to 15, coincident with meristemoid and primordium formation.

When the tissues were fed [^{14}C]-glucose, labeled shikimate and quinate were found (Beaudoin-Eagan and Thorpe, 1984). More label was incorporated into quinate than shikimate, and the incorporation into both metabolites was greater in SF tissue than in NSF tissue. Similarly, when the tissues were fed [^{14}C]-shikimate, the end products of the pathway [phenylalanine (Phe), tyrosine (Tyr) and tryptophan (Trp)] were all labeled (Beaudoin-Eagan and Thorpe, 1984; 1985a). The level of incorporation into the aromatic amino acids in SF tissue was higher than NSF tissue. Furthermore, the incorporation into Tyr in SF tissue was greater than into the other two amino acids. These studies indicated that shikimate pathway enhancement involved the turnover of tyrosine which correlated with a higher enzyme capacity of tyrosine ammonia lyase (TAL) (Beaudoin-Eagan and Thorpe, 1984; 1985a,b).

In a rice callus organogenic system, Both TAL and tyrosine aminotransferase (TAT) activities were enhanced in organ-forming cultures; however, phenylalanine ammonia lyase and phenylalanine aminotransferase were not (Kavi Kishor, 1989). This finding is interesting since the tobacco shoot-forming medium contains Tyr for optimum shoot formation, but the rice shoot-forming medium does not. Hence, it is likely that the pathway (and TAL and TAT) may be variably influenced in the different tissues during the organogenic process, but there are similarities in aromatic amino acid metabolism during development.

Studies by Tryon (1956) and Skoog and co-workers (Sargent and Skoog, 1960; Skoog and Montaldi, 1961) have implicated scopoletin and its glucoside, scopolin, in tobacco tissue growth and differentiation. The effectiveness of Tyr and other substituted phenols in stimulating organ formation in tobacco callus was attributed in part to their involvement in lignin synthesis (Lee 1962). Evidence in favor of this view has been contradictory (Dougall and Shimbayashi, 1960; Dougall, 1962), and the later study by Hasegawa et al. (1977) did not significantly increase our understanding of the role of Tyr in organogenesis. They found that [^{14}C]-

tyrosine was incorporated into polymeric substances including proteins (both cytoplasmic and cell wall), polysaccharides, and possibly lignin.

Studies on aspartate (Asp) metabolism in both radiata pine and tobacco have been recently carried out (Konschuh and Thorpe, 1997, unpubl). Asp is involved in the formation of asparagine (Asn) and is the starting amino acid for the biosynthesis of threonine (Thr), isoleucine (Ile), lysine (Lys) and methionine (Met). The activity of aspartate kinase (AspK), the enzyme catalyzing the first committed step in the biosynthesis of these latter amino acids, was similar in shoot-forming and non-shoot-forming tobacco callus, but higher activities were found during shoot primordium formation and development in radiata pine cotyledons than in non-shoot-forming cotyledons. [^{14}C]-Asp was taken up and metabolized in both tobacco and radiata pine, with much of the label being released as $^{14}CO_2$. However, label was also incorporated into amino acids (Asp, Asn, Glu, Gln, and to a lesser extent, GABA and Arg) and organic acids. There was a greater $^{14}CO_2$ release and metabolite labelling in shoot-forming radiata pine cotyledons during bud induction and promeristemoid formation, but no differences were observed in tobacco callus.

Polyamine metabolism in radiata pine cotyledons has also been investigated using [^{14}C]-putrescine and was found not to be significantly different between SF and NSF tissues except for rates of uptake during high mitotic activity at day 3 (Kumar and Thorpe, 1989). ^{14}C from labeled putrescine was also incorporated into GABA, Asp and Glu. Endogenous spermidine was the most predominant polyamine in the cultured cotyledons of radiata pine; it increased by about 60% in SF and 25% in NSF cotyledons within the first 3 days of culture. Biondi et al. (1986; 1988) found that dicyclohexylamine inhibited accumulation and biosynthesis of polyamines as well as *de novo* organogenesis in radiata pine. These data indicate that polyamine metabolism is important for both growth and *de novo* organogenesis.

In *Euphorbia esula,* putrescine did not affect organ formation in full strength B5 medium, but at lower salt concentrations it effectively inhibited both root and shoot formation (Davis and Olson, 1994). Organ formation was also inhibited by both DFMA and DFMO; however, no correlations were found between polyamine levels and organogenic capacity. Gradients of arginine decarboxylase (ADC), ornithine decarboxylase (ODC) and polyamines were found in superficial and deep tissues of tobacco stem (Altamura et al., 1993). ADC and ODC decreased in activity towards the lower portion of the plant as did morphogenic response. All of the above findings point to an important role of polyamine metabolism during organogenesis.

During organ formation the measured capacities, and presumably in vivo acitivities, of many enzymes change. Some of these enzyme changes can be correlated with pool sizes of metabolites, while others cannot. Nevertheless, these reports indicate that many areas of nitrogen metabolism, not unexpectedly, may be involved in organ formation.

2.7 Gene expression

Cyto- and histochemical changes seen during organogenesis, as well as the observed biochemical differences between shoot-forming and non-shoot-forming tissues, are clearly due to differential gene expression. Cytological studies in radiata pine demonstrated that bud formation occurs on the side of the explant in contact with the medium and that under optimal conditions a reproducible developmental time course is maintained (Aitken-Christie et al., 1985; Villalobos et al., 1984a, 1985; Yeung et al., 1981). Histochemical investigations showed increased macromolecular synthesis in areas destined to produce promeristemoids and shoot buds (Patel and Thorpe, 1984; Villalobos, 1984b). Biochemical analyses of *de novo* organogenesis in both radiata pine and tobacco callus have demonstrated enhancement of metabolism in shoot-forming tissues as compared to non-shoot-forming controls (Thorpe, 1983; 1988; 1993). A summary of the changes observed in tobacco callus in relation to shoot formation is presented in Fig. 1.

The results of such investigations all point towards shoot-forming conditions inducing differential gene expression to produce proteins and polypeptides which affect morphogenesis. Protein and polypeptide profiles have been analyzed in several species in attempts to evaluate differential expression during shoot formation. Protein analysis in bud-forming cotyledons of *Pinus ponderosa* showed that a high molecular weight protein was reduced and a low molecular weight protein appeared during organogenesis (Ellis and Judd, 1987). This study, however, used 1D SDS-PAGE which is, at best, a poor distinguishing method for total protein preparations

More recently, in a more definitive polypeptide analysis in tobacco callus, Bertrand-Garcia et al. (1992) compared polypeptide profiles in SF and NSF cultures using 2D SDS-PAGE and autoradiography. A few proteins were novel to either SF or NSF cultures, but most were found in both tissues. Two proteins had elevated levels in NSF tissue and several phosphorylated polypeptides were detected exclusively in either one or the other treatments (Bertrand-Garcia et al., 1992). It should be noted that at the time of subculture, meristemoids were already present in SF tissue. Therefore, this study is not a representation of the induction process but of the development of primordia.

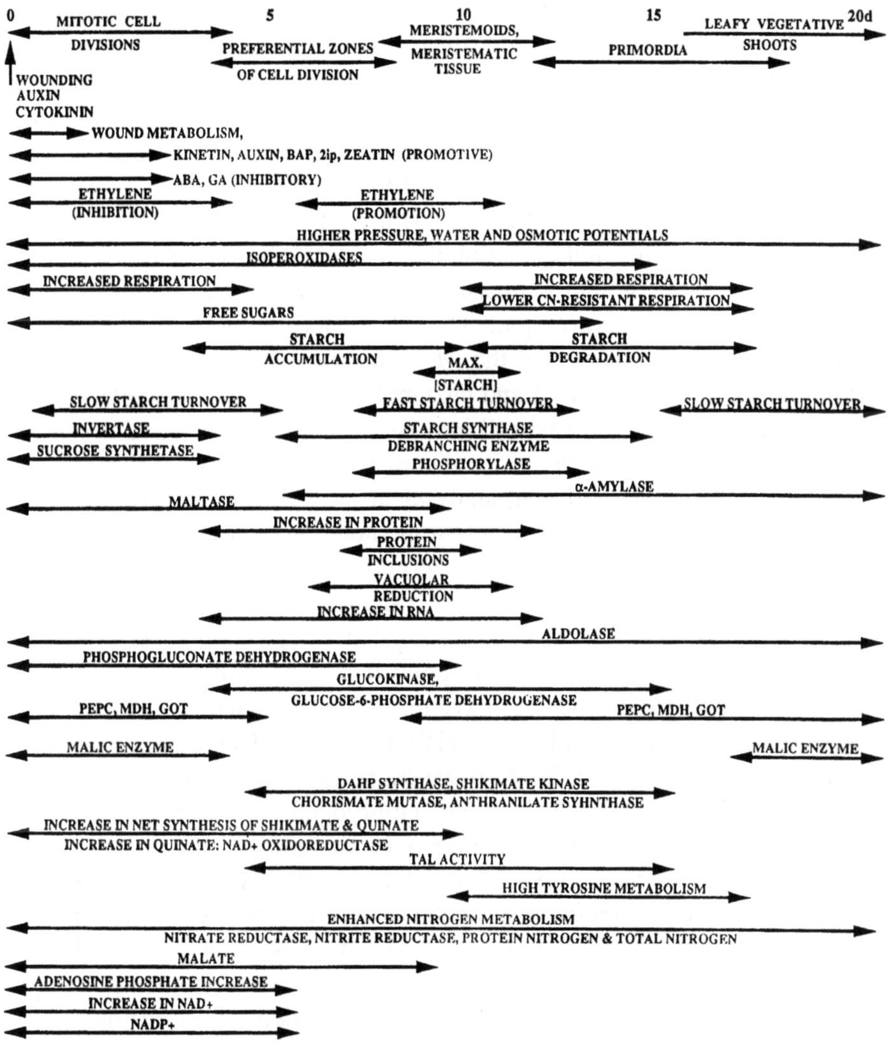

Figure 1. A summary of some sequential events taking place during *de novo* organogenesis in tobacco.

Analysis of polypeptide profiles in the *Pinus radiata* shoot-forming system indicated that at least 54 polypeptides were affected by culture conditions (Thompson and Thorpe, 1997). Seven classes were identified based on differences in cultures and appearance: I) shoot enhanced, II) BA-enhanced, III) shoot suppressed, IV) BA-suppressed, V) BA-maintained, VI) BA-enhanced in early stages, and VII) synthesis, but not accumulation, BA suppressed. The largest class consisted of polypeptides whose synthesis appeared to be specifically repressed in shoot-forming cotyledons. Further work in this area has indicated that one polypeptide was induced early in culture (by 32 hours) and an additional 8 were found to increase and one decrease by 48 hours in culture in *Psuedotsuga menziesii* (Campbell et al., 1992). In *Picea abies*, a large number of polypeptides were also found to be differentially synthesized in bud-forming embryos; however, these were not exclusive to this treatment (Stabel et al., 1990).

The conclusions that can be drawn from the above data are that total protein analyses are highly unlikely to elucidate organ-specific gene expression, as most protein expression at substantial levels is probably due to house-keeping gene expression. This can be seen in most protein analyses as the predominant presence of the *lsu* and *ssu* of Rubisco, chlorophyll a/b binding protein, and structural and enzymatic proteins, for example. As the possible number of proteins is well into 10-20 thousand, protein analysis can realistically only identify the most abundant ones those usually being of a constitutive nature. In addition, analyses at the polypeptide level are crude at best, and there is a need for different methodologies (e.g. mutant analysis, molecular genetic analysis) to help determine what are the causal factors in the initiation and maintenance of *de novo* shoot morphogenesis.

Nevertheless, these studies do show that differential gene expression takes place in tissues undergoing morphogenesis. In addition, specific genes related to morphogenic processes have recently been isolated. Yoshida et al. (1994) have been successful in the cDNA cloning of one gene specific to organogenesis in rice, two for embryogenesis and an additional two which were expressed in organogenesis and embryogenesis but not in calluses. Koornneef et al. (1987) identified and mapped a gene which controlled shoot regeneration in tomato. Also, Yasutani et al. (1994) isolated three temperature sensitive *Arabidopsis* mutants defective in the redifferentiation of shoots *in vitro*. Their analyses lead to the working hypothesis that the genes played different roles at different stages during development.

Hence, within the next few years we should anticipate a large number of morphogenic-specific genes being cloned and characterized within tissue culture systems. Indeed, this is already true of the *in planta* work proceeding with the *Drosophila* of the plant world - *Arabidopsis* (Weigel and Meyerowitz, 1994; and references therein). A large number of homeotic genes have been characterized over the past 5 years which are now being investigated for protein-protein and protein-DNA interactions.

3 CELLULAR AND MOLECULAR ASPECTS DURING ORGANIZED DEVELOPMENT

3.1 Transcription factors and control of organogenesis

Significant progress has been made in the elucidation of genes controlling organogenesis since the isolation of the first one, the 'Knotted-1' (Kn1) mutation in maize, which encodes a homeodomain protein (Hake et al., 1989; Vollbrecht et al., 1991). This pioneering work provided the first evidence that the control of plant development may be due to homeobox gene regulation specifically, and to transcription factors in general. Homeobox genes were initially discovered to regulate morphogenesis in the fruit fly (see Gehring, 1987) and have now been extended to developmental regulatory roles in most organisms (see Gehring et al., 1994, and references therein; Weigel and Meyerowitz, 1994). In plants, genetically tractable species which have been found to contain homeotic mutants include maize (Vollbrecht et al., 1991), rice (Matsuoka et al., 1993), *Arabidopsis* (Yanofsky et al., 1990), *Antirrhinum majus* (Coen et al., 1990) and petunia (Angenent et al., 1995).

Evidence to substantiate the view that transcription factors control plant morphogenesis is now quite abundant. Through the studies of homeotic mutants (mutations which result in the formation of organs in the wrong places) a clearer picture of the genetic control of development is being formed. Homeotic genes encode related DNA-binding proteins which are a class of transcription factors containing conserved regions. Homeobox and also MADS box domains code for invariable regions within their respective proteins which interact with DNA regulatory control regions. Proximal to the MADS- or homeodomain, in most cases, is a dimerization domain. Together, the protein components form homo- and heterodimers to make functional units for transcriptional regulation.

At least three classes of homeotic mutants, which control development and cell fate, have been determined: I) meristem identity genes (responsible for initial induction) which affect the initiation of floral meristems by the primary inflorescence meristem, II) cadastral genes (responsible for spatial regulation) which alter the symmetry of organs, and III) organ identity genes, controlling which organs are formed. Meristem identity genes are positive inducers of organ identity genes and their absence of activity results in partial or complete reversion of flowers to vegetative meristems. Cadastral genes set boundaries for organ identity genes by preventing their ectopic expression.

Currently, the best characterized species in regards to understanding the genetic basis of organogenesis is *Arabidopsis* (Yanofsky et al., 1990; see

Weigel and Meyerowitz, 1994, and references therein). An elegant model has been developed with *Arabidopsis* which can be extended to other species and which explains how various floral organs are formed. The model is called the 'ABC model', with each letter (A,B or C) representing a particular genetic function, controlling organ identity genes in the 4 whorls of a developing flower. In the model, positioning and expression lead to normal development; however, a mutation can lead to the ectopic expression of organs in unusual places and combinations. Thus, in this model, a mutation in 'A' corresponding to one of the genes designated 'AP' for 'apetela', eliminates the presence of sepals and petals and produces floral organs in the arrangement of carpel, stamen, stamen, carpel in the four whorls.

A number of combinations of homeotic mutants affecting the organ identity genes have been studied including double and triple mutants. Of particular interest in these latter studies is the fact that there is an apparent reversion to a vegetative state (Bowman et al., 1991; Weigel and Meyerowitz, 1994). Hence, when organ identity genes are missing, or not expressed, it appears that the vegetative state predominates, thereby producing leaf-like structures.

The above work strongly suggests that the ground state of organogenesis is the vegetative shoot which, upon appropriate signal transduction, produces a differentiated organ, in this case, floral in nature. This concept was originally proposed by Goethe in 1790 (cf. Weigel and Meyerowitz, 1994), who suggested that flowers were essentially specialized leaves.

3.2 Signal Transduction

Signal transduction still remains one of the larger black boxes in plant physiology with much of our current knowledge being phenomenologically-based. Some of the major stumbling blocks that still plague this type of research are: (1) whether certain types of signaling metabolites actually exist in higher plants; (2) difficulties in the analysis of nano- and picomole quantities; (3) whether designated signal metabolites actually produce biological responses; (4) whether the concentration of the signal metabolite is applied within a physiological range; (5) whether the enzymes of the signaling pathway exist in the plant cell; and (6) whether isolated genes and their products act in a way which is consistent with the signal molecule and pathway (i.e., produce an expected response such as a phosphorylation reaction) (Trewavas and Gilroy, 1991; Trewavas and Knight, 1994; Assmann, 1995; Stone and Walker, 1995). Even so, progress is being made, and a few areas relevant to this chapter will be discussed here.

There is strong evidence that calcium-based signal transduction controls various aspects of plant growth and development. Changes in intracellular calcium have been correlated to stomatal closure, pollen tube growth, touch, light, cold shock, fungal elicitors, salt concentration, etc. (Schroeder and Thuleau, 1991; Trewavas and Knight, 1994). Calmodulin, a calcium-binding protein, changes with changes in intracellular calcium. Plant calmodulin has been isolated and characterized in a number of plants and found to be involved in responses (similar to calcium) to wind, rain, touch, tuberization, auxin and light (Jena et al., 1989; Ling and Zielinski, 1989; Thompson et al., 1989; Braam and Davis, 1990; Trewavas and Knight, 1994).

Calcium-based signal transduction is manifested through a number of routes, with possible fluxes coming through calcium channels in the plasma membrane, organelles, and the vacuole (Schroeder and Thuleau, 1991). Through stimulus-response coupling, activation of the phosphatidylinositol pathway may also elevate intracellular calcium levels. Subsequent to calcium release, a cascade of events involving activation of protein kinase C, calmodulin, or protein kinases occurs (Boss, 1989; Trewavas and Gilroy, 1991), ultimately leading to phosphorylation cascades which provide the basis of regulatory control in growth and development (Drayer and Haastert, 1994; Trewavas and Knight, 1994; Stone and Walker, 1995).

Another source of phosphorylation cascades and one of the current focuses in signal transduction research are mitogen-activated protein (MAP) kinases and their roles in various aspects of plant growth and development. One mutant similar to a MAP kinase mutant is the ethylene triple response mutant ctr1 (constitutive triple response), which exhibits the classic response of (1) inhibition of cell elongation, (2) hypocotyl swelling, and (3) elongation of the apical hook. Because overproduction of ethylene was not seen in this mutant, it was found to be part of the ethylene response pathway. The CTR1 gene was found to encode for a putative serine/threonine protein kinase related to the raf protein kinase family (Keiber et al., 1993). As pointed out by Jonak et al. (1994), the CTR1 gene is similar to what would be considered a MAP kinase, containing conserved residues which are necessary for phosphorylation.

Plant homologues of MAP kinase have been isolated from alfalfa (*Medicago sativa*) (Duerr et al., 1993; Jonak et al., 1993), petunia (Decroocq-Ferrant et al., 1995) and *Arabidopsis* (Mizoguchi et al., 1994), for which there is evidence of multigene families with differential expression during the cell cycle and/or in different plant organs. Rapid enhancement of MAP kinase activity by 2,4-D led to the suggestion that auxins may be signal molecules which activate MAP kinase (Mizoguchi et al., 1994). Suzuki and Shinshi (1995) have shown that there is a transient activation of MAP kinase in tobacco cells treated with a fungal elicitor. These findings suggest multiple roles for various signal transduction pathways. Indeed, it appears that MAP kinases are involved in the regulation of defense-related genes, cell cycle genes, root nodule formation, and perhaps even in the regulation of morphogenesis, although no unequivocal evidence exists for their role in organogenesis. Credence is also given to this idea by the fact that it appears that transcription factors control many aspects of organogenesis, as discussed in the previous section.

3.3 Competency

The concept of competency comes from the idea that all nucleated cells may be able to form another individual from which they were originally derived (totipotency). Since most nucleated cells are considered totipotent, one would expect that appropriate treatments would result in virtually all cells in the explant forming organs. However, this rarely, if ever, occurs. More commonly, select areas of the explant produce meristemoids, and then shoots.

The reason why some cells and explants do not produce shoots may be, as Halperin (1986) suggested, genetic (i.e., cells lack totipotency due to nuclear changes such as polyteny, chromosomal rearrangements and other mutations, etc.) or epigenetic (i.e., involving stable but potentially reversible constraints in the functioning of genes required for growth and organized development), as well as physiological. If cells have epigenetic constraints to development, they must become competent to respond to an inductive signal before they can become 'determined' (also referred to as 'fixed' or 'committed') to a developmental route from which they will form organs, embryos, or specialized cells and tissues (Lyndon, 1990).

Phenocritical times in the process of competence acquisition can be examined to some extent by experimental manipulation on *in vitro* cultures. Based on media manipulations and inhibitor studies, at least 3 phases of *de novo* organogenesis in *Convolvulus* have been identified: (1) the acquisition of competence (involving dedifferentiation); (2) induction; and (3) determination leading to morphological differentiation (Christianson and Warnick, 1983; 1984; 1985; Christianson, 1987). Similar stages of *de novo* organogenesis have also been distinguished in *Populous deltoides* (Coleman and Ernst, 1990), *Nicotiana tabacum* (Attfield and Evans, 1991), and *Beta vulgaris* (Owens and Eberts, 1992). The interaction of these phases is shown schematically in Fig. 2.

Competence acquisition in tobacco thin cell layers (TCL) occurs between days 0 and 4 of induction: cycloheximide, ribose, thymidine and TIBA applied during this time disrupted events associated with morphogenesis in these explants (Kim and Ernst, 1994). Cell cycle phase, cellular location and ploidy level of cells were correlated to the competence of cells for division

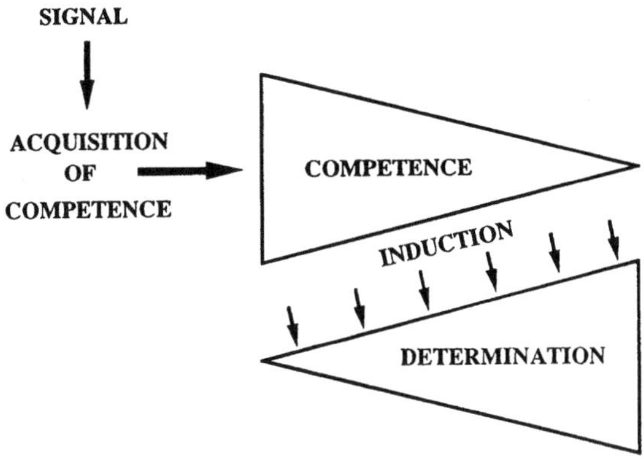

Figure 2. A scheme illustrating a hypothetical sequence of events that results in differentiation. The initiation of events starts with a signal to allow 'acquisition of competence'. After cellular competence is attained, this signal is diminished as cells become more determined, as shown by opposing triangles. Therefore, as cells become fully canalized to a particular developmental route, they lose competence to revert to a different pathway.

and primordia formation (Gilissen et al., 1994). Tobacco TCLs contain mostly nuclei with a 4C DNA content; however, the cells in epidermal and subepidermal layers had mostly 2C DNA levels. Primordia formed from the subepidermal and epidermal layers and produced predominantly normal diploid organs. Therefore, acquisition of competence apparently happened in normal diploid cells, rather than in the tetraploid cells.

Cell cycle phase (i.e. M, G_1, S, G_2, or G_0 phase) is another candidate for differences among cells for which other factors (i.e. phytohormone concentrations, metabolite levels, concentration gradients, etc.) are equal. Cells in certain stages of the cell cycle may be more competent to respond to stimuli favorable to a morphogenic response than others.

In *Dioscorea*, older leaf material was incapable of proliferation, but young leaves did not even absolutely require cytokinin in the medium to produce callus (Wernicke and Park, 1993) . This is very reminiscent of monocots such as *Zea mays*, which are highly recalcitrant to *in vitro* culture. In older, differentiated tissues of *Zea mays*, $p34^{cdc2}$ protein kinase expression was barely detectable, and there was no measurable enzyme activity (Colasanti et al., 1991). Also, susceptibility to nodule formation in legumes has been found at the G_0/G_1 phase of the cell cycle but not at G_2 (Yang et al., 1994). Thus, M-phase and S-phase may not provide an environment for competence to be expressed, but at least one of G_1 or G_2 (or G_0) may.

The asynchrony of primordia formation in tobacco also implicates cell cycle phases. Some cells would produce primordia at the beginning of shoot induction because they would be in the proper phase of the cell cycle, and would respond immediately to induction signals. Others would respond subsequently as they arrived at the same phase of the cell cycle, causing primordia to form over a window of time rather than all at once. It is likely that a similar scenario exists in other systems such as somatic embryogenesis (Komamine et al., 1992) and cytodifferentiation (Fukuda, 1992).

This, however, still does not explain why all of the cells do not eventually respond. Returning to gradients, it is quite likely that the optimal gradient conditions at the onset of the culture cease to exist after some time in culture. Phytohormone and metabolite concentrations may become depleted so that they can no longer produce a morphogenic response. Further, as the initial respondents grow and develop, they set up conditions of a 'sink-source' which draws from the surrounding environment, thereby further reducing optimal conditions so that no additional cells can respond.

Thus, the next question arises: What, during a particular phase of the cell cycle, allows the cell to respond to induction? Several lines of evidence implicate signal transduction pathways and transcription factors in the response to conditions which promote development. Hence, under appropriate conditions, the incipient respondent cell will produce a primordium provided that it is in the correct phase of the cell cycle (i.e. competent) to accept a signal (possibly a phytohormone) which, through various signal transduction pathways activates developmental genes (homeotic domains). In addition, the extracellular mileu (e.g. receptors on the plasmalemma) must be conducive to the preceding responses in order for them to take place.

3.4 The cell cycle

Considerable progress has been made in regards to our understanding of the regulation of the cell cycle. Entry into mitosis (M) and into DNA synthesis (S) require the activity of the $p34^{cdc2}$ protein kinase (reviewed in Nurse, 1990; Hayles and Nurse, 1993). Cyclin synthesis and degradation also drive the cell cycle (Evans et al., 1983; Murray and Kirshner, 1989; Murray et al., 1989). A number of different $p34^{cdc2}$ and cyclin homologues have been characterized in evolutionarily higher species, and based on their structure and function, regulatory controls of the cell cycle are being dissected

Cell cycle research in plants has progressed swiftly, using information gleaned from yeasts and other model organisms. Using PCR amplification products and antibodies against the highly conserved 'PSTAIRE' region of $p34^{cdc2}$, homologues of the $p34^{cdc2}$ and various cyclins have been isolated in a number of plant species, including *Chlamydomonas*, oats, *Arabidopsis*, red and brown algae, pea, and soybean (Feiler and Jacobs, 1990; Ferriera et al., 1991; Hirt et al., 1991; John et al., 1989; Miao et al., 1993). Unlike yeasts, higher organisms appear to regulate the cell cycle differentially at different points in the cell cycle (Hirt et al., 1993) and/or in different organs of the plant (Miao et al., 1993) through the use of different homologues of p34^{cdc2}. There have also been reports of p34 protein kinase-related genes which do not contain fully conserved consensus regions, indicating the possibility of a higher diversity of p34^{cdc2}-like genes in plants (Imajuku et al., 1992).

There are a large number of cyclins which have been characterized based on when they are expressed during the cell cycle. Hence, there are G_1, S, G_2, and M-phase cyclins. Overlap of function into more than one cell cycle phase has also been demonstrated (cf. Doerner, 1994 and references therein). Several putative cyclins with expression pattern similarity and *Xenopus* oocyte maturation induction ability have been isolated from plant species, including carrot and soybean (Hata et al., 1991), *Arabidopsis* (Hemerly et al., 1992) and alfalfa (Hirt et al., 1992). Based on expression patterns, these cyclins appear to be similar to A/B-type cyclins; however, sequence similarities are minimal (cf. Francis and Halford, 1995).

In *Arabidopsis* and *Raphanus sativa*, p34^{cdc2} homologues are highly expressed in actively dividing tissues such as primordia, apices, and floral organs (Martinez et al., 1992; Hemerly et al., 1993). Wounding, light, and hormone treatments also affected protein kinase levels. Since nonproliferating or differentiated tissues did not have detectable levels of cdc2 expression, cdc2 transcript levels may be a marker of proliferative competence in tissues. Cell division is required for proliferative and formative growth in developing organs, but it may also play a role in the induction of developmental programs.

Soybean has two distinct p34^{cdc2} protein kinases of which expression levels vary; one, cdc2-S6, is expressed at higher levels in aerial tissues, whereas the other, cdc2-S5, is more specific to roots and nodule formation. cdc2-S5 transcript levels are enhanced upon infection with *Rhizobium*; however, cdc2-S6 is unaffected (Miao et al., 1993). These data suggest that different p34^{cdc2} protein kinases may be present in different areas of the plant and respond to different signals, possibly including signals leading to organogenesis.

Cyclin expression, as mentioned above, is also cell cycle specific and in many cases a good marker for which phase of the cycle the cell is in (Hirt et al., 1992; Fobert et al., 1994). Cyclin expression may be a major control mechanism in the progression of the cell cycle, and contribute to specific phase arrest (Ferreira et al., 1994): thus it may also contribute to cell competence.

Preprophase band (PPB) formation is another hallmark of cell division. PPBs form during G_2 and demarcate the future divisional plane. p34^{cdc2} protein kinase has been shown to associate with the PPB in *Allium cepa*, *Zea mays* and *Chalmydomonas* (Colasanti et al., 1991; John et al., 1993;

Mineyuki et al., 1991). The likely function of this association is the phosphorylation signal for PPB disassembly, which marks the onset of mitosis and commitment to cell division (John et al., 1989). The 'signal(s)' to divide and differentiate may be translated to the cellular level through rearrangement of the cortical microtubules to form the PPB in a different plane. This is noteworthy as a change in the divisional plane is one of the first cyto-histological features of a future primordium, e.g., as found in radiata pine (Villalobos et al., 1985).

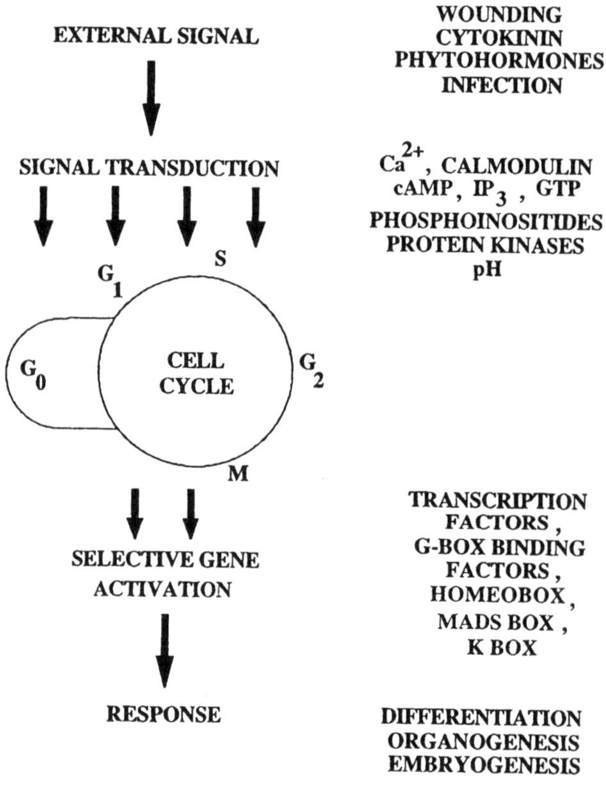

Figure 3. Hypothetical scheme illustrating connections, starting with an external signal, which may be causal in the initiation and maintenance of differentiation.

A change in the divisional direction cannot be implicated by activation of the cell cycle components alone, although their presence ($p34^{cdc2}$ and cyclins) presumably distinguishes competency towards proliferative growth. Doerner (1994) pointed out that developmental competence for division should not be confused with totipotency; thus, although cell cycle components may be markers of proliferative competence, they are not necessarily also markers of morphogenic competence. In fact, it is likely that these two events are separate as can be illustrated by direct cytodifferentiation (without intervening cell division) of mesophyll cells to tracheary elements, in which the progression of the cell cycle is not necessary for differentiation (Fukuda and Komamine, 1981). Habituated cultures can proliferate indefinitely without addition of exogenous phytohormones; however, they cannot differentiate. *De novo* organogenesis requires the activation of quiescent cells or the rechannelling of cells into proliferation into an alternative pathway of differentiation. Both of these processes require dedifferentiation and redifferentiation, as well as mitotic competence, in cells; however, cell cycle research in the future may lead to more discoveries of mechanisms required to induce competency for differentiation, as well as cell division. A model for the interaction of the various cellular and molecular aspects leading to organized development is shown in Fig. 3 (Also see the Plant Cell, Special Issue, July 1997).

4 CONCLUSIONS

We are currently at an exciting crossroads in plant biotechnology. Advances in technology over the last 5-10 years have given us even more tools to investigate fundamental questions of growth and development. The automation of *de novo* shoot morphogenesis for commercial propagation is, by necessity, developing rapidly. Within the next decade, this is likely to become a fully automated process for most economically important species (Aitken-Christie et al. 1995).

The utilization of molecular biology for which regeneration of plants from transformed cells is necessary, has allowed us to manipulate plants to our benefit. This has resulted in the production of transgenics which have given us an insight into the *in vivo* workings of the plant, usually through modifications in one or more biochemical pathways. Examples of ethylene regulation (Pua and Lee 1995), lignin modification (Atanassova et al. 1995), herbicide resistance (see Rogers and Parkes 1995, and references therein) to name just a few, abound in the literature.

To attest to the actual applied benefits of the existing technology, there were a minimum of 8 commercialized, genetically modified plants at the end of 1995. These ranged from Calgene's Flavr Savr® tomato and *laurate canola* to Pioneer Hy-Bred's transgenic corn and cotton. It is estimated that the number of commercially available transgenic crops is likely to double or triple every year from 1995 (Keith Redenbaugh, personal communication, Rogers and Parkes 1995).

We are now beginning to understand the molecular underpinnings of plant development. Through the use of morphogenic mutants in

Arabidopsis, and other genetically tractable species, there is now a clearer picture of the complicated cascade of events which are involved in the organogenesis of the flower and its component parts. These studies, in combination with *in vitro* work should, over the next few years, bring us closer to answering the bigger question: "What initiates, maintains, and controls *de novo* organogenesis?" The production of usable products (such as commercially available transgenics) is one ultimate use of the information arising from basic studies. Commercial micropropagation, today largely devoted to ornamentals, has been dependent on the basic studies of morphogenesis. The use of the model systems of tobacco callus and excised radiata pine cotyledons, as we have discussed above, will continue to increase our knowledge and understanding of basic shoot morphogenesis and contribute to the elucidation of mechanisms of plant growth and development.

5 ACKNOWLEDGMENTS

The contributions of colleagues, graduate students, post-doctoral fellows, visiting scientists and research assistants to the personal research reported here is gratefully acknowledged. Particular thanks are due to those who have contributed as yet unpublished material. The authors thank Tanya S. Hooker for her help in the preparation of this ms. T. A. Thorpe also gratefully acknowledges research funding from the Natural Sciences and Engineering Research Council of Canada (Operating/Research and Strategic Grants).

6 REFERENCES

Abdullah AA & Grace J (1987) Regeneration of carlabrian pine from juvenile needles. Plant Sci. 53: 147-155.

Aitken J, Horgan KJ & Thorpe TA (1981) Influence of explant selection on the shoot-forming capacity of juvenile tissue of *Pinus radiata*. Can. J. For. Res. 11: 112-117.

Aitken-Christie J, Kozai T & Smith MAL (eds) (1995) Automation and environmental control in plant tissue culture. Kluwer, Dordrecht.

Aitken-Christie J, Singh AP, Horgan KJ & Thorpe TA (1985) Explant developmental state and shoot formation in *Pinus radiata* cotyledons. Bot. Gaz. 146: 196-203.

Allen RD (1995) Dissection of oxidative stress tolerance using transgenic plants. Plant Physiol. 107: 1049-1054.

Altamura MM, Torrigiani P, Falasca G, Rossini P & Bagni N (1993) Morpho-functional gradients in superficial and deep tissues along tobacco stem: Polyamine levels, biosynthesis and oxidation, and organogenesis *in vitro*. J. Plant Physiol. 142: 543-551.

Angenent GC, Busscher M, Franken J, Dons HJM & van Tumen AJ (1995) Functional interaction between the homeotic genes *fbp1* and *pMADS1* during petunia floral organogenesis. Plant Cell 7: 507-516.

Angenent GC, Franken J, Busscher M, van Dijken A, van Went JL, Dons HJM & van Tumen AJ (1995) A novel class of MADS box genes is involved in ovule development in Petunia. Plant Cell 7: 1569-1582.

Asbell CW (1977) Ultrastructual modifications during shoot formation *in vitro*. In Vitro 13: 180.

Assman SM (1995) Cyclic AMP as a second messenger in higher plants. Plant Physiol. 108: 885-889.

Atanassova R, Favet N, Martz F, Chabbert B, Tollier M.-T, Monties B, Fritig B & Legrand M (1995) Altered lignin composition in transgenic tobacco expressing O-methyltransferase sequences in m .Usense and antisense orientation. Plant J. 8: 465-477.

Attfield EM & Evans PK (1991) Stages in the initiation of root and shoot organogenesis in cultured leaf explants of *Nicotiana tabacum* cv. Xanthi nc. J. Exp. Bot. 42: 59-63.

Bachs-Hüemann D & Reinert J (1970) Embryobildung durch isolierte Einzellen auk Gewebekûturen von *Daucus carota*. Protoplasma 70: 49-60.

Bagni N & Biondi S (1987) Polyamines. In: Bonga, J.M., and Durzan, D.J. (eds) Cell and Tissue Culture in Forestry. Vol. 1 pp. 113-124. Martinus Nijhoff, Dordrecht.

Beaudoin-Eagan L.D & Thorpe TA (1983) Shikimic acid pathway activity during shoot initiation in tobacco callus cultures. Plant Physiol. 73: 228-232.

Beaudoin-Eagan LD & Thorpe TA (1984) Turnover of shikimate pathway metabolites during shoot initiation in tobacco callus cultures. Plant Cell Physiol. 25: 913-921.

Beaudoin-Eagan LD & Thorpe TA (1985a) Shikimate pathway enhancement in the lower half of shoot-forming tobacco callus. J. Plant Physiol. 120: 87-90.

Beaudoin-Eagan LD & Thorpe TA (1985b) Tyrosine and phenylalanine ammonia lyase activities during shoot initiation in tobacco callus cultures. Plant Physiol. 78: 438-441.

Bender L, Joy IV RW & Thorpe TA (1987a) Studies on [^{14}C]-glucose metabolism during shoot induction in cultured cotyledon explants of *Pinus radiata*. Physiol. Plant. 69: 428-434.

Bender L, Joy IV RW & Thorpe TA (1987b) [^{14}C]-glucose metabolism during shoot bud development in cultured cotyledon explants of *Pinus radiata*. Plant Cell Physiol. 28: 1335-1338.

Bertrand-Garcia R, Walling LL & Murashige T (1992) Analysis of polypeptides associated with shoot formation in tobacco callus cultures. Amer. J. Bot. 79: 481-487.

Biondi S & Thorpe TA (1982) Growth regulator effects, metabolite changes and respiration during shoot initiation in cultured cotyledon explants of *Pinus radiata*. Bot. Gaz. 143: 20-25.

Biondi S, Bagni N & Sansovini A (1986) Dicyclohexylamine uptake and effects on polyamine content in cultured cotyledons of radiata pine. Physiol. Plant. 66: 41-45.

Biondi S, Torrigiani P, Sansovini A & Bagni N (1988) Inhibition of polyamine biosynthesis by dicyclohexylamine in cultured cotyledons of *Pinus radiata*. Physiol. Plant. 72: 471-476.

Bonnett HI & Torrey JG (1965) Chemical control of organ formation in root segments of *Convolvulus* cultured *in vitro*. Plant Physiol. 40: 1228-1236.

Bonnett HT & Torrey JG (1966) Comparative anatomy of endogenous bud and lateral root formation in *Convolvulus arvensis* roots cultured *in vitro*. Amer. J. Bot. 53: 496-507.

Bornman CH (1983) Possibilities and constraints in the regenerations of trees from cotyledonary needles of *Picea abies in vitro*. Physiol. Plant. 57: 5-16.

Boss WF (1989) Phosphoinositide metabolism: Its relation to signal transduction in plants. In: Boss, W.F., and Morre, D.F. (eds) Second Messengers in Plant Growth and Development. pp. 29-56. Alan R. Liss Inc., New York.

Bowman JL, Smyth DR & Meyerowitz EM (1991) Genetic interactions among floral homeotic genes of *Arabidopsis*. Development 112: 1-20.

Braam J & Davis RW (1990) Rain-, wind-, and touch-induced expression of calmodulin and calmodulin-related genes in *Arabidopsis*. Cell 60: 357-364.

Brown DCW, Leung DWM & Thorpe TA (1979) Osmotic requirement for shoot formation in tobacco callus. Physiol. Plant. 46: 36-41.

Brown DCW & Thorpe TA (1980a) Adenosine phosphate and nicotinamide adenine dinucleotide pool sizes during shoot initiation in tobacco callus. Plant Physiol. 65: 587-590.

Brown DCW & Thorpe TA (1980b) Changes in water potential and its components during shoot formation in tobacco callus. Physiol. Plant. 49: 83-87.

Brown DCW & Thorpe TA (1982) Mitochondrial activity during shoot formation and growth in tobacco callus. Physiol. Plant. 54: 125-130.

Button J, Kochba J & Bornman CH (1974) Fine structure of and embryoid development from embryogenic ovular callus of 'Shamouti' orange (*Citrus sinensis* Osb.). J. Exp. Bot. 25: 446-457.

Campbell MA, Gaynor JJ & Kirby EG (1992) Culture of cotyledons of Douglas-fir on a medium for the induction of adventitious shoots induces rapid changes in polypeptide profiles and mRNA populations. Physiol. Plant. 85: 180-188.

Chandler SF & Thorpe TA (1986) Hormonal regulation of organogenesis *in vitro*. In: Purohit, S.S. (ed) Hormonal Regulation of Plant Growth and Development. Vol. 3. pp. 1-27. Agro-Botanical Publ., India.

Chi G-L, Pua E-C & Goh C-J (1991) Role of ethylene on *de novo* shoot regeneration from cotyledonary explants of *Brassica campestris* sp. *Pekinensis* (Lour.) Olsson *in vitro*. Plant Physiol. 96: 178-183.

Chlyah H (1974) Formation and propagation of cell division centres in the epidermal layer of internodal segments of *Torenia fournieri* grown *in vitro*. Simultaneous surface observations of all the epidermal cells. Can. J. Bot. 52: 867-872.

Chlyah H & Tran Thanh Van M (1975) Distribution pattern of cell division centers on the epidermis of stem segments of *Torenia tournieri* during *de novo* bud formation. Physiol. Plant. 56: 28-33.

Christianson ML (1987) Causal events in morphogenesis. In: Green, C.E., Somers, D.A., Hackett, W.P., and Biesboer, D.O. (eds) Plant Tissue and Cell Culture. pp. 45-55. Liss, New York.

Christianson ML & Warnick DA (1983) Competence and determination in the process of *in vitro* shoot organogenesis. Develop. Biol. 95: 288-293.

Christianson ML & Warnick DA (1984) Phenocritical times in the process of *in vitro* shoot organogenesis. Develop. Biol. 101: 382-390.

Christianson ML & Warnick DA (1985) Temporal requirement for phytohormone balance in the control of organogenesis *in vitro*. Develop. Biol. 112: 494-497.

Citterio S, Sgorbati S, Scippa S & Sparvoli E (1994) Ascorbic acid effect on the onset of cell proliferation in pea root. Physiol. Plant. 92: 601-607.

Coen ES, Romero JM, Doyle S, Elliott R, Murphy G & Carpenter R (1990) *floricaula*: A homeotic gene required for flower development in *Antirrhinum majus*. Cell 63: 1311-1322.

Colasanti J, Tyres M & Sundaresan V (1991) Isolation and characterization of cDNA clones encoding a functional $p34^{cdc2}$ homologue from *Zea mays*. Proc. Natl. Acad. Sci. USA 88: 3377-3381.

Coleman GD & Ernst SG (1990) Shoot induction competence and callus determination in *Populus deltoides*. Plant Sci. 71: 83-92.

Constabel F, Kurz WGW, Chatson KB & Kirkpatrick JW (1977) Partial synchrony in soybean cell suspension cultures induced by ethylene. J. Cell Res. 105: 263-268.

Dalton CC & Street HE (1976) The role of the gas phase in the greening and growth of illuminated cell suspension cultures of spinach (*Spinacea oleracea* L.). In Vitro 12: 485-494

Davis DG & Olson PA (1994) Effects of putrescine and inhibitors of putrescine biosynthesis on organogenesis in *Euphorbia esula* L. In Vitro Cell. Dev. Biol. 30: 124-130.

Dey SK & Kar M (1995) Antioxidant efficiency during callus initiation from mature rice embryo. Plant Cell Physiol. 36: 543-549.

Doerner PW (1994) Cell cycle regulation in plants. Plant Physiol. 106: 823-827.

Dougall DK & Shimbayashi K (1960) Factors affecting growth of tobacco callus tissue and its incorporation of tyrosine. Plant Physiol. 35: 396-404.

Dougall DK (1962) On the fate of tyrosine in tobacco callus tissue. Aust. J. Biol. Sci. 15: 619-622.

Douglas TJ, Villalobos VM, Thompson MR & Thorpe TA (1982) Lipid and pigment changes during shoot initiation in cultured explants of *Pinus radiata*. Physiol. Plant. 55: 470-477.

Drayer AL & van Haastert PJM (1994) Transmembrane signaling in eukaryotes: a comparison between higher and lower eukaryotes. Plant Mol. Biol. 26: 1239-1270.

Duerr B, Gawienowski M, Ropp T & Jacobs T (1993) MsERK1: A mitogen-activated protein kinase from a flowering plant. Plant Cell 5: 87-96.

Earnshaw BA & Johnson MA (1987) Control of wild carrot somatic embryo development by antioxidants. Plant Physiol. 85: 273-276.

Ellis DD & Judd RC (1987) SDS-PAGE analysis of bud-forming cotyledons of *Pinus ponderosa*. Plant Cell Tissue Organ Cult. 11: 57-65.

Evans DA, Sharp WR & Flick CE (1981) Growth and behavior of cell cultures. In: Thorpe, T.A. (ed) Plant Tissue Culture - Methods and Applications in Agriculture. pp. 45-113. Academic Press, New York.

Evans T, Rosenthal ET, Youngblom J, Distel D & Hunt T (1983) Cyclin: A protein specified by material mRNA in sea urchin eggs that is destroyed at each cleavage division. Cell 33: 389-396.

Feiler HS & Jacobs TW (1990) Cell division in higher plants: A *cdc2* gene, its 34-kDa product, and histone H1 kinase activity in pea. Proc. Natl. Acad. Sci. USA 87: 5397-5401.

Ferreira PCG, Hemerly AS, de Almeida Engler J, Van Montagu M, Engler G & Inzé D (1994) Developmental expression of the *Arabidopsis* cyclin gene *cyc1At*. Plant Cell 6: 1763-1774.

Ferreira PCG, Hemerly AS, Villarroel R, Van Montagu M & Inzé D (1991) The *Arabidopsis* functional homolog of the $p34^{cdc2}$ protein kinase. Plant Cell 3: 531-540.

Flinn BS, Webb DT & Newcomb W (1988) The role of cell clusters and promeristemoids in determination and competence for caulogensis by *Pinus strobus* cotyledons *in vitro*. Can. J. Bot. 66: 1556-1565.

Flinn BS, Webb DT & Newcomb W (1989) Morphometric analysis of reserve substances and ultrastructural changes during caulogenic determination and loss of competence of Eastern White pine (*Pinus strobus*) cotyledons *in vitro*. Can. J. Bot. 67: 779-789.

Fobert PR, Coen ES, Murphy GJP & Doonan JH (1994) Patterns of cell division revealed by transcriptional regulation of genes during the cell cycle in plants. EMBO J. 13: 616-624.

Francis D & Halford NG (1995) The plant cell cycle. Physiol. Plant. 93: 365-374.

Fry SC (1990) Roles of the primary cell wall in morphogenesis. In: Nijkamp, H.J.J. et al. (eds) Progress in Plant Cellular and Molecular Biology. pp. 504-513. Kluwer, Dordrecht.

Fukuda H (1992) Tracheary element formation as a model system of cell differentiation. Int. Rev. Cytol. 136: 289-332.

Fukuda H & Komamine A (1981) Relationship between tracheary element differentiation and the cell cycle in single cells isolated from the mesophyll of *Zinnia elegans*. Physiol. Plant. 52: 423-430.

Gavidia I, Segura J & Perez-Bermudez P (1993) Effects of gibberellic acid on morphogenesis and cardenolide accumulation in juvenile and adult *Digitalis obscura* cultures. J. Plant Physiol. 142: 373-376.

Gehring WJ (1987) Homeo boxes in the study of development. Science 236: 1245-1252.

Gehring WJ, Qian YQ, Billeter M, Furukubo-Tokunaga K, Schier AF, Resendez-Perez D, Affolter M, Otting G & Wührich K (1994) Homeodomain-DNA recognition. Cell 78: 211-223.

Gilissen LJW, van Staveren MJ, Hakkert JC, Smulders MJ, Verhoeven HA & Creemers-Molenaar J (1994) The competence of cells for cell division and regeneration in tobacco explants depends on cellular location, cell cycle phase and ploidy level. Plant Sci. 103: 81-91.

Hake S, Vollbrecht E & Freeling M (1989) Cloning *Knotted*, the dominant morphological mutant in maize using Ds2 as a transposon tag. EMBO J. 8:15-22.

Halperin W & Wetherell DF (1964) Adventive embryony in tissue cultures of the wild carrot, *Daucus carota*. Amer. J. Bot. 51: 274-283.

Halperin W (1986) Attainment and retention of morphogenetic capacity *in vitro*. In: Vasil, I.K. (ed) Cell Culture and Somatic Cell Genetics of Plants. Vol. 3 pp. 3-47. Academic Press, New York.

Hammersley-Straw DRH & Thorpe TA (1987) Modulation of shoot formation in tobacco callus by tissue transfer between permissive and inhibiting media. In Vitro Cell. Dev. Biol. 23:867-872.

Hammersley-Straw DRH & Thorpe TA (1988) Use of osmotic inhibition in studies of shoot formation in tobacco callus cultures. Bot. Gaz. 149: 303-319.

Hardy EL & Thorpe TA (1990) Nitrate assimilation in shoot-forming tobacco callus cultures. In Vitro Cell. Dev. Biol. 26: 525-530.

Hasegawa PM, Murashige T & Mudd JB (1977) The fate of L-tyrosine-UL-^{14}C in shoot forming tobacco callus. Physiol. Plant. 41: 223-230.

Hasegawa PM, Yasuda T & Cheng TY (1979) Effect of auxin and cytokinin on newly synthesized proteins of cultured Douglas fir cotyledons. Physiol. Plant. 46: 211-217.

Hata S, Kouchi H, Suzuka I & Ishii T (1991) Isolation and characterization of cDNA clones for plant cyclins. EMBO J. 10: 2681-2688.

Hayles J & Nurse P (1993) The controls acting at mitosis in *Schizosaccharomyces pombe*. In: Ormrod, J.C., and Francis, D. (eds) Molecular and Cell Biology of the Plant Cell Cycle. pp. 1-7. Kluwer, Dordrecht.

Hemerly A, Bergounioux C, Van Montagu M, Inzé D & Ferreira P (1992) Genes regulating the plant cell cycle: Isolation of a mitotic-like cyclin from *Arabidopsis thaliana*. Proc. Natl. Acad. Sci. USA 89: 3295-3299.

Hemerly AS, Ferreira P, de Almeida Engler J, Van Montagu M, Engler G & Inzé D (1993) *cdc2a* expression in *Arabidopsis* is linked with competence for cell division. Plant Cell 5: 1711-1723.

Hicks GS (1980) Patterns of organ development in plant tissue culture and the problem of organ determination. Bot. Rev. 46: 1-23.

Hicks GS (1994) Shoot induction and organogenesis *in vitro*: A developmental perspective. In Vitro Cell. Dev. Biol. 30p: 10-15.

Hirt H, Páy A, Györgyey J, Bakó L, Németh K, Bögre L, Schweyen RJ, Heberle-Bors E & Dudits D (1991) Complementation of a yeast cell cycle mutant by an alfalfa cDNA

encoding a protein kinase homologous to p34[cdc2]. Proc. Natl. Acad. Sci. USA 88: 1636-1640.
Hirt H, Mink M, Pfosser M, Bögre L, Györgyey J, Jonak C, Gartner A, Dudits D & Heberle-Bors E (1992) Alfalfa cyclins: Differential expression during the cell cycle and in plant organs. Plant Cell 4: 1531-1538.
Hirt H, Páy A, Bögre L, Meskiene I & Heberle-Bors E (1993) *cdc2MsB*, a cognate *cdc2* gene from alfalfa, complements the G1/S but not the G2/M transition of budding yeast *cdc28* mutants. Plant J. 4: 61-69.
Huxter TJ, Reid DM & Thorpe TA (1979) Ethylene production by tobacco (*Nicotiana tabacum*) callus. Physiol. Plant. 46: 374-380.
Huxter TJ, Thorpe TA & Reid DM (1981) Shoot initiation in light- and dark-grown tobacco callus: The role of ethylene. Physiol. Plant. 53: 319-326.
Imajuku Y, Hirayama T, Endos H & Oka A (1992) Exon-intron organization of the *Arabidopsis thaliana* protein kinase genes CDC2a and CDC2b. FEBS Lett. 304: 73-77.
Jena PK, Reddy ASN & Poovaiah BW (1989) Molecular cloning and sequencing of a cDNA for plant calmodulin: Signal-induced changes in the expression of calmodulin. Proc. Natl. Acad. Sci. USA 86: 3644-3648.
Jensen WA (1963) Cell development during plant embryogenesis. In: Meristems and Differentiation, Brookhaven Symp. Biol. 16: 179-202.
John PCL, Sek FJ & Lee MG (1989) A homolog of the cell cycle control protein p34[cdc2] participates in the division cycle of *Chlamydomnas*, and a similar protein is detectable in higher plants and remote taxa. Plant Cell 1: 1185-1193.
Jonak C, Heberle-Bors C & Hirt H (1994) MAP kinases: Universal multi-purpose signaling tools. Plant Mol. Biol. 24: 407-416.
Jonak C, Páy A, Bögre L, Hirt H & Heberle-Bors E (1993) The plant homologue of MAP kinase is expressed in a cell cycle-dependent and organ-specific manner. Plant J. 3: 611-617.
Joy IVRW, Patel KR & Thorpe TA (1988) Ascorbic acid enhancement of organogenesis in tobacco callus. Plant Cell Tiss. Organ Cult. 13: 219-228.
Joy IVRW & Thorpe TA (1990) Non-morphogenic metabolism during shoot induction in radiata pine cotyledon explants. In Vitro Cell. Dev. Biol. 26: 643-646.
Joy IVRW, Bender L & Thorpe TA (1994) Nitrogen metabolism in cultured cotyledon explants of *Pinus radiata* during *de novo* organogenesis. Physiol. Plant. 92: 681-688.
Kavi Kishor PB (1989) Activities of phenylalanine- and tyrosine-ammonia lyases and aminotransferases during organogenesis in callus cultures of rice. Plant Cell Physiol. 30: 25-29.
Kim K-N & Ernst SG (1994) Effects of inhibitors on phenocritical events of *in vitro* shoot organogenesis in tobacco thin cell layers. Plant Sci. 103: 59-66.
Kirby EG & Schalk ME (1982) Surface structural analysis of cultured cotyledons of Douglas-fir. Can. J. Bot. 60: 2729-2733.
Kohlenbach HW (1977) Basic aspects of differentiation and plant regeneration from cell and tissue cultures. In: Barz, W. et al. (eds) Plant Tissue Culture and its Biotechnological Application. pp. 355-366. Springer, Berlin.
Komamine A, Kawahrara R, Matsumoto M, Sunabori S, Toya T, Fujiwara A, Tsukahara M, Smith J, Ito M, Fukuda H, Nomura K & Fujimura T (1992) Mechanisms of somatic embryogenesis in cell cultures: Physiology, biochemistry, and molecular biology. In Vitro Cell. Dev. Biol. 28p: 11-14.
Konshuh MN & Thorpe TA(1997) Metabolism of ^{14}C-aspartate during shoot bud formation in cultured cotyledon explants of radiate pine. Physiol. Plant. 99: 31-38.

Koornneef M, Hanhart CJ & Martinelli L (1987) A genetic analysis of cell culture traits in tomato. Theor. Appl. Genet. 74: 633-641.

Kumar PP, Bender L & Thorpe TA (1988) Activities of ribulose bisphosphate carboxylase and phosphoenolpyruvate carboxylase and ^{14}C-bicarbonate fixation during *in vitro* culture of *Pinus radiata* cotyledons. Plant Physiol. 87: 675-679.

Kumar PP & Thorpe TA (1989) Putrescine metabolism in excised cotyledons of *Pinus radiata* cultured *in vitro*. Physiol. Plant. 76: 521-526.

Lance B, Durley RC, Reid DM, Thorpe TA & Pharis RP (1976a) The metabolism of [^{3}H]-gibberellin A_{20} in light- and dark-grown tobacco callus cultures. Plant Physiol. 58: 387-392.

Lance B, Reid DM & Thorpe TA (1976b) Endogenous gibberellins and growth of tobacco callus cultures. Physiol. Plant. 36: 287-292.

Larkin PJ & Scowcroft WC (1981) Somaclonal variation-a novel source of variability from cell cultures for plant improvement. Theor. Appl. Genet. 60: 197-214.

Lee TT (1962) Effects of substituted phenols on growth, bud formation and IAA oxidase activity. Ph.D dissertation, University of Wisconsin, Madison.

Leung DWM & Thorpe TA (1985) Occurrence of α -amylases in tobacco callus: Some properties and relationships with callus induction and shoot primordium formation. J. Plant Physiol. 18: 145-151.

Ling V & Zielinski RE (1989) Cloning of cDNA sequences encoding the calcium-binding protein, calmodulin, from barley (*Hordeum vulgare* L.). Plant Physiol. 90: 714-719.

Lyndon RF (1990) Plant development: The cellular basis. Unwin Hyman, London.

Lyndon RF. (1994) Control of organogenesis at the shoot apex. New Phytol. 128: 1-18.

Maeda E & Thorpe TA (1979) Shoot histogenesis in tobacco callus cultures. In Vitro 15: 415-424.

Martinez MC, Jøgensen J-E, Lawton MA, Lamb CJ & Doerner PW (1992) Spatial pattern of *cdc2* expression in relation to meristem activity and cell proliferation during plant development. Proc. Natl. Acad. Sci. USA 89: 7360-7364.

Matsuoka M, Ichikawa H, Saito A, Tada Y, Fujimura T & Kano-Murakami Y (1993) Expression of a rice homeobox gene causes altered morphology of transgenic plants. Plant Cell 5: 1039-1048.

Miao G-H, Hong Z & Verma DPS (1993) Two functional soybean genes encoding p34^{cdc2} protein kinases are regulated by different plant developmental pathways. Proc. Natl. Acad. Sci. USA 90: 943-947.

Mineyuki Y, Yamashita M & Nagahama Y (1991) p34^{cdc2} kinase homologue in the preprophase band. Protoplasma 162: 182-186.

Mizoguchi T, Gotoh Y, Nishida E, Yamaguchi-Shinozaki K, Hayashida N, Iwasaki T, Kamada H & Shinozaki K (1994) Characterization of two cDNAs that encode MAP kinase homologues in *Arabidopsis thaliana* and analysis of the possible role of auxin in activating such kinase activities in cultured cells. Plant J. 5: 111-122.

Murashige T (1961) Suppression of shoot formation in cultured tobacco cells by gibberellic acid. Science 134: 280.

Murashige T (1974) Plant propagation through tissue culture. Ann. Rev. Plant Physiol. 25: 135-166.

Murashige T & Skoog F (1962) A revised medium for rapid growth and bioassays with tobacco tissue cultures. Physiol. Plant. 15: 473-497.

Murray AW & Kirschner MW (1989) Cyclin synthesis drives the early embryonic cell cycle. Nature 339: 275-280.

Murray AW, Solomon MJ & Kirschner MW (1989) The role of cyclin synthesis and degradation in the control of maturation promoting factor activity. Nature 339: 280-286.

Negrutiu I, Jacobs M & Cachita D (1978) Some factors controlling *in vitro* morphogenesis of *Arabidopsis thaliana*. Z. Pflanzenphysiol. 86: 113-124.

Nurse P (1990) Universal control mechanism regulating onset of M-phase. Nature 344: 503-508.

Obata-Sasamoto H, Villalobos VM & Thorpe TA (1984) ^{14}C-metabolism in cultured cotyledon explants of radiata pine. Physiol. Plant. 61: 490-496.

Okazawa Y, Katsura N & Tajawa T (1967) Effects of auxin and kinetin on the development and differentiation of potato tissue cultured *in vitro*. Physiol. Plant. 20: 862-869.

Opekarova M.& Kotyk A (1973) Uptake of sugars by tobacco callus tissue. Biol. Plant 15: 312-317.

Owens LD & Eberts DR (1992) Sugarbeet leaf disc culture: An improved procedure for inducing morphogenesis. Plant Cell Tissue Organ Cult. 31: 195-201.

Patel KR & Thorpe TA (1984) Histochemical examination of shoot initiation in cultured cotyledon explants of radiata pine. Bot. Gaz. 145: 312-322.

Plumb-Dhindsa PL, Dhindsa RS & Thorpe TA (1979) Non-autotrophic CO_2 fixation during shoot formation in tobacco callus. J. Exp. Bot. 30: 759-767.

Pua E-C & Chi G-L (1993) De novo shoot morphogenesis and plant growth of mustard (*Brassica juncea*) *in vitro* in relation to ethylene. Physiol. Plant. 88: 467-474.

Pua E-C & Lee JEE (1995) Enhanced de novo shoot morphogenesis *in vitro* by expression of antisense 1-aminocyclopropane-1-carboxylate oxidase gene in transgenic mustard plants. Planta 196: 69-76.

Pyke K (1994) Arabidopsis - its use in the genetic and molecular analysis of plant morphogenesis. New Phytol. 128: 19-37.

Reilly KJ & Washer J (1977) Vegetative propagation of radiata pine by tissue culture; plantlet formation from embryonic tissue. N. Z. J. For. Sci. 7: 199-206.

Rogers HJ & Parkes HC (1995) Transgenic plants and the environment. J. Exp. Bot. 46: 467-488.

Ross MK & Thorpe TA (1973) Physiological gradients and shoot initiation in tobacco callus cultures. Plant Cell Physiol. 14: 473-480.

Ross MK, Thorpe TA & Costerton JW (1973) Ultrastructural aspects of shoot initiation in tobacco callus cultures. Amer. J. Bot. 60: 788-795.

Sargent JA & Skoog F (1960) Effects of indoleacetic acid and kinetin on scopoletin and scopolin levels in relation to growth of tobacco tissues *in vitro*. Plant Physiol. 35: 934-941.

Schell J, Van Montague M, Holsters M, Hernalsteens JP, Dhaese P, De Greve H, Leemans J, Joos H, Inzel D, Willmitzer L, Otten L, Wostemeyer A & Schroeder J (1982) Plant cells transformed by modified Ti plasmids: A model system to study plant development. In: Jaenicke, L. (ed) Biochemistry of Differentiation and Morphogenesis, pp. 65-73. Springer, New York.

Schroeder JI & Thuleau P (1991) Ca^{2+} channels in higher plant cells. Plant Cell 3: 555-559.

Sen S, Magallanes-Cedeno ME & Kamps RH (1994) *In vitro* micropropagation of Afghan pine. Can. J. For. Res. 24: 1248-1252.

Sinnott EW (1960) Plants Morphogenesis. McGraw Hill, New York. 550 pp.

Skoog F & Miller CO (1957) Chemical regulation of growth and organ formation in plant tissues cultured *in vitro*. Symp. Soc. Exp. Biol. 11: 118-131.

Skoog F & Montaldi E (1961) Auxin-kinetin interaction regulating the scopoletin and scopolin levels in tobacco tissue cultures. Proc. Natl. Acad. Sci. USA 47: 36-49.

Skoog F & Tsui C (1948) Chemical control of growth and bud formation in tobacco stem segments and callus cultured *in vitro*. Amer. J. Bot. 35: 782-787.

Stabel P, Eriksson T & Engström P (1990) Changes in protein synthesis upon cytokinin-mediated adventitious bud induction and during seedling development in Norway Spruce, *Picea abies*. Plant Physiol. 92: 1174-1183.
Stebbins GL (1965) Some relationships between mitotic rhythm, nucleic acid synthesis and morphogenesis in higher plants. Brookhaven Symp. Biol. 18: 204-221.
Steeves TA & Sussex IM (1989) Patterns in Plant Development. Camb. Univ. Press, New York.
Steward FC, Mapes MO, Kent AE & Holsten RD (1964) Growth and development of cultured plant cells. Science 163: 20-27.
Steward FC, Mapes MO & Mears K (1958) Growth and organized development of cultured plant cells. II. Organization in cultures growth from freely suspended cells. Amer. J. Bot. 45: 705-708.
Stone JM & Walker JC (1995) Plant protein kinase families and signal transduction. Plant Physiol. 108: 451-457.
Street HE & Withers LA (1974) The anatomy of embryogenesis in culture. In: Street, H.E. (ed) Tissue culture and plant Science. pp. 71-100. Academic Press, London.
Sussex IM & Clutter ME (1967) Differentiation in tissues, free cells, and reaggregated plant cells. In Vitro 3: 3-12.
Suzuki K & Shinshi H (1995) Transient activation and tyrosine phosphorylation of a protein kinase in tobacco cells treated with a fungal elicitor. Plant Cell 7: 639-647.
Syono K (1965) Changes in organ forming capacity of carrot root callus during subcultures. Plant Cell Physiol. 6: 403-419.
Thompson MR & Thorpe TA (1980) Control of hexose monophosphate oxidation in shoot-forming tobacco callus. Plant Physiol. Suppl. 65: 89.
Thompson MR, Douglas TJ, Obata-Sasamoto H & Thorpe TA (1986) Mannitol metabolism in cultured plant cells. Physiol. Plant. 67: 365-369.
Thompson MP, Piazza GJ, Brower DP & Farrell HM Jr (1989) Purification and c haracterization of calmodulins from *Papaver somniferum* and *Euphorbia lathyris*. Plant Physiol. 89: 501-505.
Thompson MR & Thorpe TA(1997) Analysis of protein patterns during shoot initiation in cultured *Pinus radiate* cotyledons. J. Plant Physiol. 151: 724-73
Thorpe TA (1974) Carbohydrate availability and shoot formation in tobacco callus cultures. Physiol. Plant. 30: 77-81.
Thorpe TA (1978) Physiological and biochemical aspects of organogenesis *in vitro*. In: Thorpe, T.A. (ed) Frontiers of Plant Tissue Culture. pp. 49-58. Univ. of Calgary, Canada.
Thorpe TA (1979) Regulation of organogenesis *in vitro*. In: Hughes, K.W. et al. (eds) Propagation of Higher Plant Through Tissue Culture - A Bridge Between Research and Application. US Tech. Inf. Serv. pp. 87-101. Springfield, Virginia.
Thorpe TA (1980) Organogenesis *in vitro*: Structural, physiological and biochemical aspects. Int. Rev. Cytol. suppl. 11A: 71-111.
Thorpe TA (1982) Callus organization and de novo formation of shoots, roots and embryos *in vitro*. In: Tomes, D.T. et al. (eds) Application of Plant Cell And Tissue Culture and Agriculture and Industry. pp. 115-138. Univ. of Guelph, Ontario.
Thorpe TA (1983) Morphogenesis and regeneration in tissue cultures. In: Owens, L.D. (ed) Genetic Engineering: Application to Agriculture, pp. 285-303. Rowman and Allanheld, New Jersey.
Thorpe TA (1988) Physiology of bud induction in conifers in vitro. In: Hanover, J.W., and Keathley, D.E. (eds) Genetic Manipulation of Woody Plants. pp. 167-184. Plenum, New York.

Thorpe TA (1993) *In vitro* organogenesis and somatic embryogenesis: Physiological and biochemical aspects. In: Roubelakis-Angelakis, K.A., and Tran Thanh Van, K. (eds) Markers of Plant Morphogenesis. pp. 19-38. Plenum Press, New York.

Thorpe TA (1994) Morphogenesis and regeneration. In: Vasil, I.K., and Thorpe, T.A. (eds) Plant Cell and Tissue Culture. pp. 17-36. Kluwer, Dordrecht.

Thorpe TA, Bagh K, Cutler AJ, Dunstan DI, McIntyre DD & Vogel HJ (1989) A ^{14}N and ^{15}N nuclear magnetic resonance study of nitrogen metabolism in shoot-forming culture of white spruce (*Picea glauca*) buds. Plant Physiol. 91: 193-202.

Thorpe TA & Beaudoin-Eagan LD (1984) ^{14}C-metabolism during growth and shoot formation in tobacco callus. Z. Pflanzenphysiol. 113: 337-346.

Thorpe TA & Biondi S (1981) Regulation of plant organogenesis. Adv. Cell Cult. 1: 213-239.

Thorpe TA & Gaspar T (1978) Changes in isoperoxidases during shoot formation in tobacco callus. In Vitro 14: 522-526

Thorpe TA, Harry IS & Kumar PP (1991) Application of micropropagation to forestry. In: Debergh, P.C., and Zimmerman, R.H. (eds) Micropropagation. pp. 311-336. Kluwer, Dordrecht.

Thorpe TA, Joy IVRW & Leung DWM (1986) Starch turnover in shoot-forming tobacco callus. Physiol. Plant. 66: 58-62.

Thorpe TA & Laishley EJ (1973) Glucose oxidation during shoot initiation in tobacco callus cultures. J. Exp. Bot. 24: 1082-1089.

Thorpe TA & Meier DD (1972) Starch metabolism, respiration and shoot formation in tobacco callus cultures. Physiol. Plant. 27: 365-369.

Thorpe TA & Meier D (1973) Effects of gibberellic acid and abscisic acid on shoot formation in tobacco callus cultures. Physiol. Plant. 29: 121-124.

Thorpe TA & Meier DD (1974) Starch metabolism in shoot-forming tobacco callus. J. Exp. Bot. 25: 288-294.

Thorpe TA & Murashige T (1968) Starch accumulation in shoot-forming tobacco callus culture. Science 160: 421-422.

Thorpe TA & Murashige T (1970) Some histochemical changes underlying shoot initiation in tobacco callus culture. Can. J. Bot. 48: 277-285.

Thorpe TA & Patel KR (1986) Comparative morpho-histological studies on the sites of shoot initiation in various conifer explants. N. Z. J. For. Sci. 16: 257-268.

Thorpe TA, Tran Thanh Van M & Gaspar T (1978) Isoperoxidases in epidermal layers of tobacco and changes during organ formation *in vitro*. Physiol. Plant. 44: 388-394.

Torrigiani P, Altamura MM, Pasqua G, Monacelli B, Serafini-Fracassini D & Bagni N (1987) Free and conjugated polyamines during *de novo* floral and vegetative bud formation in thin cell layers of tobacco. Physiol. Plant. 70: 453-460.

Torrigiani P, Altamura MM, Capitani F, Serafini-Fracassini D & Bagni N (1989) *De novo* root formation in thin layers of tobacco: Changes in free and bound polyamines. Physiol. Plant. 77: 294-301.

Tran Thanh Van K, Toubart P, Cousson A, Darvill AJ, Gollin DG, Chelf P & Albersheim P (1985) Manipulation of morphogenetic pathways of tobacco explants by oligosaccharines. Nature 314: 615-617.

Tran Thanh Van K & Trinh TH (1990) Organogenic differentiation. In: Bhojwani, S.S. (ed) Plant Tissue Culture: Applications and Limitations. pp. 34-53. Elsevier, Amsterdam.

Trewavas A & Gilroy S (1991) Signal transduction in plant cells. Trends Genet. 7: 356-361.

Trewavas A & Knight M (1994) Mechanical signaling, calcium and plant form. Plant Mol. Biol. 26: 1329-1341.

Tyron K (1956) Scopoletin in differentiating and nondifferentiating cultured tobacco tissue. Science 123: 590.

Villalobos VM, Leung DWM & Thorpe TA (1984a) Light cytokinin interactions in shoot formation in cultured cotyledon explants of radiata pine. Physiol. Plant. 61: 497-504.
Villalobos VM, Oliver MJ, Yeung EC & Thorpe TA (1984b) Cytokinin-induced switch in development in excised cotyledons of radiata pine cultured *in vitro*. Physiol. Plant. 61: 483-489.
Villalobos VM, Yeung EC, Biondi S & Thorpe TA (1982) Autoradiographic and ultrastructural examination of shoot initiation in radiata pine cotyledon explants. In: Fujiwara, A. (ed) Plant Tissue Culture. pp. 41-42. Japanese Assoc. Plant Tissue Cult., Tokyo.
Villalobos VM, Yeung EC & Thorpe TA (1985) Origin of adventitious shoots in excised radiata pine cotyledons cultured *in vitro*. Can. J. Bot. 63: 2172-2176.
Vollbrecht E, Veit B, Sinha N & Hake S (1991) The developmental gene *Knotted-1* is a member of a maize homeobox gene family. Nature 350: 241-243.
Walker KA, Yu PC, Sato SJ & Jaworski EG (1978) The hormonal control of organ formation in callus of *Medicago sativa* L. cultured *in vitro*. Amer. J. Bot. 65: 654-659.
WeigelD & Meyerowitz EM (1994) The ABCs of floral homeotic genes. Cell 78: 203-209.
Wernicke W & Park H-Y (1993) The apparent loss of tissue culture competence during leaf differentiation in yams (*Dioscorea bublifera* L.). Plant Cell Tiss. Organ Cult. 34: 101-105.
Wetmore RH & Reir JP (1963) Experimental induction of vascular tissues in callus of angiosperms. Amer. J. Bot. 50: 418-430.
Winkler MN, Mawson BT & Thorpe TA (1994) Alternative and cytochrome pathway respiration during shoot bud formation in cultured *Pinus radiata* cotyledons. Physiol. Plant. 90: 144-151.
White PR (1939) Controlled differentiation in a plant tissue culture. Bull. Torrey Bot. Club 66: 507-513.
White PR (1943) Nutrient deficiency studies and an improved inorganic nutrient for cultivation of excised tomato roots. Growth 7: 53-65.
Yang W-C, de Blank C, Meskiene I, Hirt H, Bakker J, van Kammen A, Franssen H & Bisseling T (1994) *Rhizobium* nod factors reactivate the cell cycle during infection and nodule primordium formation, but the cycle is only completed in primordium formation. Plant Cell 6: 1415-1426.
Yanofsky MF, Ma H, Bowman JL, Drews GN, Feldmann KA & Meyerowitz EM (1990) The protein encoded by the *Arabidopsis* homeotic gene *agamous* resembles transcription factors. Nature 346: 35-39.
Yasuda T, Hasegawa PM & Cheng TY (1980) Analysis of newly synthesized proteins during differentiation of cultured Douglas fir cotyledons. Physiol. Plant. 48: 83-87.
Yasutani I, Ozawa S, Nishida T, Sugiyama M & Komamine A (1994) Isolation of temperature-sensitive mutants of *Arabidopsis thaliana* that are defective in the redifferentiation of shoot. Plant Physiol. 105: 815-822.
Yeung EC, Aitken J, Biondi S & Thorpe TA (1981) Shoot histogenesis in cotyledon explants of radiata pine. Bot. Gaz. 142: 494-501.
Yoshida KT, Naito S & Takeda G (1994) cDNA cloning of regeneration-specific genes in rice by differential screening of randomly amplified cDNAs using RAPD primers. Plant Cell Physiol. 35: 1003-1009.

Chapter 8

FLORAL AND VEGETATIVE DIFFERENTIATION IN VITRO AND IN VIVO

K. TRAN THANH VAN
Institut de Biotechnologie des Plantes, Université de Paris-Sud, Bâtiment 630, URA 1128, CNRS, 91405 Orsay Cedex, France

1 INTRODUCTION

Plant morphogenesis proceeds through a multitude of events which occur according to a spatial and temporal sequence, some of them at specific time and sites (cell / tissue / organ), some gradually or permanently intertwined, some constitutively, and some with switches on and off which are unpredictable by the investigator although "programmed" in the plant. Unlike the animals, a plant which from its birth to its death is confined to grow and to develop at the geographical site of its birth, has to adapt itself to predictable and unpredictable seasonal environmental conditions. Those adaptive mechanisms acquired through the evolution of the species will be switched on or off according to the perception of the signals by the plant. Unlike the animals, the plant signal perception and transduction mechanisms are not as specific. Therefore, using mutants, developmental processes have been more clearly elucidated in animals than in plants.

Moreover, for the topic we deal with in this chapter, there are more floral organ identity mutants than floral meristem identity mutants (mutation affecting precisely the decision to shift to flower) found in *Arabidopsis* and *Antirrhinum* ; non-flowering mutants have not been identified so far.

The unpredictable variable background found *in vivo* in the context of an entire plant, even in mutants, is a serious handicap for the analysis of growth and development mechanisms in plants. However, higher plants offer an advantage over higher animals in that transgenic plants can easily be constructed at least for certain species and gametogenesis can be precociously obtained. These considerations have led us to propose that the analysis of the mechanisms of the control of morphogenesis be conducted on *in vitro* as well as *in vivo* systems.

In the first case, *in vitro* simplified model system such as Thin Cell Layer (TCL) system (Tran Thanh Van, 1981, 1991) of wild type and of developmental mutants (if available) should be used ; simplified system because it reduces the variable background due to interactions between organ/tissue/cell and environment to a minimum, and *in vitro* for the simplification of transport of nutrients metabolites, growth regulators and signal perception from other plant parts.

In vitro simplified model systems allow the investigator to programme separately specific morphological pattern and to identify gene(s) supposedly involved in each programme. Comparison between different morphogenetic programmes will lead to the identification of specific genes.

For the second case, we can use *in vivo* systems from transgenic plants transformed with sense/antisense/ transposon constructs. However, there are several limitations in the use of transgenic plants. One morphogenetic programme results from the activity (positive and/or negative) of a set of gene(s) and rarely from a single gene. Even in the latter case, neither the spatial and temporal sequence of its expression within the plant body, nor its interrelation with other gene(s) can be mimicked. However, some important progress has been made with the tools presently available, such as genetic manipulation and mutants.

In this chapter, in order to underline particularity of the flower programme, we first present some selected significant morphological changes illustrating the limitations in the plasticity of *in vivo* wild-type and mutant systems. Then, the experimentally induced changes in *in vitro* systems will be examined and compared to *in vivo* systems. Key-factors controlling the morphogenetic programme, common to both *in vitro* and *in vivo* systems will be underlined.

2 IN VIVO SYSTEMS : LIMITATIONS IN MORPHOGENETIC CHANGES

2.1 Normal Pattern of Growth and Development

In zygotic or somatic embryogenesis (with some exceptions, such as protocorms in Orchids), the formation of an embryo body results from the differentiation of an apical meristem (the future shoot meristem) and a basal meristem (the future root meristem) at each pole of an axis (the future stem). During normal plant growth and development (classically defined as an acquisition of the reproductive phase ; see section 2.2), the basal meristem develops into root meristem in which growth pattern is limited to the initiation of adventitous root ; this growth pattern is maintained throughout the life cycle of the plant. In contrast, the apical meristem (shoot meristem) develops from its juvenile structure to the adult structure with the same function: initiation of internodes and leaves of different shapes which sign the chronological / physiological phases of growth of the shoot meristem.

This function ceases abruptly or progressively according to the species. A decision is taken after integrating an array of signals: the vegetative meristem now changes into a floral meristem. Flowering, a crucial event during plant development, could be divided into: (i) an induction phase, during which the floral stimulus is perceived but the functioning of the meristem has not changed as yet, (ii) a flower initiation phase, leading to a prefloral meristem during which modified leaves or bracts are formed, and (iii) a flower development phase during which floral organs are formed. Therefore, shoot meristem is functionally different from root meristem. Shoot meristem can develop into floral meristem and *vice versa*.

2.2 Morphogenic Changes at the Site of Organogenesis

2.2.1 Root meristem

In order to outline the characteristics of floral meristem we briefly report some morphogenic changes (*in vivo* and *in vitro*) recorded at the site of root meristem itself or of root tissue.

It has been reported that in *Nasturtium*, the adventitous roots formed at the stem node can be converted into shoot meristem. This conversion occurs at the site of the root meristem itself, following kinetin application to the node, provided the root was at an appropriate stage of growth. As far as we know, there is no conversion of root meristem into a floral meristem. Cortical cells of mature roots can be induced to form vegetative buds in several species (Trinh, 1978 ; for review see Aeschbacher, 1994) or somatic embryos (Sticklen, 1991).

2.2.2 Mature shoot meristem

A large number of works were devoted to the shift from vegetative to floral meristem (Bernier, 1988; Bernier et al., 1993; Tran Thanh Van, 1985). We only report here some characteristic traits which introduced concepts different from the classical one.

2.2.2.1 Extreme case of recalcitrance to flowering

In a perennial rosette plant, *Geum urbanum* (Rosaceae), requiring vernalization, not all mature vegetative shoot meristems are transformed in to floral meristems after a cold treatment; vernalization is perceived differently by the terminal meristem and the different axillary meristems. The terminal meristem keeps its vegetative functioning during the whole life cycle of the plant, thus insuring the perennial habit and the rosette type of growth while certain types of axillary meristem develop into floral meristem (Fig. 1) after active cell divisions in the corpus zone. Figure 2 illustrates the histological stage of an axillary bud before cold treatment (Tran Thanh Van-Le Kiem Ngoc 1965, Tran Thanh Van 1975 ; 1985).

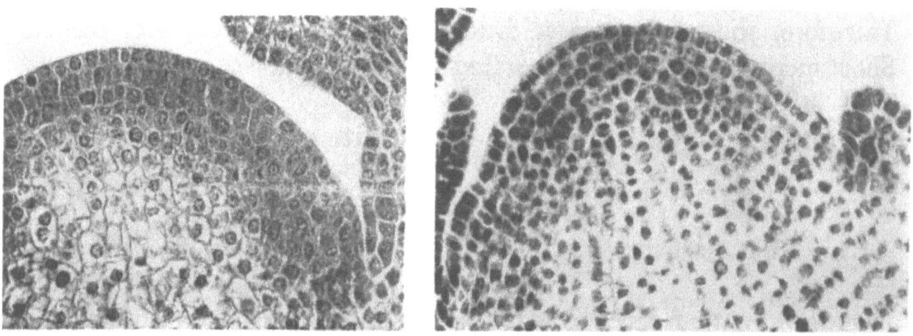

Figure 1-2. Geum urbanum. Axillary bud at the floral stage, showing sepal and petal initiation(Fig.1, X=560). Axially bud before vernalization (Fig.2, X=378).

We have analyzed the histological changes in the terminal bud during a cold treatment for 8 weeks, sufficient for the floral induction of certain type of axillary buds. Unexpectedly, we found an activation in cell division in the corpus and medullary zones but this activation was not strong enough for this meristem to be converted into a floral meristem.

(a) Substitution of vernalization. This led us to work out two methods to trigger flowering on this recalcitrant meristem. One method was to prolong the cold treatment up to 50 weeks instead of 8 weeks during which the corpus and the medullary zones have reached their appropriate activation level for its shift to a floral meristem. The second method was to substitute the 50 weeks of cold treatment by a combination treatment of 13 weeks of cold with kinetin and gibberellins for activating cell division in corpus and medullary zones, respectively.

With the second treatment, the terminal bud meristem underwent progressive change in its functioning: instead of multilobed leaves, which are characteristic of the the vegetative phase, trilobed and stipulated leaves, characteristic of the prefloral meristem were formed indicating a progressive shift from vegetative to floral functioning (Fig. 3). This was a unique example in the literature of such a strong reluctance to flowering being overcome. The perennial and polycarpic habits were changed into annual and monocarpic ; the rosette type of growth changed into an erect type of growth. As for the axillary meristems which absolutely required 8 weeks of

Figure 3. Geum urbanum. Multilobed (vegetative) leaf (left), trilobed and stipulated (prefloral) bract (right) and intermediate stipulated and multilobed leaf (middle)

cold treatment for flowering, we have shown that this treatment can be substituted either by the removal of the terminal bud thus suppressing the inhibition by apical dominance (hence related to, among other factors, auxin transport/metabolism) or by the application of kinetin in the presence of the terminal bud.

(b) New concept of floral differentiation. These observations have led to a new concept of floral differentiation. Flowering occurs normally unless inhibited. It results from the suppression of an inhibition. If the induction requires vernalization and/or photoperiodism, it is to overcome the inhibition. This concept was different from the classical ones in which floral induction was attributed to the synthesis of a specific "florigen". Up to now, there is no report of gene(s) coding for a florigen; instead, gene required for vegetative shoot development, inhibiting flowering was found. Since the first publication of the new concept on flowering, the florigen theory was modulated into "florigen/antiflorigen" theory.

This concept has been adapted successfully for controlling flowering in less reluctant species, such as *Bambusa* or Orchids; precocious flowering was induced in these plants. Flowering in most *Bambusa* species occurs once or twice in a century. Young plants from seeds can flower in vivo after regular removal of leaves (Tran Thanh Van, unpublished results) and in vitro (Nadgauda et al., 1990). *Phalaenopsis* plants, which flower 5-6 years after development from protocorms, flowered within 2 years after the suppression of the terminal bud (Tran Thanh Van, 1974a, 1975). Terminal bud of floral stalks (*Phalaenopsis*), or pseudo-bulb (*Cymbidium*, *Odontonia*, *Odontoglossum*) which normally senesced and necrosed, developed into inflorescences when the nutritional and light regimes were optimal (Tran Thanh Van, 1974b; 1975).

According to our working hypothesis, flowering occurs when the meristem is not inhibited by vegetative differentiation or by limiting factors or limiting level of growth. We will see in the next section that in EMF *Arabidopsis* mutants, flowering occurred in the absence of EMF gene product regulating vegetative differentiation (Sung et al 1992).

2.2.2.2 Early / Late flowering mutants

Developmental mutant phenotypes provide information on the role of gene(s) in the wild type. Two species have been extensively used for screening morphogenetic mutants : *Arabidopsis* and *Antirrhinum*. It turned out that the mutants isolated until now are more related to the floral organs rather than to the floral meristem, i.e. regulation of shift from vegetative differentiation to floral differentiation (Haugh et al., 1995, Schultz et al., 1991, Schwarz-Sommer et al., 1990, Shannon., 1991, Sommer et al., 1990, Alvarez et al., 1992).

(a) Pin formed mutants. For *Arabidopsis*, until now only early or late flowering mutants have been obtained but not non-flowering mutants. The most drastic mutant in the inflorescence axis structure is the *pin* 1-1 mutant with no flower nor floral related structure; the top of the "*pin*" has no meristematic structure (Okada et al., 1991). Wild type plants cultured in the presence of polar auxin transport inhibitors morphactin (HFCA, 9-hydroxyfluorene-9-carboxylic acid) or NPA (N-(1- naphthyl) phthalamic acid), exhibited phenotype similar to the *pin* 1-1 phenotype. TIBA (2,3,5-triiodobenzoic acid) was less effective and the CPIB (2-(p-chlorophenoxy)-isobutyric acid), an antagonist of auxin action (and not of an auxin transport), had no effect. In *emf* mutants, endogenous level of IAA was about 8% and IAA polar transport 10% of the wild type.

The mutation might effect both IAA synthesis and its polar transport. It was hypothesized that ‘cell to cell communication networks that share the auxin polar transport’ could also be required. In the *in vitro* model system of tobacco Thin Cell Layer (TCL), data were presented in relation to the cell wall oligosaccharides as informational molecules released upon treatment with growth substances (Tran Thanh Van 1973, Tran Thanh Van et al., 1985, 1990b, Mutafstchief et al., 1987).

(b) Other morphogenetic mutants. Other mutations related to floral organ identity in *Arabidopsis* were found more or less related to floral meristem identity. It is the case of mutation at loci TFL (Terminal Flower), EFL 1 (Early Flowering) and EFL 2 (Shannon, 1991, Zagotta et al., 1992). The TFL mutant develops a terminal flower from the inflorescence shoot which remains indeterminate in the wild type. TFL, EFL1 and EFL2 gene product(s), still unknown, could activate the vegetative programme and therefore repress the floral programme as does the EMF (Embryogenic Flower) gene (see next section).

Another single copy gene, LEAFY (LFY) regulates the formation of determinate floral meristem in *Arabidopsis*. Its transcripts were reported to be expressed only in floral meristem and floral organs (Schultz et al., 1991). Kelly et al. (1995), therefore, hypothesized that its homologs NFL1 and NFL2 transcripts would also be expressed only in the floral meristem / organ in tobacco, a determinate (in its main axis) flowering plant . Unexpectedly, they found the expression not confined to the floral meristems but also in indeterminate vegetative meristems. This divergence could be explained, according to the authors, by a difference in the growth habit, an indeterminate racemose plant, and a determinate cymose plant (*Arabidopsis*). We think that the difference between these two species also lies in the difference in the requirement for environmental conditions for flowering: the photoperiodism and vernalization for *Arabidopsis*, no photoperiodism nor vernalization required for tobacco.

These findings bring some limitation to the assertion that since "*Arabidopsis* and *Antirrhinum* present homology" for certain gene(s), "they are representative of all Angiosperms" (Haugh et al., 1995). Reservations have also to be made relative to the specificity of expression of the presumed floral organ identity genes. Jokufu et al. (1994) found that, in *Arabidopsis* itself, APETALA2 (AP2) gene transcripts were expressed, not only in floral organs as previously reported, but in inflorescence meristem and in non floral organs (leaf, stem). They assumed that AP2 gene is expressed throughout most of *Arabidopsis* development.

2.2.3 Juvenile shoot- / floral meristem

The *emf* (Embryogenetic Flowering) *Arabidopsis* mutant develops a short determinate inflorescence immediately after germination; this occurs independently of the environmental conditions "bypassing the vegetative shoot" (Sung et al.,1992). This corroborates with our concept on flowering presented in section 2.2.2.1. In fact, these authors also hypothesized that flowering is the obligate developmental programme of the apical meristem unless it is inhibited by the vegetative programme regulated by EMF gene, which function is not known as it is the case with most of the morphogenetic gene(s).

Other cases of embryogenetic flowering were observed in *A. hypogaea* and *Bidens radiatus*, (Tran Thanh Van unpublished results) and *Panax ginseng*, (Chang et al., 1980). In *Geum urbanum*, continuous suppression of floral buds developed after vernalization from axillary floral stalks, triggered the development of lateral branches up to the fifth or the sixth degree of ramification. These lateral branches initiated juvenile meristems from which flowers of very small size developed (Fig. 4). This observation together with the embryogenetic flowering indicates that the enlargement of the meristem is not a prerequisite condition for flowering to occur.

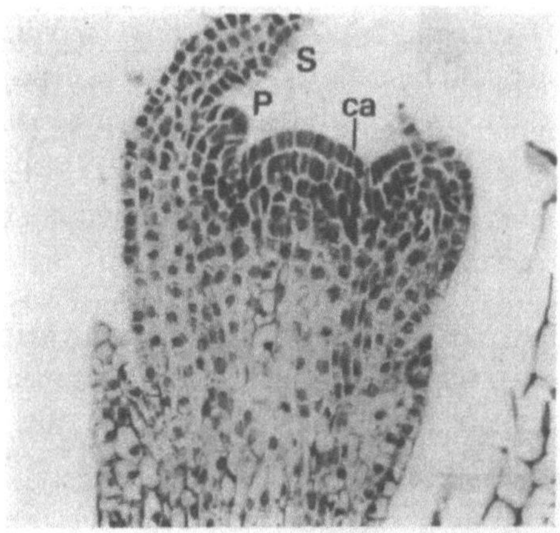

Figure 4. Geum urbanum . Miniaturized floral meristem. X=235

2.3 Conclusion

There are limitations in the conversion from one morphogenetic programme to the other. These limitations were found in *Geum urbanum* where the absolute recalcitrance to flowering has been overcome by activation of cell division in the corpus zone.

All the examples of morphogenetic changes in *in vivo* systems are related to changes occurring at the site of pre-existing meristems ; that is to say that floral differentiation seems to require for its occurrence a pre-existing shoot meristem (juvenile or adult), i.e. a group of more or less actively dividing corpus-, medullary- and flank-zone cells covered by one or several layers of tunica. In these meristems, activation of cell division in the corpus zone is a prerequisite condition for floral differentiation. This activation can be induced by cytokinins.

There are other limitations in morphogenetic changes in *in vivo* systems. For example, the shoot meristem originates from an embryonic apical meristem itself built during zygotic embryogenesis and is, therefore, subjected to ontogenetical/physiological constraints. Similarly, the structure of the root meristem does not lend itself to a conversion into a floral meristem but only into a shoot meristem in a few exceptional cases (where root originates from superficial tissues as in *Nasturtium*).

One can ask first about the identity of gene(s) regulating the differentiation of meristematic structures itself, and next the ones regulating shoot-, root-, floral-, vegetative- and embryo-, differentiation. In *in vivo* systems, as stressed in the introduction, developmental gene(s) might have been expressed or inhibited following a spatial and temporal sequence preimposed during ontogenetic processes from the embryogenic phase to the juvenile, mature and reproductive phases. Some gene(s), such as AP2, believed to be expressed specifically in whorl 1 and 2 of *Arabidopsis* floral primordia was found to be expressed throughout the development. Therefore, even though morphogenetic and especially developmental mutants are of great value, because of the common developmental background inherent to entire organisms, and even the mutants, the analysis of morphogenesis gains when carried out in parallel with *in vitro* systems in which different morphogenetic patterns (somatic embryo, root, flower, vegetative bud) can be programmed *directly*, and *separately* from differentiated cells and not from embryogenic or meristematic cells. It is a case of the *in vitro* model system of Thin Cell Layer (TCL).

3 IN VITRO SYSTEMS: PROGRAMMABLE DIFFERENTIATION OF FLORAL AND VEGETATIVE MERISTEMS DIRECTLY FROM CELL LAYERS AND NOT FROM PRE-EXISTING MERISTEMS

In vitro systems for differentiation of root or shoot meristem from a plant organ or from a callus generated through cambium activity are largely used for the study of the corresponding morphogenesis (see Chapters 6 and 7). Somatic embryo differentiation from a single cell isolated from a callus is also a largely used in vitro system (see Chapter 5). In contrast, the control of direct differentiation of floral meristem without an intermediate callus or vegetative phase is up to now a unique example in literature.

The model system of tobacco Thin Cell Layer has been already described (Tran Thanh Van, 1973a, b, 1980, 1981, 1991; Tran Thanh Van et al., 1990a, b, 1992, 1993, 1995; Bridgen et al., 1988; Compton et al., 1992; Rajeevan et al., 1993; Tran Thanh Van et al., 1999) and extended to other economically important species (Mulin et al., 1989; Pélissier et al., 1990; Jullien et al., 1994). We have introduced it to other research groups for different economically useful species (Pua et al., 1987, 1989; Goh et al., 1994, 1995; Lakshmanan et al., 1995).

We will briefly report in the following pages the key-factors controlling floral / vegetative differentiation in longitudinal (TCL) and transverse Thin Cell Layer (tTCL) systems in relation to some key-factors presented in section 2.

3.1 The morphogenetic patterns programmed in tobacco TCL

Tobacco TCL are small explants (1mm x 5mm) of 3 to 6 layers of epidermal and subepidermal cells excised from floral branches. They are programmed to differentiate directly, on the surface of the epidermis, approximately 30 to 50 functional flowers per TCL, at its proximal pole (towards the stem apex of the donor plant). Alternatively, they can be programmed to differentiate directly 10 to 20 roots or 500 to 800 vegetative buds with no polarity, in contrast to the flower programme. Flowers-, roots- and callus- programmes are represented in Figures 5, 6, 7, respectively.

All these organs differentiated from the same layer, the subepidermal layer in which cell division was arrested before the excision of TCLs from the donor plant . The time interval was very short: 10-12 days for flower and vegetative bud programmes, and 16-18 days for root programme. Such a rapid and direct differentiation of organs raised the question of the difference in the nature of cell division itself in the subepidermal cells leading to the direct differentiation of specific organ. Flowers were functional, i.e. meiosis occurred resulting in fruit setting and seed formation after 3 to 5 weeks from the Thin Cell Layer culture. The high frequency of organ formation, the short time interval required for their differentiation conferred to the TCL system distinct advantages for the study of floral related gene(s) expressed in transgenic plants during floral differentiation and after meiosis (Ammirato, 1987; Trinh et al 1987).

The TCL systems allow the study of cytological, physiological, biochemical and molecular changes occurring in each morphogenetic programme (Tran Thanh Van et al., 1992). To assess if these changes are specific to a given morphogenetic programme, a comparison between these programmes will permit selection for specific markers. Their advantage over

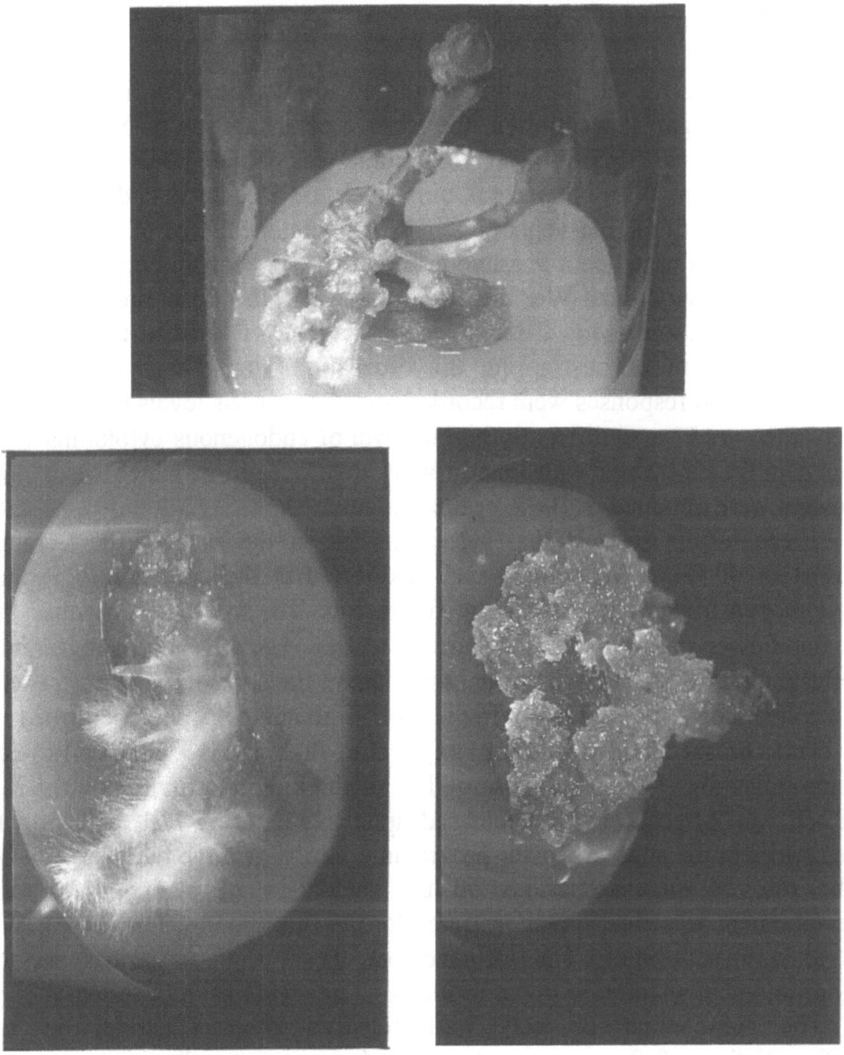

Figure 5-7. Tabacco Thin Cell Layer: flower programme (Fig. 5 above), root programme (Fig. 6 left) and callus programme (Fig. 7 right)

morphogenetic mutants resides in the fact that the selected programme, flower for example, is *not sustained* by the other *unavoidable programmes* such as root, stem, leaf with their intertwined functions thus forming a background which does not lend itself to be dissected!

Once morphogenetic markers are identified with the in vitro model system allowing to programme *separately* different and pure (in 100 % of TCLs) morphogenetic pattern, a test of their expression has to be conducted with *in vivo* system.

3.1.1 Influence of cytokinins

The key-factors for the floral differentiation in tobacco were light intensity and quality, sugar concentration, the relative concentration of NAA or IBA versus cytokinins. A pure flower programme was obtained only with kinetin as the cytokinin. With other cytokinins, such as BA, zeatin and 2iP, mixed programme of floral and vegetative buds was obtained. One of our working hypothesis was based on a possible difference in catabolism between kinetin *versus* other cytokinins. We, therefore, compared the floral and vegetative bud differentiation using different concentrations of kinetin, zeatin and a substituted form of zeatin, the dihydrozeatin (Tran Thanh Van 1991). Morphological responses were recorded and endogenous levels of cytokinins were measured : a significantly higher level of endogenous cytokinins were found in TCLs treated with kinetin and dihydrozeatin in which cases only flowers were obtained in 100% of the explants.

These results suggest that cytokinins (as in the case of the terminal meristem of *Geum urbanum*) and auxins (as in the case of *pin* formed *Arabidopsis* mutant) are among the key-factors. The strict polarity observed in the flower programme, in contrast to the vegetative bud programme, reflects the importance of a normal polar auxin transport if we compare with the conversion of the wild-type to the *pin* 1-1 mutant when it is perturbed.

This polar auxin transport might be in relation with cell-to-cell communication. This is on the same line of thinking and finding on the role of cell wall oligosaccharides released upon treatment of TCL with growth regulators in the shift of flower programme into vegetative bud programme. The role of oligosaccharides on the regulation of plant growth and development was clearly demonstrated for the first time by Tran Thanh Van et al. (1985, 1990c). The influence of these oligosaccharides on the morphological restoration in *pin* formed mutants remains to be tested.

The implication of cell wall released oligosaccharides in the differentiation of flowers, vegetative buds, roots or callus suggests an implication of cell wall hydrolases. As a matter of fact, gene(s) identified as being involved in flower differentiation in tobacco TCL were : basic b 1-3 glucanase, basic chitinase , extensine, a cell wall protein (Neale et al., 1990) and thaumatin. This identification was made through a comparison between 7 day-TCLs programmed separately for floral or vegetative bud differentiation. However, their transcripts were found strongly expressed also in root (Meeks-Wagner et al., 1989). This result could be related to the transcripts homology level or to the expression (which could be stopped at the transcription level) of the genes. In fact, thaumatin-like proteins (47 and 42 KDa) were found in the 7 day-TCLs programmed for flowers and also in *in vivo* floral organs (anther, ovary and bracts). These proteins were absent in vegetative organs (stem, shoot meristem, root or leaf) (Richard et al., 1992). It was reported that thaumatin shared homology with osmotin (26 KDa) and was found in roots and in cells of water-stressed tobacco plants (King et al., 1986). However, their compartmentalization in the cell was different: osmotin in vacuoles and thaumatin-like proteins in the cytoplasm.

(a) Preinduction or induction. With tobacco, a day neutral species, flower programme can only be obtained on TCLs excised from floral branches and not from the vegetative part, the basis of the stem (Tran Thanh Van, 1973a). This raises the question whether the TCLs were preinduced to form flower. Even if it was the case, this would not weaken the "morphogenetic control ": for whatever the cells were prepared to, it was possible to programme all patterns, not only flowers, but also root which normally do not originate from superficial tissue such as the subepidermal cells. But the TCLs were not preinduced for floral differentiation: the transcripts of genes identified in TCLs-programme for flower formation were not expressed in the TCLs at the time of their excision. However, with photoperiodic *Nicotiana* species, flowers can be obtained provided that the appropriate photoperiod be applied to the TCLs in culture. These considerations suggest strongly that the flower differentiation on TCLs results from an expression of gene(s) during *in vitro* TCL culture under specific *in vitro* conditions and not from a preinduced phase "inheritated" from the donor plant. The requirement for floral phase of the donor plant could be interpreted as the low level of vegetative gene(s) products (similar to EMB gene product) after the onset for flowering in the main axis. In 7 day-vegetative programme-TCLs another set of genes was expressed (Meeks-Wagner et al., unpublished results). Thus, the floral programme could be considered as resulting from the expression of a set of genes and the non-activation of another set of genes.

Transgenic plants with selectively manipulated sense / antisense gene(s) will be constructed and used for programming morphogenesis on TCLs. Several serious limitations may arise due to: (i) the choice of gene to be inserted, (ii) the expression / suppression of inserted genes, and (iii) their interference with genes used for the selection of the transgenic plants.

In a preliminary study (Tran Thanh Van, 1990a) since auxin was shown to be one of the key- factors, we compared morphogenetic differentiation in the cultures of leaf discs and TCLs from Ri transformed tobacco plants. While TCLs differentiated morphogenetic programmes accordingly to the medium established for different morphogenic programme, leaf discs formed hairy root irrespective of the medium used. One can raise the question of differential expression of Ri in different organs / tissues. An elegant method was to use gene construct with transposon in transgenic plants where mosaic cells expressing the inserted gene upon transposon excision are surrounded by cells in which the inserted gene was not expressed (Estruch et al 1993).

3.1.2 Ectopic vegetative/floral expression in mosaic transgenic tobacco with *ipt* gene

According to the previous results, cytokinins are an important key-factor in flower differentiation. Tobacco plants transformed with the 35S-Ac-ipt construct (maize Ac transposon inserted into the gene encoding for isopentenyl transferase) showed, *in vivo*, differentiation of adventitous shoots on leaves (epiphylly) upon transposon excision (Estruch et al 1993). After flowering of the main inflorescence, the new epiphyllous meristems formed at the leaf tip, differentiated two leaves first and then normal or abnormal floral organs, while the adventitous shoot formed prior to the onset of flowering remain vegetative.

The normal epiphyllous flowers contained 140 +/- 100 p.mol.g.fw of zeatin riboside as compared to in 28000 +/- 6000 p.mol.g.fw in the abnormal flowers, 39 +/- 13 p.mol.g.fw in the flower buds of 35S-Ac-*ipt* transgenic plants and 55 +/- 25 p.mol.g.fw in flowers of nontransformed plants. These data indicate that a high cytokinin level altered the development of floral organs.

As for the relation between the onset of flowering and the differentiation of epiphyllous normal and abnormal flowers, one can again hypothesize that the activity of vegetative gene(s) declined and thus favored floral differentiation ; unless the rate of transposon excision changed with the phase change of the plant, causing a decline in the endogenous level of cytokinins in the mosaic cell population to a level favouring floral differentiation.

3.2 Vegetative / floral differentiation on transverse TCL of juvenile meristem/ tissue

tTCLs 1 mm thick from *Soybean* cotyledonary nodes cultured *in vitro* in the presence of 1 mM BA produced multiple shoot clusters (Tran Thanh Van, unpublished results). BA used as a pretreatment of seedlings enhanced somatic embryogenesis on tTCLs excised from seedling epicotyl of the monocot plants *Sorghum* and *Digitaria* and from hypocotyl or cotyledon tTCL of *Panax ginseng* (Ahn et al., 1995). The same treatment (20 mM BA) to 2 mm thick tTCLs from the cotyledonary nodes of soybean seedlings (pretreated with 5 mM BA) induced floral differentiation on 30% of the tTCLs; the other tTCLs formed vegetative shoots as described above.

juvenile meristem situated at the cotyledonary node (tTCLs of 1 mm would be better than of 2 mm for a more homogenous response). Cell division would occur at different frequencies, leading to vegetative or floral meristem differentiation. Similar observations were previously described in the corpus zone of the terminal shoot meristem of *Geum urbanum* depending upon its vegetative functioning or its shift to a floral functioning (section 2.2.2.1.).

Pretreatment by a cytokinin had an important impact on the subsequent morphognenetic response of the tTCLs. As the cytokinin transport is not polar, one can hypothesize that it is transported to different parts of the seedlings and thus increased its endogenous level prior to the exogenous supply to the tTCLs *via* the culture medium.

These results confirm the data recorded from morphogenic mutants, that cytokinins (Tran Thanh Van, 1992, for review see Binns, 1994), auxins and oligosaccharides (Tran Thanh Van et al., 1985; Lerouge et al., 1990; Marfa et al., 1991) play important role in determining the differentiation of flowers or vegetative buds in several plant species. It is interesting to point out that other biochemical markers such as polyamines are involved in floral organ differentiation. However gene(s) regulating floral organ differentiation isolated from *Arabidopsis* and *Antirrhinum* mutants were not related to these molecules.

3.3 Conclusion

With tobacco TCLs, all morphogenetic patterns can be programmed directly, and in a sequence desired by the investigator himself and not as determined by the ontogenetic processes inherited from the embryo. For example, root programme can be initiated first, or flower programme first followed by the other one without any stem or leaf. This will allow the investigator to "erase" the unpredictable background inherent to all integrated systems i.e entire plants (mutants or wild type).

4 GENERAL CONCLUSIONS

With the blossoming in the literature of morphogenetic mutants especially in *Arabidopsis* and *Antirrhinum*, more than 20 genes related to early or late flowering (but not to non-flowering) have been isolated, two genes have been sequenced (for review see Haugh et al., 1995). Their function is not yet fully elucidated. However, study at the physiological level of *in vivo* systems pointed out to some functions related to, but may be not causal to, the decision of a meristem to shift from the vegetative pattern to the floral pattern. Moreover, the isolation of specific gene(s) from a complex background of unpredictable changes is not an easy task.

With *in vitro* systems, one can ask how a group of differentiated cells has taken the decision to enter cell division and to programme themselves directly (and not to shift) into flower, vegetative bud, root or embryo formation.

While more appropriate morphogenetic mutants, especially non-flowering mutants, are yet to be identified for the study of floral / vegetative differentiation, in *in vitro* simplified systems, *flowering or non-flowering* can be programmed through the use of well defined key-factors. Although their molecular mechanisms of action remain to be more clearly elucidated in plant system, gene(s) which were switched on or off in the early phase of floral differentiation were identified in simplified model systems. New information will emerge with progress being made in the knowledge of gene expression and its regulation and of gene once inserted in transgenic plants which however can still bear its resident genes. Furthermore, analysis should be carried out on both morphogenetic mutants and on simplified model systems using wild type and transgenic plants constructed with gene(s) identified through morphogenetic mutation (if available) and through morphogenetic changes that the investigator can programme.

Lastly, we should bear in mind that the reproductive organ differentiation is the ultimate morphogenetic programme in phanerogams. Therefore, floral / vegetative differentiation raises another ultimate question : what controls mitosis and meiosis.

5 ACKNOWLEDGEMENTS

Nicole Bachala and Aida Cherkaoui are greatly acknowledged for their help in the preparation of the manuscript.

6 REFERENCES

Aeschbacher RA, Schiefelbein JW & Benfey Ph N (1994) The genetic and molecular basis of root development. Ann. Rev. Plant Physiol. Plant Mol. Biol. 45: 25-45.

Alvarez J, Gulf CL, Lu X-H & Smyth DR (1992) Terminal flower: a gene affecting inflorescence development in *Arabidopsis thaliana*. Plant J. 2: 103-116.

Ammirato PV (1987) Speeding transgenic Plants. Bio/Technology 5: 1015

Bernier G (1988) The control of floral evocation and morphogenesis. Ann. Rev. Plant Physiol. 39: 175-219.

Bernier G, Havelange A, Houssa C, Petitjean A & Lejeune P (1993) Physiological signals that induce flowering. Plant Cell 5: 1147-1155.

Binns AN (1994) Cytokinin accumulation and action: biochemical, genetic and molecular approaches. Ann.Rev.Plant.Physiol. Plant.Mol.Biol. 45: 173-96.

Bridgen MP & Veuilleux RE (1988) A comparison of *in vitro* flowers to *in vivo* flowers of haploid and diploid *Nicotiana tabacum* L. Plant Cell Tiss. Organ Cult. 13: 3-13.

Chang WC & Hsing YI (1980) *In vitro* flowering of embryoids derived from mature root callus of ginseng. Nature 284: 341-342.

Compton ME & Veilleux RE (1992) Thin Cell Layer morphogenesis, Hort. Rev 14: 239-264.

Estruch JJ, Granell A, Hansen G, Prinsen E, Redig P, Van Onckelen H, Schwarz-Sommer Z, Sommer H & Spena A (1993) Floral development and expression of floral homeotic genes are influenced by cytokinins. Plant J. 4: 379-384.

Goh CJ, Lakshmanan P & Loh CS (1994) High frequency direct shoot bud regeneration from excised leaves of mangosteen (*Garcinia mangostana* L.). Plant Sci. 101: 173-180.

Goh CJ, Nathan MJ & Kumar PP (1995) Direct organogenesis and induction of morphogenic callus through thin section culture of *Heliconia psittacorum*. Sci. Hortic. 62: 113-120.

Haugh GW, Schultz EA & Martinez-Zapater JM (1995) The regulation of flowering in *Arabidopsis thaliana*: meristems, morphogenesis, and mutants. Can. J. Bot. 73: 959-981.

Jofuku KD, den Boer BGW, Van Montagu M & Okamuro JK (1994) APETALA2: regulation of *Arabidopsis* flower development and homeotic gene expression by a new class of regulatory protein. Plant Cell 6: 1211-1225.

Jullien F & Tran Thanh Van K (1994) Micropropagation and embryoid formation from young leaves of *Bambusa glaucescens* 'Golden goddess'. Plant Sci. 98: 199-207.

Jullien F & Wyndaele R (1992) Precocious *in vitro* flowering of soybean cotyledonary nodes. J. Plant Physiol. 140: 251-253.

Kelly AJ, Bonnlander MB & Meeks-Wagner DR (1995) NFL, the tobacco homolog of floricaula and leafy, is transcriptionaly expressed in both vegetative and floral meristems. Plant Cell 7: 225-234.

King GL, Hussey CE & Turner VA (1986) Protein induced by NaCl in suspension cultures of *Nicotiana tabacum* accumulates in whole plant roots. Plant Mol. Biol. 7: 441-449.

Lakshmanan P, Loh C-S & Goh C-J (1995) An *in vitro* method for rapid regeneration of a monopodial orchid hybrid *Aranda deborah* using thin section culture. Plant Cell Rep. 14: 510-514.

Lerouge P, Roche P, Faucher C, Fabienne M, Truchet G, Prom JC & Der J (1990) Symbiotic host specificity of *Rhizobium meliloti* is determined by a sulphated and acylated glucosamine oligosaccharide signal. Nature 344: 781-784.

Marfa V, Gollin D, Eberhard S, Mohnen D, Darvill A & Abersheim P (1991) Oligogalacturonides are able to induce flowers to form on tobacco explants. Plant J. 1: 217-225.

Meeks-Wagner D, Denis E, Tran Thanh Van K & Peacock W (1989) Tobacco genes expressed during *in vitro* floral initiation and their expression during normal plant development. Plant Cell 1: 25 - 35.

Mulin M & Tran Thanh Van K (1989) Obtention of *in vitro* flowers from epidermal cell layers of Petunia hybrida Hort. Plant Sci. 62: 113-121.

Mutafstchiev S, Cousson A & Tran Thanh Van K (1987) Modulation of cell growth and differentiation by pH and oligosaccharides. In British Plant Growth Regulator Group, Monograph 16.

Nadgauda SM, Paarasharapi VA & Mascarenhas AF (1990) Precocious flowering and seedling behaviour in tissue-cultured bamboos. Nature 344: 335-336.

Neale A, Walhleithner J, Lund M, Bonnet KA, Meeks-Wagner D, Peacock W & Dennis E (1990) Chitinase, b -1,3-Glucanase, osmotin, and extension are expressed in tobacco explants during flower formation. Plant Cell 2: 673-684.

Okada K, Ueda J, Komaki MK, Bell CJ & Shimura Y (1991) Requirement of the auxin polar transport system in early stages of *Arabidopsis* floral bud formation. Plant Cell 3: 677-684.

Pua EC, Mehra-Palta A, Nagy F & Chua NH (1987) Transgenic plants of *Brassica napus*. Bio/Technology 5: 815-817.

Pua EC, Trinh TH & Chua NH (1989) High frequency plant regeneration from stem explants of Brassica alboglabra Bailey *in vitro*. Plant Cell Tiss. Organ Cult. 17: 143-152.

Pélissier B, Bouchefra O, Pépin R & Freyssinet G (1990) Production of isolated somatic embryos from sunflower thin cell layers. Plant Cell Rep. 9: 47-50.

Rajeevan M & Lang A (1993) Flower bud formation in explants of photoperiodic and day-neutral *Nicotiana biotypes* and its Bearing on the regulation of flower formation. Proc. Natl. Acad. Sci. USA. 90: 4636-4640.

Richard L, Arro M, Hoebeke J, Meeks-Wagner DR & Tran Thanh Van K (1992) Immunological evidence of thaumatin-like proteins during *in vitro* and *in vivo* tobacco floral differentiation. Plant Physiol. 98: 337-342.

Schultz EA & Haughn GW (1991) LEAFY, a homeotic gene that regulates inflorescence development in *Arabidopsis*. Plant Cell 3: 771-781.

Schwarz-Sommer Z, Huijser P, Nacken W, Saedler H & Sommer H (1990) Genetic control of flower development by homeotic genes in *Antirrhinum majus*. Science 250: 931-936.

Shannon S & Meeks-Wagner DR (1991) A mutation in the *Arabidopsis TFL1* gene affects inflorescence meristem development. Plant Cell 3: 877-892.

Sommer H, Beltran J-P, Huijser P, Pape H, Lonnlg W-E, Saedler H & Schwarz-Solnmer Z (1990) *Deficiens*, a homeotic gene involved in the control of flower morphogenesis in *Antirrhinum majus*: The protein shows homology to transcription factors. EMB0 J. 9. 605-613.

Sticklen MB (1991) Direct somatic embryogenesis and fertile plants from rice root cultures. J. Plant Physiol. 138: 577-580.

Sung, ZR, Belachew A, Shunong B & Bertrand-Garcia R (1992) EMF, an *Arabidopsis* gene required for vegetative shoot development. Science 258: 1645-1647.

Tran Thanh Van - Le Kiem Ngoc (1975) Comment faire fleurir l'infleurissable ? Conférence au Palais de la Découverte, Paris. Sciences et Techniques 20: 1-8.

Tran Thanh Van - Le Kiem Ngoc M (1965) La vernalisation du *Geum urbanum* L. Etude expérimentale de la mise à fleur chez une plante vivace en rosette exigeant le froid vernalisant pour fleurir. Ann. Sci. Nat., Bot., Paris. 12éme série, 6: 373-594.

Tran Thanh Van K & Trinh TH (1990a) Organogenic differentiation. In: Bhojwani SS (ed) Plant Tissue Culture: Applications and Limitations. pp. 34-54. Elsevier, Amsterdam.

Tran Thanh Van K (1973a) *In vitro* and *de novo* flower, bud, root and callus differentiation from excised epidermal tissue. Nature 246: 44-45.

Tran Thanh Van K (1980) Control of morphogenesis by inherent and exogenously applied factors in thin cell layers.Int. Rev. Cytol. 11 A: 175-194.

Tran Thanh Van K (1981) Control of morphogenesis in *in vitro* cultures. Ann. Rev. Plant. Physiol. 32: 291-311.

Tran Thanh Van K (1985) Control of developmental pattern in *Geum urbanum*. In: Handbook of Flowering. Vol. 3. pp. 53-70. CRC, New York.

Tran Thanh Van K (1991) Molecular aspects of flowering. In: Harding J et al. (eds) Genetics and Breeding of Ornamental Plants, Plenum, New York.

Tran Thanh Van K & Gendy C (1992) Cytological, Biochemical and Molecular Markers of Plant Morphogenesis. In: Roubelakis-Angelakis K & Tran Thanh Van K (eds) Molecular markers of plant morphogenesis. NATO Workshop, Crete.

Tran Thanh Van K & Gendy C (1993) Biochemical and molecular markers in programmed plant pattern differentiation. In: Soh W-Y et al. (eds) Advances in Developmental Biology and Biotechnology of Higher Plants. PP.255-263. Korea Soc. Plant Tissue Cult., Korea.

Tran Thanh Van K & Gendy C (1995) Thin Cell Layer (TCL) method to programme morphogenetic differentiation In: Plant Tissue Culture Handbook, Kluwer Academic Publishers, Dordrecht.

Tran Thanh Van K & Mutaftschiev S (1990c) Signals influencing cell elongation, cell enlargement, cell division and morphogenesis. In: Nijkamp HJJ(eds) Progress in Plant Cellular and Molecular Biology. pp. 514-519. Kluwer, Dordecht.

Tran Thanh Van M (1973b) Direct flower neoformation from superficial tissues of small explant of *Nicotiana tabacum* L. Planta 115: 87-92.

Tran Thanh Van M (1974a) Methods of acceleration of growth and flowering in a few species of Orchids. Amer. Orchid Soc. Bull. 43: 699-707.

Tran Thanh Van M (1974b) Growth and flowering of *Cymbidium* buds normally inhibited by apical dominance. J. Am. Soc. Hortic. Sci. 99: 450-453.

Tran Thanh Van K, Richard L & Gendy C (1990b) An experimental model for the analysis of plant / cell differentiation: Thin cell layer concept, strategy, methods, records and potential. In: Rodriguez R et al. (eds) Plant Aging, Basic and Applied Approaches. Plenum Press, New York.

Tran Thanh Van K & Bui Van Le (1999) Current status of thin cell layer method for the induction of organogenesis or somatic embryogenesis. In: Jainsm, Gupta PK and Newton RJ (eds) Somatic embryogenesis in woody plants. Vol. 6 (in press)

Trinh TH (1978) Organogenèse induite *in vitro* sur des fragments de racines de *Nicotiatna tabacum. L.* Can. Bot. J. 56: 2370-2374.

Trinh TH, Mante S, Pua EC & Namhai C (1987) Rapid production of transgenic flowering shoots and F1 progeny from epidermal peel of *Nicotiana plumbaginifolia.* Bio/Technology 5: 1032-1038.

Zagotta MT, Shannon S, Jacobs C & Meeks-Wagner DR (1992) Early-flowering mutants of *Arabidopsis thaliana.* Aust. J. Genet. 19: 411-418.

Chapter 9

DEVELOPMENTAL AND STRUCTURAL PATTERNS OF IN VITRO PLANTS

Meira ZIV
Department of Agricultural Botany and The Warburg Centre for Biotechnology in Agriculture, Faculty of Agriculture, The Hebrew University of Jerusalem, Rehovot 76100, Israel

1 INTRODUCTION

The onset of developmental and morphogenic events in an isolated explant *in vitro* is associated at the very initial stages with the loss of coordinated, correlative and integrated control. Isolation and wounding of the explanted tissue, exposure to new signals from the medium and the formation of new gradients, induce a series of molecular and physiological events which lead to dedifferentiation. In most plants studied the early events after explant isolation, during dedifferentiation, result in unorganized cell division and growth, followed by the formation of callus.

As new gradients form and different signals function, reorganization begins with a more coordinated cell division, and the formation of meristematic growth centres. When new correlation among cells in the meristem are established, organized morphogenetic expression and redifferentation occur. This is followed by the formation of unipolar or bipolar structures, i.e. either organs through organogenesis or embryos through somatic embryogenesis. These events are summarized in Figure 1.

Both organogenesis and somatic embryogenesis take place through dedifferentiation and redifferentiation events. These events depend on the renewal of meristematic activity in mature differentiated cells or in an unorganized callus tissue.

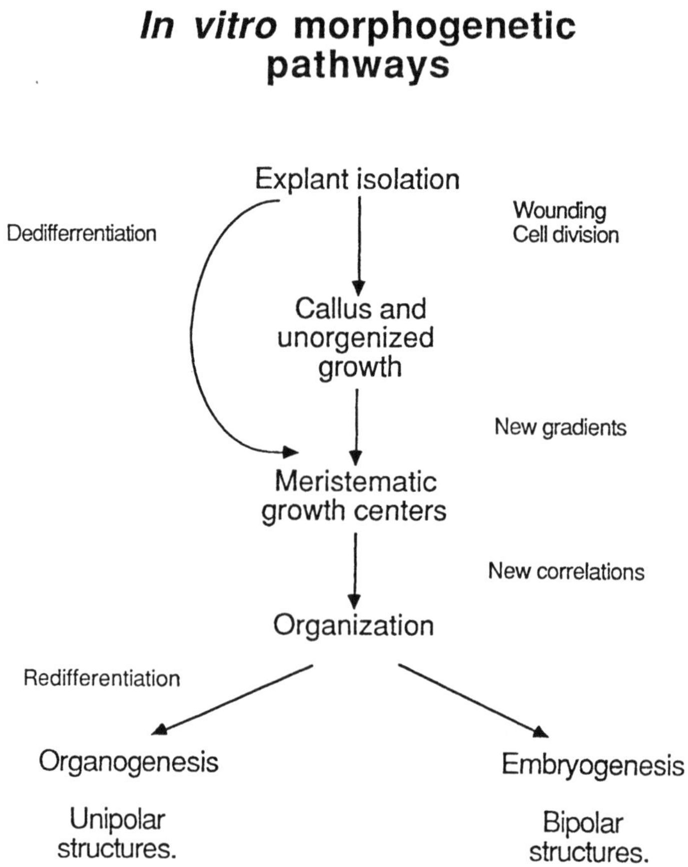

Figure 1. Morphogenetic pathways in isolated explants in vitro

Plant regeneration through organogenesis and somatic embryogenesis has been established during the last 40 years in a large number of species, belonging to almost all plant families. For most of these species regeneration has been achieved by the transfer of isolated explants or dedifferentiated callus tissue through a sequence of media, in which the major changes were variation in the organic compounds and in the growth regulators level, as well as in the physical environment.

The main problem confronted in plant regeneration is that in most cases, information is available on how to manipulate culture conditions to induce the sequence of events which lead to morphogenetic expression *in vitro.* However, information on what is the exact sequence of events at the cellular and subcellular level, or why specific events take place during morphogenesis in culture, is to date sketchy, limited and inadequate.

Information is available on the type of explants, culture conditions, growth regulators levels, the various ways to manipulate *in vitro* protocols, to name just a few subjects which cover a large volume of literature (George,

1996). Our incomplete understanding of why or when a somatic cell undergoes the transition to become competent as an organogenic or an embryogenic cell, is disappointing from both, the basic and applied aspects of plant morphogenesis. In many cases our limited, or even lack of understanding of the control mechanism of events *in vitro,* may lead to abnormal growth and development, resulting in aberrant structures and organs and to the development of plants that cannot survive the *ex vitro* environment.

Manipulation of culture conditions to promote proliferation and growth, often results in the loss of integrated control diverting developmental pathways from normal patterns. This chapter will cover morphogenetic patterns deviating from normal organogenesis and somatic embryogenesis (discussed in other chapters in this book) emphasizing their importance to plant development *in vitro* and survival *ex vitro*.

2 PATHWAYS IN DEVELOPMENTAL EVENTS IN VITRO

One of the most fundamental issues in plant regeneration from somatic cells is the understanding and elucidation of the mechanisms controlling developmental events. Plant morphogenesis *in vivo* and *in vitro* is a complex developmental event. Morphogenesis, as defined by Wardlaw (1968), is a physiological and/or morphological change allowing the specialization of a cell, tissue, organ or an entire plant during its development. Morphogenesis involves the development of cells differing from their mother cell, and the differentiation and arrangement of these cells in a defined order as a result of cell division according to a specific plan and sequence. An isolated organ, explanted from the whole organism, is an organized multicellular structure, controlled by both extrinsic and intrinsic correlative signals. It is made of an orderly assortment of cell types, which develop under both separated and integrated morphogenetic control.

What are the events which take place in plant morphogenesis *in vitro*? How are they regulated and can be better manipulated to improve our knowledge and, thus, provide better means to control normal patterns of plant regeneration and development *in vitro?*

Two main phases are involved in the morphogenetic pathway *in vitro*: the induction phase and the expression phase (Ammirato, 1985): inducing the cells to acquire organogenic or embryogenic competence and the expression of this potential in the development of organs or embryos. Competence is defined by the ability of a cell or tissue to respond to morphogenetic signals. Only competent cells can be induced as they have the ability to respond to

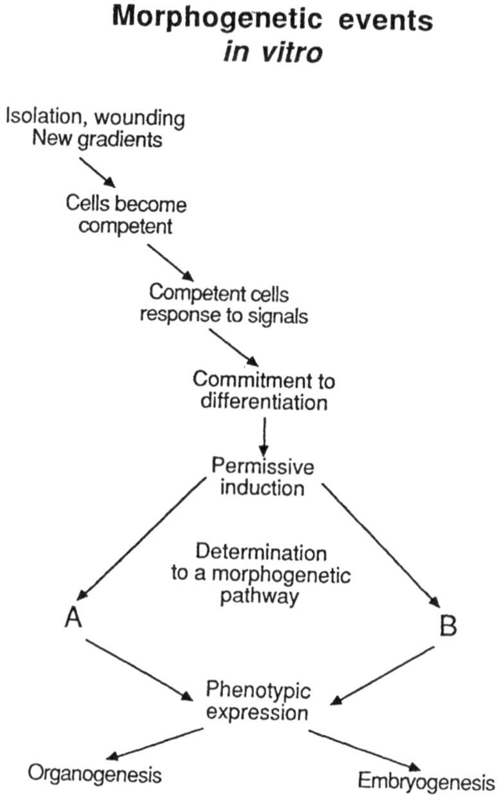

Figure 2. Signals and events in organogenesis and somatic embryogenesis *in vitro*

one or several inductive factors and become capable to express morphogenesis (Christianson, 1987).

Activation of a tissue makes it predisposed to receiving the appropriate signals. These signals reprogramme somatic cells which have a defined morphogenetic commitment, to revert *in vitro* to a meristematic state, divide, differentiate, reorganize and express totipotency. Once induced, the cells have a specific new constellation of biochemical and physiological components which will redirect them to a particular developmental event, at which stage they become determined in a specific pattern. A determined cell can be further induced to express its morphogenetic path by additional inductive signals and further express the morphogenetic potential in either organogenesis or somatic embryogenesis pathway, as outlined in Figure 2.

Various developmental signals are known to affect morphogenesis *in vitro* as they do *in vivo*. They include nutrient levels, physiological and electrical gradients, environmental factors, receptors in target tissue and physical forces such as directional stress (Steward et al., 1970). What are the immediate effects of these signals and in which sites of an isolated explant, organ, callus or tissue, they act to induce the response, are still unknown

3 ABNORMAL MORPHOGENETIC PATTERNS

Throughout the organogenic or embryogenic pathways *in vitro*, we continuously observe deviation from the normal course of *in vivo* morphogenetic events. Culture requirements for rapid regeneration and proliferation lead to metabolic and physiologic changes which are manifested in anatomical and morphological aberration. Some of the observed morphologic abnormalities, such as hyperhydricity and malformed leaves, secondary embryogenesis, anatomical aberration, precocious germination of embryos, and other disorders are apparently the result of interruption and faulty timing of the inducing signals in the normal sequence of organizational events.

4 DEVELOPMENTAL ABERRATION IN SOMATIC EMBRYOGENESIS

Changes in developmental patterns of somatic embryogenesis which deviate from normal embryo development, especially in embryos originating from cell suspension cultures, were observed in many species. In some cases the abnormal embryos can eventually develop into normal plants. In many others, these morphological abnormalities result in malformed structures which cannot revert to normal plants and which can not survive *ex vitro*.

A malformed cucumber somatic embryo can be seen in Figure 3a. The embryonal axis has two shoots, the leaves are hyperhydrous and they lack a cuticular epidermal tissue.

Several malformed embryos can be seen in Figure 3b: embryos with shoots lacking cotyledons, while some were multicotyledonous (Fig. 3c). Other malformation observed included multiple shoots on a single root system or accessory embryos as reported for carrot and caraway (Ammirato, 1977, 1983a, b, 1985) and can also be seen on the cotyledon of a somatic embryo in *Nerine* (Fig. 4; Lilien-Kipnis et al., 1994). Unlike zygotic embryos which enter dormancy after maturation, somatic embryos did not enter a rest or dormancy period, germinated precociously, continued to grow and formed plantlets with elongated roots, and malformed cotyledons and leaves as reported for carrot and celery (Ammirato, 1985,1986; Nadel et al., 1989 a, b). Frequently, somatic embryos of cucumber in liquid cultures

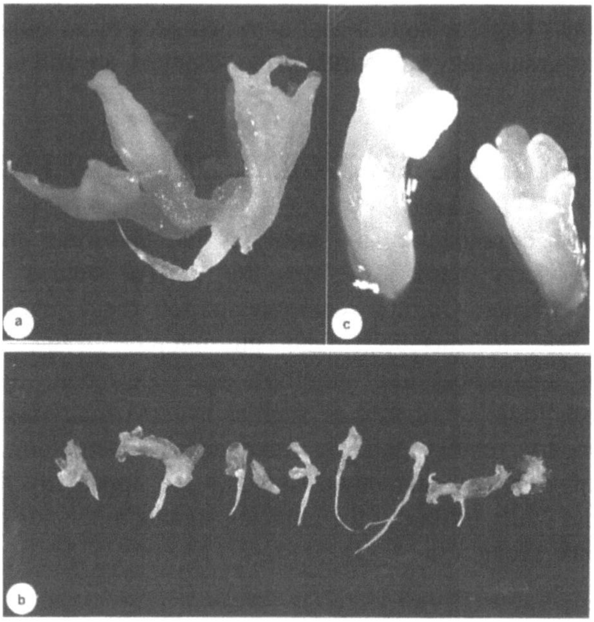

Figure 3. A malformed cucumber somatic embryo with two embryonal shoot axes (a), several malformed cucumber somatic embryos (b) and multicotyledonous Nerine somatic embryos (c).

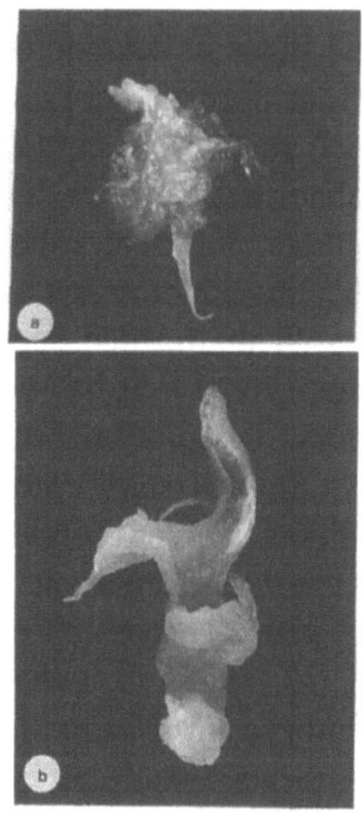

Figures 4-5

Figure.4(left). Accessory embryos developing on a senescing cotyledon (ct) of *Nerine* with a bulblet developing from the embryonal axis (bl). *Figure 5(right).* Cucumber somatic embryos with a dedifferentiated shoot showing several meristematic centers and leaf proimordia (a) and a dedifferentiated root meristem (b).

failed to differentiate a normal shoot and the apical meristem dedifferentiated and formed a callus with several meristematic centres, connected to an elongated root (Fig 5a). In several other embryos the root meristem dedifferentiated and formed a callus and the cotyledons showed severe hyperhydricity (Fig. 5b).

The addition of ABA to carrot, caraway (Ammirato, 1974, 1977, 1985), cucumber (Ziv and Gadasi, 1986) and mango (Mansalus et al., 1995) prevented abnormal embryo development, and in mango reduced hyperhydricity. ABA appears to act as a signal for normal morphogenetic expression as well as for the induction of dormancy and growth arrest as reported for white spruce somatic embryos (Dunstan et al., 1991) and other species. In soybean a morphogenetic callus produced abnormal embryos after the withdrawal of 2,4-D. However, the manipulations of auxins, cytokinins, reduced nitrogen levels and changes in the medium pH provided signals which promoted normal embryo development (Christianson, 1985).

Krikorian and Kahn (1981) observed in daylily abnormal somatic embryos with malformed shoots and roots which they termed "Neomorphs". Embryogenic cells, usually those on the outer enveloping layers of the developing embryo continued to divide and formed new meristematic centres, which redifferentiated and produced additional-secondary embryos described as recurrent somatic embryogenesis in *Nerine* (Lilien-Kipnis et al., 1994) and other species. The cells remain determined and continue to express a morphogenetic pathway which was induced at an earlier stage. This occurred even though the medium and hormonal signals were changed. A Nerine somatic embryo subcultured from liquid to a semi-solid bulb-inducing medium, developed secondary embryos on the senescing cotyledon, while the embryonal axis developed into a bulb (Fig 4).

Secondary embryos were observed in carrot (Steward et al., 1970; Kamada and Harada, 1981; Ammirato 1987), celery (Al Abta and Collin, 1978; Nadel et al., 1989a,b) and *Panicum maximum* and *Pennisetum purpureum* (Vasil & Vasil 1982; Karlson and Vasil, 1986).

It is possible that a concentration gradient of a specific substance or signal originating in a competent source cell is transmitted to other cells, which remain meristemaic, thus inducing recurrent embryogenesis in the cells of a mature embryo which remain totipotent.

5 PRECOCIOUS GERMINATION AND HYPERHYDRICITY IN SOMATIC EMBRYOS

In many plants the induction of somatic embryogenesis can be manipulated by culture conditions and growth regulator levels, and a normal pattern of development results in faultless and complete embryos. However, often, once the embryos mature they do not always enter dormancy. They start to germinate precociously *in vitro*, without following normal zygotic ontogeny, continue to grow and develop into mature plantlets. Precocious germination delays or obstructs growth-arrest and the dormancy period which is typical of zygotic embryo development. The unbalanced development of shoot and root systems of somatic embryos results in an entangled plant biomass, which makes it impossible to produce singulated propagation units for encapsulation, and the production of synthetic seeds or for transplanting and further plant growth.

In several species, precocious germination was reduced by elevated osmolality in the culture medium (Ammirato and Steward, 1971), or with the addition of ABA (Ammirato, 1974, 1972, 1983a, b; Vasil and Vasil, 1982; Dunstan et al., 1991).

Physical manipulation of the culture improved embryo maturation in yellow poplar, and precocious germination was prevented by transferring embryos to a layer of filter paper overlaid on semi-solid basal medium (Merkle et al., 1990).

The phenomenon of hyperhydricity in carrot, cucumber and celery somatic embryos was observed in agar cultures but was much more severe in liquid embryogenic cultures used for large scale cloning (Ammirato et al., 1984; Ammirato, 1985; Ziv and Gadasi 1986; Nadel et al., 1989a, b,; 1990). Liquid medium which is used as an efficient large-scale culture system for production of embryo biomass is the major cause for malformation and hyperhydricity in developing embyos as reported in celery, carrot and other species (Nadel et al., 1989; Nickle and Yeung, 1993; 1994). In liquid cultured embryos the epidermal tissue was damaged and appeared irregular in shape. In carrot and *Nerine* the cortex tissue comprised of enlarged parenchyma cells interspaced with large air spaces and the vascular tissue was incomplete (Ammirato, 1987; Ziv and Azizbekova, unpublished).

6 HYPERHYDRICITY IN DEVELOPING SHOOTS

Anomalous development in existing or newly formed meristems in many plant species studied *in vitro*, were reported to result in the appearance of hyperhydrous glassy plants (Gaspar et al., 1987; Ziv, 1991a). The first organs to exhibit hyperhydricity were the leaves which were brittle, translucent, often chlorotic and succulent (Werker and Leshem, 1987; Ziv and Ariel, 1994). Histological examination of malformed leaves in several species revealed an unorganized hyperhydrous mesophyll, characterized by a disordered palisade tissue and a spongy parenchyma with large intercellular spaces. The mesophyll cells had thin walls and contained a relatively poor, largely vacuolated cytoplasm (Brainard et al., 1981; Fabbri et al., 1986; Grout and Aston 1978; Reuther 1988; Vieitez et al., 1985; Leshem 1983; Vieth et al., 1983).

In sweetgum many of the chloroplasts lacked normal organization into grana and stroma (Wetzstein and Sommer, 1982, 1983). In carnation shoots cultured in liquid medium, the chloroplasts contained large starch grains (Ziv and Ariel, 1994). In plum, carnation and raspberry, the chlorophyll content was lower in hyperhydrous as compared to normal leaves (Phan and Letouze, 1983; Ziv et al., 1983; Donnelly and Vidaver, 1984a, b).

In several species, leaf hyperhydricity was associated with defective epidermal tissue. Faulty deposition of epicuticular waxes on vitreous leaves was manifested both in the quantity and quality of waxes (Grout, 1975; Grout and Aston, 1977,1978; Sutter, 1985; Sutter and Langhans, 1979; Ziv et al., 1981, 1983). Epicuticular waxes, deposited thinly on hyperhydrous leaves, were significantly different in composition as compared to normal leaves. In vitreous cabbage and several ornamental plants, a higher proportion of polar components in the wax, was formed (Sutter, 1984, 1985). Ziv (1986) observed in cultured carnation shoots an inverse correlation between the amount of epicuticular wax and the relative humidity in the culture vessels.

The guard cells in the epidermal tissue of several species did not function properly and remained open in darkness, or even under water stress conditions. In apple, carnation and cauliflower, stomata did not close in response to treatment with ABA, high CO_2, Ca^{++} or hypertonic solutions of mannitol or sucrose (Brainard et al., 1981; Fuchigami et al., 1981; Wardle et al., 1979; Ziv et al., 1987). CO_2-free air, but not darkness or ABA, induced partial closure of stomata from tissue cultured *Chrysanthemum* plants (Wardle and Short, 1983). In geranium and rose abnormally large stomata were observed (Reuther, 1988) and this was also reported for *Solanum laciniatum* plants (Conner and Conner, 1984) and apple leaves (Blanke and Belcher, 1989). Histochemical studies of guard cells from hyperhydrous carnation leaves revealed lower levels of cutin, pectins and cellulose (Ariel, 1987; Ziv and Ariel, 1992,1994; Werker and Leshem, 1987; Koshuchowa et al., 1988). Callose deposits in guard cells from hyperhydrous leaves were reported in cherry and carnation. The cell walls lacked the normal orientation of cellulose microfibrils typical to guard cells (Ziv et al., 1987; Marin et al., 1988; Ziv and Ariel, 1992,1994), which may have contributed to their failure to close. Abnormal stomata had damaged walls around the stomatal pore which appear to result from a deformed cell plate during the division of the primary stomata mother cells in liquid medium (Ziv and Ariel, 1992; Ziv, 1995a).

Reduced lignification, thin cell walls, large intercellular air spaces and reduced vascular tissue were also observed in hyperhydric stems. In chestnut, geranium and apple, hyperhydrous shoots were smaller in diameter, the stems lacked sclerenchymatous tissue and the cortical and pith cells were hyperhydrated (Vieitez et al., 1985; Vieth et al., 1983). In hyperhydrous carnation plants procambial strands were not observed and the vascular bundles lacked normal organization (Leshem, 1983). In cauliflower plants the vascular connection between the stem and the roots was incomplete (Grout and Aston, 1977).

7 EMBRYO AND ORGAN MORPHOGENESIS IN LIQUID CULTURES

The development of mature viable plants through organogenesis or embryogenesis in liquid cultures is impeded by morphogenetic divergence, which later leads to distorted hyperhydric shoot and root systems (Ziv 1991a). Various attempts were made to control abnormal developmental pathways through the modification of different media components: minerals, organic substances and growth regulators. Controlling the osmoticum and viscosity of the medium improved normal development (Ziv 1991a, b; 1995b). One of the major factors involved in embryogenic or organogenic development was the source of nitrogen. Reduced nitrogen given as $NH^{+}{}_{4}$, as a single amino acid or as a mixture of various amino acids reduced 'neomorph' formation in alfalfa, carrot and carnation (Stuart et al., 1985; Kamada and Harada, 1981; Christianson, 1985; Ziv and Ariel, 1994). Reduced nitrogen is apparently mandatory at the early stages of somatic embryo development as they appear to lack nitrate reductase.

Various types and levels of sugars, as well as the use of non-metabolic carbohydrates, such as mannitol or sorbitol, affected the medium osmoticum and contributed significantly to the control of embryo development in carrot, alfalfa, celery and papaya (Ammirato, 1985; Stuart et al., 1985; Nadel et al., 1989a; Litz and Conover, 1982). In celery proembryogenic suspension cultures, 3% mannitol in addition to sucrose in the medium, enhanced normal embryo development and increased the number of singulated embryos (Nadel et al., 1989a). Singulated embryos are an important aspect in 'synseeds' (synthetic seeds) production and the application of automation to micropropagation.

The role of abscisic acid (ABA) reported in zygotic embryo and seed development prompted its use in somatic embryogenesis in the early seventies (Ammirato, 1974). ABA added to differentiating caraway and carrot suspension cultures, reduced the elongation of the embryo axis and controlled a more normal cotyledon growth than in its absence (Ammirato, 1977; Kamada and Harada, 1981; Fuji et al., 1990). The role of ABA in regulating normal somatic embryo development in carrot was further demonstrated by Nickle and Yeung (1994) in a series of detailed experiments using both ABA and Fluridone (an inhibitor of ABA synthesis). ABA prevented precocious germination and induced maturation in white spruce somatic embryos (Dunstan et al., 1991). In cucumber, ABA and activated charcoal (AC) had a positive effect on plantlet development from somatic embryos. ABA when added as a second liquid layer on top of an agar medium with AC, reduced hyperhydricity and controlled shoot malformation (Ziv and Gadasi, 1987). Ziv used various polymers to control precocious germination and prevent dedifferentiation to callus in the embryonal shoot apex of cucumber and asparagus proembryos. The most effective polymer used for the control of normal embryo development was Gum-xanthan added to 0.8% agar medium at concentrations in the range of 0.5-1.0%. The presence of Gum-xanthan inhibited embryo aberration and induced normal shoot growth (Ziv, unpublished).

Problems of malformation, delayed maturation and the enhancement of normal embryo development, which were reported also for woody species, were overcome by treatments with ABA and osmoticum changes in the medium (see review by Merkle, 1995).

In several ornamental and vegetable crops, the severity of malformation and hyperhydricity in proliferating organogenic liquid cultures, was regulated by the use of growth retardants and inhibitors of gibberellin biosynthesis (Ziv, 1992). Three growth retardants tested, Paclobutrazol (ICI), ancymidol and flurprimidol (Elanco Eli Lilly), were found to reduce shoot and leaf expansion and to induce meristematic or bud cluster formation (Ziv, 1996b, 1992; Ziv et al., 1994; Ziv and Ariel, 1991; Ziv and Shemesh, 1996). In gladiolus, lilium and *Nerine,* geophytes which form corms or bulbs as perennating organs, the meristematic clusters (Fig. 6a, b, c) developed corms or bulbs in the presence of 0.5-2.5 ppm paclobutrazol added to a storage organ induction medium with elevated sucrose level. In banana (Fig. 6d), philodendron (Ziv and Ariel, 1991) and potato (Ziv and Shemesh, 1996), 1.0-5.0 ppm ancymidol reduced the developing hyperhydric shoots into bud clusters and enhanced biomass proliferation in bioreactor cultures. Reduced shoot aberration with limited leaf development and cluster formation were obtained also in *Ornithogalum dubium* (Ziv and Lilien-Kipnis, 1997) and in *Brodiaea* (Ilan et al., 1995). Histological examination of *Nerine* and gladiolus clusters revealed that in both species the morphogenetic expression in the cluster started in the outer layers after active cell division followed by the formation of meristematic centres(Lilien-Kipnis et al., 1994; Ziv and Azizbekova, unpublished).

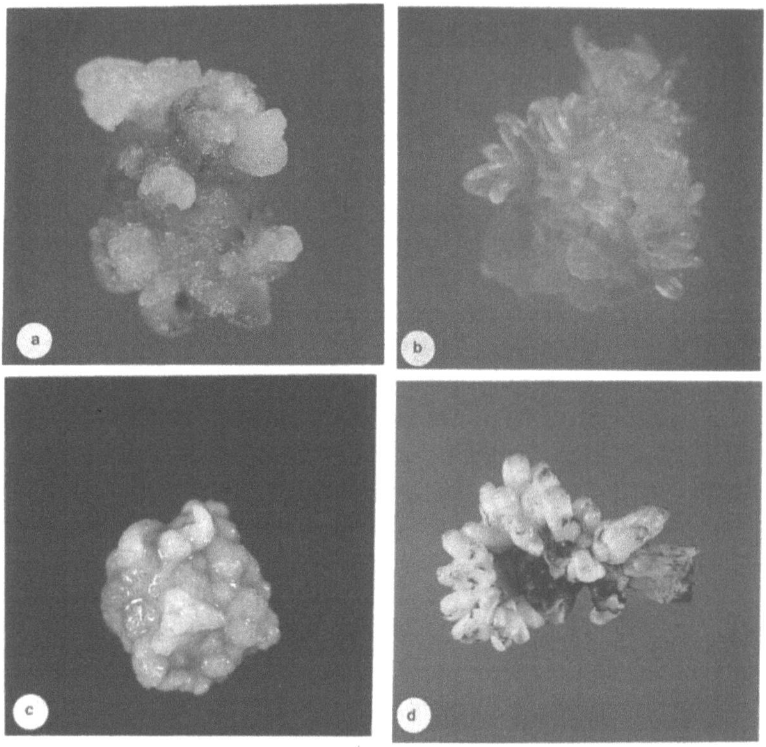

Figure 6. Meristematic and bud clusters from liquid cultured gladiolus (a), Nerine (b), lilium (c) and banana (d)

However, two different morphogenetic pathways were observed depending on the species: in *Gladiolus grandiflora* cv. 'Eurovision' the developing meristems were unipolar forming a shoot (Fig. 7a), while in *Nerine sarniensis* the meristems organized into a bipolar structure from which embryos developed (Fig. 7b). In both cases the triggering signal was paclobutrazol which upon its removal trigered the formation of a uni- or bipolar meristem. In *Nerine* the development of somatic embryos was depended also on the substitution of benzyl adenine by isopentenyl adenine (2ip; Lilien-Kipnis et al., 1994).

In banana the meristems developed into shoots after the removal of ancymidol and changes in growth regulators levels (Ziv et al., 1997). A similar response was observed in potato bud clusters, which elongated into shoots upon the removal of ancymidol (Ziv and Shemesh, 1996). However,

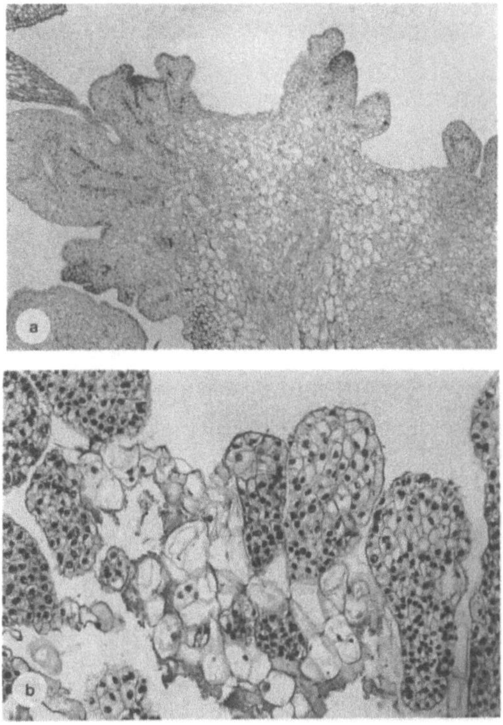

Figure 7. Histological sections of a gladiolus cluster showing shoot and leaf primordia development (a; x70) and a Nerine cluster showing bipolar development of somatic embryos (b; x150)

the buds on the clusters could be induced to form microtubers after separation from the clusters, in the presence of ancymidol and elevated sucrose (6-8%) in the medium. Storage organ formation can be induced from meristem or bud clusters in corm and bulb producing plants, which is an advantage for both the transplanting stage and for long term storage. In gladiolus, the clusters appear to accumulate sugars which are stored as starch and this stimulates corm production in a humidified culture vessel in the total absence of medium (Ziv et al., 1997; Fig. 8a). In lilium clusters, the meristematic centres developed bulblets upon the removal of PAC and when the sucrose level was increased to 6% (Fig. 8b).

The ability of embryogenic and organogenic cells to restrict or control cell expansion in culture appears to reside in the cell wall and in metabolic activity associated with cell wall enzymes (Fry, 1990). The use of growth retardants appears to reduce water uptake during cell proliferation, inhibit shoot elongation, decrease vacuolation and intercellular spaces and induce

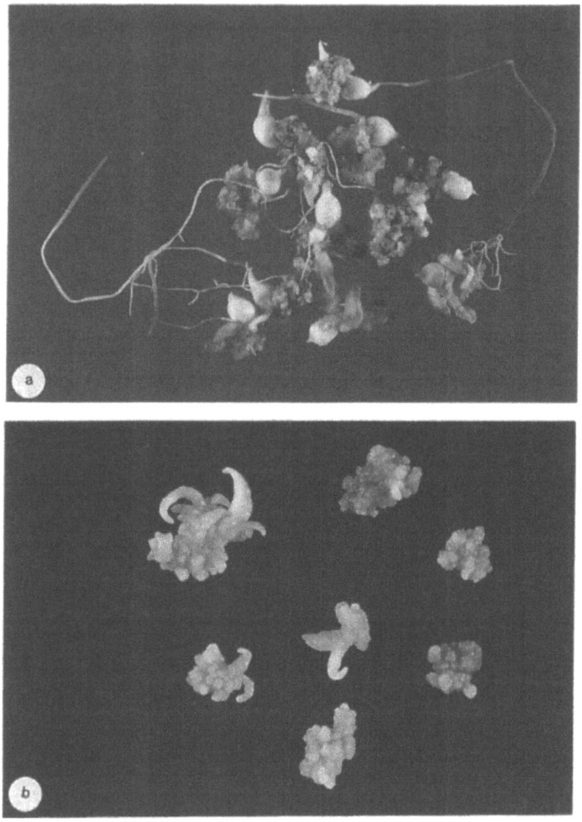

Figure 8. Corm development from gladiolus meristematic clusters in a medium-free culture vessel (a), and lilium bulblets developing from meristematic clusters in an elevated sucrose medium (b).

the formation of a protodermal enveloping layer around the meristematic centres. This layer may be necessary for controlled meristematic activity and compartmentation in the inner layers which, thus, decreases the variation patterns in culture.

8 CONCLUSION

The morphogenetic expression of meristems *in vitro* as it is *in vivo,* is determined by distinctive patterns of gene control which characterize early developmental events under inductive conditions. Morphogenesis is affected by continuous changes of various signals from the immediate environment in culture. *In vitro* culture conditions foster a high degree of cell plasticity which in many plant species can lead to developmental aberration.

Developmental variation *in vitro* may result from the intrinsic potential of the cells to differentiate and their competence to respond to extrinsic factors. As yet we do not fully understand the signals that permit or direct the developmental chain of reactions.

As developmental changes must follow a sequence of events coordinated in time and space, any interruption in the signal-transduction chain may result in a deranged morphogenetic expression. Furthermore, the isolation and culture of an explant in a defined medium does not only change internal existing gradients, it also alters the sensitivity of the tissue to triggering signals which can shift meristematic activity to a pattern unlike the one initiated prior to these changes, or to the one which exsists in the plant *in vivo*.

In view of the role plant growth bioregulators play in morphogenesis, attempts should be made to enlighten the understanding of the response of the cells to the interactions among of the various plant bioregultos and to the balance between them and the culture microenvironment on organ or embryo development *in vitro*.

Although recent research in molecular biology has contributed to further our understanding of morphogenetic events and expression *in vitro* (de Jong et al., 1992), several questions remain open. Is the same meristematic centre capable of forming both types of morphogenetic patterns, i.e. unipolar and bipolar structures? Do competence, determination and expression follow separate pathways in organogenesis and somatic embryogenesis -one that leads to organ formation and the other to embryos? Can one be diverted to the other at one or several stages of development or is the determination irreversible? Do similar or different gradients and signals exist for each one of the pathways? Is there a single competent source cell which transmits the signal to other cells in the meristematic centre, or does a group of competent cells, organized spatially through newly formed gradients, respond morphogenetically? Does the gradient start from one cell at the periphery of the explant and then a sequence of changes are transmitted to other inner cells? At what stage of the organogenic or embryogenic pathways, variation in the developmental potential of the constituent cells results in irreversible aberration which lead to abnormal plants?

At present, our inadequate understanding of the flexibility of the developmental systems hinders our ability to control normal patterns of morphogenesis in many plant species. A substantial amount of research at the molecular level is required to provide us with the knowledge and understanding of the events that precede the actual differentiation and growth observed in cultured plants.

9 REFERENCES

Al-Abta S & Collin HA (1978) Cell differentiation in embryoids and plantlets of celery tissue cultures. New Phytol. 80: 517-521.
Ammirato PV (1974) The effects of abscisic acid on the development of somatic embryos from cells of caraway (*Carum carvi* L.). Bot. Gaz. 135: 328-337.
Ammirato PV (1977) Hormonal control of somatic embryo development from cultured cells of caraway: Interactions of abscisic acid, zeatin and gibberellic acid. Plant Physiol. 59: 579-586.
Ammirato PV (1983a) The regulation of somatic embryo development in plant cell cultures: Suspension culture techniques and hormone requirements. Bio/Technology 1: 68-74.
Ammirato PV (1983b) Embryogenesis. In: Evans DA et al. (eds) The Handbook of Plant Cell Culture. Vol. 1, Techniques for Propagation and Breeding. (pp. 82-123) Macmillan, New York.
Ammirato PV (1985) Patterns of development in culture. In: Henke RR et al. (eds) Tissue Culture in Forestry and Agriculture. (pp. 9-29) Plenum, New York.
Ammirato PV (1986) Control and expression of morphogenesis *in vitro*. In: Withers LA & Alderson PG (eds) Plant Tissue Culture and its Agricultural Application. (pp. 23-46) Butterworths, London.
Ammirato PV (1987) Organizational events during somatic embryogenesis. In: Green CE et al. (eds) Plant Tissue and Cell Culture. (pp. 57-81), Alan R. Liss Inc., New York.
Ammirato PV, & Steward FC (1971) Some effects of the environment of embryos from cultured free cells. Bot. Gaz. 132: 149-158.
Ammirato PV, Evans DA, Flick CE, Whitaker RJ & Sharp WR (1984) Biotechnology and agricultural improvement. Trends Biotechnol. 2: 1-6.
Ariel T (1987) The effect of culture condition on development and acclimatization of *Philodendron* 'Burgundy' and *Dianthus caryophyllus* (pp. 1-96), M.Sc. Thesis, The Hebrew University of Jerusalem, Israel (Hebrew with English summary)
Blanke MM & Belcher AR (1989) Stomata of apple leaves cultured *in vitro*, Plant Cell Tissue Organ Cult. 19: 85-89.
Brainard KE, Fuchigami LH, Kwiatkowski S & Clark CS (1981) Leaf anatomy and water stress of aseptically cultured 'Pixy' plum grown under different environments, Hortic. Sci. 16: 173-175.
Christianson ML (1985) An embryogenic culture of soybean: towards a general theory of somatic embryogenesis. In: Henke RR et al. (eds) Tissues Culture in Forestry and Agriculture. (pp. 83-103) Plenum, New York.
Christianson ML (1987) Causal events in morphogenesis. In: Green CE et al. (eds) Plant Tissue and Cell Culture. (pp. 45-56) Allan R. Liss Inc., New York.
Conner LN & Conner AJ (1984) Comparative water loss from leaves of *Solanum laciniatum* plants cultured *in vitro* and *in vivo*. Plant Sci. Lett. 36: 241-246.
de Jong AJ, Schmidt EDL & Vries SC (1992) Early events in higher-plant embryogenesis. Plant Mol. Biol. 22: 367-377
Donnely DJ & Vidaver WE (1984a) Leaf anatomy of red raspberry transferred from *in vitro* culture to soil. J. Am. Soc. of Hortic. Sci. 109: 172-176.
Donnely DJ & Vidaver WE (1984b) Pigment content and gas exchange of red raspberry *in vitro* and *ex vitro*. J. Am. Soc. Hortic. Sci. 109: 177-181.
Dunstan DI, Bethune TD & Abrams SR (1991) Racemic abscisic acid and abscisyl alcohol promote maturation of white spruce (*Picea glauca*) somatic embryos. Plant Sci. 76: 219-228.

Fabbri A, Sutter E & Dunston SK (1986) Anatomical changes in persistent leaves of tissue cultured strawberry plants after removal from culture. Sci. Hortic. 28: 331-337.
Fry SC (1990) Roles of the primary cell wall in morphogenesis. In: Nijkamp HJJ et al. (eds) Progress in Plant Cellular and Molecular Biology. (pp. 504-513) Kluwer, Dordrecht.
Fuchigami LH, Cheng TY & Soeldner A (1981) Abaxial transpiration and water loss in aseptically cultured plums. J. Am. Soc. Hortic. Sci. 106: 519-522.
Fujii JAA, Slade D, Olsen R et al. (1990) Alfalfa somatic embryo maturation and conversion to plants. Plant Sci. 72: 72-100.
Gaspar T, Kevers C, Debergh P, Maene L, Paques M & Boxus P (1987) Vitrification: morphological, physiological and ecological aspects. In: Bonga JM & Durzan DJ (eds) Cell and Tissue Culture in Forestry, Vol. 1. (pp 152-166) Martinus Nijhoff, Dordrecht.
George EF(1996) Plant Propagation by Tissue Culture, Part 2. Exgetics, Edington.
Grout BWW & Aston H (1977) Transplanting of cauliflower plants regenerated from meristem culture I. Water loss and water transfer related to changes in leaf wax and to xylem regeneration. Hortic. Res. 17: 1-7.
Grout BWW & Aston H (1978) Modified leaf anatomy of cauliflower plantlets regenerated from meristem culture. Ann. Bot. 42: 993-995.
Grout BWW (1975) Wax development on leaf surface of *Brassica oleracea botrytis cv* Currawon regenerated from meristem culture. Plant Sci. Lett. 5: 401- 405.
Ilan A, Ziv M & Halevy AA (1995) In vitro propagation in liquid cultures and acclimatization of *Brodiaea*. Sci. Hortic. 63: 101-112
Kamada H & Harada H (1981) Changes in the endogenous level and effects of abscisic acid during somatic embryogenesis of *Daucus carota*. Plant Cell Physiol. 22:1423-1429.
Karlson SB & Vasil IK (1986) Morphology and ultrastructure of embryogenic cell suspension cultures of *Panicum maximum* and *Pennisetum purpureum*. Am. J. Bot. 73: 894-901.
Koshuchowa S, Bottchcr I, Zogaucr K & Goring H (1988) Avoidancc of vitrification of *in vitro* cultured plants. In: Proceedings 6th Congress of RESPP (Abstract p 1418) Split, Yugoslavia.
Krikorian AD & Kann RP (1981) Plantlet production from morphogenetically competent cell suspensions of daylily. Ann. Bot. 47: 679-686.
Leshem B (1983) Growth of carnation meristems *in vitro*: anatomical structure of abnormal plantlets and the effect of agar concentration in the medium on their formation. Ann. Bot. 52: 413-415.
Lilien-Kipnis H, Azizbekova N & Ziv M (1994) Scale-up proliferation and regeneration of Nerine in liquid cultures. Part II. Ontogeny of somatic embryos and bulblet regeneration. Plant Cell Tissue Organ Cult. 39: 117-123.
Litz RE & Conover RA (1982) *In vitro* somatic embryogenesis and plant regeneration from *Carica papaya* L. ovular callus. Plant Sci. Lett. 26: 153-158.
Mansalus MJ, Mathews H, Litz RE & Grey DJ (1995) Control of hyperhydricity of mango somatic embryos. Plant Cell Tissue Organ Cult. 42: 195-206.
Marin JA, Gella R & Herrero M (1988) Stomatal structure and functioning as a response to environmental changes in acclimatized micropropagated *Prunus cerasus* L. Ann. Bot. 62: 663-670.
Merkle SA (1995) Strategies dealing with limitations of somatic embryogenesis in hardwood trees. Plant Tissue Cult. Biotechnol. 1: 112-121.
Merkle SA, Wiecko AT, Sotak RJ & Sommer HE (1990) Maturation and Conversion of *Liriodendron tulipifera* somatic embryos. In Vitro Cell. Dev. Biol. 26p : 1086-1093.
Nadel BL, Altman A & Ziv M (1989a) Regulation of somatic embryogenesis in celery cell suspension: 1. Promoting effects of mannitol on somatic embryo development. Plant Cell Tissue Organ Cult. 18: 181-189.

Nadel BL, Altman A & Ziv M (1989b) Regulation of somatic embryogenesis in celery cell suspension: 1. Early detection of embryogenesis potential and the induction of synchronized cell cultures. Plant Cell Tissue Organ Cult. 20:119-124.

Nadel BL, Altman A & Ziv M (1990) Regulation of large scale embryogenesis in celery Acta Hortic. 280: 75-82.

Nickle TC & Yeung EC (1993) Failure to establish a functional shoot meristem may be a cause of conversion failure in somatic embryos of *Daucus carota*. Am. J. Bot. 80: 1284-1291.

Nickle TC & Yeung EC (1994) Further evidence of a role for abscisic acid in conversion of somatic embryos of *Daucus carota*. In Vitro Cell. Dev. Biol. 309: 96-108.

Phan CT & Letouze R (1983) A. comparative study of chlorophyll, phenolic and protein contents and of hydroxycinnamate: Coa ligase activity of normal and vitreous plants (*Prunus avium* L.) obtained *in vitro*. Plant Sci. Lett. 31: 323-327.

Reuther G (1988) Comparative anatomical and physiological studies with ornamental plants under *In vitro* and greenhouse conditions. Acta Hortic. 226: 91-98.

Steward FC, Ammirato PV & Mapes MO (1970) Growth and development of totipotenet cells: some problems procedures and perspectives. Ann. Bot. 34:761-787.

Stuart DA, Nelsen J, McCall CM, Strickland SG & Walker KA (1985) Physiology of the development of somatic embryos in cell cultures of alfalfa and celery. In: Zaitlin M, Day P & Hollaender A (eds) Biotechnology in Plant Science Relevance to Agriculture in the Eighties in the (pp. 35-47) Academic Press, New York.

Sutter EG & Langhans RW (1979) Epicuticular wax formation on carnation plantlets regenerated from shoot tip culture, J. Am. Soc. Hortic. Sci. 104: 493-496.

Sutter EG (1984) Chemical composition of epicuticular wax in cabbage plants grown *in vitro*. Can. J. Bot. 62: 74-77.

Sutter EG (1985) Morphological, physical and chemical characteristics of epicuticular wax on ornamental plants regenerated *in vitro*. Ann. Bot. 55: 321-329.

Vasil V & Vasil IK (1982) Characterization of an embryogenic cell suspension culture derived from cultured inflorescences of *Pannisetum americanum* (Pearl Millet, L. Gramineae). Am. J. Bot. 69: 1441-1449.

Vieitez AM, Ballester A, San-Jose MC & Vieitez E (1985) Anatomical and chemical studies of vitrified shoots of chestnuts regenerated *in vitro*. Physiol. Planta. 65: 177-184.

Vieth J, Morisset C & Lamand M (1983) Histologie de plantules vitreuses de *Pyrus malus cv* M-26 et de *Pelargonium peltatum* cv Chester Frank issues de la culture *in vitro*. Rev. Can. Biol. Exp. 42: 29-32.

Wardlaw CW (1968) Morphogenesis in Plants. (see pp. 11-24) Methuen & Co. Ltd., London.

Wardle K & Short KC (1983) Stomatal responses of *in vitro* cultured plantlets I. Responses in epidermal strips of *Chrysanthemum* to environmental factors and growth regulators. Biochem. Physiol. Pflanzen. 178: 619-624.

Wardle K, Quinlan A & Simpkins I (1979) Abscisic acid and the regulation of water loss in plantlets of *Brassica aleracea* L. var *botrytis* regenerated through apical meristem culture. Ann. Bot. 43: 745-752.

Werker E & Leshem B (1987) Structural changes during vitrification of carnation plantlets. Ann. Bot. 59: 377-385.

Wetzstein HY & Sommer HE (1982) Leaf anatomy of tissue cultured *Liquidambar styraciflua* (Hamamelidaceae) during acclimatization. Am. J. Bot. 69: 1579-1586.

Wezstein HY & Sommer HE (1983) Scanning electron microscopy of *in vitro* cultured *Liquidambar styraciflua* plantlet during acclimatization. J. Am. Soc. Hortic Sci. 108: 475-480.

Ziv M & Ariel T (1991) Bud proliferation and plant regenreation in liquid-cultured *Philodendron* treated with ancymidol and paclobutrazol. J. Plant Growth Regul. 10: 53-57.
Ziv M & Ariel T (1992) On the relation between vitrification and stomatal cell wall deformity in carnation leaves *in vitro*. Acta Hortic. 314: 121-129
Ziv M & Ariel T (1994) Vitrification in relation to stomatal deformation and malfunction in carnation leaves *in vitro*. In: Lumsden PJ et al. (eds) Physiology, Growth and Development of Plants in Culture. (pp. 143-154), Kluwer , Dordrecht.
Ziv M & Gadasi D (1986) Enhanced embryogenesis and plant regeneration from cucumber (*cucumis sativus* L.) callus by activated charcoal in solid/liquid double-layer culture. Plant Sci. 47: 115-122.
Ziv M & Lilien-Kipnis H (1997) Bud cluster proliferation and bulb formation from bioreactor cultures of *Ornithogalum dubium*. Acta Hortic. 430 (in Press)
Ziv M & Shemesh M (1996) Propagation and tuberization of potato bud clusters from bioreactor culture. In Vitro Cell Dev. Biol. 32p: 31-36.
Ziv M (1986) *In vitro* hardening and acclimatization of tissue culture plants. In:. Withers LA & Alderson PG (eds) Plant Tissue Culture and its Agricultural Application. (pp. 187-196). Butterworths, London.
Ziv M (1991a) Vitrification: morphological and physiological disorders of *in vitro* plants. In: Debergh PC & Zimmerman RH (eds) Micropropagation Technology and Application. (pp. 45-69) Kluwer, Dordrecht.
Ziv M (1991b) Quality of micropropagated plants-Vitrification. In Vitro Cell. Dev. Biol. 27p: 64-69.
Ziv M (1992) The use of growth retardants for the regulation and acclimatization of in vitro plants In: Karsen CM van Loon LC & Vreugdenhil D(eds) Progress in Plant Growth Regulation. (pp. 807-817) Kluwer, Dordrecht.
Ziv M (1995a) *In vitro* acclimatization. In: Aitkcn-Christic J ct al. (cds) Automation and Enviromental Control in Plant Tissue Culture. (pp. 493-516) Kluwer, Dordrect.
Ziv M (1995b) The control of bioreactor environment for plant propagation in liquid cultures Acta Hortic. 393: 25-38.
Ziv M, Kahany S & Lilien-Kipnis H (1994) Scale-up proliferation and regeneration of Nerine in liquid cultures. Part I. The induction and maintenance of proliferating meristematic clusters by paclobutrazol in bioreactors. Plant Cell Tissue Organ Cult. 39: 109-115.
Ziv M, Meir G & Halevy AH (1981) Hardening carnation plantlets regenerated from shoot tips cultured *in vitro*. Environ. Exp. Bot. 21: 423.
Ziv M, Meir G & Halevy AH (1983) Factors influencing the production of hardened glaucous carnation plantlets *in vitro*. Plant Cell Tissue Organ Cult. 2: 55-60.
Ziv M, Ronen G & Raviv M (1998) Proliferation of meristematic clusters in disposable presterilized plastic bioreactors for the large-scale micropropagation of plants. In Vitro Cell. Dev. Biol. 34p:152-158.
Ziv M, Schwarts A & Fleminger D (1987) Malfunctioning stomata in vitreous leaves of carnation (*Dianthus caryophyllus*) plants propagated *in vitro*; implications for hardening. Plant Sci. 52: 127-134.

Chapter 10

MORPHOGENESIS IN CELL AND TISSUE CULTURES:
ROLE OF ETHYLENE AND POLYAMINES

Eng-Chong Pua
Plant Genetic Engineering Laboratory, Department of Biological Sciences, Faculty of Science, National University of Singapore, 10 Kent Ridge Crescent, Singapore 119260, Republic of Singapore

1 INTRODUCTION

An unique plant cell and tissue culture system has long been recognized as a powerful tool for study of the fundamental aspect of plant biology to provide a better understanding of the process and control of developmental pathways, including cell differentiation and morphogenesis. The system is also crucial for plant biotechnology, through which plant resources can be tailored to generate industrial processes and novel plant materials for the use in agriculture, forestry and horticulture. The rapid advances in plant biotechnology for the last two decades have stimulated much interest in developing efficient tissue culture systems for high frequency plant regeneration from cultured cells and tissues. To date, plant regeneration from cultured cells and tissues, via either organogenesis or somatic embryogenesis, has been reported for a wide range of species, but these systems have been developed based on studies that are mainly empirical in nature. The mechanism underlying the initiation of cell differentiation leading to organ or plant formation *in vitro* is not well understood, although regulation of various morphogenetic events can be physiological, biochemical, epigenetical and genetical in nature (Halperin, 1986; Thorpe and Kumar, 1993). In recent years, there has been increasing evidence showing that both ethylene and polyamines (PAs) are involved in plant cell growth, differentiation and morphogenesis *in vitro* (Biddington, 1992; Pua,

1993, 1996; Buddendorf-Joosten and Woltering, 1994; Matthys et al. 1995). The purpose of the present review is to summarize the recent development in plant morphogenesis *in vitro* in relation to ethylene, PAs and their possible regulatory mechanisms.

2 HISTORICAL BACKGROUND

A cell theory proposed during the 1800s has a profound influence on the advance and our current understanding of plant cell and tissue culture. The theory states that all living things are composed of cells, which can arise from other cells and are capable of maintaining its vitality independent of the rest. This cell doctrine coupled with the fact that all cells, except gametes, in a multicellular organism possess an identical genetic blueprint as the fertilized egg from which they originated, have led to the formulation of the concept of totipotency. This concept implies that most living cells possess an ability for growth and differentiation, as fertilized egg, leading to regeneration of whole multicellular organisms. In plants, numerous early attempts to demonstrate totipotency were not successful (Haberlandt, 1902; Kotte, 1922), although significant progress was made in the 1930s that plant tissues and organs could be cultured for a prolonged period (White, 1934, 1939; Gautheret, 1938; Nobercourt, 1939). It took another two decades before the concept of totipotency was demonstrated by Steward and co-workers (1958a, 1958b), who successfully obtained complete plantlet regeneration from cell clumps grown in suspension initiated from root callus of carrot (*Daucus carota*). Several remarkable achievements were made in the 1960s and early 1970s. One of which was the formulation of a tissue culture medium developed by Murashige and Skoog (1962). This medium, known as MS medium, has been most widely used in plant tissue culture even today. Another accomplishment was regeneration of fertile plants from single somatic cells of tobacco (*Nicotiana tabacum*) grown in isolation from other cells in a defined medium (Vasil and Hildebrandt, 1965). These, together with the first successful regeneration of plants from tobacco protoplasts (Takebe et al. 1971), clearly demonstrate that plant cells are indeed totipotent. The success of plant regeneration from protoplasts is attributable, at least in part, to an innovative approach to produce a large number of protoplasts by enzymatic digestion of plant cells (Cocking, 1960). Apart from somatic cells, Guha and Maheshwari (1964, 1966) also described the first androgenesis from anther culture of *Datura innoxia*, and shortly after haploid plants were successfully regenerated from tobacco anther culture (Bourgin and Nitch, 1967). These earlier achievements not only are the important milestones in plant cell and tissue culture but also provide a significant impetus to the research activity in plant cell differentiation and morphogenesis *in vitro* for the next three decades.

Stewart and Freebaire (1967) were first reported ethylene production by plant tissues in culture. It was later confirmed by Gamborg and LaRue (1968), who observed a variation in ethylene production by suspension culture of different plant species. Since then, there has been increasing report on cultured plant cells and tissues producing ethylene (Mackenzie and Street, 1970; LaRue and Gamborg, 1971; Thomas and Murashige, 1979; Gavinlertvatana et al. 1982; De Proft et al. 1985; Sauerbrey et al. 1988). However, the role of ethylene in culture was not understood until recently.

3 ETHYLENE BIOSYNTHESIS AND REGULATION

Ethylene is a gaseous plant hormone that is synthesized from the precursor methionine, which derived mainly from recycling of 5'-methylthioadenosine, as a precursor. Methionine is first converted to S-adenosylmethionine (SAM) by SAM synthetase, followed by converting SAM to 1-aminocyclopropane-1-carboxylate (ACC) by ACC synthase that is the rate-limiting step (Yang and Hoffman, 1984). ACC oxidase(ethylene-forming enzyme) catalyzes the last step of ethylene biosynthesis by converting ACC to ethylene (Fig. 1).

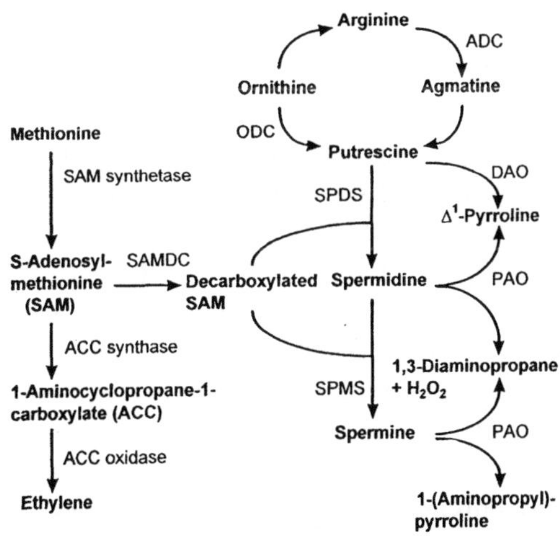

Figure 1. The pathways of ethylene and polyamine biosynthesis and metabolism. ADC, arginine decarboxylase; DAO, diamine oxidase; ODC, ornithine decarboxylase; PAO, polyamine oxidase; SAMDC, S-adenosyl-L-methionine decarboxylase; SPDS, spermidine synthase; SPMS, spermine synthase

Ethylene is involved in regulation of various physiological processes, including seed and dormancy release, flower initiation and senescence, leaf abscission and fruit ripening (Yang and Hoffman, 1984; Reid, 1987). The amount of ethylene produced by the plant varied with tissues, but the level is generally low in most stages of growth and development. However, high levels of ethylene are produced at certain plant developmental or physiological stages such as fruit ripening, leaf and flower senescence and abscission. Other factors known to stimulate ethylene production include wounding, pathogen attacks, auxin treatment and environmental stresses (Abeles, 1973; Yang and Hoffman, 1984). In the tissue culture system, ethylene production is rapid and its synthesis capacity can reach to a peak as early as one week after culture (Chi et al. 1991; Pua 1993). This may be due in part to the wounding effect. It has been shown that an accumulation of ACC synthase (Rottmann et al. 1991) and ACC oxidase mRNA (Holdsworth et al. 1987) in tomato (*Lycopersicon esculentum*) leaves or fruits can be detected as early as 45 min to 8 h after wounding treatment. In addition, the presence of auxin, a common component in culture medium, may also promote ethylene production (see below).

Our current knowledge on ethylene physiology has been derived mainly from studies of plant physiological processes in response to changes in levels of ethylene in the environment and the correlation between the plant response and endogenous levels of key enzymes and substrates of ethylene biosynthesis. These studies rely heavily on various approaches of ethylene modulation, which are discussed below.

3.1 Chemical approach

Modulation of ethylene production and action using the chemical approach has contributed significantly to our understanding of physiological basis of various ethylene-induced responses in plants. Several chemicals are known to inhibit ethylene synthesis and its action (Table 1). The most common inhibitors are aminoethoxyvinylglycine (AVG) and aminooxyacetic acid (AOA) that inhibit the activity of ACC synthase, while salicylic acid inhibits ACC oxidase. The use of these inhibitors usually results in reduced ethylene production. Alternatively, the level of ethylene can also be reduced using ethylene absorbent, e.g. mercuric perchlorate or $KMn0_4$. Another group of chemicals known as ethylene antagonists are Ag^+ ions, 2,5-norbornadiene (NBD) and CO_2, which inhibit ethylene action but not its synthesis. The action of these chemicals are thought to interfere or compete with ethylene binding. On the contrary, plant tissues can be induced to produce high levels of ethylene by the use of ethylene precursor ACC or ethylene-releasing compound such as 2-chloroethylphosphonic acid (CEPA) or ethrel, or direct application of ethylene gas. Although the chemical approach is commonly used because of its convenience, it suffers from some inherent problems such as inefficient uptake and/or transport of chemicals to the target tissue. Non-specificity of inhibitors may also be problematic. For instance, both AVG and AOA inhibit ACC synthase by inhibiting pyridoxal phosphate, which is required for the ACC synthase activity. In view of the fact that pyridoxal phosphate is a common coenzyme in a living system, inhibition of pyridoxal phosphate may also interrupt biochemical processes apart from ethylene biosynthesis.

Table 1. Inhibitors for enzymes and/or action of ethylene and PA biosynthesis

Enzyme/action	**Inhibitor**
ACC synthase	Aminoethoxyvinylglycine (AVG), aminooxyacetic acid (AOA)
ACC oxidase	Salicylic acid
Ethylene action	CO_2, Ag^+, norbornadiene (NBD)
SAM decarboxylase	Methylglyoxal-bis-guanylhydrazone (MGBG)
Arginine decarboxylase (ADC)	α-Difluoromethylarginine (DFMA)
Ornithine decarboxylase (ODC)	α-Difluoromethylornithine (DFMO)
Spermidine synthase (SPMS)	Cyclohexylamine (CHA), dicyclohexylammonium sulphate (DCHA)
Spermine synthase (SPDS)	S-Methyl-5-methlythioadenosine
Diamine oxidase (DAO)	Aminoguanidine (AG)
Polyamine oxidase (PAO)	Guazatine, 2-hydroxyethylhydrazines

3.2 Genetic and molecular approaches

More recently, genetic and molecular approaches have been employed to produce mutants or transgenic plants with specific ethylene responses, which may circumvent the problem encountered using the chemical approach. In *Arabidopsis thaliana*, various ethylene-response mutants such as ethylene-insensitive (*ein*), ethylene-resistant (*etr*), ACC-insensitive (*ain*), constitutive triple responses (*ctr*) and ethylene-overproducing (*eto*) have been isolated (Zarembinski and Theologis, 1994) and these mutants are being used to elucidate the signal transduction pathway of ethylene (Ecker, 1995). However, unlike *A. thaliana*, whose genome size (~100,000 kb) is one of the smallest among higher plants and 80% of it is protein-coding DNA sequence (Meyerowitz, 1992), genomes of other plant species are substantially larger and mutants are not readily obtained. Alternatively, ethylene synthesis can be manipulated by downregulation of ethylene biosynthesis genes using antisense RNA approach (Gray et al. 1992) or decreasing the pool of precursors, e.g. SAM and ACC, by expression of genes coding for precursor-degrading enzymes (Klee et al. 1991; Good et al. 1994). The former strategy can be achieved by, first, cloning genes encoding ethylene biosynthesis enzymes. To date, genes encoding SAM synthetase (Wen et al. 1995), ACC synthase (Wen et al. 1993; Zarembinski and Theologis, 1994) and ACC oxidase (Pua et al. 1992; Zarembinski and Theologis, 1994) have been isolated from various plant species. The cloned gene can be constructed to generate antisense RNA to inactivate expression of the corresponding endogenous gene in transgenic plants. This has been demonstrated in transgenic tomato and mustard (*Brassica juncea*) plants overexpressing antisense ACC synthase (Oeller et al. 1991) or ACC oxidase RNA (Hamilton et al. 1990; Pua and Lee, 1995). These plants showed a marked reduction in corresponding mRNA and enzyme synthesis, and a decrease in ethylene production by 80-99%. In precursor degradation strategy, transgenic tomato plants expressing a gene encoding bacteriophage T3 SAM hydrolase, which degrades SAM into methylthioadenosine and homoserine, displayed a marked reduction in the SAM pool in tissues and a decrease in ethylene synthesis (Good et al. 1994). A similar decrease in the endogenous ACC pool and ethylene production has also been reported in transgenic plants expressing a bacterial ACC deaminase gene (Klee et al. 1991). ACC deaminase catalyzes the conversion of ACC to α-ketobutyrate and ammonia (Honma, 1985).

4 ETHYLENE ACTION

The mechanism of how ethylene exerts its action in various plant physiological processes is not well understood, although steps in ethylene biosynthetic pathway have been well established for some time. It has been proposed that the mode of ethylene action may be mediated through its binding to receptor, which may be a Zn^{2+}- or Cu^{2+}-containing metalloprotein (Burg and Burg, 1967). The existence of ethylene-binding receptor has been confirmed in several plant species, including tobacco (Sisler, 1979), bean (*Phaseolus vulgaris*), citrus and *Ligustrum japonicum* (Goren and Sisler,

1986), pea (*Pisum sativum*) (Sisler and Yang, 1984) and rice (*Oryza sativa*) (Sanders et al. 1990). In addition, ethylene-binding proteins of 26 and 28 kDa have been purified from bean cotyledons (Harpham et al. 1996). However, attempts to isolate genes encoding the receptor have not been successful until recently. Molecular genetic analysis of several *Arabidopsis* ethylene-response mutants has led to the successful isolation of two putative receptor/sensor genes, *ETR1* (Chang et al. 1993; Schaller and Bleecker, 1995) and *ERS* (Hua et al. 1995), both of which encode a putative histidine protein kinase.

The amino acid sequence of both ETR1 and ERS shows a high degree of similarity to proteins in the prokaryotic two-component system that consists of a sensor and an associated response regulator. The two-component system is an efficient communication mechanism of prokaryotes in response to signals as a result of changes in their environment such as nitrogen availability, osmotic stress and oxygen tension (Parkinson, 1993). The signal is usually perceived by the input domain that is located at the N-terminal of the sensor, and its perception triggers phosphorylation of histidine residue, which in turn phosphorylates aspartate residue in the receiver domain of the response regulator. The activity of the downstream output domain, which are mainly transcriptional regulators in bacteria, is regulated by activation of aspartate residue (Chang, 1996). The structural similarities between the ethylene receptor/sensor and two-component system proteins suggest that the signaling mechanism for ethylene in plants may be similar to signal transduction pathway in bacteria. Apart from *ETR1* and *ERS*, Ecker and co-workers (Kieber et al. 1993) also cloned an *Arabidopsis CTR1* gene, which is a downstream negative regulator of ethylene response. *CTR1* encodes a putative *Raf*-like serine-threonine protein kinase. *Raf* proteins are mitogen-activated protein kinase kinase kinases (MAPKKKs), which activate the downstream MAPKKs by phosphorylation. MAPKKs in turn phosphorylate MAPKs. These proteins are involved in regulation of cell growth and differentiation in animals by transduction of signals from cell-surface receptors to cytoplasmic transcription factors (Blumer and Johnson, 1994). The unique feature of CTR1 protein has led to the speculation that a similar phosphorylation cascade might be involved in ethylene signaling in plants (Zarembinski and Theologis, 1994: Chang, 1996), although evidence from genetic analysis indicates that *CTR1* might be under the control of *ETR1* and/or *ERS* (Kieber et al. 1993).

Based on the current knowledge of ethylene signal transduction pathway, it can be assumed that ethylene action may begin with ethylene perceived at the plasma membrane by ETR1 and/or ERS receptors, whose ethylene binding is likely to involve metal ions that has yet to be demonstrated. The signal resulting from the binding may be transduced to other molecules via phosphorylation of the histidine residue, as a result, alters expression of *CTR1* and other downstream cascade genes leading to plant physiological response (Chang, 1996; Fluhr and Mattoo, 1996). In view of the evidence showing the implication of protein phosphorylation and dephosphorylation in tobacco pathogenesis response (Raz and Fluhr, 1993), it was suggested that the similar reactions might also be involved in other ethylene-induced responses in plants.

5 EFFECT OF ETHYLENE

5.1 Growth and development

Evidence from several lines of study have shown that high levels of ethylene markedly influence plant growth and development in culture. In rapid-cycling *Brassica campestris*, plants grown in the sealed vessel for three weeks accumulated ethylene up to 0.25 ppm/plant, causing abnormal growth and development of the plant, e.g. hypocotyl swelling and inhibition of leaf expansion and flowering (Lentini et al. 1988). However, plant growth was reverted to normal after treated with an ethylene antagonist, 250-500 ppm NBD. The similar adversary effect of accumulated ethylene has also been reported in potato (*Solanum tuberosum*), whose shoots grown in the sealed vessel were retarded (Hussey and Stacey, 1981; 1984). Shoot culture maintained in the sealed vessel is also a poor source of protoplasts, as demonstrated in potato (Perl et al. 1988) and tomato (Rethmeier et al. 1991), whose yield and plating efficiency of protoplasts are low. However, the plating efficiency and growth of protoplasts can be markedly improved by maintaining cultured shoots in the presence of Ag^+. Unlike *B. campestris* and potato, mustard plants grown in the absence or presence of AVG for four weeks in the sealed vessel were phenotypically normal, although growth and development of those maintained in the presence of 20 μM $AgNO_3$ were severely inhibited (Pua and Chi, 1993). $AgNO_3$ also promoted ethylene production, which was thought to be responsible for growth inhibition, as high levels of ethylene could lead to a decrease in photosynthetic rate and assimilate allocation (Subrahmanyam and Rathore, 1992). In addition, growth may be adversely affected by the phytotoxic effect exerted by $AgNO_3$ because Ag^+ is a heavy metal.

5.2 Shoot organogenesis and proliferation

Apart from growth and development of cultured shoots or plants, high levels of ethylene also inhibit shoot organogenesis from various cultured tissues or organs of a wide range of plant species (Table 2). In tobacco callus, there is an inverse relationship between the capacity of shoot formation and ethylene production. Callus with higher shoot forming capacity contained lower levels of endogenous ACC (Grady and Bassham 1982) and produced less ethylene (Huxter et al. 1981) than non-shoot forming callus. The inhibitory effect of ethylene is further exemplified by studies, in which shoot regeneration from callus of wheat (*Triticum aestivum*) and tobacco (Purnhauser et al. 1987) and sunflower (*Helianthus annuus*) (Robinson and Adams, 1987) was promoted by inhibition of ethylene production or action using AVG or $AgNO_3$. These findings suggest that high levels of ethylene in culture may be responsible for inhibition of shoot regeneration from these calli. The similar promoting effect of ethylene inhibitors has also been demonstrated in mustard and *B. campestris*, which are usually recalcitrant in culture (Murata and Orton, 1987; Jain et al. 1988: Narasimhulu and Chopra, 1988). Cultured explants grown on medium supplemented with 5-10 μM AVG or 10-30 μM $AgNO_3$ gave rise to adventitious shoots at high frequencies after 3-4 weeks, whereas those grown without ethylene inhibitors were poorly regenerative (Chi and Pua, 1989, Chi et al. 1990, 1991; Pua and Chi, 1993). Measurements for the ethylene content in culture, where cotyledons or leaf disc explants were grown, showed that the level of ethylene increased rapidly after three days and reached a peak after 10-14 days (Pua, 1993). Because initiation of shoot primordium in mustard occurred at 4-8 days after culture (Sharma and Bhojwani, 1990) that is coincided with the time of rapid ethylene synthesis, high levels of ethylene may cause inhibition to shoot primordium differentiation and subsequent regeneration. In recent years, the implication of ethylene in shoot organogenesis has also been reported in a range of plant species. These include various members in Cruciferae such as oilseed *B. campestris* (Palmer, 1992), Brussels sprouts (*B. oleracea* var. *gemmifera*) (Williams et al. 1990), oilseed rape (*B. napus*) (De Block et al. 1989), *B. rapa* (Burnett et al. 1994), Chinese kale (*B. alboglabra*) (Pua et al. 1996b), *A. thaliana* (Márton and Browse, 1991) and Chinese radish (*Raphanus sativus* var. *longipinnatus*) (Pua et al. 1996a).

Although both AVG and $AgNO_3$ are equally effective in promoting shoot regeneration, their effects on ethylene production are divergent. The use of AVG has been shown to reduce ethylene production (Robinson and Adams, 1987; Chi et al. 1991; Pua and Chi, 1993). This is conceivable because AVG inhibits ethylene synthesis by blocking ACC synthase activity (Yang and Hoffman, 1984). On the contrary, explants grown in the presence of $AgNO_3$ showed a marked increase in endogenous ACC and ACC synthase activity and produced large amounts of ethylene. An overproduction of ethylene as a result of Ag^+ treament has also been reported in cultured sunflower cotyledons (Chraibi et al. 1991), tomato fruits (Penarrubia et al. 1992) and an ethylene-overproducing *A. thaliana* mutant (Guzman and Ecker, 1990). This $AgNO_3$-stimulated ethylene production is intriguing, because Ag^+ inhibits ethylene action and, theoretically, has no effect on ethylene production, as demonstrated in rice callus (Adkins et al. 1994). According to Theologis (1992), ethylene overproduction may be explained by interference of ethylene receptor by Ag^+. He assumed that amount of ethylene produced by plant cells might be controlled by the level of receptor binding to ethylene. A high level of reception might be perceived by the cell as the amount of ethylene had reached to a high level, as a result, decreased ethylene production. This might be achieved by reducing synthesis and/or activity of ACC synthase. The interference of Ag^+ with the receptor system that resulted in a lower level of reception might be interpreted by the cell as an absence of ethylene, thereby leading to ethylene overproduction. However, the contradictory results of $AgNO_3$ effect on ethylene production have also been reported. For examples, cultured cucumber cotyledons (Roustan et al. 1992) and cauliflower (*B. oleracea* var. *botrytis*) hypocotyls (Sethi et al. 1990) grown in the presence of $AgNO_3$ showed decreased ethylene production. This discrepancy may be attributed to genotypic variability and a differential physiological stage and Ag^+-sensitivity of the explant.

Although various ethylene inhibitors have been used to promote shoot regeneration *in vitro*, not all inhibitors are equally effective. We found that AOA and 2,4-dinitrophenol, an uncoupler of oxidative phosphorylation (Murr and Yang, 1975), had no effect on shoot regeneration from cultured cotyledons of *B. campestris* ssp. *chinensis*, although both $AgNO_3$ and AVG were effective (Chi and Pua, 1989). This inconsistency of the inhibitor effect suggests that the role of ethylene on shoot regeneration cannot be determined conclusively based on the inhibitor study alone. To search for more direct evidence to determine the role of ethylene, we employed an antisense RNA approach to produce transgenic plants with impaired ethylene synthesis. This has been achieved by cloning a gene encoding ACC oxidase (Pua et al. 1992) and optimization of a gene transfer methodology based on Ti plasmid vector for mustard plants (Barfield and Pua, 1991).

Table 2. The promoting effect of ethylene inhibitors on cell growth, viability and morphogenesis of various plant species *in vitro*

Plant species	Tissue / explant	Basal medium*	Ethylene inhibitor	Enhanced growth/ morpho-genesis	Reference
Monocotyledons					
Hordeum vulgare	Anthers	MMS + 1 $mg.l^{1}$ PAA + 0.2 $mg.l^{-1}$K	1 $mg.l^{-1}Ag_2S_2O_3$	Androgenesis	Evans and Patty, 1994
Oryza sativa	Caryopses	MS + 10 $mg.l^{-1}$ 2,4-D	25 µM AVG; 50 µM $AgNO_3$	Callus	Adkins et al., 1993
Triticum aestivum	Callus	MS + 0.5 -1 $mg.l^{-1}$ 2,4-D	10 $mg.l^{-1}$ $AgNO_3$	Shoots	Purnhauser et al., 1987
Zea mays	Callus	D medium	100 µM $AgNO_3$; 250 µM NBD	Plants	Songstad et al., 1988
Z. mays	Immature embryos	MS/N6 + 1 $mg.l^{-1}$ 2,4-D + 6 mM proline + 100 $mg.l^{-1}$ CH	5.9-59 µM $AgNO_3$; 0.5-5 µM AVG	Type II callus	Vain et al., 1989
Z. mays.	Immature embryos	N6 + 1$mg.l^{-1}$ 2,4-D + 25 mM proline +100 $mg.l^{-1}$CA	10-100 µM $AgNO_3$	Type II callus	Songstad et al., 1991
Dicotyledons					

* AC, activated charcoal; AOA, aminooxyacetic acid; AVG, aminoethoxyvinylglycine; BA, benzylaminopurine; 1-MCP, 1-methyl-cyclopropene ; TCZ, thidiazuron; BDS medium (see Dunstan and Short, 1977); CA, casamino acid; CH, casein hydrolysate; D medium (See Duncan et al., 1985) ; 2,4-D, 2,4-dichlorophenoxyacetic acid; GA_3, gibberellic acid; HaR medium (see Robinson and Adams, 1987); IPA, isopentenyladenine; IPAR, isopentenyladenosine; K, kinetin; MG5 medium (see Keller et al., 1975); MHI medium (see Carron and Enjalric, 1985); NAA, α-naphthaleneacetic acid; NBD, norbornadiene; PAA, phenylacetic acid; MMS medium(see Evans and Patty, 1994); MS_0 medium (see Vain et al., 1989); N6 medium (see Chu et al., 1975); RMOP medium (see Sidorov et al., 1981)

Plant species	Tissue / explant	Basal medium*	Ethylene inhibitor	Enhanced growth/ morpho-genesis	Reference
Arabidopsis Thaliana	Root segments	MS + 4 mg.l^{-1} IPA + 2 mg.l^{-1} NAA/MS + 2mg.l^{-1} IPA + 0.1 mg.l^{-1} NAA + 1 mg.l^{-1} IPAR	25 mg.l^{-1} $AgNO_3$	Shoots	Márton and Browse, 1991
Brassica spp.	Cotyledons; hypocotyls	MS + 8.8 µM BA + 5.4 µM NAA	5-30 µM $AgNO_3$; 1-10 µM AVG	Shoots	Chi et al., 1990
B. alboglabra	Hypocotyls	MS + 2 mg.l^{-1} BA + 1 mg.l^{-1} NAA + 0.4% agarose	10 µM $AgNO_3$; 5 µM AVG	Shoots	Pua et al., 1996b
B. campestris	Cotyledons	MS + 2 mg.l^{-1} BA + 1 mg.l^{-1} NAA	30/60 µM $AgNO_3$	Shoots	Palmer, 1992
B. campestris	Hypocotyls ; cotyledons	MS + 5 mg.l^{-1} BA + 3 mg.l^{-1} NAA	25-100 µM $AgNO_3$	Shoots	Mukhopadhya y et al., 1992
B. campestri ssp. *chinensis*	Cotyledons	MS + 2 mg.l^{-1} BA + 1 mg.l^{-1} NAA	0.1-10 µM AVG; 7.5-60 µM Ag_2SO_4; 7.5-60 µM $AgNO_3$	Shoots	Chi and Pua, 1989
B. campesrtis ssp. *Pekinensis*	Cotyledons	MS + 2 mg.l^{-1} BA + 1 mg.l^{-1} NAA	30 µM $AgNO_3$; 5 µM AVG	Shoots	Chi et al., 1991
B. juncea	Somatic embryos	MS + 1-2 mg.l^{-1} BA	5 µM $AgNO_3$	Shoots /plants	Pua, 1994
B. juncea	Leaf discs; petioles	MS + 10 µM BA + 5 µM NAA	20 µM $AgNO_3$; 5 µM AVG	Shoots	Pua and Chi, 1993
B. juncea	Hypocotyls	MS + 2 mg.l^{-1} BA + 1 mg.l^{-1} NAA or MS + 2 mg.l^{-1} BA + 0.05 mg.l^{-1} 2,4-D	20 µM $AgNO_3$	Shoots	Pental et al., 1993
B. oleracea var. botrytis	Callus	B5 + 1 mg.l^{-1} BA + 1 mg.l^{-1} NAA	10 µM $AgNO_3$; 1 µM $CoCl_2$; 1-100 µM AVG	Shoot differentiation	Sethi et al., 1990

Plant species	Tissue / explant	Basal medium*	Ethylene inhibitor	Enhanced growth/ morpho-genesis	Reference
B. oleracea var. *gemmifera*	Anthers	B5 + 10% sucrose + 0.1 $mg.l^{-1}$ NAA + 0.1 $mg.l^{-1}$ 2,4-D	10 $mg.l^{-1}$ $AgNO_3$	Androgenesis	Biddingdon et al., 1988
B. oleracea var. *gemmifera*	Anthers	MG5 + 0.1-0.3 $mg.l^{-1}$ 2,4-D	5 $mg.l^{-1}$ $AgNO_3$	Androgenesis	Ockendon and McClenaghan, 1993
B. oleracea var. *gemmifera*	Callus	B5 + 0.1 $mg.l^{-1}$ 2,4D + 0.1 $mg.l^{-1}$ K	1-3 $mg.l^{-1}$ $AgNO_3$	Shoots	Williams et al., 1990
B. oleracea & *B. napus*	Hypocotyls	MS + 1 $mg.l^{-1}$ BA + 0.1 $mg.l^{-1}$ NAA + 0.01 $mg.l^{-1}$ GA_3	5-10 $mg.l^{-1}$ $AgNO_3$	Shoots	De Block et al., 1989
B. rapa	Hypocotyls	B5 + 3 $mg.l^{-1}$BA + 0.1 $mg.l^{-1}$ zeation	5/10 $mg.l^{-1}$ $AgNO_3$	Shoots	Radke et al., 1992
B. rapa	Cotyledons	MS + 4.4 µM BA + 5.4 µM NAA	1 µM AVG	Shoots	Burnett et al., 1994
Capsicum annuum	Cotyledins	BDS (+ / -2 $mg.l^{-1}$ GA + 1 $mg.l^{-1}$ BA	29.4 µM $AgNo_3$	Shoot proliferation and elongation	Hyde and Phillips, 1996
Chrysanthemum coronarium	Leaf discs	MS + 2.5 µM BA + 2.5 µM NAA	1 µM $AgNO_3$	Shoots	Lee et al., 1997
Cucumis melo	Cotyledons	MS + 0.54 µM NAA + 2.2 µM BA	60-120 µM $AgNO_3$	Shoots	Roustan et al., 1992
Daucus carota	Cell suspension	B5 + 4.5 µM 2,4-D	10-50 µM $CoCl_2$; 100 µM $NiCl_2$	Somatic embryos	Roustan et al., 1989
Garcinia mangostana	Leaves	MS + 20 µM BA	0.5 µM AVG; 10-20 µM $AgNO_3$	Shoot buds (in airtight vessels)	Goh et al., 1997

Plant species	Tissue / explant	Basal medium*	Ethylene inhibitor	Enhanced growth/ morpho-genesis	Reference
Glycine max	Cotyledons	MS salts + B5 vit + 6% sucrose + 40 $mg.l^{-1}$ 2,4-D	2 $mg.l^{-1}$ AVG	Normal embryoids and plants	Santo et al., 1997
Helianthus annuus	Hypocotyl callus	HaR	10 µM AVG	Plants	Robinson and Adams 1987
H. annuus	Cotyedons	MS + 4.4 µM BA + 5.4 NAA	10-25 µM $AgNO_3$; 20 µM $CoCl_2$	Shoots	Chraibi et al., 1991
Hevea brasiliensis	Callus	MH1 + 4.5 µM BA + 4.5 µM 2,4-D	5/50 µM AOA; 5.9/58.9 µM $AgNO_3$	Somatic embryos	Auboiron et al., 1990
H. brasiliensis	Callus	MH1 + 4.5 µM BA + 4.5 µM 2,4-D	250 µM AOA	Somatic embryos	Housti et al., 1992
Lycopersicon esculentum	Cultured shoots	Hormone-free MS + 1.8% sucrose	2.5 $mg.l^{-1}$ $Ag_2S_2O_3$	Protoplast yield and viability	Rethmeier et al., 1991
Malus × Domestica	Shoot culture	MS salts + 1/2 B5 vit. + 0.5 mg/l^{-1} IBA	3-10 µM $AgNO_3$; 1-3 µM AVG; 1-30 µM $CoCl_2$	Root formation and growth	Ma et al., 1998
Nicotiana plumbaginifolia	Callus	RMOP + /- 8.25 µM NH_4-succinate	10 $mg.l^{-1}$ $AgNO_3$	Shoots	Purnhauser et al., 1987
Pelargonium × Hortorum	Hypocotyls	MS salts + B5 vit + 10 µM TDZ	1.2 ppm 1-MCP	Somatic embryos	Hutchinson et al., 1997
Raphanus sativus var. longipinnatus	Hypocotyls	MS + 2 $mg.l^{-1}$ BA+ 1 $mg.l^{-1}$ NAA + 0.4% agarose	30 µM $AgNO_3$/AVG + 10-25 mM putrescine	Shoots	Pua et al., 1996a
Solanum tuberosum	Leaves	MS + 1% sucrose	2 $mg.l^{-1}$ $Ag_2S_2O_3$	Protoplast yield	Perl et al., 1988
S. tuberosum	Anthers	MS + 0.5% AC + 1 $mg.l^{-1}$ BA	54 µM $AgNO_3$; 47 µM n-propygallate	Androgenesis	Tiainen, 1992

Subsequently, an antisense ACC oxidase gene was constructed and transferred into mustard plants using *Agrobacterium tumefaciens*-mediated transformation (Pua and Lee, 1995). Several transgenic plants were selected and characterized. These R0 transgenic plants showed low levels of endogenous 1.4-kb ACC oxidase transcript and a marked decrease in corresponding enzyme activity and ethylene production (Fig. 2). More

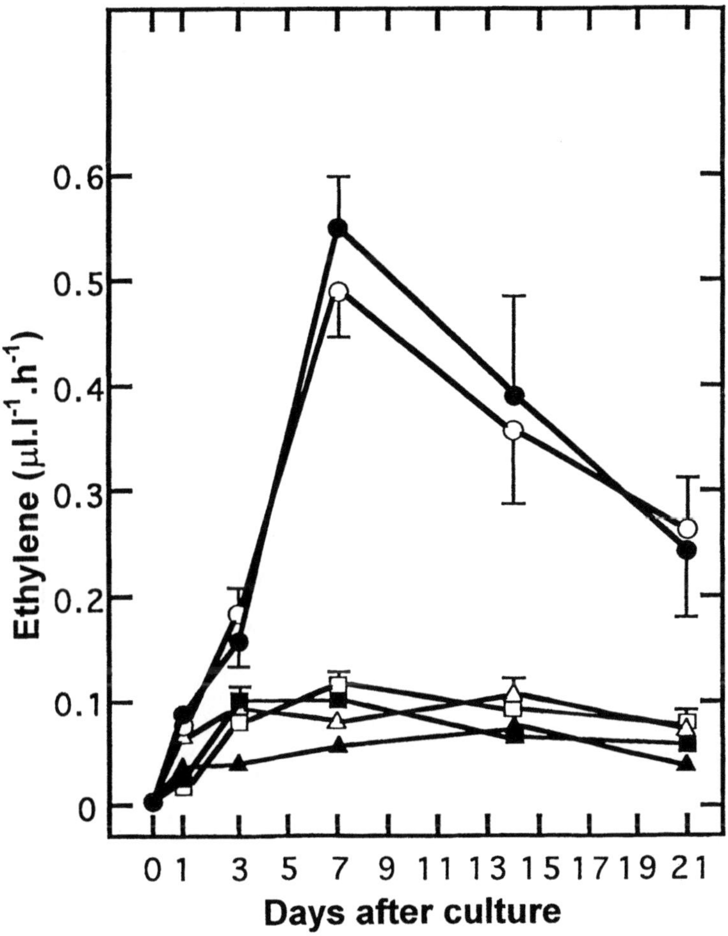

Figure 2. Ethylene production in cultured tissues of mustard during three weeks of culture. Leaf disc explants grown on MS medium supplemented with 4 mg.l^{-1} BA and 1 mg.l^{-1} NAA and 0.8% Difco-Bacto agar. Each culture flask containing five explants was sealed with an airtight serum stopper. The ethylene level in the flask was measured at different intervals using gas chromatography according to method described previously (Pua and Chi, 1993). ●, negative control (non-transformed plant); ○, positive control (transgenic mustard plants carrying pROA93); □, ■, Δ, ▲; individual transgenic plants overexpressing antisense ACC oxidase RNA (Pua and Lee 1995).

strikingly, leaf disc explants excised from transgenic plants grown on medium in the absence of ethylene inhibitor were highly regenerative. On the contrary, control tissues grown on the same medium produced high levels of ethylene and formed mostly callus. This inverse relationship between the capacity of ethylene synthesis and shoot regenerability is also manifested in R1 progeny. This is evidenced by high frequency shoot regeneration from transgenic R1 hypocotyls grown in the absence of ethylene inhibitor (Fig. 3). The transgenic explant contained lower

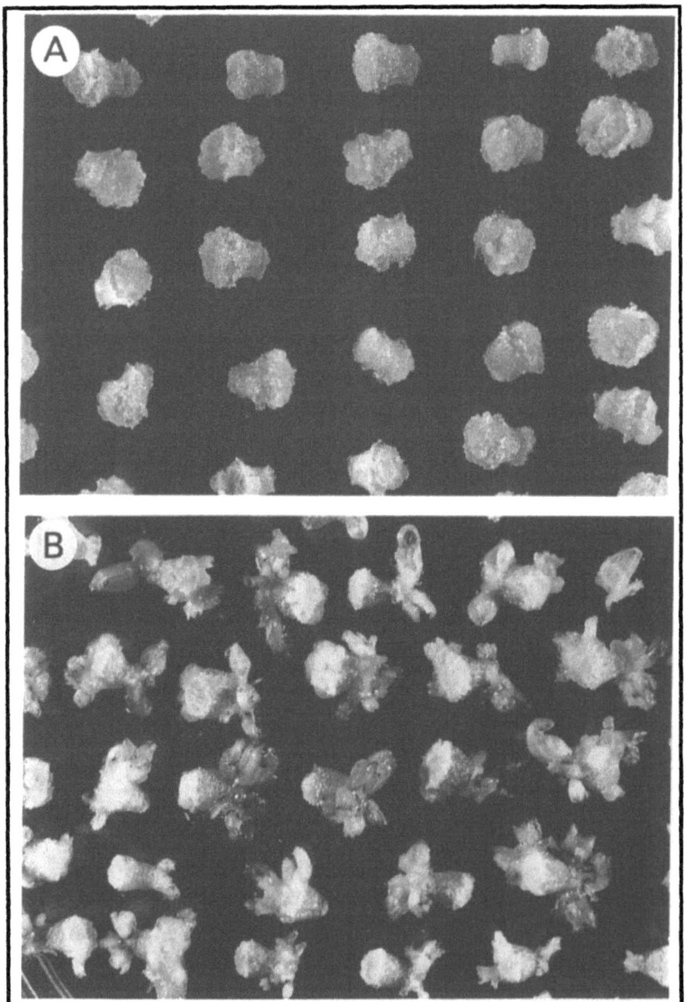

Figure 3. Shoot regeneration response of hypocotyl explants of control (A) and R1 transgenic (B) mustard seedlings grown for two weeks on MS medium containing 2 mg.l^{-1} BA and 1 mg.l^{-1} NAA solidified with 0.4% agarose

Table 3. The promoting effect of ethylene on growth and morphogenesis of plant cells and tissues *in vitro*.

Plant species	**Tissue/ explant**	**Basal medium**	**Chemical** *	**Growth/ morphogenesis**	**Reference**
Monocotyledons					
Heliconia psittacorum	Morphogenic callus	Hormine-free MS	≥20 µM $AgNO_3$; ≥5µ AVG	Inhibited plant regeneration	Kumer et al., 1996
Hordeum vulgare	Anthers	BAC1 + 2 mg/l 2,4-D + 0.5 mg/l ZR + 750 mg/l glutamine	10^{-5} M ethrel; 10^{-4}-10^{-5} M ACC	Enhanced androgenesis (genotype dependent)	Cho and Kasha, 1989
H. vulgare	Anthers	Hormone-free liquid FHG	1 µM $CoCl_2$ +/- 0.5 µM Cl	Enhanced androgenesis	Cho and Kasha, 1995
Lilium speciosum	Bulb-scales segments	MS+5 µM NAA	10-50 µM ACC; 1-10 ppm C_2H_4 ; 0.5 µM Cl	Enhanced shoot proliferation	Van Aartrijk et al., 1985
Triticum aestivum	Anthers	Hormone-free liquid FHG	1 µM $CoCl_2$ +/- 0.5 µM Cl	Enhanced androgenesis	Cho and Kasha, 1995

* ABA, abscisic acid; ACC, 1-aminocycloropane-1-carboxylate; BAC1 medium (see Marsolai and Kasha, 1985); CEPA, 2-chloroethylphosphonic acid; CI, calcium inophore; FHG medium (see Hunter, 1977); K(h) medium (see Cheng, 1975); MLT medium, modified Lu and Thorpe (1987); SF/NSF, shoot forming/non-shoot forming edium (see Huxter et al., 1981); QP medium (see Quoirin and LePoivre, 1977); TDZ, thidiazuron; TIBA, 2,3,5- triiodobenzoic acid; WPM medium (see Lloyd and McCown, 1980); SH, Schenk and Hildebrandt medium (see Aitken et al. 1981)

Plant species	Tissue/ explant	Basal medium	Chemical *	Growth/ morphogenesis	Reference
Dicotyledons					
Corylus avellana	Cotyledon segments	1/2 K(h) + 50 µM IBA + 5 µM K	5 µM CEPA	Enhanced rooting	González et al., 1991
Daucus carota	Embryogenic cells	B5 + 7% sucrose + 0.45 µM 2,4-D	10 µM ACC	Increased embryoid yield	Nissen, 1994
Lonicera nitida	Leaf discs	LS + 2.3 µM TDZ + 2.9 µM IAA	8.8 µM ACC + 20 µM TIBA	Increased adventitious bud formation	Cambecedes et al., 1992
Medicago sativa	Petiole callus	Hormone-free MS	1.3 x 10^{-4} M NBD	Inhibited callus growth and embryoid maturation	Kepczynski et al., 1992
Nicotiana tabacum	Callus	SF/NSF	8 µM ethrel	Earlier shoot emergence	Huxter et al., 1981
N. tabacum	Thin-layer pendicels	MS + 1 µM BA + 1-2.2 µM NAA	1 µM ACC	Earler flower bud formation	Smulder et al., 1990
Petunia hybrid	Shoots	MS + 1 µM BA	0.01-10 ppm C_2H_4	Enhanced shoot proliferation	Dimasi-Theriou et al., 1993
Picea glauca	Embryogenic tissues	MLT + 10 µM 2,4-D + 2 µM BA + 1 µM kinetin	5 µM AVG/10 µM $AgNO_3$ + 10/50 µM ABA	Embryoid production and quality	Kong and Yeung, 1994

Plant species	Tissue/ explant	Basal medium	Chemical *	Growth/ morphogenesis	Reference
Pinus radiata	Cotyledons	SH + 25 µM BA	0.25 M $Hg(ClO_4)_2$	Inhibited shoot bud differentiation	Kumar et al., 1987
Prunus persica × *P. amygdalus*	Shoots	Hormone-free WPM	0.1 ppm C_2H_4	Enhanced shoot growth and proliferation	Dimasi-Theriou and Economou, 1995
Rosa hybrida	Shoots	MS + 0.49 µM + 6.69 µM BA+ 0.29 µM GA_3	5 ppm C_2H_4	Enhanced shoot proliferation	Kevers et al., 1992
Solanum carolinense	Anthers	Liquid MS	0.001% ethrel; 10^{-4} M ACC	Enhanced androgenesis	Reynolds, 1987
Thuja occidentalis	Embryos	1/2 QP + 1 µM BA	0.25 M $Hg(ClO_4)_2$	Inhibited bud formation and shoot elongation	Nour and Thorpe, 1994
Tulipa sp.	Floral stems	MS + 1 $mg.l^{-1}$ BA + 1 $mg.^{-1}$ NAA	10 ppm C_2H_4; 1-100 µM ACC	Enhanced shoot production	Taeb and Alderson, 1990

endogenous ACC oxidase activity produced less ethylene than the poorly regenerative control explant (Pua and Lee, 1995). Theses finding s are in agreement with results of the inhibitor study that ethylene indeed plays a pivotal role on shoot morphogenesis *in vitro* and high levels of ethylene produced by cultured tissues are responsible for suppression of cell differentiation leading to regeneration.

The inhibitory effect of ethylene on shoot organogenesis is not an universal response of cultured cells and tissues in plants. Ethylene has been shown to be required for somatic embryogenesis, androgenesis, and shoot organogenesis and proliferation of several plant species *in vitro* (Table 2). In lavandin, the capacity of shoot formation was positively correlated with the level of ethylene produced by the explant (Panizza et al. 1988). An elevated level of ethylene in culture also promoted formation of bulb primordia from floral stem explants of tulip (*Tulipa* sp.) (Taeb and Alderson, 1990) and regeneration of flower buds from thin-layer explants of tobacco (Smulders et al. 1990). In radiata pine (*Pinus radiata*), ethylene was shown to be, at least in part, responsible for shoot bud differentiation from cotyledonary explants (Kumar et al. 1987). Apart from organogenesis, ethylene is also promotive to shoot proliferation *in vitro*. Shoot culture of peach rootstock (*Prunus persica* x *P. amygdalus*) that grown in the vessel containing higher amounts of ethylene showed a significant higher number of shoots with longer shoot length (Dimasi-Theriou et al. 1995). In addition, proliferation of petunia (*Petunia hybrida*) shoots was markedly enhanced by exposure of explants to 0.01-10 ppm exogenous ethylene, whereas the presence of $KMnO_4$, $CoCl_2$ or $AgNO_3$ inhibited shoot proliferation (Dimasi-Theriou et al. 1993). The similar beneficial effect of ethylene has also been reported in rose (*Rosa hybrida*) shoot culture, whose multiplication increased as a result of addition of ACC to the medium or pulse treatment of culture with 5 ppm ethylene gas (Kevers et al. 1992). Furthermore, the presence of ethylene was also important for rapid shoot proliferation of eastern white cedar (*Thuja occidentalis*) (Nour and Thorpe, 1994). Application of mercuric perchlorate to remove ethylene in culture suppressed axillary bud development.

5.3 Somatic embryogenesis

The mechanism underlying the process of somatic embryo formation and development is complex (Thorpe, 1995). In some species, the embryogenic capacity is associated with the rate of ethylene production by the cultured tissue. The embryogenic tissue of conifers has been shown to produce 10-100-fold less ethylene than the undifferentiated tissues (Wann et al. 1987). Furthermore, inhibition of ethylene production or action using inhibitors also enhanced somatic embryogenesis from callus culture of rubber (*Hevea brasiliensis*) (Auboiron et al. 1990), carrot (Roustan et al. 1989, 1990) and maize (Songstad et al. 1988; Vain et al. 1989). The promoting effect of ethylene inhibitors may be related to decreased tissue browning, as demonstrated in rubber, whose callus culture treated with AOA showed less necrosis and a decrease in polyphenol oxidase, peroxidase and NADH-quinone reductase and an increase in superoxide dismutase, catalase and ascorbate peroxidase (Housti et al. 1992). On the contrary, the presence of ethylene appears to be beneficial to somatic embryogenesis in some species. Somatic embryogenesis of carrot was promoted by application of low levels of ACC and CEPA, but higher concentrations were inhibitory (Nissen, 1994), suggesting that low levels of ethylene are important to differentiation and growth of embryogenic tissues. A study on ethylene synthesis and growth of embryogenic tissues of Norway spruce (*Picea abies*) also indicates the possible implication of ethylene or ACC in the induction of embryogenic tissue and in the early stages of embryoid maturation (Kvaalen, 1994). Unlike the species discussed above, ethylene may not be involved in somatic embryogenesis of alfalfa (*Medicago sativa*). This is evidenced by the lack of response of cultured tissues to NBD (Kepczynski et al. 1992) and cobalt and nickel (Meijer, 1989). However, NBD inhibited embryoid maturation of alfalfa, indicating a differential requirement for ethylene in embryoid production and maturation. This may explain as to why mustard protoplasts could grow and differentiate into embryoids in the absence of ethylene inhibitor but these embryoids required $AgNO_3$ for further growth and development into shoots and plants (Pua, 1994). As alfalfa, the use of ethylene inhibitors had no effect on embryoid induction in soybean (*Glycine max*), but the presence of AVG significantly increased the frequency of normal embryoid formation and plant regeneration from embryoids (Santo et al. 1997).

5.4 Androgenesis

Anthers grown in culture have been shown to produce ethylene (Horner et al. 1977) and pollen of many plant species also contains high levels of ACC (Hill et al. 1987). Evidence has shown that the level of ethylene present in anther culture may be associated with androgenesis. In *Solanum carolinense*, pollen embryogenesis was enhanced by application of 10^{-4} M ACC, 0.001% ethrel or, to a greater extent, by 6×10^{-5} M IAA (Reynolds, 1987). It was speculated that IAA-induced embryogenesis might be attributed to auxin-induced ethylene production. Apart from auxin, the addition of 0.5 μM calcium ionophore A23187 or 1 mM $CoCl_2$ to the medium also stimulated embryoid production from anther culture of barley (*Hordeum vulgare*) and wheat (Cho and Kasha, 1995) with a concomitant increase in ethylene production. This promoting effect of Ca^{2+} on ethylene production has also been reported in mung bean hypocotyls (Lau and Yang, 1974; Arteca, 1984). It was therefore suggested that Ca^{2+}-stimulated androgenesis might be mediated through enhanced ethylene synthesis. On the contrary, enhanced androgenesis by decreasing ethylene production or action has also been reported in some plant species. $AgNO_3$ was stimulatory to androgenesis of Brussels sprouts (Biddington et al. 1988; Ockendon and McClenaghan, 1993). Apart from $AgNO_3$, androgenesis of potato was also promoted by n-propyl-gallate and exogenous PAs (Tiainen, 1992). These divergent effects of ethylene on androgenesis may be related to endogenous levels of ACC and ethylene production of the plant tissues. In barley anther culture, cultivars that contained lower amounts of ACC and produced lower levels of ethylene showed decreased androgenesis, which could be enhanced by addition of 0.1 mM ethrel or exogenous ACC to the medium (Cho and Kasha, 1989). For those cultivars with higher levels of endogenous ACC and produced higher amounts of ethylene, the presence of 0.1 mM exogenous putrescine was beneficial to androgenesis. These indicate that androgenesis may require an optimal level of ethylene, which can be modulated by ethylene promoter and inhibitor. A similar genotypic variability of androgenic response to ethylene has also been reported in barley, but $Ag_2S_2O_3$ is required for plant regeneration from anther culture of all cultivars (Evans and Batty, 1994). It suggests that there may be a differential ethylene requirements for androgenesis and plant regeneration.

5.5 Rhizogenesis

Effect of ethylene on rhizogenesis has been controversial. Some studies have shown that ethylene stimulated adventitious root formation *in vivo*, e.g. mung bean (Robbins et al. 1985) and sunflower (Liu et al. 1990). The stimulatory effect of ethylene is thought to be attributed to enhanced degradation of cytokinins, as a result, lowering cytokinin/auxin ratios, thereby favoring root formation (Bollmark and Eliasson, 1990). On the other hand, others reported that ethylene had no effect to rhizogenesis (Batten and Mullins, 1978; Mudge and Swanson, 1978) or that it was inhibitory (Geneve and Heuser, 1983; Nordstr m and Eliasson, 1984). In *Brassica*, there is a genotypic variation in rooting *in vitro* in response to ethylene inhibitors. Both $AgNO_3$ and AVG had no effect on root formation of cultured shoots of Chinese cabbage (*B. campestris* ssp. *pekinensis*), whereas $AgNO_3$ but not AVG decreased rooting of mustard cv. Indian Mustard by 50% (Chi et al. 1990). The inhibitory effect of $AgNO_3$ may be explained in part by the phytotoxic effect of Ag^+. Furthermore, application of CEPA significantly reduced root elongation and number of roots of *in vitro*-grown plants (Pua and Chi, 1993). The similar rooting response was also observed in shoot culture of cheery (*Prunus avium*) (Biondi et al. 1990) and apple (*Malus* x *domestica*) (Ma et al. 1998). Root formation of these shoots was inhibited by exogenous application of ACC but promoted by $AgNO_3$, AVG or $CoCl_2$, indicating that ethylene is not required for rooting of these species, while exogenous ethylene may be inhibitory. Nevertheless, unlike mustard, apple and cherry, low levels of ethylene provided by 5 μM CEPA are beneficial to rooting of hazelnut (*Corylus avellana*) cotyledons *in vitro* (González et al. 1991). This discrepancy may be due in part to the use different genotypes and CEPA concentrations.

6 FACTORS AFFECTING CULTURE RESPONSE IN RELATION TO ETHYLENE

6.1 Gaseous environment

Apart from ethylene, cultured plant cells and tissues also emanate other gaseous substances, including CO_2, ethanol, acetaldehyde and acetic acid (Thomas and Murashige, 1979; Righetti et al. 1990; Demeester et al. 1995). Considerable efforts have been devoted to investigate whether CO_2 plays a role in culture growth and differentiation (Cornejo-Martin et al. 1979; Kumar et al. 1987, 1989; Nour and Thorpe, 1994), and its progress has recently been reviewed (Buddendorf-Joosten and Woltering, 1994). However, unlike ethylene and CO_2, the role of other gases in culture is less clear.

An accumulation of headspace components, including ethylene, is affected both quantitatively and qualitatively by the type of culture vessel used (Mensuari-Sodi et al. 1992; Demeester et al. 1995). In general, plant cells and tissues grow vigorously in the vessel sealed with cotton plug or other closure devices that allow gas exchange with environment to occur. On the contrary, the airtight vessel is poor and sometime detrimental to growth of cultured cells and tissues. This has been demonstrated in peach suspension culture, whose growth was arrested after 72 h and thereafter most cells showed a progressive browning and death (Marino et al. 1995). Analysis of the headspace components revealed that both CO_2 and ethylene increased considerably but O_2 decreased Furthermore, peach shoots maintained in the sealed vessel had shifted from aerobic to anaerobic growth phase by releasing various fermentation products such as ethanol, acetaldehyde and acetic acid (Demeester et al. 1995). As a result, shoot growth was arrested that was followed by necrosis and death of culture. Although it is not clear as to how changes in these gases can lead to growth arrest and cell death, these results indicate that accumulation of fermentation products may be detrimental to culture growth.

6.2 Auxin

Among growth regulators interacting with ethylene, auxin has been most extensively studied. Auxin stimulating ethylene biosynthesis has been well documented and most of its effects may be mediated by ethylene (Yang and Hoffman, 1984). Because auxin is a common hormone used in plant tissue medium, its presence may affect the amounts of ethylene produced in culture. In wheat and tobacco, poor regeneration of callus culture was thought to be due to accumulation of auxin-stimulated ethylene production (Purnhauser et al. 1987). High levels of ethylene in the presence of auxin may also be implicated in the capacity of shoot formation in tobacco callus cultures. Grady and Basham (1982) thought that callus with lower levels of endogenous ACC and ethylene production showing high capacity of shoot formation might be attributed to the absence of NAA in the medium, on which it was growing, as compared to non-forming callus maintained on NAA-containing medium. However, we have previously shown that lowering the level of ethylene in culture by decreasing the NAA concentration (0.1 $mg.l^{-1}$) in the medium was not beneficial to shoot regeneration, irrespective of the absence or presence of ethylene inhibitor (Pua et al. 1996b). Explants grown on medium containing 1 $mg.l^{-1}$ NAA produced higher amounts of ethylene but its regenerability was greatly enhanced in the presence of AVG or $AgNO_3$, indicating that an optimal level of auxin together with low levels of ethylene are important to high frequency shoot regeneration of Chinese kale. The importance of auxin in the medium is also illustrated in tobacco thin-layer explants. It has been shown that flower bud regeneration from explants grown at lower NAA concentrations (1 μM) could be completely inhibited by exogenous ethylene, whereas exogenous ethylene had no effect on those grown in the presence of higher NAA concentration (4.5 μM) (Smulders et al. 1990). These results were interpreted as a reduction in tissue sensitivity to auxin and in regenerability by ethylene. The type of auxin present in the medium can also affect the mode of morphogenesis of the cultured tissue. In *Solanum carolinense*, anthers grown in IAA-containing medium resulted in direct regeneration of embryoids and plantlets from pollen grains (Reynolds, 1986). This morphogenetic event was subsequently shown to be attributed to IAA-induced ethylene (Reynolds, 1987). Unlike IAA, pollen grains formed callus in the presence of 2,4-D (Reynolds, 1984), whose effect could not be substituted by exogenous ethylene (Reynolds, 1989). The author concluded that the effect of 2,4-D was not mediated through ethylene synthesis and endogenous ethylene was not involved in pollen callus formation.

6.3 Abscisic acid

Abscisic acid (ABA) is a naturally occurring plant hormone that is involved in various physiological and developmental processes of cultured cells and tissues, including freezing resistance (Chen and Gusta, 1983), desiccation tolerance (Kim and Janick, 1989) and somatic embryo development (Roberts, 1991; Kong and Yeung, 1995). Because exogenous ABA stimulates ethylene production, it has led to a speculation that some ABA-induced responses in culture, e.g. callus formation (Goren et al. 1979) and somatic embryogenesis (Kochba et al. 1978) in Shamouti orange (*Citrus sinensis*), may be mediated through ethylene. However, this appears not to be the case in conifer somatic embryo development. In white spruce (*Picea glauca*), embryoid production was markedly enhanced by 10-50 μM ABA but no ethylene elevation was observed (Kong and Yeung, 1994). The ABA-induced embryoid production could be suppressed by 70 μM ethephon. Similar to ABA, embryoid development could also be enhanced, but to a lesser extent, by 100 μM $AgNO_3$ or 2.5 μM $CoCl_2$. Analysis of embryogenic tissues grown in the presence of $AgNO_3$ showed an elevated level of ABA in the tissue, suggesting that the promoting effect of $AgNO_3$ might be attributed, at least in part, to increased levels of endogenous ABA (Kong and Yeung, 1995). This may also explain why ABA is essential for normal embryoid development in conifers (Duncan et al., 1988; Roberts et al. 1990; Attree and Fowke, 1993). However, information regarding the interaction between ethylene and ABA on plant morphogenesis *in vitro* has been limited and the mechanism of ABA action on plant morphogenesis *in vitro* is not well understood.

6.4 Gelling agent

Agar and gellan gum are two common gelling agents used in plant tissue culture medium (Smith and Spomer, 1995) but both are capable of releasing ethylene (Mensuali-Sodi et al. 1992). The amounts of ethylene produced can be varied with brand of agar, due to its differences in biological origin and nature of production. Some brands of agar, when exposed to light, have been shown to release considerable amounts of ethylene (Leonhardt and Kandeler, 1987), which increased with an increase in agar concentration (Mensuali-Sodi et al. 1990). Therefore, selection of an appropriate brand and concentration of gelling agent may be important for the success of the culture, as demonstrated in tomato anther callus (Jaramillo and Summers,

1990). A comparative study on gelling agent shows a general superiority of Gelrite, a bacterial gellan gum, with respect to proliferation, rooting and vigor of cultured shoots and callus development, as compared to tissue culture-agar (Huang et al. 1995). The exact cause of this differential effect of gelling agent is not clear, but Gelrite has been shown to release less ethylene than Difco-Bacto-agar, which was thought to contain more impurities of organic substances, particularly phenolic compounds (Mensuali-Sodi et al. 1992). Difco-Bacto agar was also shown to contain a relatively higher level of Na^+ and Cu^{2+} (Deberge, 1983), which may affect growth and differentiation of some tissues or species *in vitro*.

A recent study has shown that the best agar contained a high content of trace elements and low levels of salts and impurities, especially chlorine (Scholten and Pierik, 1998). It was suggested that chlorine may be used as a marker for agar quality and purity. In our laboratory, we found that agarose that is a component of agar was the best gelling agent for shoot regeneration from cultured hypocotyls of Chinese kale, whereas other gelling agents such as Gelrite, Oxoid and Difco-Bacto agar were not effective (Pua et al. 1996b). Agarose is also essential for high frequency shoot regeneration from hypocotyl explants of mustard, which regenerated poorly on agar medium (Barfield and Pua, 1991). However, leaf tissues responded very differently to agarose, which resulted in an enlargement and poor regeneration of mustard leaf disc explants (Pua and Chi, 1993) and cotyledons of several *Brassica* genotypes (unpublished results), but this can be overcome by replacing agarose with agar (Pua and Chi, 1993). Although the physiological basis of the requirement for specific gelling agent by different explants is not clear, these findings indicate that the culture performance may not be determined solely by the purity of gelling agent, and explant source may also be a decisive factor.

6.5 Source and age of explants

The capacity of shoot regeneration from cultured explant is affected by their source, age and/or position. In *B. campestris* ssp. *chinensis*, cotyledons originating from three day-old seedlings are more regenerative than five day-old explants (Chi and Pua, 1989). The similar results have also been reported in *B. rapa* ssp. *oleifera* (Burnett et al. 1994). However, younger tissues are not always more regenerative. In Ethiopia mustard (*B. carinata*), shoot regenerability of hypocotyls derived from 6-7 day-old seedlings was higher than that of younger and older seedlings (Yang et al. 1991), indicating genotypic variability of age preference. Nevertheless, shoot regenerability is associated with explant age in response to ethylene inhibitor. This has been demonstrated in *B. campestris*, whose regeneration from cotyledons derived from seedlings up to seven day-old was enhanced by $AgNO_3$ but regenerability decreased by half for explants from older (eight day-old) seedlings (Palmer, 1992). However, despite of seedling age, $AgNO_3$ had little effect on hypocotyls, which is in line with our previous findings (Chi and Pua, 1989; Chi et al. 1990). Nevertheless, we later found poor regeneration of *B. campestris* hypocotyls could be overcome by replacing agar with agarose (unpublished results). While the exact cause of the differential regenerability between different ages and sources of the explant remains obscure, our previous studies show that ethylene and, perhaps, PA metabolisms may be involved.

In Chinese kale, shoots could be regenerated from stem explants at high frequencies in the absence of ethylene inhibitor (Pua et al. 1989), but ethylene inhibitor or exogenous putrescine was required for high frequency regeneration from hypocotyls, which were otherwise recalcitrant in culture (Pua et al. 1996b). Apart from explant source and age, we have also observed a strong acropetal polarity in shoot regenerability along the hypocotyl axis in *Brassica*. Explants derived from <20 mm proximal to the shoot apex are most regenerative, while regenerability diminishes along the axis towards the base. In lily (*Lilium speciosum*), the process of bud formation from cultured explants has a strong basipetal polarity, which can be suppressed by wounding of explants and application of exogenous ethylene, NAA and ACC (Van Aartrijk and Blom-Barnhoorn, 1983; Van Aartrijk et al. 1985). This implies that the polarity may be related to auxin distribution and the capacity of ethylene synthesis along the axis.

7 PA BIOSYNTHESIS AND REGULATION

There is a close relationship between ethylene and PAs because both share a common precursor SAM for their synthesis(Fig.1). PAs are essential for cell viability of both prokaryotes and eukaryotes (Heby and Persson, 1990) and, as ethylene, also implicated in an overwhelming array of growth and developmental processes in plants (Evans and Malmberg 1989; Walden et al. 1997). The most abundant PAs in plants are the diamine putrescine, the triamine spermidine and the tetramine spermine, which present in amounts

Table 4. The promoting effect of polyamines on plant morphogenesis *in vitro*

Plant species	Tissue/explant	Basal medium *	Medium* additive	Growth/ morphogenesis	Reference
Monocotyledons					
Lilium longiflorum	Bulb-scales	MS + 4% sucrose + 1 µM NAA	10-100 µM spm; 100 µM spd	Promoted bulblet formation	Tanimoto and Matsubara,1995
Oryza sativa	one year-old callus	MS + 0.5 mg.l^{-1} BA	1-5 mM spd	Enhanced plant regeneration	Bajaj and Rajam, 1995
Zea mays	Callus	MS + 9 µM 2,4-D	5-10 mM DFMA	Promoted shoot bud formation	Guergué et al., 1997
Dicotyledons					
Brassica campestris ssp. *pekinensis*	Cotyledons	MS + 1 mg.l^{-1} NAA + 2 mg.l^{-1} BA	1-20 mM put; 0.1-2.5 mM spd; 0.1- 1 mM spm	Enhanced shoot regeneration	Chi et al., 1994
Daucus carota	Cultured cells	MS + 0.2 mg.l^{-1} 2,4-D	0.1-0.5 mM MGBG	Inhibited somatic embryogenesis but reversed by exogenous spd	Minocha et al., 1991
Nicotiana tabacum	Mid-binucleate pollen	Miller's macro-elements + 0.4 M mannitol	1.5 mM DFMO or DFMA	Inhibited embryoid formation	Garrido et al., 1995
N. tabacum	Thin-layer stem	MS + 1 µM NAA + 10 µM K	0.5-5 mM spd	Promoted floral bud formation	Kaur-Sawhney et al., 1988

Plant species	Tissue/explant	Basal medium *	Medium* additive	Growth/ morphogenesis	Reference
N. tabacum	Thin layer stem	MS + 10 μM IBA + 0.1 μM K	1 mM MGBG; 5mM CHA; 1 mM DFMA or DFMO	Inhibited rhizogenesis	Altamura et al., 1991
Picea glauca × *P. engelmannii*	Embryogenic cell line	1/2 LM + 1 % sucrose + 10 μM 2,4-D	40 μM ABA + 1 μM IBA	Promoted embryoid maturation	Amarasinghe et al., 1996
Populus tremula × *P. tremuloides*	Cultured shoots	1/2 MS + 0.3 $mg.1^{-1}$ NAA	100 μmol. $^{-1}$ put or CHA	Promoted rhizogenesis	Hausman et al., 1995
Prunus avium	Cultured shoots	1/2 MS + 5 μM IBA	0.5-5 μM DFMO; 0.1 –1 mM DCHA or MGBG	Inhibited rhizogenesis	
Solanum tuberosum	Cultured plants	1/2 MS + 6% sucrose + 2 $mg.1^{-1}$ BA + 1 $mg.1^{-1}$ NAA	0.2 mM put	Increased tuberization	Mader, 1997
S. tuberosum	Leaves (Basal region)	MS + 10.73 μM NAA	0.5 mM put	Enhanced somatic embryogenesis	Yadav and Rajam, 1998

* ABA, abscisic acid; BA, benzyladenine; CHA, cyclohexylamine; 2,4-D, 2,4-dichlorophenoxyacetic acid; DCHA, dicyclohexylammonium sulphate; DFMA, difluoromethylarginine; DFMO, difluoromethylonithine, IBA, indole-3-butyric acid; K, kinetin; 1/2 LM., half-strength LM medium (see Tremblay 1990); MGBG, methylglyoxal-bis-guanylhydrazone; MS, Murashige and Skoog's medium; 1/2 MS, half-strength MS medium; NAA, α-naphthaleneacetic acid; put, putrescine; spd, spermidine; spm, spermine

varying from micromolar to more than millimolar (Galston and Sawhney 1990). Changes in PAs have been related to environmental stresses (Flores 1990; Kuehn et al. 1990) and various plant physiological processes. These include rhizogenesis (Jarvis et al. 1985), development of flowers (Gerats et al. 1988) and fruits (Galston and Sawhney 1990), senescence (Muhitch et al. 1983) and plant morphogenesis *in vitro* (Table 4). However, the regulatory mechanism of PA action is not clear.

In PA biosynthesis, putrescine is synthesized from ornithine and arginine through ornithine decarboxylase (ODC) and arginine decarboxylase (ADC) activities. Putrescine is then used as a precursor, together with the aminopropyl group derived from decarboxylated-SAM, which is synthesized from SAM via the catalysis of SAM decarboxylase (SAMDC), to synthesize spermidine and spermine by spermidine synthase and spermine synthase, respectively (Fig. 1). These PAs can be further metabolized by other enzymes in plant cells. One of these enzymes is diamine oxidase that converts putrescine to Δ^1-pyrroline. Another type of enzyme is polyamine oxidase that degrades spermidine and spermine to give rise to the common products of hydrogen peroxide and 1,3-diaminopropane, and either Δ^1-pyrroline for spermidine or 1-(aminopropyl)-pyrroline for spermine (Phillips and Kuehn, 1991). 1,3-Diaminopropane may be used as a substrate for synthesis of uncommon PAs, e.g. norspermidine and norspermine (Bagga et al. 1991), which may be related to drought tolerance (Rodriduez-Garay et al. 1989). Nevertheless, it is speculated that PAs and ethylene may regulate each other s synthesis by competing for SAM. Results from several lines of study using exogenous PAs and its inhibitors appear to support the precursor competition hypothesis. This is evidenced by the inhibitory effect of exogenous PAs on ethylene production in several types of plant tissues, including *Malus* fruits, bean and tobacco (Apelbaum et al. 1981) and *Tradescantia* petals and mung bean (*Vigna radiata*) hypocotyls (Suttle 1981). Furthermore, the use of PA biosynthesis inhibitors has been shown to promote ethylene production and flower senescence of carnation (*Dianthus caryophyllus*) (Roberts et al. 1983). However, the findings of other studies are not in line with the hypothesis. In Chinese cabbage, exogenous PAs resulted in an increase in ethylene production in culture (Chi et al. 1994). The similar ethylene increase has also been reported in apple fruits (Wang and Kramer, 1990) and cut carnation flowers (Downs and Lovell, 1986) after PA treatments.

Apart from the chemical approach, recent research has focused on cloning and expression of PA biosynthesis genes in transgenic plants to explore the interactive role of PA and ethylene in plant growth and development. Several genes encoding SAMDC have been cloned from potato (Taylor et al. 1992), *Catharanthus roseus* (Schroder and Schroder

1995), carnation (Lee et al. 1997a), spinach (Bolle et al. 1995) and mustard (Lee et al. 1997c). In potato, SAMDC gene was highly expressed in young and growing tissues, but expression was low in the mature and non-dividing tissues (Mad Arif et al. 1994). This is in line with previous findings, in which SAMDC activity and PAs are present at high levels in actively growing and differentiating tissues (Evans and Malmberg 1989; Santanen and Simola 1992). Kumar and co-workers (1996) have demonstrated that it is possible to manipulate metabolic partitioning of PAs in plants by overexpression of sense and antisense SAMDC gene. Transgenic potato plants expressing the homologous antisense SAMDC gene under the control of tetracyclin inducible promoter showed a general decline of endogenous PAs and a marked increase in ethylene production. On the contrary, the level of PAs in transgenic plants overexpressing sense SAMDC RNA was considerably higher than that in the control tissue.

The cellular PAs can also be modulated through manipulation of ADC or ODC gene expression in transgenic plants. Genes encoding ADC have been cloned from several plant species, including oat (*Avena sativa*) (Bell and Malmberg 1990), tomato (Rastogi et al. 1993), pea (P rez-Amador et al. 1995), rice (Chattopadhyay et al. 1997) and soybean (Nam et al. 1997). Transgenic tobacco plants expressing an oat ADC gene under the control of tetracyclin inducible promoter showed a marked increase in endogenous ADC activity and putrescine content by up to 20- and 4.6-fold, respectively, upon tetracyclin induction (Masgrau et al. 1997). However, the phenotype of these transgenic plants was abnormal when ADC activation was induced at the vegetative growth stage of the plant. The growth abnormality includes short internodes, thin stems and leaves, and reduced root growth and development. Interestingly, ADC activation induced at the flowering stage had no effect on plant growth and development. The cause of this differential response between the two growth stages is not clear. nlike ADC, ODC genes have not been cloned from plants until recently, e.g. *Datura stramonium* (Michel; et al. 1996), although it has been isolated from several vertebrates, invertebrates and fungi. In *Nicotiana rustica*, transformed roots expressing the yeast ODC gene showed an accumulation of putrescine and nicotine, which is a putrescine-derived alkaloid (Hamill et al. 1990). Similarly, callus derived from transgenic tobacco plants expressing mouse ODC cDNA also accumulated high levels of endogenous putrescine but not spermidine and spermine (DeScenzo and Minocha 1993). It was thought that the synthesis of spermidine and spermine might be limited by the availability of precursor, decarboxylated SAM. However, unlike tobacco, carrot cells expressing the mouse ODC cDNA showed a marked increase in spermidine and spermine, in addition to putrescine (Andersen et al. 1998). This discrepancy may be attributable to genotypic variability. Although increased production of putrescine in transformed carrot cells was accompanied by higher rates of putrescine conversion to spermidine and spermine, cells fed with exogenous putrescine resulted in higher rates of putrescine degradation. Because high levels of endogenous PAs are inhibitory to plant growth and development, as demonstrated in transgenic tobacco plants expressing an oat ODC gene, putrescine degradation may serve to prevent further increase in endogenous PAs. It suggests that PA synthesis in these transformed cells may be controlled by the endogenous regulatory mechanism. Nevertheless, these studies clearly demonstrate the powerful system of transgenic plants to be used for study of the regulatory role of ethylene and PA metabolisms in plant morphogenesis *in vitro*.

8 EFFECT OF PAS AND ITS RELATION WITH ETHYLENE

8.1 Shoot organogenesis

PAs have been associated with shoot organogenesis of several plant species *in vitro* (see Table 4). In maize, short treatment of young callus with high dosages of difluoromethylarginine (DFMA), an irreversible inhibitor of ADC (Table 1), significantly increased the number of regenerated buds with a concomitant decrease in total protein and PA content, ADC and ODC activities (Guergu et al. 1997). In some plant species, accumulation of a specific PA in cultured cells or explants appears to play an important role in shoot morphogenesis. Kaur-Sawhney and co-workers (1988) reported that thin-layer explants of tobacco giving rise to flower buds possessed high levels of cellular spermidine. The importance of spermidine was further exemplified by a study showing induction of floral-bud differentiation in the presence of 0.5-5 mM exogenous spermidine. Noh and Minocha (1994) have shown that tobacco tissues originated from transgenic plants that overexpressing human SAMDC RNA showed a 2-4-fold increase in SAMDC activity and cultures with higher levels of endogenous spermidine were capable of forming shoots on callus-inducing medium. In radiata pine, although spermidine was the most predominant PA accumulated in cultured cotyledons, there was no difference between shoot-formation and non-shoot-formation explants (Kumar and Thorpe 1989). nlike tobacco, spermine has been shown to be most effective for promoting bulblet formation from bulb-scale explants of lily (Tanimoto and Matsubara 1995). However, the regulatory role of PAs on plant morphogenesis *in vitro* in these studies is not clear. We found that the promoting effect of AVG and $AgNO_3$ on shoot regeneration from Chinese kale hypocotyls (Pua et al. 1996b) and Chinese cabbage cotyledons (Chi et al. 1994) can be mimicked by the use of exogenous PAs, indicating the possible regulatory role of PAs in regeneration of these species. This is based on the speculation that inhibition of ethylene synthesis and action by ethylene inhibitors may lead to an accumulation of SAM, which may trigger metabolic shift favoring PA metabolism. The implication of PAs on shoot organogenesis is supported by evidence, in which shoot regeneration in Chinese kale can be suppressed by 0.1 mM DFMA, whose inhibitory effect can be abolished by application of exogenous 10 mM putrescine (Pua et al. 1996b). In Chinese cabbage, ethylene inhibitors generally did not affect endogenous free PAs of the cultured explant. Similarly, levels of ethylene production, endogenous ACC synthase activity and ACC of the explant were not affected by exogenous PAs (Chi et al. 1994). These suggest that the promoting effect of PAs on shoot regeneration of Chinese cabbage may not be due to inhibition of ethylene synthesis.

8.2 Somatic embryogenesis

Changes in PA metabolism in cultured cells and tissues are associated with at least some stages during initiation, growth and development of somatic embryos of some plant species (see Table 4). An optimal level of endogenous PAs has been shown to be important for somatic embryogenesis of carrot (Montague et al. 1978; Fierer et al. 1984; Fienberg et al. 1984), alfalfa (Meijer and Simmonds, 1988), grape (Faure et al. 1991), mango (Litz and Shaffer 1987) and eggplant (Sharma and Rajam 1995). Furthermore, PAs are also required for embryoid maturation, e.g. interior spruce (Amarasinghe et al. 1996). In tobacco pollen culture, PAs are beneficial only to embryo formation but not to embryo induction (Garrido et al. 1995).

The implication of PAs in somatic embryogenesis has been investigated by adjusting the cellular PA content in the tissue using exogenous PAs and/or PA synthesis inhibitors. In eggplant, a high frequency somatic embryogenesis from the apical region of the leaf was closely linked to high levels of endogenous ADC activity and PAs, especially putrescine (Yadav and Rajam 1998). This was exemplified by decreasing the embryogenic capacity by treatment of explants with DFMA, which significantly decreased ADC activity and PA content. Furthermore, embryogenesis from basal region of the leaf with low embryogenic capacity was found to possess low levels of PAs, but the capacity increased markedly by treatment of explants with exogenous putrescine that resulted in increased cellular PAs. Somatic embryogenesis of some species appears to be linked to particular PA in the tissue. In cultured cells of wild carrot, the embryogenic capacity is associated with high levels of endogenous spermidine (Fienberg et al. 1984;

Fierer et al. 1984). Enhanced synthesis of spermidine may also play a role during globular embryo development of Norway spruce (Santanen and Simola 1992). In rice, application of exogenous spermidine also promoted plant regeneration from long-term callus via somatic embryogenesis (Bajaj and Rajam 1995). However, it is not clear whether ethylene is also involved in these cultures. The interaction between PAs and ethylene has been studied in carrot somatic embryogenesis. The presence of 2,4-D was inhibitory to cell differentiation and somatic embryogenesis of carrot with a concomitant increase in ethylene production. The inhibitory effect of 2,4-D could be overcome by application of 1-10 mM difluoromethylornithine (DFMO), a potent inhibitor of ODC (Table 1), which resulted in an increase in endogenous PAs and a decrease in ethylene production (Robie and Minocha, 1989). These results imply that enhanced PA synthesis may be important for carrot somatic embryogenesis. The implication is substantiated by a subsequent study using methylglyoxal bis(guanylhydrazone) (MGBG), an inhibitor of SAMDC (Table 1). Somatic embryogenesis was inhibited in the presence of 0.1-0.5 mM MGBG, which also inhibited spermidine and spermine synthesis but resulted in an accumulation of putrescine and ACC in the auxin-free medium (Minocha et al. 1991). Application of exogenous 0.1-0.2 mM spermidine effectively suppressed the inhibitory effect of MGBG, indicating that inhibition of somatic embryogenesis in the presence of MGBG may be resulted from inhibition of converting putrescine to spermidine and spermine due to the lack of decarboxylated SAM. In a transgenic study, carrot cells expressing the mouse ODC gene showed increased putrescine synthesis but spermidine was not affected (Bastola and Minocha 1995). These transformed cells were more embryogenic than non-transformed counterpart, suggesting that some catabolites of putrescine, in addition to spermidine, may play an important role in improved somatic embryogenesis in carrot.

8.3 Rhizogenesis

PAs have been implicated in rhizogenesis both *in vitro* (Biondi et al. 1990; Hausman et al. 1995a) and *in vivo* (Jarvis et al. 1985; Friedman et al. 1985; Wang and Faust 1986). Cultured shoots during rhizogenesis usually showed an increase in endogenous levels of PAs (Desai and Mehta, 1985; Chriqui et al. 1986; Biondi et al. 1990), particularly putrescine (Tiburcio et al. 1989). Analysis of tobacco thin cell-layer explants revealed that both free and bound putrescine and spermidine increased when root meristemoids emerged but PAs decreased when root formation was completed (Torrigiani et al. 1989). Application of PA biosynthesis inhibitors (DFMA + DFMO, cyclohexylamine (CHA) or MGBG) blocked both PA accumulation and rhizogenesis (Altamura et al. 1991). The cytological evidence indicates that the inhibition may be related to high incidence of nucleolar extrusion as well as cell expansion, thinning of cell wall and inhibition of cell division (Altamura et al. 1993). Although these findings suggest the possible PA involvement in root formation *in vitro*, the question of how PAs exert their effects on rhizogenesis remains to be addressed The role of PAs on root formation has also been explored using cultured poplar shoots, which usually formed roots in the presence of NAA and did not root in the absence of auxin. It was found that shoots rooted in the presence of NAA accumulated putrescine but not spermidine and spermine during the inductive phase, while exogenous putrescine but not other PAs promoted root formation from shoots grown on NAA-free medium (Hausman et al. 1994). Furthermore, application of CHA, an inhibitor of spermidine synthase that catalyzes the conversion of putrescine to spermidine, promoted root formation, whereas DFMA and aminoguanidine (AG), which inhibits the reaction of putresine→ δ-pyrroline by inhibiting diamine oxidase (Table 1), were inhibitory to rooting. These findings have led to a postulation that rooting of poplar shoots *in vitro* might involve the Δ^1-pyrroline pathway (Hausman et al. 1994). Putrescine accumulation has also been shown to correlate well with a transient increase in endogenous levels of IAA and its conjugated IAA aspartate (Hausman et al. 1995a), which appears to be in line with the requirement for high levels of auxin for root induction (Gaspar et al. 1994). In addition to auxin, an increase in endogenous peroxidase activity was detected preceded rooting (Hausman et al. 1995b), indicating that oxidative burst may also be implicated in root formation *in vitro*. However, the role of increased peroxidase activity in rhizogenesis remains to be elucidated

9 CONCLUSIONS

The number of species that are responsive to ethylene and PAs, with respect to organogenesis, embryogenesis and rhizogenesis *in vitro*, has been increasing since 1987. These clearly indicate the pivotal role of ethylene and PAs in regulation of cell differentiation and morphogenesis in plants, particularly for species recalcitrant in culture, e.g. monocots and some members in Cruciferae. This is not surprising because both ethylene and PAs are ubiquitous in plants and the DNA sequence for genes encoding key enzymes of ethylene and PA biosynthesis among species are highly homologous. These genes include ACC synthase (Zarembinski and Theologis 1994), ACC oxidase (Pua 1993), SAMDC (Bolle et al. 1995) and ADC (Pérez-Amador et al. 1995). However, their regulatory mechanisms leading to cell differentiation and morphogenesis remain poorly understood.

The regulatory role of auxin, cytokinin and ABA on organogenesis (Thorpe, 1994) and somatic embryogenesis (Merkle et al. 1995; Nomura and Komamine 1995) has been well documented. It is therefore speculated that ethylene and PAs may act in concert with these growth regulators to trigger the morphogenetic event. Nevertheless, the capacity of plant morphogenesis *in vitro* can be controlled genetically, as demonstrated in petunia (Dulieu et al. 1991), cucumber (Nadolska-Orczyk and Malepszy 1989), alfalfa (Reich and Bingham 1980) and maize (Tomes and Smith 1985). A gene controlling shoot regeneration from tomato root explants has also been located and mapped in the genome using morphological and restriction fragment length polymorphism markers (Koornneef et al. 1993). However, the DNA sequence of the gene or its product responsible for plant morphogenesis *in vitro* has not been isolated Evidence from transgenic plants overexpressing ODC and antisense ACC oxidase genes suggests that genetic control of morphogenesis may be related to genes responsible for regulation of ethylene and PA biosynthesis.

10 REFERENCES

Abeles FB (1973) Ethylene in Plant Biology, Academic Press, New York.

Adkins SW, Kunanuvatchaidach R, Gray SJ & Adkins AL (1993) Effect of ethylene and culture environment on rice callus proliferation. J. Exp. Bot. 44: 1829-1835.

Aitken J, Horgan KJ & Thorpe TA (1981) Influence of explant selection on the shoot-forming capacity of juvenile tissue of *Pinus radiata*. Can. J. For. Res. 11: 112-117.

Altamura MM, Capitani F, Cerchia R, Falasca G & Bagni N (1993) Cytological events induced by the inhibition of polyamine biosynthesis in thin cell layers of tobacco. Protoplasma 175: 9-16.

Altamura MM, Torrigiani P, Capitani F, Scaramagli S & Bagni N (1991) *De novo* root formation in tobacco thin layers is affected by inhibition of polyamine biosynthesis. J. Exp. Bot. 42: 1575-1582.

Amarasinghe V, Dhami R & Carlson JE (1996) Polyamine biosynthesis during somatic embryogenesis in interior spruce (*Picea glauca* x *Picea engelmannii* complex). Plant Cell Rep . 15: 495-499.

Amor MB, Guis M, Latché A, Bouzayen M, Pech J-C & Roustan J-P (1998) Expression of an antisense 1-aminocyclopropane-1-carboxylate oxidase gene stimulates shoot regeneration in *Cucumis melo*. Plant Cell Rep. 17: 586-589.

Andersen SE, Bastola DR & Minocha SC (1998) Metabolism of polyamines in transgenic cells of carrot expressing a mouse ornithine decarboxylase cDNA. Plant Physiol. 116: 299-307.

Apelbaum A, Burgoon AC, Anderson AD, Lieberman M, Ben-Arie R & Mattoo AK (1981) Polyamines inhibit biosynthesis of ethylene in higher plant tissue and fruit protoplasts. Plant Physiol. 68: 453-456.

Arteca RN (1984) Ca^{2+} acts synergistically with brassinosteroid and indole-3-acetic acid in stimulating ethylene production in etiolated mung bean hypocotyl segments. Physiol. Plant. 62: 102-104.

Attree SM & Fowke LC (1993) Somatic embryogenesis and synthetic seeds of conifers. Plant Cell Tiss. Organ Cult. 35: 1-35.
Auboiron E, Carron MP & Michaux-Ferriere N (1990) Influence of atmospheric gases, particularly ethylene, on somatic embryogenesis of *Hevea brasiliensis*. Plant Cell Tiss. Organ Cult. 21: 31-37.
Bagga S, Dharma A, Phillips GC & Kuehn GD (1991) Evidence for the occurrence of polyamine oxidase in the dicotyledonous plant *Medicago sativa* L. (alfalfa). Plant Cell Rep. 10: 550-554.
Bajaj S & Rajam MV (1995) Efficient plant regeneration from long-term callus cultures of rice by spermidine. Plant Cell Rep. 14: 717-720.
Barfield DG & Pua EC (1991) Gene transfer in plants of *Brassica juncea* using *Agrobacterium tumefaciens*-mediated transformation. Plant Cell Rep. 10: 308-314.
Bastola DR & Minocha SC (1995) Increased putrescine biosynthesis through transfer of mouse ornithine decarboxylase cDNA in carrot promotes somatic embryogenesis. Plant Physiol. 109: 63-71.
Batten DJ & Mullins MC (1978) Ethylene and adventitious root formation in hypocotyl segments of etiolated mung bean seedlings. Planta. 138: 193-197.
Biddingdon NL (1992) The influence of ethylene on plant tissue culture. Plant Growth Regul. 11: 173-187.
Biddington NL, Sutherland RA & Robinson HT (1988) Silver nitrate increases embryo production in anther culture of Brussels sprouts. Ann. Bot. 62: 181-185.
Biondi S, Diaz T, Iglesias I, Gamberini G & Bagni N (1990) Polyamines and ethylene in relation to adventitious root formation in *Prunus avium* shoot cultures. Physiol. Plant. 78: 474-483.
Blumber KJ & Johnson GL (1994) Diversity in function and regulation of MAP kinase pathways. Trends Biochem. Sci. 19: 236-240.
Bolle C, Herrmann RG & Oelmuller R (1995) A spinach cDNA with homology to *S*-adenosylmethionine decarboxylase. Plant Physiol. 107: 1461-1462.
Bollmark M & Eliasson L (1990) Ethylene accelerates the breakdown of cytokinins and thereby stimulates rooting in Norway spruce hypocotyl cuttings. Physiol. Plant. 80: 534-540.
Bourgin JP & Nitsch JP (1967) Obtention de *Nicotiana* haploides à partir d'étamines cultivées *in vitro*. Ann. Physiol. Veg. 9: 377-382.
Buddendorf-Joosten JMC & Woltering EJ (1994) Components of the gaseous environment and their effects on plant growth and development *in vitro*. Plant Growth Regul. 15: 1-16.
Burg SP & Burg EA (1967) Molecular requirements for the biological activity of ethylene. Plant Physiol. 42: 144-152.
Burnett L, Arnoldo M, Yarrow S & Huang B (1994) Enhancement of shoot regeneration from cotyledon explants of *Brassica rapa* ssp. *oleifera* through pretreatment with auxin and cytokinin and use of ethylene inhibitors. Plant Cell Tiss. Organ Cult. 37: 253-256.
Cambecedes J, Duron M & Decourtye L (1992) Interacting effects of 2,3,5-triiodobenzoic acid, 1-aminocyclopropane-1-carboxylic acid, and silver nitrate on adventitious bud formation from leaf explants of the shrubby honeysucker, *Lonicera nitida* Wils. 'Maigrun'. J. Plant Physiol. 140: 557-561.
Carron MP & Enjalric F (1985) Embryogenèse somatique à partir du tégument interne de la graine '*Hevea*. CR Acad. Sci. Sér. III 300: 653-658.
Chang C (1996) The ethylene signal transduction pathway in *Arabidopsis*: an emerging paradigm? Trends Biochem. Sci. 21: 129-133.
Chang C, Kwok SF, Bleecker AB & Meyerowitz EM (1993) *Arabidopsis* ethylene-response gene *ETR1*: a similarity of product to two component regulators. Science 262: 539-544.

Chattopadhyay MK, Gupta S, Senguptam DN & Ghose B (1997) Expression of arginine decarboxylase in indica rice (*Oryza sativa* L.) cultivars as affected by salinity stress. Plant Mol. Biol. 34: 477-483.

Chen THH & Gusta LV (1983) Abscisic acid induced freezing resistance in cultured plant cells. Plant Physiol. 73: 71-74.

Cheng TY (1975) Adventitious bud formation in culture of Douglas fir (*Pseudotsuga menziesii* Mirb. Franco). Plant Sci. Lett. 5: 97-102.

Chi GL, Barfield DG, Sim GE & Pua EC (1990) Effect of $AgNO_3$ and aminoethoxyvinylglycine on *in vitro* shoot and root organogenesis from seedling explants of recalcitrant *Brassica* genotypes. Plant Cell Rep. 9: 195-198.

Chi GL, Lin WS, Lee JEE & Pua EC (1994) Role of polyamines on *de novo* shoot morphogenesis from cotyledons of *Brassica campestris* ssp. *pekinensis* (Lour) Olsson *in vitro*. Plant Cell Rep. 13: 323-329.

Chi GL &Pua EC (1989) Ethylene inhibitors enhanced de novo shoot regeneration from cotyledons of *Brassica campestris* ssp. *chinensis* (Chinese cabbage). Plant Sci. 64: 243-250.

Chi GL, Pua EC & Goh CJ (1991) Role of ethylene on *de novo* shoot regeneration from cotyledonary explants of *Brassica campestris* ssp. *pekinensis* (Lour) Olsson *in vitro*. Plant Physiol. 96: 178-183.

Cho UH & Kasha KJ (1989) Ethylene production and embryogenesis from anther cultures of barley (*Hordeum vulgare*). Plant Cell Rep. 8: 415-417.

Cho UH & Kasha KJ (1995) The effect of calcium on ethylene production and microspore-derived embryogenesis in barley (*Hordeum vulgare* L.) and wheat (*Triticum aestivum* L.) anther cultures. J. Plant Physiol. 146: 677-680.

Chraibi BKM, Latche A, Roustan JP & Fallot J (1991) Stimulation of shoot regeneration from cotyledons of *Helianthus annuus* by ethylene inhibitors, silver and cobalt. Plant Cell Rep. 10: 204-207.

Chriqui D, O'Orazi D & Bagni N (1986) Ornithine and arginine decarboxylase and polyamine involvement during *in vivo* differentiation and *in vitro* dedifferentiation of *Datura innoxia* leaf explants. Physiol. Plant. 68: 589-596.

Chu CC, Wang CC, Sun CS, Hsu C, Yin KC & Chu CY (1975) Establishment of an efficient medium for anther culture of rice through comparative experiments on nitrogen source. Sci. Sin. 18: 659-668.

Cocking EC (1960) A method for the isolation of plant protoplasts and vacuoles. Nature 187: 962-963.

Cornejo-Martin MJ, Mingo-Castel AM & Primo-Millo E (1979) Organ redifferentiation in rice callus: Effect of C_2H_4, CO_2 and cytokinin. Z. Pflanzenphysiol. 94: 117-123.

Debergh PC (1983) Effect of agar brand and concentration on the tissue culture medium. Physiol. Plant. 59: 270-276.

De Block M, de Brouwer D & Tenning T (1989) Transformation of *Brassica napus* and *Brassica oleracea* using *Agrobacterium tumefaciens* and the expression of the *bar* and *neo* genes in the transgenic plants. Plant Physiol. 91: 694-701.

Demeester JJ, Matthijs DG, Pascat B & Debergh PC (1995) Toward a controllable headspace composition – growth, development, and headspace of a micropropagated *Prunus* rootstock in different containers. In Vitro Cell. Dev. Biol. 31: 105-112.

De Proft MP, Maene LJ & Deberge PC (1985) Carbon dioxide and ethylene evolution in the culture atmosphere of *Magnolia* cultured *in vitro*. Physiol. Plant. 65: 375-379.

Desai HV & Mehta AR (1985) Changes in polyamine levels during shoot formation, root formation and callus induction in cultured *Passiflora* leaf discs. J. Plant Physiol. 119: 45-53.

DeScenzo RA & Minocha SC (1993) Modulation of cellular polyamines in tobacco by transfer and expression of mouse ornithine decarboxylase cDNA. Plant Mol. Biol. 22: 113-127.

Dimasi-Theriou K & Economou AS (1995) Ethylene enhances shoot formation in cultures of peach rootstock GF-677 (*Prunus persica* x *P. amygdalus*). Plant Cell Rep. 15: 87-90.

Dimasi-Theriou K, Economou AS & Sfakiotakis EM (1993) Promotion of petunia (*Petunia hybrida* L.) regeneration *in vitro* by ethylene. Plant Cell Tiss. Organ Cult. 32: 219-225.

Downs CG & Lovell PH (1986) The effect of spermidine and putrescine on the senescence of cut carnation. Physiol. Plant. 66: 679-684.

Dulieu H (1991) Inheritance of the regeneration capacity in the genus *Petunia*. Euphytica 53: 173-118.

Duncan DI, Bekkaoui F, Pilon M, Fowke LC & Abrams SR (1988) Effects of abscisic acid and analogues on the maturation of white spruce (*Picea glauca*) somatic embryos. Plant Sci. 58: 77-84.

Duncan DR, Williams ME, Zehr BE & Widholm JM (1985) The production of callus capable of plant regeneration from immature embryos of numerous *Zea mays* genotypes. Planta 165: 322-332.

Dunstan DI & Short KC (1977) Improved growth of tissue culture of the onion, *Allium cepa*. Physiol. Plant. 41: 70-72.

Ecker JR (1995) The ethylene signal transduction pathway in plants. Science 268: 667-675.

Evans JM & Batty NP (1994) Ethylene precursors and antagonists increase embryogenesis of *Hordeum vulgare* L. anther culture. Plant Cell Rep. 13: 676-678.

Evans PT & Malmberg RL (1989) Do polyamines have a role in plant development? Annu. Rev. Plant Physiol. Plant Mol. Biol. 40: 235-269.

Faure O, Mengoli M, Nougarede A & Bagni N (1991) Polyamine pattern and biosynthesis in zygotic and somatic embryo stages of *Vitis vinifera*. J. Plant Physiol. 138: 545-549.

Feirer RP, Magnon G & Litvay JD (1984) Arginine decarboxylase and polyamines required for embryogenesis in the wild carrot. Science 223: 1433-1435.

Fienberg A, Choi JH, Lubich WP & Sung ZR (1984) Developmental regulation of polyamine metabolism in growth and differentiation of carrot culture. Planta 162: 532-539.

Flores HE (1990) Polyamines and plant stress. In: Alscher RG, Cumming JR & Allen NS (eds) Stress Responses in Plants: Adaptation and Acclimation Mechanisms. pp. 217-239 Wiley-Liss, J. Wiley Sons Inc. Publ, New York.

Fluhr R & Mattoo AK (1996) Ethylene - biosynthesis and perception. Critical Rev. Plant Sci. 15: 479-523.

Friedman R, Altmann A & Bachrach U (1995) Polyamines and root formation in mung bean hypocotyl cuttings. II. Incorporation of precursors into polyamines. Plant Physiol. 79: 80-83.

Galston AW & Sawhney RK (1990) Polyamines in plant physiology. Plant Physiol. 94: 406-410.

Gamborg OL & LaRue TAG (1968) Ethylene production by plant cells in suspension cultures. Nature 220: 604-605.

Garrido D, Chibi F & Matilla A (1995) Polyamines in the induction of *Nicotiana tabacum* pollen embryogenesis by starvation. J. Plant Physiol. 145: 731-735.

Gaspar T, Kevers C, Hausman,JF & Ripetti V (1994) Peroxidase activity and endogenous free auxin during adventitious root formation. In: Lumsden PJ, Nicholas JR & Davies WJ (eds) Physiology, Growth and Development of Plants in Culture. pp. 289-298. Kluwer Academic Publ., Dordrecht.

Gautheret RJ (1938) Sur la repiquage des cultures de tissues cambial de *Salix capraea*. CR Acad. Sci. Paris 206: 125-127.

Gavinlertvatana P, Read PE, Wilkins HF & Hein R (1982) Ethylene levels in flask atmospheres of *Dahlia pinata* Cav. leaf segments and callus cultured *in vitro*. J. Amer. Soc. Hortic. Sci. 107: 3-6.
Geneve RL & Heuser CW (1983) The relationship between ethephon and auxin on root initiation in cuttings of *Vigna radiata* (L.) R. Wilcz. J. Amer. Soc. Hortic. Sci. 108: 330-333.
Gerats AGM, Kaye C, Collins C & Malmberg RL (1988) Polyamine level in *Petunia* genotypes with normal and abnormal morphologies. Plant Physiol. 86: 390-393.
Goh CJ, Ng SK, Lakshmanan P & Loh CS (1997) The role of ethylene on direct shoot bud regeneration from mangosteen (*Garcinia mangostena* L.) leaves cultured in vitro. Plant Sci. 124: 193-202.
González A, Rodríguez R & Tamés RS (1991) Ethylene and in vitro rooting of hazelnut (*Corylus avellana*) cotyledons. Physiol. Plant. 81: 227-233.
Good X, Kellogg JA, Wagoner W, Langhoff D, Matsumura W & Bestwick RK (1994) Reduced ethylene synthesis by transgenic tomatoes expressing *S*-adenosylmethionine hydrolase. Plant. Mol. Biol. 26: 81-790.
Goren R, Altman A & Giladi I (1979) Role of ethylene in abscisic acid-induced callus formation in citrus bud cultures. Plant Physiol. 63: 280-282.
Goren R & Sisler EC (1986) Ethylene-binding characteristics in *Phaseolus*, *Citrus* and *Ligustrum* plants. Plant Growth Regul. 4: 43-54.
Grady KL & Bassam JA (1982). 1-Aminocyclopropane-1-carboxylic acid in shoot forming and non-shoot forming tobacco callus cultures. Plant Physiol. 70: 919-921.
Gray J, Picton S, Shabbeer J, Schuch W & Grierson D (1992) Molecular biology of fruit ripening and its manipulation with antisense genes. Plant Mol. Biol. 19: 69-87.
Guergué A, Claparols I, Santos M & Torné JM (1997) Modulator effect of DL-α-difluoromethylarginine treatments on differentiation processes of young maize calluses. Plant Growth Regul. 21: 7-14.
Guha S & Maheshwari SC (1964) *In vitro* production of embryos from anthers of *Datura*. Nature 204: 497.
Guha S & Maheshwari SC (1966) Cell division and differentiation of embryos in the pollen grains of *Datura in vitro*. Nature 212: 97-98.
Guzman P & Ecker JR (1990) Exploiting the triple response of *Arabidopsis* to identify ethylene-related mutants. Plant Cell 2: 513-523.
Haberlandt G (1902) Culturversuche mit isolierten Planzenzellen. Sitzber Kaiser Akad Wiss Berlin Math Naturw KI, Abt. I. 111: 69-92.
Halperin W (1986) Attainment and retention of morphogenetic capacity *in vitro*. In: Vasil IK (ed) Cell Culture and Somatic Cell Genetics of Plants. pp. 1-47 Academic Press, Orlando, Florida.
Hamill JD, Robins RJ, Parr AJ, Evans DM, Furze JM & Rhodes MJC (1990) Over-expressing a yeast ornithine decarboxylase gene in transgenic roots of *Nicotiana rustica* can lead to enhanced nicotine accumulation. Plant Mol. Biol. 15: 27-38.
Hamilton AJ, Lycett GW & Grierson D (1990) Antisense gene that inhibits synthesis of the hormone ethylene in plants. Nature 346: 284-287.
Harpham NVJ, Berry AW, Holland MG, Moshkov IE, Smith AR & Hall MA (1996) Ethylene binding sites in higher plants. Plant Growth Regul. 18: 71-77.
Hausman J-F, Kevers C & Gaspar T (1994) Involvement of putrescine in the inductive rooting phase of poplar shoots raised *in vitro*. Physiol. Plant. 92: 201-206.
Hausman JF, Kevers C & Gaspar T (1995a) Auxin-polyamine interaction in the control of the rooting inductive phase of poplar shoots *in vitro*. Plant Sci. 110: 63-71.

Hausman JF, Kevers C & Gaspar T (1995b) Putrescine control of peroxidase activity in the inductive phase of rooting in poplar shoots *in vitro*, and the adversary effect of spermidine. J. Plant Physiol. 146: 681- 685.

Heby O & Persson L (1990) Molecular genetics of polyamine synthesis in eukaryotic cells. TIBS 15: 153-158.

Hill JE, Stead AD & Nichols R (1987) Pollination-induced ethylene and production of 1-aminocyclopropane-1-carboxylic acid by pollen of *Nicotiana tabacum* cv White Burley. J. Plant Growth Regul. 6: 1-13.

Holdsworth MJ, Bird CR, Ray J, Schuch W & Grierson D (1987). Structure and expression of an ethylene-related mRNA from tomato. Nucleic Acid Res. 15: 731-739.

Honma M (1985) Chemically reactive sulfhydryl groups of 1-aminocyclopropane-1-carboxylic deaminase. Agric. Biol. Chem. 49: 567-571.

Horner M, McComb JA, McComb AJ & Street HE (1977) Ethylene production and plantlet formation by *Nicotiana* anthers cultured in the presence and absence of charcoal. J . Exp. Bot. 28: 1365-1372.

Housti F, Coupe M & d'Auzac J (1992) Effect of ethylene on enzymatic activities involved in the browning of *Hevea brasiliensis* callus. Physiol. Plant. 86: 445-450.

Hua J, Chang C, Sun Q & Meyerowitz EM (1995) Ethylene insensitivity conferred by *Arabidopsis ERS* gene. Science 269: 1712-1714.

Huang LC, Kohashi C, Vangundy R & Murashige T (1995) Effects of common components on hardness of culture media prepared with Gelrite. In Vitro Cell. Dev. Biol. 31: 84-89.

Hunter B (1987) Plant regeneration method. European patent application #87200773.7.

Hussey G & Stacey NJ (1981) In vitro propagation of potato (*Solanum tuberosum* L.). Ann. Bot. 48: 787-796.

Hussey G & Stacey NJ (1984) Factors affecting the formation of in vitro tubers of potato (*Solanum tuberosum* L.). Ann. Bot. 53: 565-578.

Hutchinson MJ, Murr D, Krishnaraj S, Senaratna T & Saxena PK (1997) Does ethylene play a role in thidiazuron-regulated somatic embryogenesis of geranium (*Pelargonium* x *Hortorum* Bailey) hypocotyl cultures? In Vitro Cell. Dev. Biol. 33: 136-141.

Huxter T, Trevor TA & Reid DM (1981) Shoot initiation in light- and dark-grown tobacco callus: the role of ethylene. Physiol. Plant. 53: 319-326.

Hyde C & Phillips GC (1996) Silver nitrate promotes shoot development and plant regeneration of chile pepper (*Capsicum annuum* L.) via organogenesis. In Vitro Cell. Dev. Biol. 32: 72-80.

Jain RK, Chowdhury JB, Sharma DR & Friedt W (1988). Genotypic and media effects on plant regeneration from cotyledon explant cultures of some *Brassica* species. Plant Cell Tiss. Organ Cult. 14: 197-206.

Jaramillo J & Summers WL (1990) Tomato anther culture callus production: solidifying agent and concentration influence induction of callus. J. Amer. Soc. Hort Sci. 115: 1047-1050.

Jarvis BC, Yasmin S & Coleman MT (1985) RNA and protein metabolism during adventitious root formation in stem cuttings of *Phaseolus aureus* cultivar berkin. Physiol. Plant. 64: 53-59.

Kaur-Sawhney R, Tiburcio AF & Galston AW (1988) Spermidine and flower-bud differentiation in thin layer explants of tobacco. Planta 173: 282-284.

Keller WA, Rajhathy T & Lacapra J (1975) *In vitro* production of plants from pollen in *Brassica campestris*. Can. J. Genet. Cytol. 17: 655-666.

Kepczynski J, McKersie BD & Brown DCW (1992) Requirement of ethylene for growth of callus and somatic embryogenesis in *Medicago sativa* L. J. Exp. Bot. 43: 1199-1202.

Kevers C, Boyer N, Courduroux JC & Gaspar T (1992) The influence of ethylene on proliferation and growth of rose shoot cultures. Plant Cell Tiss. Organ Cult. 28: 175-181.

Kieber JJ, Rothenberg M, Roman G, Feldmann KA & Ecker JR (1993) *CTR1*: a negative regulator of the ethylene response pathway in *Arabidopsis*, encodes a member of the Raf family of protein kinase. Cell 72: 427-441.

Kim YH & Janick J (1989) ABA and polyox-encapsulation or high humidity increases survival of desiccated somatic embryos of celery. Hort Sci. 24: 674-676.

Klee HJ, Hayford MB, Kretzmer KA, Barry GF & Kishore GM (1991) Control of ethylene synthesis by expression of a bacterial enzyme in transgenic tomato plants. Plant Cell 3: 1187-1193.

Kochba J, Spiegal-Roy P, Neumann H & Saad S (1978) Stimulation of embryogenesis in *Citrus* ovular callus by ABA, ethephon, CCC and Alar and its suppression by GA. Z. Pflanzenphysiol. 89: 427-432.

Kong L. & Yeung EC (1994) Effects of ethylene and ethylene inhibitors on white spruce somatic embryo maturation. Plant Sci. 104: 71-80.

Kong L & Yeung EC (1995) Effects of silver nitrate and polyethylene glycol on white spruce (*Picea glauca*) somatic embryo development: enhancing cotyledonary embryo formation and endogenous ABA content. Physiol. Plant. 93: 298-304.

Koornneef M, Bade J, Hanhart C, Horsman K, Schel J, Soppe W, Verkerk R & Zabel P (1993) Characterization and mapping of a gene controlling shoot regeneration in tomato. Plant. J. 3: 131-141.

Kotte W (1922) Kulturversuche mit isolierten Wurzels pitzen Beitr. Allg. Bot. 2: 413-434.

Kuehn GD, Rodriguez-Garay B, Bagga S & Phillip GC (1990) Novel occurrence of uncommon polyamines in higher plants. Plant Physiol. 94: 855-857.

Kumar A, Taylor MA, Mad Arif SA & Davies HV (1996) Potato plants expressing antisense and sense *S*-adenosylmethionine decarboxylase (SAMDC) transgenes show altered levels of polyamines and ethylene: antisense plants display abnormal phenotypes. Plant. J. 9: 147-158.

Kumar PP, Joy IV RW & Thorpe TA (1989) Ethylene and carbon dioxide accumulation and growth of cell suspension cultures of *Picea glauca* (white spruce). J. Plant Phyiol. 135: 592-596.

Kumar PP, Nathan MJ & Goh CJ (1996) Involvement of ethylene on growth and plant regeneration in callus cultures of *Heliconia psittacorum* L. J. Plant Growth. Regul. 19: 145-151.

Kumar PP, Reid DM & Thorpe TA (1987) The role of ethylene and carbon dioxide in differentiation of shoot buds in excised cotyledons of *Pinus radiata in vitro*. Physiol. Plant. 69: 244-252.

Kumar PP & Thorpe TA (1989) Putrescine metabolism in excised cotyledons of *Pinus radiata* cultured *in vitro*. Physiol. Plant. 76: 521-526.

Kvaalen H (1994) Ethylene synthesis and growth in embryogenic tissue of Norway spruce: Effect of oxygen, 1-aminocyclopropane-1-carboxylic acid, benzyladenine and 2,4-dichlorophenoxyacetic acid. Physiol. Plant. 92: 109-117.

LaRue TAG & Gamborg OL (1971) Ethylene production by plant cell cultures: Variations in production during growing cycle and in different plant species. Plant Physiol. 48: 394-398.

Lau O & Yang SF (1974) Synergistic effects of calcium and kinetin on ethylene production by the mung bean hypocotyl. Planta 118: 1-6.

Lee MM, Lee SH & Park KY (1997a) Characterization and expression of two members of the *S*-adenosylmethionine decarboxylase gene family in carnation flower. Plant Mol. Biol. 34: 371-382.

Lee T, Huang MEE & Pua EC (1997b) High frequency shoot regeneration from leaf disc explants of garland chrysanthemum (*Chrysanthemum coronarium* L.) *in vitro*. Plant Sci. 126: 219-226.

Lee T, Liu JJ & Pua EC (1997c) Molecular cloning of two cDNAs (Accession Nos. X95729 and U80916) encoding *S*-adenosyl-L-methionine decarboxylase in mustard (*Brassica juncea* [L.] Czern & Coss) (PGR 97-157). Plant Physiol. 115: 1287.

Lentini Z, Mussell H, Mutschler MA & Earle ED (1988) Ethylene generation and reversal of ethylene effects during development *in vitro* of rapid-cycling *Brassica campestris* L. Plant Sci. 54: 75-81.

Leonhardt W & Kandeler R (1987) Ethylene accumulation in culture vessels - a reason for vitrification? Acta. Hort. 212: 223-227.

Litz RE & Shaffer B (1987) Polyamines in adventitious and somatic embryogenesis in mango (*Mangifera indica* L.). J. Plant Physiol. 128: 251-258.

Liu J, Mukherjee I & Reid DM (1990) Adventitious rooting in hypocotyls of sunflower (*Helianthus annuus*) seedlings. III. The role of ethylene. Physiol. Plant. 78: 268-276.

Lloyd GB & McCown BH (1980) Commercially feasible micropropagation of mountain laurel, *Kalmia latifolia*, by use of shoot-tip culture. Proc. Int. Plant Prop. Soc. 30: 421-437.

Lu C-Y & Thorpe TA (1987) Somatic embryogenesis and plantlet regeneration in cultured immature embryos of *Picea glauca*. J. Plant Physiol. 128: 297-302.

Ma J-H, Tao J-L, Cohen D & Morris B (1998) Ethylene inhibitors enhance in vitro root formation from apple shoot culture. Plant Cell Rep. 17: 211-214.

Mackenzie IA & Street HE (1970) Studies on the growth in culture of plant cells. VIII. The production of ethylene by suspension cultures of *Acer pseudoplatanus*. J. Exp. Bot. 21: 824-834.

Mad Arif SA, Taylor MA, George LA, Butler AR, Burch LR, Davies HV, Stark MJR & Kumar A (1994) Characterization of the *S*-adenosylmethionine decarboxylase (SAMDC) gene of potato. Plant Mol. Biol. 26: 327-338.

Mader JC (1997) Studies on polyamines in *Solanum tuberosum in vitro*: Effects of DFMO, DFMA, chlorogenic acid and putrescine on the endogenous distribution of polyamines, tuberization and morphology. J. Plant Physiol. 150: 141-152.

Marino G, Berardi G & Ancherani M (1995) The effect of the type of closure on the gas composition of the headspace and the growth of GF 677 peach X almond rootstock cell suspension cultures. In Vitro Cell. Dev. Biol. 31: 207-210.

Marsolais AA & Kasha KJ (1985) Callus induction from barley microspores. The role of sucrose and auxin in a barley anther culture medium. Can. J. Bot. 63: 2209-2212.

Marton L & Browse J (1991) Facile transformation of *Arabidopsis*. Plant Cell Rep. 10: 235-239.

Masgrau C, Altabella T, Farra R, Flores D, Thompson AJ, Besford R & Tiburcio AF (1997) Inducible expression of oat arginine decarboxylase in transgenic tobacco plants. Plant J. 11: 465-473.

Matthys D, Gielis J & Debergh P (1995) Ethylene. In: Aitken-Christie K, Kozai T & Smith MAL (eds) Automation and Environmental Control in Plant Tissue Culture, pp. 473-491. Kluwer Academic Publ., Dordrecht.

Meijer EGM (1989) Developmental aspects of ethylene biosynthesis during somatic embryogenesis in tissue culture of *Medicago sativa*. J. Exp. Bot. 40: 479-484.

Meijer EGM & Simmonds J (1988) Polyamine levels in relation to growth and somatic embryogenesis of *Medicago sativa*. J. Exp. Bot. 203: 787-794.

Mensuari-Sodi A, Panizza M & Tognoni F (1992) Quantitation of ethylene losses in different container-seal systems and comparison of biotic and abiotic contributions to ethylene accumulation in cultured tissues. Physiol. Plant. 84: 472-476.

Merkle SA, Parrott WA & Flinn BS (1995) Morphogenic aspects of somatic embryogenesis. In: Thorpe, T.A. (ed) *In Vitro* Embryogenesis in Plants, pp. 155-203. Kluwer Academic Publ., Dordrecht.

Meyerowitz EM (1992) Introduction to the *Arabidopsis* genome. In: Koncz C, Chua N-H & Schell J (eds) Methods in *Arabidopsis* Research, pp. 100-118. World Scientific Publ., Singapore.

Michael AJ, Furze JM, Rhodes MJ & Burtin D (1996) Molecular cloning and functional identification of a plant ornithine decarboxylase cDNA. Biochem. J. 314: 241-248.

Minocha SC, Papa NS, Khan AJ & Samuelsen AI (1991) Polyamines and embryogenesis in carrot. III. Effects of methyl glyoxal bis(guanylhydrazone). Plant Cell Physiol. 32: 395-402.

Montague MJ, Koppenbrink JW & Jaworski EG (1978) Polyamine metabolism in embryogenic cells of *Daucus carota*. I. Changes in intracellular content and rates of synthesis. Plant Physiol. 62: 430-433.

Mudge KW & Swanson BT (1978) Effect of ethephon, indolebutyric acid and treatment solution pH on rooting and on ethylene levels within mung bean cuttings. Plant Physiol. 61: 271-273.

Muhitch MJ, Edwards LA & Fletcher JS (1983) Influence of diamines and polyamines on the senescence of plant suspension culture. Plant Cell Rep. 2: 82-84.

Mukhopadhyay A, Arumugam N, Nandakumar PBA, Pradhan AK, Gupta V & Pental D (1992) *Agrobacterium*-mediated genetic transformation of oilseed *Brassica campestris*: Transformation frequency is strongly influenced by the mode of shoot regeneration. Plant Cell Rep. 11: 506-513.

Murashige T & Skoog F (1962) A revised medium for rapid growth and bioassays in tobacco tissue culture. Physiol. Plant. 15: 473-493.

Murata M & Orton TJ (1987) Callus initiation and regeneration capacities in *Brassica* species. Plant Cell Tiss. Organ Cult. 11: 111-123.

Murr DP & Yang SF (1975) Inhibition of *in vitro* conversion of methionine to ethylene by L-canaline and 2,4-dinitrophenol. Plant Physiol. 66: 79-82.

Nadolska-Orczyk A & Malepszy S (1989) *In vitro* culture of *Cucumis sativus* L. 7. Genes controlling plant regeneration. Theor. Appl. Genet. 78: 836-840.

Nam KH, Lee SH & Lee JH (1997) Differentiation expression of ADC mRNA during development and upon acid stress in soybean (*Glycine max*) hypocotyls. Plant Cell Physiol. 38: 1156-1166.

Narasimhulu SB & Chopra VL (1988) Species specific shoot regeneration response of cotyledonary explants of Brassicas. Plant Cell Rep. 7: 104-106.

Nissen P (1994) Stimulation of somatic embryogenesis in carrot by ethylene: Effects of modulators of ethylene biosynthesis and action. Physiol. Plant. 92: 397-403.

Nobercourt P (1939) Sur la perennité et l'augmantation de volumes des cultures de tissues végétaux. CR Soc. Biol. Paris 130: 1270.

Noh EW & Minocha SC (1994) Expression of a human *S*-adenosylmethionine decarboxylase cDNA in transgenic tobacco and its effects on polyamine synthesis. Transgenic Res. 3: 26-35.

Nomura K & Komamine A (1995) Physiological and biochemical aspects of somatic embryogenesis. In: Thorpe TA (ed) *In Vitro* Embryogenesis in Plants. pp. 249-265. Kluwer Academic Publ., Dordrecht.

Nordström AC & Eliasson L (1984) Regulation of root formation by auxin-ethylene interaction in pea stem cuttings. Physiol. Plant. 61: 298-302.

Nour KA & Thorpe TA (1994) The effect of gaseous state on bud induction and shoot multiplication in vitro in eastern white cedar. Physiol. Plant. 90: 163-172.

Ockendon DJ & McClenaghan R (1993) Effect of silver nitrate and 2,4-D on anther culture of Brussels sprouts (*Brassica oleracea* var. *gemmifera*). Plant Cell Tiss. Organ Cult. 32: 41-46.

Oeller PW, Lu MW, Taylor LP, Pike DA & Theologis A (1991) Reversible inhibition of tomato fruit senescence by antisense RNA. Science 254: 437-439.

Palmer EE (1992) Enhanced shoot regeneration from *Brassica campestris* by silver nitrate. Plant Cell Rep. 11: 541-545.

Panizza M, Mensuali-Sodi A & Tognoni F (1988) "In vitro" propagation of lavandin: ethylene production during plant development. Acta . Hort. 227: 334-339.

Parkinson JS (1993) Signal transduction schemes of bacteria. Cell 73: 857-871.

Penarrubia L, Aguilar M, Margossian L & Fischer RL (1992) An antisense gene stimulates ethylene hormone production during tomato fruit ripening. Plant Cell 4: 681-687.

Pental D, Pradhan AK, Sodhi YS & Mukhopadhyay A (1993) Variation amongst *Brassica juncea* cultivars for regeneration from hypocotyl explants and optimization of conditions for *Agrobacterium*-mediated genetic transformation. Plant Cell Rep. 12: 462-467.

Pérez-Amador M, Carbonell J & Granell A (1995) Expression of arginine decarboxylase is induced during early fruit development and in young tissues of *Pisum sativum* (L.). Plant Mol. Biol. 28: 997-1009.

Perl A, Aviv D & Galun E (1988) Ethylene and *in vitro* culture of potato: suppression of ethylene generation vastly improves yield, plating efficiency and transient expression of an alien gene. Plant Cell Rep. 7: 403-406.

Phillips GC & Kuehn GD (1991) Uncommon polyamines in plants and other organisms. In: Slocum RD & Flores HE (eds) Biochemistry and Physiology of Polyamines in Plants. pp. 121-136. Chemical Rubber Co Press, Uniscience, Boca Raton, Florida.

Pua EC (1993) Cellular and molecular aspects of ethylene on plant morphogenesis of recalcitrant *Brassica* species *in vitro*. Bot. Bull. Acad. Sin. 34: 191-209.

Pua EC (1994). Regeneration of plants from protoplasts of *Brassica juncea* (L.) Czern & Coss (brown mustard). In: Bajaj YPS (ed) Biotechnology in Agriculture and Forestry Vol. 29: Plant Protoplasts and Genetic Engineering V, pp. 38-51. Springer-Verlag, Berlin-Heidelberg.

Pua EC (1996) The regulatory role of ethylene on cell differentiation and growth. In: Xu Z-H & Chen ZH (eds) Plant Biotechnology for Sustainable Development of Agriculture, pp. 129-135. China Forestry Publ. House, Beijing.

Pua EC & Chi GL (1993) *De novo* shoot morphogenesis and plant growth of mustard (*Brassica juncea*) *in vitro* in relation to ethylene. Physiol. Plant. 88: 467-474.

Pua EC & Lee JEE (1995) Enhanced de novo shoot morphogenesis in vitro by expression of antisense 1-aminocyclopropane-1-carboxylate oxidase gene in transgenic mustard plants. Planta 196: 69-76.

Pua EC, Sim GE, Chi GL & Kong LF (1996a) Synergistic effect of ethylene inhibitors and putrescine on shoot regeneration from hypocotyl explants of Chinese radish (*Raphanus sativus* L. var. *longipinnatus* Bailey) *in vitro*. Plant Cell Rep. 15: 685-690.

Pua EC, Sim GE & Chye ML (1992) Isolation and sequence analysis of a cDNA clone encoding ethylene-forming enzyme in *Brassica juncea* (L.) Czern & Coss. Plant Mol. Biol. 19: 541-544.

Pua EC, Teo SH & Loh CS (1996b) Interactive role of ethylene and polyamines on shoot regenerability of Chinese kale (*Brassica alboglabra* Bailey) *in vitro*. J. Plant Physiol. 149: 138-148.

Pua EC, Trinh TH & Chua NH (1989) High frequency plant regeneration from stem explants of *Brassica alboglabra* Bailey *in vitro*. Plant Cell Tiss. Organ Cult. 17: 143-152.

Purnhauser L, Medgyesy P, Czekó M, Dix PJ & Márton L (1987) Stimulation of shoot regeneration in *Triticum aestivum* and *Nicotiana plumbaginifolia* Viv. tissue culture using the ethylene inhibitor $AgNO_3$. Plant Cell Rep. 6: 1-4.

Quoirin M & LePoivre P (1977) Etude de milieux adaptes aux cultures in vitro de prunus. Acta. Hort. 78: 437-442.

Radke SE, Turner JC & Facciotti D (1992) Transformation and regeneration of *Brassica rapa* using *Agrobacterium tumefaciens*. Plant Cell Rep. 11: 499-505.

Rastogi R, Dulson J & Rothstein SJ (1993) Cloning of tomato (*Lycopersicon esculentum* Mill.) arginine decarboxylase gene and its expression during fruit ripening. Plant Physiol. 103: 829-834.

Raz V & Fluhr R (1993) Ethylene signal is transducted via protein phosphorylation events in plants. Plant Cell 5: 523-530.

Reich B & Bingham ET (1980) The genetic control of bud formation from callus cultures of diploid alfalfa. Plant Sci. Lett. 20: 71-77.

Reid MS (1987) Ethylene in plant growth, development, and senescence. In: Davies, P.J. (ed) Plant Hormones and Their Role in Plant Growth and Development, pp. 257-279. Martinus Nijhoff Publ., Dordrecht.

Rethmeier NOM, Jansen CE, Snel EAM, Nijkamp HJJ & Hille J (1991) Improvement of regeneration of *Lycopersicon pennillii* by decreasing ethylene production. Plant Cell Rep. 9: 539-543.

Reynolds TL (1984) Callus formation and organogenesis in anther cultures of *Solanum carolinense* L. J. Plant Physiol. 11: 157-161.

Reynolds TL (1986) Pollen embryogenesis in anther cultures of *Solanum carolinense* L. Plant Cell Rep. 5: 273-275.

Reynolds TL (1987) A possible role for ethylene during IAA-induced pollen embryogenesis in anther cultures of *Solanum carolinense* L. Amer. J. Bot. 74: 967-969.

Reynolds TL (1989) Ethylene effects on pollen callus formation and organogenesis in anther cultures of *Solanum carolinense* L. Plant Sci. 61: 131-136.

Righetti B, Magnanini E, Infante R & Predieri S (1990) Ethylene, ethanol, acetaldehyde and carbon dioxide released by *Prunus avium* shoot cultures. Physiol. Plant. 78: 507-510.

Roberts DR (1991) Abscisic acid and mannitol promote early development, maturation and storage protein accumulation in somatic embryos of interior spruce. Physiol. Plant. 83: 247-254.

Roberts DR, Flinn BS, Webster FB & Sutton CS (1990) Abscisic acid and indole-3-acetic acid regulation of maturation and accumulation of storage proteins in somatic embryos of interior spruce. Physiol. Plant. 78: 355-360.

Roberts DR, Walker MA, Thompson JE & Dumbroff EB (1983) The effects of inhibitors of polyamines and ethylene biosynthesis on senescence, ethylene production and polyamine levels in cut carnation. Plant Cell Physiol. 25: 315-322.

Robbins JA, Reid MS, Paul JL & Rost TJ (1985) The effect of ethylene on adventitious root formation in mung bean (*Vigna radiata*) cuttings. J. Plant Growth Regul. 4: 147-157.

Robie AC & Minocha SC (1989) Polyamines and somatic embryogenesis in carrot. I. The effects of difluoromethylornithine and difluoromethylarginine. Plant Sci. 65: 45-54.

Robinson KEP & Adams DO (1987) The role of ethylene in the regeneration of *Helianthus annuus* (sunflower) plants from callus. Physiol. Plant. 71: 151-156.

Rodriguez-Garay B, Phillips GC & Kuehn GD (1989) Detection of norspermidine and norspermine in *Medicago sativa* L. (alfalfa). Plant Physiol. 89: 525-529.

Rottmann WH, Peter GF, Oeller PW, Keller JA, Shen NF, Nagy BP, Taylor LP, Campbell AD & Theologis A (1991) 1-Aminocyclopropane-1-carboxylate synthase in tomato is encoded by a multigene family whose transcription is induced during fruit and floral senescence. J. Mol. Biol. 222: 937-961.

Roustan JP, Latche A & Fallot J (1989) Stimulation of *Daucus carota* somatic embryogenesis by inhibitors of ethylene synthesis: cobalt and nickel. Plant Cell Rep. 8: 182-185.

Roustan J-P, Latche A & Fallot J (1992) Enhancement of shoot regeneration from cotyledons of *Cucumis melo* by $AgNO_3$, an inhibitor of ethylene action. J. Plant Physiol. 140: 485-488.

Sabapathy S & Nair H (1992) In vitro propagation of taro, with spermine, arginine, and ornithine. I. Plantlet regeneration from primary shoot apices and axillary buds. Plant Cell Rep. 11: 290-294.

Sabapathy S & Nair H (1995) In vitro propagation of taro, with spermine, arginine and ornithine. II. Plantlet regeneration via callus. Plant Cell Rep. 14: 520-524.

Sanders IO, Ishizawa K, Smith AR & Hall MA (1990). Ethylene binding and action in rice seedlings. Plant Cell Physiol. 31: 1091-1099.

Santanen A & Simola LK (1992) Changes in polyamine metabolism during somatic embryogenesis in *Picea abies*. J. Plant Physiol. 140: 475-480.

Santos KGB, Mundstock E & Bodanese-Zanettini MH (1997) Genotype-specific normalization of soybean somatic embryogenesis through the use of an ethylene inhibitor. Plant Cell Rep. 16: 859-864.

Sauerbrey E, Grossmann K & Jung J (1988) Ethylene production by sunflower cell suspensions. Plant Physiol. 87: 510-513.

Schaller GE & Bleecker AB (1995) High-affinity binding sites for ethylene are generated in yeast expressing the *Arabidopsis ETR1* gene. Science 270: 1809-1811.

Scholten HJ & Pierik RLM (1998) Agar as a gelling agent: chemical and physical analysis. Plant Cell Rep. 17: 230-235.

Schroder G & Schroder J (1995) cDNAs for S-adenosyl-L-methionine decarboxylase from *Catharanthus roseus*, heterologous expression, identification of the proenzyme-processing site, evidence for the presence of both subunits in the active enzyme, and a conserved region in the 5' messenger-RNA leader. Euro. J. Biochem. 228: 74-78.

Sethi U, Basu A & Guha-Mukherjee S (1990) Control of cell proliferation and differentiation by modulators of ethylene biosynthesis and action in *Brassica* hypocotyl explants. Plant Sci. 69: 225-229.

Sharma KK & Bhojwani SS (1990) Histological aspects on *in vitro* root and shoot differentiation from cotyledon explants of *Brassica juncea* (L.) Czern. Plant Sci. 69: 207-214.

Sharma P & Rajam MV (1995) Spatial and temporal changes in endogenous polyamine levels associated with somatic embryogenesis from different hypocotyl segments of eggplant (*Solanum melongena* L.). J. Plant Physiol. 146: 658-664.

Sidorov VA, Menczel L & Maliga P (1981) Isoleucine-requiring *Nicotiana* plant deficient in threonine deaminase. Nature 294: 87-88.

Sisler EC (1979) Measurement of ethylene binding in plant tissue. Plant Physiol. 64: 538-542.

Sisler EC & Yang SF (1984) Anti-ethylene effects of cis-2-butene and cyclic olefins. Phytochemistry 23: 2765-2768.

Smith MAL & Spomer LA (1995) Vessels, gels, liquid media, and support systems. In: Aitken-Christie J, Kozai T & Smith MAL (eds) Automation and Environmental Control in Plant Tissue Culture pp. 371-404. Kluwer Academic Publ., Dordrecht.

Smulders MJM, Kemp A, Barendse GWM, Croes AF & Wullems GJ (1990) Role of ethylene in auxin-induced bud formation in tobacco explants. Physiol. Plant. 78: 167-172.

Songstad DD, Armstrong CL & Petersen WL (1991) $AgNO_3$ increases type II callus production from immature embryos of maize inbred B73 and its derivatives. Plant Cell Rep. 9: 699-702.

Songstad DD, Duncan DR & Widholm JM (1988) Effect of 1-aminocyclopropane-1-carboxylic acid, silver nitrate, and norbornadiene on plant regeneration from maize callus cultures. Plant Cell Rep. 7: 262-265.

Steward FC, Mapes MO & Smith J (1958a) Growth and organized development of cultured cells. I. Growth and division of freely suspended cells. Amer. J. Bot. 45: 693-703

Steward FC, Mapes MO, Mears K (1958b) Growth and organized development of cultured cells. II. Organization in cultures grown from freely suspended cells. Amer. J. Bot. 45: 705-708.

Stewart ER & Freebaire HT (1967) Ethylene production in tobacco pith cultures. Plant Physiol. 42: S-30.

Subrahmanyam D & Rathore VS (1992) Influence of ethylene on $^{14}CO_2$ assimilation and partitioning in Indian Mustard. Plant Physiol. Biochem. 30: 81-86.

Suttle JC (1981) Effect of polyamines on ethylene production. Phytochemistry 20: 1477-1480.

Taeb AG & Alderson PG (1990) Shoot production and bulbing of tulip *in vitro* related to ethylene. J. Hortic Sci. 65: 199-204.

Takebe L, Labib G & Melchers G (1971) Regeneration of whole plants from isolated mesophyll protoplasts of tobacco. Naturwissen. 58: 318-320.

Tanimoto S & Matsubara Y (1995) Stimulating effect of spermine on bulblet formation in bulb-scale segment of *Lilium longiflorum*. Plant Cell Rep. 15: 297-300.

Taylor MA, Mad Arif SA, Kumar A, Davies HV, Scobie LA, Pearce SR & Flavell AJ (1992) Expression and sequence analysis of cDNAs induced during the early stages of tuberisation in different organs of the potato plant (*S. tuberosum* L.). Plant Mol. Biol. 641-651.

Theologis A (1992) One rotten apple spoils the whole bushel: The role of ethylene in fruit ripening. Cell 70: 181-184.

Thomas DS & Murashige T (1979) Volatile emissions of plant tissue culture. I. Identification of the major components. In Vitro 15: 654-658.

Thorpe TA (1994) Morphogenesis and regeneration. In: Vasil IK & Thorpe TA (eds) Plant Cell and Tissue Culture, pp. 17-36. Kluwer Academic Publ., Dordrecht.

Thorpe TA (1995) *In Vitro* Embryogenesis in Plants. Kluwer Academic Publ., Dordrecht.

Thorpe TA & Kumar PP (1993) Cellular control of morphogenesis. In: Ahuja MR (ed) Micropropagation of Woody Plants, pp. 11-29. Kluwer Academic Publ., Dordrecht.

Tiainen T (1992) The role of ethylene and reducing agents on anther culture response of tetraploid potato (*Solanum tuberosum* L.). Plant Cell Rep. 10: 604-607.

Tiburcio AF, Gendy CA & Tran Thanh Van K (1989) Morphogenesis in tobacco subepidermal cells: Putrescine as marker of root differentiation. Plant Cell Tiss. Organ Cult. 19: 43-54.

Tomes DT & Smith OS (1985) The effect of parental genotype on initiation of embryogenic callus from elite maize (*Zea mays* L.) germplasm. Theor. Appl. Genet. 70: 505-509.

Torrigiani P, Altamura MM, Capitani F, Serafini-Fracassini D & Bagni N (1989) De novo root formation in thin cell layers of tobacco: changes in free and bound polyamines. Physiol. Plant. 77: 294-310.

Torrigiani P, Altamura MM, Pasqua G, Monacelli B, Serafini-Fracassini D & Bagni N (1987) Free and conjugated polyamines during de novo floral and vegetative bud formation in thin cell layers of tobacco. Physiol. Plant. 70: 453-460.

Tremblay FM (1990) Somatic embryogenesis and plant regeneration from embryos isolated from stored seeds of *Picea glauca*. Can. J. Bot. 68: 236-242.

Vain P, Flament P & Soudain P (1989) Role of ethylene in embryogenic callus initiation and regeneration in *Zea mays* L. J. Plant Physiol. 135: 337-340.

Van Aartrijk J & Blom-Barnhoorn GJ (1983) Adventitious bud formation from bulb-scale explants of *Lilium speciosum* Thunb. *in vitro*. Effect of wounding, TIBA, and temperature. Z. Pflanzenphysiol. 110: 355-363.

Van Aartrijk J, Blom-Barnhoorn GJ & Bruinsma J (1985) Adventitious bud formation from bulb-scale explants of *Lilium speciosum* Thunb. *in vitro*. Effects of aminoethoxyvinyl-glycine, 1-aminocyclopropane-1-carboxylic acid, and ethylene. J. Plant Physiol. 117: 410-410.

Vasil V & Hildebrandt AC (1965) Differentiation of tobacco plants from single isolated cells in microculture. Science 150: 889-892.

Walden R, Cordeiro A & Tiburcio AF (1997) Polyamines: small molecules triggering pathways in plant growth and development. Plant Physiol. 113: 1009-1013.

Wang CY & Kramer GF (1990) Effect of polyamine treatment on ethylene production of apples. In: Flores HE, Arteca RN & Shannon JC (eds) Polyamines and Ethylene: Biochemistry, Physiology and Interactions, pp. 411-413. Amer. Soc. Plant Physiologists, Maryland.

Wang SY & Faust M (1986) Effects of growth retardants on root formation and polyamine content in apple seedlings. J. Amer. Soc. Hortic Sci. 111: 912-917.

Wann SR, Johnson MA, Noland TL & Carlson JA (1987) Biochemical differences between embryogenic and nonembryogenic callus of *Picea abies* (L.) Karst. Plant Cell Rep. 6: 39-42.

Wen CM, Wu M, Goh CJ & Pua EC (1993) Nucleotide sequence of a cDNA clone encoding 1-aminocyclopropane-1-carboxylate synthase from mustard (*Brassica juncea* [L.] Czern & Coss). Plant Physiol. 103: 1019-1020.

Wen CM, Wu M, Goh CJ & Pua EC (1995) Cloning and nucleotide sequence of a cDNA encoding a *S*-adenosyl-L-methionine synthetase from mustard (*Brassica juncea* [L.] Czern & Coss). Plant Physiol. 107: 1020-1022.

White PR (1934) Potentially unlimited growth of excised tomato root tips in a liquid medium. Plant Physiol. 9: 585-600.

White PR (1939) Potentially unlimited growth of excised plant callus in an artificial nutrient. Amer. J .Bot. 26: 59-64.

Williams J, Pink DAC & Biddington NL (1990) Effect of silver nitrate on long-term culture and regeneration of callus from *Brassica oleracea* var. *gemmifera*. Plant Cell Tiss. Organ Cult. 21: 61-66.

Yadav JS & Rajam MV (1998) Temporal regulation of somatic embryogenesis by adjusting cellular polyamine content in eggplant. Plant Physiol. 116: 617-625.

Yang MZ, Jia SR & Pua EC (1991) High frequency of plant regeneration from hypocotyl explants of *Brassica carinata* A. Br. Plant Cell Tiss. Organ Cult. 24: 79-81.

Yang SF & Hoffman NE (1984) Ethylene biosynthesis and its regulation in higher plants. Annu Rev Plant Physiol. 35: 155-189.

Zarembinski TI & Theologis A (1994) Ethylene biosynthesis and action: a case of conservation. Plant Mol. Biol. 26: 1579-1597.

Chapter 11

REGULATION OF MORPHOGENESIS BY BACTERIAL AUXIN AND CYTOKININ BIOSYNTHESIS TRANSGENES

Ann C. SMIGOCKI and Lowell D. OWENS
Molecular Plant Pathology Laboratory, Agricultural Research Service, U.S. Department of Agriculture, Beltsville, Maryland 20705, U.S.A.

1 INTRODUCTION

Many physiological studies have shown that the plant growth regulators, auxin and cytokinin, havc fundamental roles as modulators of plant growth and development. Virtually every aspect of development appears to be affected by these hormones based on observations gathered from hormone applications to cells, tissues or organs. Propagation of plant cells in tissue culture requires both auxin and cytokinin, and in many cases, organogenesis is regulated by the overall ratio of cytokinin-to auxin ; a higher ratio induces shoots and a lower ratio, roots. Limitations of exogenously applied hormones for deciphering the mechanisms of hormone action have recently been supplemented by molecular genetic approaches. The use of hormone mutants and the expression of bacterial genes involved in auxin and cytokinin biosynthesis in transgenic plants have enlarged our understanding of hormone biology and biochemistry.

This review will focus on the physiological effects of transgenes that have been used to alter auxin and cytokinin concentrations and sensitivity *in planta*. Other recent reviews summarizing molecular approaches for determining the mode of action of auxin and cytokinin are also recommended (Hamill, 1993; Hobbie et al., 1994; Gaudin et al., 1994; Costacurta and Vanderleyden, 1995).

2 EFFECTS OF AUXIN AND CYTOKININ

Plant hormones have been undisputedly recognized as modulators of plant growth and development. Physiological studies have indicated that almost all aspects of plant development are influenced by plant hormones. The nature, occurrence, transport and effects of the major plant hormones have been elucidated mainly from observations gathered from exogenous applications to cultured cells, tissues or organs. The importance of two of these hormones, cytokinin and auxin, in promoting the development of shoots and roots from undifferentiated cells was first demonstrated with cultured tobacco callus cells by Skoog and Miller (1957). As a result, auxin and cytokinin have been considered to be the two major classes of hormones involved in control of plant morphogenesis. However, over the years it became apparent that complex interactions among the various hormones exist and that growth and development are the result of the net effect of the hormonal balance (Davies, 1987). For example, it is the ratio of cytokinin-to-auxin in the growth medium that ultimately determines whether shoots (higher ratio) or roots (lower ratio) will form from undifferentiated tobacco cells (Skoog and Miller, 1957). In addition, it has been documented that the morphogenic response can also be modulated by other hormones such as gibberellic acid in development of somatic embryos of soybean, ethylene in inhibition of shoot organogenesis from cultured potato leaf discs, and abscisic acid in the prevention of precocious germination in culture of very young embryos (Norstog, 1979; DeBlock, 1988; Christou and Yang, 1989).

Despite years of research, our knowledge of the mechanisms of hormone action is limited. An important aspect of elucidating the mode of action of hormones has been the need to determine their exact concentration and location within individual, cells, tissues and organs. Until recently, little progress was made in that respect because the overall hormone concentrations at which plant physiological processes are influenced are extremely low, and sophisticated equipment and sensitive methods for their quantitation were lacking. With the advent of new technology, determination of minute quantities of hormones in differentiating cells is now feasible. Coupling this technology with physiological and molecular genetic approaches will most certainly advance the current understanding of the biology and biochemistry of plant hormones (Klee and Estelle, 1991; Hobbie et al., 1994).

Although, as mentioned earlier, our current knowledge regarding hormonal mode of action has been deduced from exogenous applications of hormones, this approach has its limitations regarding how plant cells perceive the added hormones in terms of uptake, sequestration and metabolism. Currently, several laboratories are utilizing genetics as a tool to address the problem of hormone action (Binns, 1994; Hobbie and Estelle, 1994). Identification of plants with hormonal mutations has provided information not only on phenotypic effects but also hormone biochemistry and, in some cases, has led to the analysis of the defective genes involved in modulation of hormonal processes.

Mutants requiring hormones are potentially useful for deciphering the biosynthetic and physiological pathways, whereas those with altered responses to hormones could provide information on how their action and interaction regulates development. Most progress has been made in identifying auxin-related mutants which display a surprisingly wide variety of phenotypes (Hobbie and Estelle, 1994). One approach, that utilizes screening of seedlings for root elongation on concentrations of auxins normally inhibitory to normal root growth, has yielded genetic lesions, some of which are being molecularly characterized (Leyser et al., 1993; Estelle and Klee, 1994; Hobbie and Estelle, 1994).

Similar studies with cytokinin yielded cytokinin-resistant mutants and, in one case, five independent recessive alleles of a new locus, *CKR1*, were identified (Su and Howell, 1992). One cytokinin-resistant mutant, that was originally identified by its resistance to cytokinin during seedling development, its reduced root branching and wilted shoots, was later shown to be deficient in the biosynthesis of another hormone, abscisic acid (Parry et al., 1991). Other cytokinin mutants have been described that have phenotypes correlated with elevated levels of cytokinin or are associated with cytokinin effects (Chaudhury et al., 1993; Chory et al., 1994; Estelle and Klee, 1994).

Gibberellic acid-deficient mutants have been extensively studied and their characteristic dwarf phenotypes have been attributed to particular blocks in the gibberellic acid biosynthetic pathway (Stoddart, 1987; Klee and Estelle, 1991). Little progress, however, has been made in isolating auxin, cytokinin and ethylene-deficient mutants. Whether this is an indication that these hormones are absolutely essential for survival of plant cells or that their absence produces no phenotypic abnormalities awaits further studies.

Based on the work with exogenous applications of hormones, it is not surprising that the hormone response in almost all of the mutants identified to date appears to be modulated by one or more of the other hormones. This phenomenon emphasizes the complexity of action and interaction of hormones, and foreshadows the challenges that lie ahead for developing novel approaches to decipher the mechanisms of hormone action.

One of the most promising new approaches, activation T-DNA tagging, generates mutants in which the expression of the tagged gene allows growth under selective conditions. This approach has been employed to isolate mutant tobacco cells that can grow in the absence of auxin or cytokinin in the culture medium, or under selective levels of an inhibitor of polyamine biosynthesis (Hayashi et al., 1992; Walden et al., 1994). By using multiple enhancer sequences in a T-DNA-derived plasmid that was introduced into tobacco protoplasts, clones in which insertion of the plasmid conferred hormone-independent growth were identified. Isolation of the tagged plant genes by plasmid rescue has in some cases been achieved. In addition, transfer of these genes to plant cells, followed by selection for growth under the same conditions as the mutant lines, confirmed the function of the cloned genes. A variety of mutants were recovered using this novel approach that may represent: 1) activation of a specific developmental pathway, 2) overexpression of the target of selective pressure, as well as, 3) detoxification, or the increased turnover or reduced uptake of the selective compound. These mutants no doubt will facilitate isolation of genes involved in complex biochemical and morphological pathways.

2.1 Plant hormone-induced organogenesis

Many plant species have been grown *in vitro* for purposes of micropropagation and genetic engineering. Numerous empirically determined methods for regeneration of specific plants have been published, however, a major problem exists in that many of them are not applicable to other species or cultivars, even when the plants are closely related genetically. Exogenously added auxin and cytokinin facilitate the process of organogenesis in many cultured cells, but some of the most important plant species do not readily respond to such treatments. Lack of recognition, uptake, proper targeting or metabolism of the exogenously added hormones by the plant cells may contribute to their ineffectiveness.

Manipulation of auxin and cytokinin levels *in vivo* circumvents to a certain degree the problems associated with exogenous applications. Molecular manipulation of genes encoding key enzymes in the auxin or cytokinin biosynthetic pathway is one approach that would allow the endogenous hormone levels to be increased or decreased. However, these enzymes have not been purified and, therefore, no corresponding plant genes have been cloned. Recently, three cytokinin-metabolizing enzymes have been purified to homogeneity, and the cDNA clones for two of them have been isolated (Mok and Mok, 1994). These genes may provide valuable information on cytokinin accumulation but do not provide information on cytokinin synthesis *in planta*. To the best of our knowledge, genes for auxin degrading enzymes have not been cloned.

2.2 Synthesis of auxin and cytokinin by pathogenic bacteria

The discovery that microorganisms can also synthesize plant hormones has been exploited for the study of auxin and cytokinin biology in plants. The role of phytohormone biosynthesis by microbes is not fully elucidated, but, in several cases of pathogenic fungi and bacteria, these compounds have been shown to play a role in disease development. Among the pathogens, bacteria are the most studied.

Crown gall and hairy root diseases of plants are caused by *Agrobacterium tumefaciens* and *A. rhizogenes,* respectively. Induction of either shooty teratomas or undifferentiated tumours, usually at the crown of plants, by *A. tumefaciens* is the result of the expression of bacterial genes that are transferred to the plant during the infection process. Similarly, *A. rhizogenes* induces the proliferation of roots from a wound site, or it can cause formation of tumours upon roots. The genes that are expressed in the transformed plant cells carry a portion of either the Ti (tumour-inducing) or Ri (root-inducing) plasmids from *A. tumefaciens* or *A. rhizogenes,* respectively. Through mutational analysis of the DNA that gets transferred to the plant genome (T-DNA), auxin and cytokinin biosynthesis genes were identified. Mutations at different loci induced a different type of tumour that was associated with auxin or cytokinin overproduction. Although a number of different genes are transferred, only some of them function in hormone metabolism, while others have no clearly established functions. The hormone-specifying genes contain eukaryotic regulatory sequences at the 5′ and 3′ flanking-regions and, as a result, are regulated by mechanisms similar to all plant genes.

In *A. tumefaciens,* three hormone biosynthetic genes were identified (Table 1; Garfinkel and Nester, 1980; Garfinkel et al., 1981; Leemans et al., 1982). Mutations in two genes *(tms1 or iaaM* and *tms2 or iaaH*) induced shooty tumours and, therefore, were determined to code for enzymes in the auxin biosynthetic pathway. The other gene was identified as the *ipt* gene coding for isopentenyl transferase, a key enzyme in the cytokinin biosynthetic pathway. Mutations in the *ipt* gene induced root growth from tumours. Mutation at yet another locus induced large tumours and was termed the *tml* locus. The *tml* locus is comprised of two genes, *6a* and *6b*. Secretion of opines produced by these bacteria is believed to be encoded by gene 6a (Messens et al., 1985) and gene *6b* is considered to be a tumour inducing gene (Hooykaas et al., 1988; Spanier et al., 1989; Tinland et al., 1990).

Six genes involved in tumour production have also been identified in a similar fashion in the T-DNA region of the Ri plasmid of *A. rhizogenes* (Table 1; White et al., 1985). Two genes, *aux1* and *aux2*, show 60 and 71%

homology to the *tms1 (iaaM)* and *tms2 (iaaH)* genes, respectively, and have been shown to complement mutations in those genes (Offringa et al., 1986; Camilleri and Jouanin, 1991). The other four, *rolA, rolB, rolC* and *rolD,* play a role in root induction. Although *rolC* confers cytokinin-like effects, no homology to the cytokinin biosynthesis gene *ipt* was found in the T-DNA of the Ri plasmid.

Table 1. Bacterial genes associated with auxin and cytokinin biosynthesis

Pathogenic Bacteria	Genes	
	Auxin-like	Cytokinin-like
Agrobacterium tumefaciens	*iaaM(tms1)* *iaaH (tms2)*	*ipt*
Agrobacterium rhizogenes	*aux1* (60%)[1] *aux2* (71%) *rolA* *rolB* *rolD*	*rolC*
Pseudomonas syringae subsp. *savastanoi*	*iaaM* (50%) *iaaH* (30%)	*ptz* (40%) *iaaL*[2]
Pseudomonas syringae subsp. *solanacearum*		*tzs*
Erwinia herbicola pv. gypsophilae	*iaaM* *iaaH*	cytokinin locus
Rhodococcus fascians	-	cytokinin gene (limited)

1 Percent homology to the corresponding genes of *A. tumefaciens.*

2 Appears to reduce the levels of active auxin and shifts the hormonal balance towards cytokinin

In addition to the *A. tumefaciens* and *A. rhizogenes* hormone biosynthetic genes, similar genes have been detected in phytopathogenic *Pseudomonas syringae* subsp. *savastanoi* bacteria that cause olive knot disease in olive trees *(Olea europa* L.) and oleander *(Nerium oleander* L.) (Table 1). Genes for auxin biosynthesis (*iaaM* and *iaaH)* share approximately 50% and 30% homology to the *A. tumefaciens tms1 (iaaM) and tms2 (iaaH)* genes, respectively (Gielen et al., 1984; Camilleri and Jouanin, 1991). The source of the *P. savastanoi* strain appears to determine whether these genes are plasmid encoded or are found on the bacterial chromosome. Genes encoding *iaaM* and *iaaH* are plasmid-encoded in the case of oleander-pathogenic strains, while those from olive-specific strains are most likely on the chromosome, since loss of auxin production was not accompanied by plasmid loss (Comai and Kosuge, 1980; Costacurta and Vanderleyden, 1995).

High cytokinin levels are also characteristic of *P. savastanoi* cultures, and a plasmid encoded gene *ptz,* sharing approximately 40%

homology with the *ipt* gene, has been cloned (Morris, 1986). A study of strains cured of their plasmids, however, revealed that cytokinin production is only in part encoded by this gene. In strains of *P. solanacearum*, another gene, *tzs*, has also been found to have homology to the *ipt* gene and to a *tzs* gene located outside the T-DNA region in nopaline type strains of *A. tumefaciens* (Table 1; Akyoshi et al., 1989).

Phytohormones appear to play a major role in the pathogenic nature of *Erwinia herbicola* pv. *gvpsophilae* that induces galls on baby's breath *(Gypsophila paniculata)* (Table 1). *E. herbicola* possesses two major pathways for auxin biosynthesis. The indole-3-pyruvate route is found in all nonpathogenic and pathogenic isolates while the indoleacetamide (IAM) route has been found only in pathogenic strains (Manulis et al., 1991b). The auxin genes in the pathogenic strains reside on a plasmid and are similar to the *tms1 (iaaM) and tms2 (iaaH)* genes of *A. tumefaciens and P. savastanoi* (Manulis et al., 1991a). Insertional inactivation of these genes results in induction of smaller galls (Clark et al., 1993). A locus conferring cytokinin biosynthesis has been localized to the same plasmid, but no homology was found between that locus and the *ipt*, *tzs* or *ptz* genes (Lichter et al., 1995).

Cytokinin biosynthesis genes have also been found in gram-positive bacteria. A gene with limited homology to the *ipt* and *ptz* genes has been isolated from *Rhodococcus fascians,* a plant pathogen that induces leafy galls in dicotyledonous plants (Table 1; Crespi et al., 1992). The gene is carried on a large, linear plasmid, pFi, and appears to be only partially responsible for the observed phenotype. Numerous other organisms have been shown to secrete cytokinins (van Staden and Davey, 1978; Elzen, 1983; Morris, 1986). They include *Rhizobium, Chromobacterium, Azotobacter* and *Arthrobacter* bacteria. In addition, ectomycorrhizal fungi and pathogenic *Plasmodiophora brassicae*, as well as some insects, are also part of this group for which little information is available concerning the role these hormones play in the organisms' physiology or their interactions with host organisms (van Staden and Davey, 1978; Elzen, 1983; Morris, 1986).

Many of the cloned bacterial genes encoding enzymes that lead to the synthesis of auxin and cytokinin have been introduced into plant cells for *in planta* manipulation of hormone levels (Klee et al., 1987b; Smigocki and Owens, 1988; Schmulling et al., 1989). Using well established molecular techniques that allow for overexpression, underexpression, as well as tissue-specific and temporal expression of genes, the steady-state levels of auxin and cytokinin were altered in order to determine their biological effects on plant structure.

2.3 Bacterial auxin biosynthesis genes in plants

Several auxin biosynthetic pathways have been identified in plants. Tryptophan appears to be the precursor to most of the rather well conserved auxin pathways and is converted most frequently to indole-3-acetaldehyde, a direct precursor of indole-3-acetic acid (IAA). Interestingly, a new pathway has been suggested recently based on work with a tryptophan auxotroph that does not require tryptophan as a precursor (Baldi et al., 1991; Wright et al., 1991). In bacteria and crown galls, however, the IAA pathway is well known because it involves only two simple steps: 1) the conversion of L-tryptophan into IAM by tryptophan monooxygenase and 2) the transformation of IAM into IAA by an indoleacetamide hydrolase (Table 2). The *iaaM* and *iaaH* genes code for the tryptophan monooxygenase and indoleacetamide hydrolase enzymes, respectively (Schroder et al., 1984; Thomashow et al., 1986). In some of the initial molecular approaches to alter auxin balances *in planta,* Klee et al. (1987b) engineered the two bacterial auxin biosynthetic genes and introduced them into *Petunia hybrida.*

Table 2. Biochemical functions of some of the bacterial genes associated with auxin and cytokinin biosynthesis

Gene	Biochemical function
iaaM; tms1; aux1	-trytophan monoxygenase; converts tryptophan to indole-3-acetamide (IAM)
iaaH; tms2; aux2	-indoleacetamide hydrolase; converts IAM to indole-3-acetic acid (IAA)
rolB	-indole glucoside; converts indole-3-acetyl- glucoside to IAA
iaaL	-IAA lysine synthetase; converts IAA to indole-3-acetyl-lysine
ipt	-isopentenyl transferase; condenses isopentenylpyrophosphate and AMP to isopentenyl AMP, a cytokinin precursor
rolC	-cyotkinin-beta-glucosidase; converts cytokinin-N-glucosides to cytokinins

Plants transformed with either the *iaaM* or *iaaH* gene under the control of their own regulatory sequence were morphologically normal (Table 3). The expression of the *iaaM* gene increased the levels of IAM. As an intermediate, IAM has been determined not to have any auxin activity and to be inefficiently converted to IAA. Surprisingly, however, fusion of the *iaaM* gene to a stronger promoter, the cauliflower mosaic virus (CAMV) 19S, did significantly alter the growth pattern of regenerated plants (Table 3; Klee et al., 1987a) Transgenic plants exhibited elongated stems, extreme apical dominance and small, thick, downward - curled leaves. Adventitious root

Table 3. Phenotypes of plants transformed with bacterial genes associated with auxin biosynthesis

Auxin-associated gene	Phenotype
iaaM	-normal; higher levels of IAM (Klee et al., 1987b)
iaaH	-normal (Klee et al., 1987b)
CaMV19S-*iaaM*	-elongated stems; strong apical dominance; small, thick, downward-curled leaves; adventitious root emergence from stems and leaves (Klee et al., 1987a)
CaMV35S-*iaaM*	-pale calli; normal plants; abnormal plants similar to CaMV19S-*iaaM* with higher lignin content (Sitbon et al., 1992)
aux1 + aux 2	-as for CaMV35S-*iaaM*, but shorter internodes and smaller epinastic leaves; some with highly branched and stunted root systems (Tepfer, 1984; McInnes et al., 1991)
rolA + B + C	-hairy root syndrome: reduced apical dominance wrinkled leaves, reduced internodal distance, reduced seed production, small flowers (Schmulling et al., 1988)
rolA	-wrinkled leaves; condensed inflorescences; reduced internodal distance; root induction in leaf tissue (Schmulling et al., 1988)
rolB	-altered leaves and flowers; heterostyly; adventitious root proliferation on stems (Schmulling et al., 1988)
CaMV35S-*rolB*	-rounded, necrotic leaves; heterostyly (Schmulling et al., 1988)

emergence from stems and leaves and numerous adventitious root primordia that frequently aborted were also noted. These abnormalities are symptomatic of high auxin or ethylene treatment and, in fact, the transformed tissues contained a 10-fold excess of free IAA and produced three times more ethylene than normal plants (Medford and Klee, 1989).

Similar experiments in different plants yielded a number of different phenotypic effects. Transformation of *Nicotiana* with both the *iaaM* gene fused to another strong viral promoter, CaMV35S, and the *iaaH* gene under regulation of its own promoter induced either pale calli with compact morphology or plants with normal and abnormal phenotypes (Table 3; Sitbon et al., 1992). The abnormal plants were phenotypically similar to the transgenic petunia, but their stem segments had 4 to 5 fold higher lignin content.

In nightshade *(Solanum dulcamara)* and cucumber *(Cucumis sativus)*, the *aux1* and *aux2* genes from *A. rhizogenes* induced similar developmental effects as in tobacco but, in addition, transgenic *S. dulcamara* plants exhibited shorter internodes and smaller, epinastic leaves that curled downwards (Tepfer, 1984; McInnes et al., 1991). Although the root system was normal in *S. dulcamara,* transgenic cucumber roots were characterized by a thick, highly branched and stunted root system characteristic of auxin overproduction (Tepfer, 1984).

Introduction of the rol genes from the Ri plasmids into several different species of plants have also been found to induce developmental changes in plants that appear to be associated with altered auxin physiology (Table 3). To ascribe a role to each of the rol genes within the very complex hairy root syndrome, single *rol* genes have been introduced into plant cells. *rolB* alone was shown fully capable of inducing hairy roots on several plants, while either *rolA* or *rolC* alone could only induce roots on a very susceptible host (Spena et al., 1987; Capone et al., 1989). Expression of the *rolB* gene also induced abnormal leaf morphology, large flowers, heterostyly and increased adventitious root formation on stems (Schmulling et al., 1988). It has been proposed that the function of the *rolB* gene is to hydrolyze inactive auxin-glucoside conjugates into free and active auxins resulting in a hormonal balance favouring auxins (Table 2; Estruch et al., 1991a). Such a shift in hormone ratio would account for some of the observed phenotypic characteristics of *rolB* transformed plants. However, no corresponding increase in IAA content has yet been demonstrated (Nilsson et al., 1993).

In *rolC* transformed plants, a reduction in apical dominance and internodal distance has been noted, along with altered leaf morphology, small flowers and reduced seed production (Schmulling et al., 1988; Oono et al., 1990). Except for the effect on the root system, these characteristics generally are associated with elevated levels of cytokinin. Strong constitutive overexpression of *rolC* gene fused to the CaMV35S viral promoter resulted in drastically reduced apical dominance, pale green, small, lanceolate leaves, increased and highly branched roots and very small male sterile flowers (Schmulling et al., 1988). Estruch et al., (1991b) have shown that *rolC* codes for an enzyme with cytokinin beta-glucosidase activity that can potentially cleave weakly active cytokinin-N-glucosides into highly active cytokinins *in planta* (Table 2). But, it is still unclear how this gene can, on one hand, stimulate hairy root and root growth but, at the same time, increase the pool of active cytokinins that are known to inhibit root initiation and growth.

Function of the *rolA* gene is, as yet, unknown but it has been shown to increase sensitivity of cells to auxin during the flowering stage (Vansuyt et al., 1992). Some of the phenotypic traits of *rolA* transgenic plants include wrinkled leaves, condensed inflorescences, reduced internodal distance and large flowers (Schmulling et al., 1988).

Interesting ways in which the auxin-modulated transgenic plants are being utilized to answer questions concerning auxin's role in plant growth and development are in crosses with different auxin mutants. For example, the effects of auxin genes on apical dominance and leaf morphology could not be unequivocably attributed to the expression of these genes since auxin synthesis has been shown to be associated with a rapid increase in ethylene levels. Ethylene is known to cause morphological changes in plants making it difficult to dissociate auxin from ethylene effects (Reid, 1987, Estelle and Klee, 1994). But, by crossing transgenic *iaaM* plants to ethylene insensitive mutants, *ein1* and *ein2*, or to another transgenic line expressing an 1-aminocyclopropane-l-carboxylic acid (ACC) deaminase gene that degrades ACC, the immediate precursor of ethylene, Romano et al. (1993) demonstrated that the observed changes in development were due to auxin alterations and not ethylene. Similar work with an auxin resistant mutant, *axr1*, indicated that *axr1* is an auxin-response mutation and does not reflect a change in auxin uptake, transport or metabolism (Estelle and Klee, 1994). In yet another mutant that was believed to be deficient in auxin transport, Okada et al. (1991) determined that, most likely, the phenotype is not simply the consequence of too much or not enough auxin.

Future perspectives for auxin transgenes lie in applications such as the ones described above for deciphering the mechanism of auxin action in *planta*. In addition, reduction of free IAA levels *in vivo* may also provide useful information. The effects of reducing the pool of active auxin have been studied in transgenic plants that overexpress a *P. savastanoi* gene coding for an enzyme that converts IAA to an inactive, conjugated form, IAA-lysine (Table 2; Romano et al., 1991; Spena et al., 1991). Distinct phenotypic alterations in transgenic tobacco and potato plants included severely wrinkled leaves with shorter midribs, reduced root systems, fewer and larger xylem elements and a lower degree of lignification. In potato, unlike results in tobacco, an increase in internode length was observed characteristic of a shift towards lower auxin content. Results from crosses with transgenic plants overexpressing the *iaaM* gene suggest that the *iaaL* gene does bring about its effects on development primarily by reducing the IAA levels *in vivo* - suggesting that endogenous auxin-conjugated activities are of physiological importance (Romano et al., 1991).

2.4 Bacterial cytokinin biosynthesis genes in plants

The involvement of cytokinins in a variety of plant developmental processes, such as control of senescence, promotion of cell proliferation, chloroplast development, flowering and plant defense responses, have been reported (Davies, 1987; Smigocki et al., 1993; Mok and Mok, 1994). Most studies have utilized exogenous applications to gain a better understanding of the events that lead to the induction of physiological responses by cytokinin. Several laboratories, however, have exploited the use of the *ipt* gene from *A. tumefaciens* to manipulate endogenous cytokinin levels in transformed plant cells. The *ipt* gene codes for the enzyme isopentenyl transferase that catalyzes the condensation of adenosine 5'-monophosphate and isopentenyl pyrophosphate to form isopentenyl adenosine 5'-monophosphate (IPA), a precursor of most other cytokinins (Table 2; Akiyoshi et al., 1983). Plants are also believed to synthesize cytokinins from IPA because enzymatic activity similar to that encoded by the bacterial *ipt* gene has been demonstrated in plants, but, to date, the plant enzyme has not been purified to homogeneity (Chen and Melitz, 1979; Blackwell and Horgan, 1994). The apparent lack of homology between the *ipt* gene and plant genomic DNA has further hampered the cloning of the corresponding plant gene. Therefore, the *Agrobacterium ipt* gene has been principally used to manipulate cytokinin concentrations *in planta* in order to learn more about cytokinin's role in regulation of plant growth and development.

Molecular manipulations of the *ipt* gene for expression in transgenic plants have revealed interesting phenomena (Table 4). In most cases, the introduction of the *ipt* gene under control of its own promoter or the CaMV35S promoter has been found to promote shoot development and inhibit root formation (Smigocki and Owens, 1988; Schmulling et al., 1989). Rooted plants were regenerated only from some of the *ipt* transformed tissues when certain medium conditions were used (Ooms et al., 1991; Yusibov et al., 1991). In *Nicotiana* spp., constitutive overexpression of the *ipt* gene with the CaMV35S promoter was correlated with more frequent, rapid and profuse shoot development than that observed with the unmodified *ipt* gene (Smigocki and Owens, 1988). In *Cucumis sativa,* this chimeric gene uniquely induced shoots, but in similar studies with petunia and peach, no shoots were regenerated from transformed cells (Klee et al., 1987a; Smigocki and Owens, 1988; Smigocki and Hammerschlag, 1991). The transgenic cucumber shoots could be rooted but they were stunted and did not survive transfer to soil.

Transformed tissues of *Nicotiana* spp. and cucumber were analyzed for the cytokinins zeatin and zeatinriboside known to be overproduced by *ipt* gene expression. Cytokinin levels were elevated an average of 300-fold in the CaMV35S-*ipt* transformed shoots and tissues as compared to normal plants (Smigocki and Owens, 1988, 1989). In contrast, only a sevenfold increase over normal levels was observed in tissues transformed with the native *ipt* gene. Since it is the ratio of cytokinin-to-auxin that is believed to regulate morphogenesis, free and total IAA levels were also determined for these tissues. Unexpectedly, the CaMV35S-*ipt* gene construct appeared to negatively effect IAA levels (Smigocki and Owens, 1989). In general, free IAA levels were reduced by about two-thirds and total IAA by about one-third in two of the three species of *Nicotiana* and cucumber. Taken together, the cytokinin and auxin data revealed that cytokinin-to-auxin ratios were elevated an average of some 700-fold in tissues transformed with the CaMV35S-*ipt* gene and sevenfold in those transformed with the native gene, as calculated on the basis of free IAA. Using total IAA, the respective increases were about 400- and 12-fold.

Generally, these high ratios were associated with shoot organogenesis of transformed cells, however, in a few cases, the regenerated shoots began to proliferate mainly as undifferentiated tissue. These calli differentiated dark green areas that were similar in appearance to shoot buds but failed to develop further. Interestingly, both the cytokinin levels and the cytokinin-to auxin ratios in these transformed calli were the highest observed within a particular group of transformed tissues. From these observations, we speculated that plant cells are induced to differentiate shoots by elevated endogenous hormone ratios; but that still higher ratios or, perhaps, absolute levels of cytokinin become detrimental either to initiation of differentiation or to continued development of the shoot bud.

Another subtle form of differentation was noted in *ipt* transformants. All of the *ipt*-transformed *Nicotiana* spp. and cucumber tissues grew on media devoid of auxin, a trait not exhibited by non-transformed tissues. Since the IAA levels were actually reduced in these tissues, it was suggestive that cytokinin-producing cells provided a signal for auxin-autonomous growth. Perhaps induction of expression of the native genes for endogenous auxin-receptor or signal molecules would sensitize the tissues to low endogenous levels of IAA which in non-transformed tissue would be inadequate for growth.

It became obvious from the above work that the cytokinin gene under control of constitutive promoters was of limited practical value from the standpoint of regenerating whole plants. Recently, this has been overcome by spatial and temporal regulation of cytokinin gene expression in transgenic plants. Transient overproduction of cytokinins using gene promoters regulated by environmental or developmental factors did not prevent root growth on *ipt* transformed shoots. In a number-of reports, rooted plantlets were obtained from shoots transformed by the *ipt* gene under control of promoters from various bacterial and plant genes (Table 4). Regenerants transformed with the *ipt* gene fused to the heat-, auxin- and light-inducible promoters all exhibited characteristics associated with higher endogenous cytokinin concentration (Medford et al., 1989; Schmulling et al., 1989; Smart et al., 1991; Smigocki, 1991; Li et al., 1992; Ainley et al., 1993; Thomas et al., 1995). They were generally greener, shorter, had a less developed root system, reduced leaf width, increased growth of axillary buds, and in some cases, abnormal flower development. This was expected due to the documented low levels of transcription from these promoters in the uninduced state during plant growth and development. In addition, plants transformed with the *ipt* gene under the heat-shock promoter had a

Table 4. Phenotypes of plants transformed with the reconstructed cytokinin biosynthesis gene *ipt.*

Cytokinin-associated gene	Phenotype
ipt (constitutive)[1]	-shoot formation; inhibition of root formation; teratomas; normal phenotype on certain media (Smigocki and Owens; 1988; Schmulling et al., 1989; Ooms et al., 1991; Yusibov et al., 1991)
CaMV35S-*ipt* (constitutive)	-shoot organogenesis; inhibition of root formation; auxin autonomous growth (Smigocki and Owens, 1988)
hs-*ipt* (heat inducible)	-slightly modified growth; shorter plants; underdeveloped root system; release of axillary buds; generally higher chlorophyll content; delayed senescence; reduced leaf width (Medford et al, 1989; Schmulling et al., 1989; Smart et al., 1991; Smigocki et al., 1991; Ainley et al., 1993)
SAUR-*ipt* (auxin inducible)	-loss of apical dominance; reduction in root growth; adventitious shoot formation; delayed senescence (Li et al., 1992)
PI-II-*ipt* (wound inducible)	-early bolting; pronounced apical dominance; underdeveloped root system; taller plants; larger, thinner leaves; reduced chlorophyll content early in development; delayed senescence associated with root expression of the cytokinin gene; enhanced insect resistance (Smigocki et al., 1993; Smigocki, 1995)
2A11-*ipt* (fruit specific)	-red fruit mottled with green islands (Martineau et al., 1994)
rbc-*ipt* (light inducible)	- teratomas; reduced apical dominance and internodal distance; reduced root development; wilting and browning of leaves and reduction in chlorophyll content in high light (Thomas et al., 1995)

[1]Indicates how the promoter is regulated in plant cells.

less developed vascular system, with considerable reduction in the xylem (Medford et al., 1989).

Conversely, specific expression of the *ipt* gene construct in tomato fruit yielded phenotypically normal plants, except for some fruit which were mottled with green islands on otherwise deep red, ripe fruit (Table 4; Martineau et al., 1994). The green islands were areas of cytokinin gene expression and corresponded to greatly elevated cytokinin levels.

Plants transformed with the wound-inducible *ipt* gene construct were characterized by early bolting, pronounced apical dominance, increased height and leaf size, thinner leaves, reduced chlorophyll content and an underdeveloped root system (Table 4; Smigocki et al, 1993; Smigocki, 1995). Except for the underdeveloped root system and reduced chlorophyll content, these phenotypic characteristics have not previously been associated with the expression of the ipt gene or increased endogenous cytokinin levels (Li et al., 1992; Ainley et al., 1993).

In all regenerated plants, substantial increases in cytokinin concentration were observed. Zeatin, zeatinriboside and zeatinriboside 5'-monophosphate levels were elevated from several fold to over 200-fold above uninduced levels. Implications from these data are that the expression of the *ipt* gene increased the levels of IPA, a rate limiting intermediate, and that IPA appears to be efficiently converted to the more active cytokinins in *ipt* transgenic plants.

The variability in the observed phenotypes of *ipt*-transformed plants stresses the importance that organ-, tissue- and cell-specific, as well as temporal, cytokinin production have on plant growth and development. It is also quite clear from these studies that regulated expression of cytokinin biosynthesis genes will not prevent whole plant regeneration and indicates that this approach can potentially be used to manipulate commercially important plant physiological processes.

One of the postulated roles for cytokinin is in control of the physiological process of senescence (Davies, 1987). It is well established that leaf cytokinin levels decline in senescing leaves and that external applications of cytokinins often delay senescence (Nooden and Letham, 1993). Studies on slow-senescing mutants of several crops, including sorghum and tomato, have demonstrated the importance of delayed senescence in increasing plant yield (Ambler et al., 1987). Directed expression of *ipt* transgenes has been shown to at least partially delay senescence in plants. In many of the studies with the *ipt* gene, the senescence process was not well controlled since expression was not targeted to any particular tissue or organ and the utilized promoters exhibited a leaky nature. Delayed onset of senescence in individually heat treated leaves of plants transformed with the *ipt* gene fused to a soybean heat shock promoter HS6871 was observed by Smart et al. (1991). This was a localized response that subsequently, however, induced earlier death of adjacent, non-heat shocked leaves.

Gan and Amasino (1995), using a promoter from a senescence-associated gene from *Arabidopsis,* were able to develop a system in which cytokinin production was specifically targeted to senescing leaves. Sufficient cytokinin was produced in the leaves of transgenic tobacco plants to retard senescence and, at the same time, to suppress the senescence-specific promoter so as to prevent hormone overproduction. Plants showed increased yield of both biomass and seed which correlated with an enhanced post-harvest longevity of the leaves.

A general view has emerged that it is the root-derived cytokinins that are responsible for the control of senescence in nontransformed plants (Nooden and Letham, 1993). Therefore, control of *ipt* gene expression by a developmentally regulated promoter from a gene that is, for example, activated in the roots upon cessation of growth would potentially extend the life of the whole plant. This has been achieved using the potato proteinase inhibitor gene promoter (Smigocki et al., 1993; Smigocki, 1995). The expression of a potato proteinase inhibitor gene is known to be wound-inducible in leaves and constitutively expressed in tubers and flowers in later stages of plant growth (Lorbeth et al., 1992; Ryan, 1992). Flowering *N. plumbaginifolia* plants transformed with the *ipt* gene fused to this wound-inducible promoter showed delayed whole-plant senescence (Smigocki et al., 1993; Smigocki, 1995).

Based on mRNA analyses of unwounded plants, the appearance of cytokinin effects seemed to be correlated to the root and, to a lesser extent, stem expression of the *ipt* gene. When the *ipt* gene transcripts were detected early in development as in rosette plants of *N. tabacum,* a number of characteristic effects associated with elevated cytokinin levels were observed throughout development. However, when the expression was delayed until after the plants bolted as in *N. plumbaginifolia,* only few cytokinin effects such as leaf greening and release of shoots from the basal stem were induced when plants matured. These results suggest that alteration of root-derived cytokinin production can be used to manipulate physiological processes such as the onset of senescence in some economically important crops.

In addition to morphological effects, cytokinins have been shown to influence the production of a number of secondary metabolites, some of which are known to have insecticidal properties (Hino et al., 1982; Kayser and Gemmrich, 1984; Nakagawa et al, 1984; Mothes et al., 1985; Lees, 1986; Ozeki and Komamine, 1986; Merillon et al., 1991; Decendit et al., 1992). Since current interests are focusing on combined effects of natural defense mechanisms of plants and biotechnology for crop improvement, we (Smigocki et al., 1993; Smigocki, unpublished) evaluated cytokinin's role in bringing about physiological changes that may protect plants from pests and pathogens. Plants transformed with the wound-inducible *ipt* gene were exposed to the herbivorous pest, tomato hornworm *(Manduca sexta)* and a virus transmitting pest, the green peach aphid *(Myzus persicae).* All transformants exhibited enhanced insect resistance which was correlated with the appearance of cytokinin effects. On the average, hornworm larvae fed on transgenic leaf material ate 60% less and weighed 40% less than those feeding on control plants. Only 30 to 40% of newly hatched green peach aphids survived to adulthood and 75% of them reproduced as compared to 75% survival and almost 100% rate of reproduction of the controls. A combination of wound induction of the cytokinin gene and normal developmental regulation appeared to raise cytokinin levels to that needed for increased resistance. The insecticidal activity has been localized to the surface extracts of transgenic plants, and efforts are underway to identify the active compound (Smigocki et al., unpublished).

3 CONCLUDING REMARKS

In the last few years, the manipulation of plant growth regulator levels in numerous transgenic plants has provided new insights into the physiological roles of auxin and cytokinin during plant growth and development. However, several major limitations still impede our understanding of auxin and cytokinin biology. It is quite clear that in order to obtain further information about how auxin and cytokinin modulate specific aspects of plant growth and development, a more precise regulation of expression of bacterial auxin and cytokinin genes in transgenic plants will be necessary. This will only be possible if well defined cell- and tissue-specific gene promoters become available.

In addition, the present lack of cloned plant genes involved in auxin and cytokinin biosynthesis further hampers the progress. Mutational approaches have so far failed to provide auxin- and cytokinin-deficient mutant plants. This has led to speculation that the genes required for auxin and cytokinin action are also essential for plant viability. Therefore, a need to develop novel genetic approaches is imperative for the future cloning and genetic manipulation of the endogenous plant genes involved in auxin and cytokinin biosynthesis. A combination of new biochemical, genetic and molecular approaches will no doubt further expand our current knowledge of the biological role and mechanism of action of auxin and cytokinin.

4 REFERENCES

Ainley WM, NcNeil KJ, Hill JW, Lingle WL, Simpson RB, Brenner ML, Nagao RT & Key JL (1993) Regulatable endogenous production of cytokinins up to 'toxic' levels in transgenic plants and plant tissues. Plant Mol. Biol. 22: 13-23.

Akiyoshi DE, Morris RO, Hinz R, Mischke BS, Kosuge T, Garfinkel D, Gordon MP & Nester EW (1983) Cytokinin/auxin balance in crown gall tumors is regulated by specific loci in the T-DNA. Proc. Natl. Acad. Sci. USA 80: 407-411.

Akiyoshi DE, Regier DA & Gordon MP (1989) Nucleotide sequence of the *tzs* gene from *Pseudomonas solanacearum* strain K60. Nucl. Acids Res. 17: 8886.

Ambler JR, Morgan PW & Jordan WR (1987) Genetic regulation of scienescence in a tropical grass. In: Thomson WW et al. (eds) Plant senescence: its biochemistry and physiology. (pp. 43-53) The American Society of Plant physiologists, Rockville, MD.

Baldi BG, Maher BR, Slovin JP & Cohen JD (1991) Stable isotope labelling, *in vivo*, of D- and L-tryptophan pools in *Lemna gibba* and the low incorporation of label into indole-3-acetic acid. Plant Physiol. 95: 1203-1208.

Binns AN (1994) Cytokinin accumulation and action: biochemical, genetic, and molecular approaches. Annu. Rev. Plant Physiol. Plant Mol. Biol. 45: 173-196.
Blackwell JR & Horgan (1994) Cytokinin biosynthesis by extracts of *Zea mays*. Phytochemistry 35:339-342.
Camilleri C & Jouanin L (1991) The TR-DNA region carrying the auxin synthesis genes of *A. rhizogenes* agropine type plasmid pRiA4: nucleotide sequence analysis and introduction into tobacco plants. Mol. Plant Microbe Inter. 4: 144-162.
Capone I, Cardarelli M, Trovato M & Costantino P (1989) Upstream non-coding region which confers polar expression to Ri plasmid root inducing gene *rolB*. Mol. Gen. Genet. 216: 239-244.
Chaudhury AM, Letham S, Craig S, Dennis ES (1993) AMPl-a mutant with high cytokinin levels and altered embryonic pattern, faster vegetative growth, constitutive photomorphogenesis and precocious flowering. Plant J. 4: 907-916.
Chen C-M & Melitz DK (1979) Cytokinin biosynthesis in a cell-free system from cytokinin-autotrophic tobacco tissue cultures. FEBS Lett. 107: 15-20.
Chory J, Reinecke D, Sim S, Washburn T & Brenner M (1994) A role for cytokinins in de-etiolation in *Arabidopsis*. Plant Physiol. 104: 339-347.
Christou P & Yang N-S (1989) Developmental aspects of soybean *(Glycine max)* somatic embryogenesis. Ann. Bot. 64: 225-234.
Clark E, Manulis , Ophir Y, Barash I & Gafni Y (1993) Cloning and characterization of *iaaM* and *iaaH* from *Erwinia herbicola* pathovar *gypsophila*. Phytopathology 83: 234-240.
Comai L & Kosuge T (1980) Involvement of plasmid deoxyribonucleic acid in indoleacetic acid synthesis in *Pseudomonas savastanoi*. J. Bacteriol. 143: 950-957.
Costacurta A & Vanderleyden J (1995) Synthesis of phytohormones by plant-associated bacteria. Cri. Rev. Microbiol. 21: 1-18.
Crespi M, Messers E, Capala AB, Van Montagu M & Desomer J (1992) Fasciation induction by the phytopathogen Rhodococcus fascians depends upon a linear plasmid encoding a cytokinin synthase gene. EMBO J. 11: 795-804.
Davies PJ (1987) Plant Hormones and Their Role in Plant Growth and Development. Kluwer, Boston.
De Block M (1988) Genotype-independent leaf.disc transformation of potato *(Solanum tuberosum)* using *Agrobacterium tumefaciens*. Theor. Appl. Genet. 76: 767-774.
Decendit A, Liu D, Ouelhazi L, Doireau P, Merillon J-M & Rideau M (1992) Cytokinin-enhanced accumulation of indole alkaloids in *Catharanthus roseus* cell cultures - the factors affecting the cytokinin response. Plant Cell Rep. 11: 400-403.
Elzen GW (1983) Cytokinins and insect galls. Comp. Biochem. Physiol. 76A: 17-19.
Estelle M & Klee HJ (1994) Auxin and cytokinin in *Arabidopsis*. Cold Spring Harbor Monogr. Ser. 27: 555-578.
Estruch JJ, Chriqui D, Grossmann K, Schell J & Spena A (1991b) The plant oncogene *rolC* is responsible for the release of cytokinins from glucoside conjugates. EMBO J. 10: 2889-2895.
Estruch JJ, Schell J & Spena A (1991a) The protein encoded by the *rolB* plant oncogene hydrolyzes indole glucosides. EMBO J. 10: 3125-3128.
Gan S & Amasino RM (1995) Inhibition of leaf senescence by autoregulated production of cytokinin. Science 270:1986-1988.
Garfinkel DJ & Nester EW (1980) *Agrobacterium tumefaciens* mutants affected in crown gall tumorigenesis and octopine catabolism. J. Bacteriol. 144: 732-743.
Garfinkel DJ, Simpson RB, Ream LW, White FF, Gordon MP & Nester EW (1981) Genetic analysis of crown gall: fine structure map of the T-DNA by site-directed mutagenesis. Cell 27: 143-153.

Gaudin V, Vrain T & Jouanin L (1994) Bacterial genes modifying hormonal balances in plants. Plant Physiol. Biochem. 32: 11-29.

Gielen J, De Beuckeleer M, Seurinck J, Deboeck F, De Greve H, Lemmers M, Van Montagu M & Schell J (1984) The complete nucletide sequence of the TL-DNA of the *Agrobacterium tumefaciens* plasmid pTiAch5. EMBO J. 3: 835-846.

Hamill JD (1993) Alterations in auxin and cytokinin metabolism of higher plants due to expression of specific genes from pathogeneic bacteria: a review. Aust. J. Plant Physiol. 20: 405-423.

Hayashi H, Czaja T, Lubenow H, Schell J & Walden R (1992) Activation of a plant gene by T-DNA tagging: auxin-independent growth *in vitro*. Science 258: 1350-1353.

Hino F, Okazaki M & Miura Y (1982) Effects of kinetin on formation of scopoletin and scopolin in tobacco tissue cultures. Agric. Biol. Chem. 46: 2195-2202.

Hobbie L & Estelle M (1994) Genetic approaches to auxin action. Plant Cell and Environ. 17: 525-540.

Hobbie L, Timpte C & Estelle M (1994) Molecular genetics of auxin and cytokinin. Plant Mol. Biol. 26: 1499-1519.

Hooykaas PJJ, den bulk-Ras H & Schilperoort RA (1988) The *Agrobacterium tumefaciens* T-DNA gene 6b is an *onc* gene. Plant Mol. Biol. 11: 791-794.

Kayser H & Gemmrich AR.(1984) Hormone induced changes in carotenoid composition in *Ricinus* cell cultures. I. Identification of rhodoxanthin. Z. Naturforsch. 39: 50-54.

Klee H & Estelle M (1991) Molecular genetic approaches to plant hormone biology. Annu. Rev. Plant Physiol. 42: 529-551.

Klee HJ, Horsch R & Rogers S (1987a) *Agrobacterium* mediated plant transformation and its further applications to plant biology. Annu. Rev. Plant Physiol. 38: 467-486.

Klee HJ, Horsch RB, Hinchee MA, Hein MB & Hoffmann NL (1987b) The effects of overproduction of two *Agrobacterium tumefaciens* T-DNA auxin biosynthctic gcnc products in transgenic petunia plants. Genes Dev. 1: 86-96.

Leemans J, Deblaere R, Willmitzer L, De Greeve H & Hernalsteens JP (1982) Genetic identification of functions of TL-DNA transcripts in octopine crown galls. EMBO J. 1: 147-152.

Lees GL (1986) Condensed tannins in the tissue culture of sainfoin *(Onobrychis viciifolia* Scop.) and birdsfoot trefoil *(Lotus corniculatus* L.). Plant Cell Rep. 5: 247-251.

Leyser HMO, Lincoln CA, Timpte C, Lammer D, Turner J & Estelle M (1993) *Arabidopsis* auxin-resistance gene AXR1 encodes a protein related to ubiquitin-activating enzyme El. Nature 364: 161-164.

Li Y, Hagen G & Guilfoyle TJ (1992) Altered morphology in transgenic tobacco plants that overproduce cytokinins in specific tissues and organs. Dev. Biol. 153: 386-395.

Lichter A, Manulis S, Sagee 0, Gafni Y, Gray J, Meilan R, Morris RO & Barash I (1995) Production of Cytokinins by *Erwinia herbicola* pv. *gypsophilae* and isolation of a locus conferring cytokinin biosynthesis. MPMI 8: 114-121.

Lorberth R, Dammann C, Ebneth M,. Amati S & Sanchez-Serrano J (1992) Promoter elements involved in environmental and developmental control of potato proteinase inhibitor II expression. Plant J. 2: 477-486.

Manulis S, Gafni Y, Clark E, Zutra D, Ophir Y & Barash I (1991a) Identification of a plasmid DNA probe for detection of *Erwinia* herbicola pathogenic on *Gypsophila paniculata.* Phytopathology 81: 54-57.

Manulis S, Valinski L, Gafni Y & Hershenhorn J (1991b) Indole-3-acetic acid biosynthesis in *Erwinia herbicola* in relation to pathogenicity on *Gypsophila paniculata.* Physiol. Mol. Plant Pathol. 39: 161-171.

Martineau B, Houck CM, Sheehy RE & Hiatt WR (1994) Fruit-specific expression of the *A. tumefaciens* isopentenyl transferase gene in tomato: effects on fruit ripening and defense-related expression in leaves. Plant J. 5: 11-19.

McInnes E, Morgan AJ, Mulligan BJ & Davey MR (1991) Phenotypic effects of isolated pRiA$_4$ TL-DNA *rol* genes in the presence of intact TR-DNA in transgenic plants of *Solanum dulcamara L.* J. Exp. Bot. 42: 1279-1286.

Medford J & Klee H (1989) Manipulation of endogenous auxin and cytokinin levels in transgenic plants. In: Goldberg RB (ed.) The Molecular Basis of Plant Development, UCLA Symp. on Mol. and Cell. Biol., New Ser. 92. (pp. 211-220) Alan Liss, New York.

Medford JI, Horgan R, El-Sawi Z & Klee HJ (1989) Alterations of endogenous cytokinins in transgenic plants using a chimeric isopentenyl transferase gene. Plant Cell 1: 403-413.

Merillon J-M, Liu D, Huguet F, Chenieux J-C & Rideau M (1991) Effects of calcium entry blockers and calmodulin inhibitors on cytokinin-enhanced alkaloid accumulation in *Catharanthus roseus* cell cultures. Plant Physiol. Biochem. 29: 289-296.

Messens E, Lenaerts A, Van Montagu M & Hedges RW (1985) Genetic basis for opine secretion from crown gall tumour cells. Mol. Gen. Genet. 199: 344-348.

Mok DW & Mok MC (1994) Cytokinins: Chemistry, Activity, and Function. CRC Press, Boca Raton.

Morris RO (1986) Genes specifying auxin and cytokinin biosynthesis in phytopathogens. Annu. Rev. Plant Physiol. 37: 509-538.

Mothes K, Schutte HR & Luckner M (1985) Biochemistry of Alkaloids. VEB Beutscher Verlag der Wissenschaften, Berlin.

Nakagawa K, Konagai A, Fukui H & Tabata M (1984) Release and crystallization of berberine in the liquid medium of *Thalictrum minus* cell suspension cultures. Plant Cell Rep. 3: 254-257.

Nilsson O, Crozier A, Schmulling T, Sandberg G & Olsson O (1993) Indole-3-acetic acid homeostasis in transgenic tobacco plants expressing the *Agrobacterium rhizogenes rolB* gene. Plant J. 3:681-689.

Nooden LD & Letham DS (1993) Cytokinin metabolism and signalling in the soybean plant. Aust. J. Plant Physiol. 20:639-653.

Norstog K (1979) Embryo culture as a tool in the study of comparative and developmental morphology. In: Sharp WR et al. (eds) Plant CE11 and Tissue Culture (pp. 179-202) Ohio State Univ. Press, Columbus.

Offringa IA, Melchers LS, Regensburg-Tuink AJG, Costantino P, Schilperoort RA & Hooykaas PJJ (1986) Complementation of *Agrobacterium tumefaciens* tumor-inducing *aux* mutants by genes from the TR-region of the Ri plasmid of *Agrobacterium rhizogenes.* Proc. Natl. Acad. Sci. USA 83: 6935-6939.

Okada K, Ueda J, Komaki MK, Bell CJ & Shimura Y (1991) Requirements of the auxin polar transport system in early stages of *Arabidopsis* floral bud formation. Plant Cell 3: 677-684.

Ooms G, Risiott R, Kendall A, Keys A, Lawlor D, Smith S, Turner J & Young A (1991) Phenotypic changes in T-*cyt*-transformed potato plants are consistent with enhanced sensitivity of specific cell types to normal regulation by root-derived cytokinin. Plant Mol. Biol 17: 727-743.

Oono Y, Kanaya K & Uchimiya H (1990) Early flowering in transgenic tobacco plants possessing the *rolC* gene of *Agrobacterium rhizogenes* Ri plasmid. Jap. J. Genet. 65: 7-16.

Ozeki Y & Komamine A (1986) Effects of growth regulators on the induction of anthocyanin synthesis in carrot suspension cultures. Plant Cell Physiol. 27: 1361-1368.

Parry AD, Blonstein AD, Babiano MJ, King PJ & Horgan R (1991) Abscisic acid metabolism in a wilty mutant of *Nicotiana plumbaginifolia.* Planta 183: 237-243.

Reid MS (1987) Ethylene in plant growth, development and senescence. In: Davies PJ (ed.) Plant hormones and their role in plant growth and development. (pp. 257-279) Martinus Nijhoff, Dordrecht.

Romano CP, Cooper ML & Klee HJ (1993) Uncoupling auxin and ethylene effects in transgenic tobacco and *Arabidopsis* plants. Plant Cell 5: 181-189.

Romano CP, Hein MB & Klee HJ (1991) Inactivation of auxin in tobacco transformed with the indoleacetic acid-lysine synthetase gene of *Pseudomonas savastanoi.* Genes Dev. 5: 438-446.

Ryan CA (1992) The search for the proteinase inhibitor-inducing factor, PIIF. Plant Mol. Biol. 19: 123-133.

Schmulling T, Beinsberger S, DeGreet J, Schell J, Van Onckelen H & Spena A (1989) Construction of a heat inducible chimeric gene to increase cytokinin content in transgenic plant tissue, FEBS Lett. 249: 401-406.

Schmulling T, Schell J & Spena A (1988) Single genes from *Agrobacterium rhizogenes* influence plant development. EMBO J. 7: 2621-2629.

Schroder G, Waffenschmidt S, Weiler EW & Schroder J (1984) The Tregion of Ti plasmids codes for an enzyme synthesizing indole-3-acetic acid. Eur. J. Biochem. 138: 387-391.

Sitbon F, Hennion S, Sundberg B, Anthony Little CH, Olsson H & Sandberg G (1992) Transgenic tobacco plants coexpressing the *Agrobacterium tumefaciens iaaM* and iaaH genes display altered growth and indoleacetic acid metabolism. Plant Physiol. 99: 1062-1069.

Skoog F & Miller CO (1957) Chemical regulation of growth and organ formation in plant tissues cultured *in vitro.* Symp. Soc. Exp. Biol. 11: 118-130.

Smart CM, Scofield SR, Bevan MW & Dyer TA (1991) Delayed leaf senescence in tobacco plants transformed with *tmr,* a gene for cytokinin production in *Agrobacterium.* Plant Cell 3: 647-656.

Smigocki AC (1991) Cytokinin content and tissue distribution in plants transformed by a reconstructed isopentenyl transferase gene. Plant Mol. Biol. 16: 106-115.

Smigocki AC (1995) Expression of a wound-inducible cytokinin biosynthesis gene in transgenic tobacco: correlation of root expression with induction of cytokinin effects. Plant Sci. 109: 153-163.

Smigocki AC & Hammerschlag FA (1991) Regeneration of plants from peach embryo cells infected with a shooty mutant strain of *Agrobacterium.* J. Am. Soc. Hortic. Sci. 116:1092-1097.

Smigocki AC & Owens LD (1988) Cytokinin gene fused with a strong promoter enhances shoot organogenesis and zeatin levels in transformed plant cells. Proc. Natl. Acad. Sci. USA 85: 5131-5135.

Smigocki AC & Owens LD (1989) Cytokinin-to-auxin ratios and morphology of shoots and tissues transformed by a chimeric isopentenyl transferase gene. Plant Phyiol. 91: 808-811.

Smigocki AC, Neal JW, McCanna I & Douglass L. (1993) Cytokinin-mediated insect resistance in *Nicotiana* plants transformed with the *ipt* gene. Plant Mol. Biol. 23: 325-335.

Spanier K, Schell J & Schreier PH (1989) A functional analysis of T-DNA gene *6b*: the fine tuning of cytokinin effects on shoot development. Mol. Gen. Genet. 219: 209-216.

Spena A, Prinsen E, Fladung M, Schulze SC & Van Onckelen H (1991) The indoleacetic acid-lysine synthetase gene of *Pseudomonas syringae* subsp. *savastanoi* induces developmental alterations in transgenic tobacco and potato plants. Mol. Gen. Genet. 227: 205-212.

Spena A, Schmulling T, Koncz C & Schell J (1987) Independent and synergistic activity of the *rolA, B* and *C* loci in stimulating abnormal growth in plants. EMBO J. 6: 3891-3899.

Stoddart JL (1987) Genetic and hormonal regulation of stature. In: Thomas H & Grierson D (eds) Developmental Mutants in Higher Plants Vol. 32. (pp. 155-180) Cambridge Univ. Press, Cambridge.

Su W & Howell S (1992) A single genetic locus, CKRK, defines Arabidopsis mutants in which root growth is resistant to low concentrations of cytokinin. Plant Physiol. 99: 1569-1574.

Tepfer D (1984) Transformation of several species of higher plants by *Agrobacterium rhizogenes:* sexual transmission of the transformed genotype and phenotype. Cell 37: 959-967.

Thomas JC, Smigocki AC & Bohnert HJ (1995) Light-induced expression of *ipt* from *Agrobacterium tumefaciens* results in cytokinin accumulation and osmotic stress symptoms in transgenic tobacco. Plant Mol. Biol. 27: 225-235.

Thomashow MF, Hughly S, Buchholz WG & Thomashow LS (1986) Molecular basis for the auxin-independent phenotype of crown gall tumor tissue. Science 231: 616-618.

Tinland B, Rohfritsch O, Michler P & Otten L (1990) *Agrobacterium tumefaciens* T-DNA gene 6b stimulates rol-induced root formation, permits growth at high auxin concentrations and increases root size. Mol. Gen. Genet. 223: 1-10.

van Staden J & Davey JE (1978) Endogenous cytokinins in the larvae and galls of *Erythina latissima* leaves. Bot. Gaz. 139: 36-41.

Vansuyt G, Vilaine F, Tepfer M & Rossignol M (1992) *rolA* modulates the sensitivity to auxin of the proton translocation catalyzed by the plasma membrane H+-ATPase in transformed tobacco. FEBS Lett. 298: 89-92.

Walden R, Fritze K, Hayashi H, Miklashevichs E, Harling H & Schell J (1994) Activation tagging: a means of isolating genes implicated as playing a role in plant growth and development. Plant Mol. Biol. 26: 1521-1528.

White FF, Taylor BH, Huffman GA, Gordon MP & Nester EW (1985) Molecular and genetic analysis of the transferred DNA regions of the root-inducing plasmid of *Agrobacterium rhizogenes.* J. Bacteriol. 164: 33-44.

Wright AD, Sampson MB, Neuffer MG, Michalczuk L, Slovin JP & Cohen JD (1991) Indole-3-acetic acid biosynthesis in the mutant maize *orange pericarp,* a tryptophan auxotroph. Science 254: 998-1000

Yusibov VM, Chun II P, Andrianov M & Piruzian ES (1991) Phenotypically normal transgenic T-*cyt* tobacco plants as a model for the investigation of plant gene expression in response to phytohormonal stress. Plant Mol. Biol. 17: 825-836.

PART II

APPLICATIONS

Chapter 12

IN VITRO INDUCED HAPLOIDS IN PLANT GENETICS AND BREEDING

Han HU & Xiangrong GUO
State Key Laboratory of Plant Cell and Chromosome Engineering(PCCEL) Institute of Genetics Chinese Academy of Sciences, Beijing, 100101,China

1 INTRODUCTION

Haploids have attracted great interest of plant physiologists, embryologists, geneticists and breeders since the first discovery of haploid plants in *Datura stramonium* as early as 1922 (Blakeslee et al, 1922). In the beginning, haploid plants were regarded as a special biological phenomenon. The occurrences of haploids were reported in 71 species of angiosperms belonging to 39 genera and 14 families up to early 1960's (Kimber & Riley, 1963). However, the low frequency of spontaneously arising haploid plants severely limited the utilization of haploids for crop improvement and genetic studies. In the last three decades many efficient and simple techniques especially *in vitro* culture have been developed to produce haploid plants in large numbers. For instance, using anther culture, haploids were induced in 247 species of angiosperms, including some hybrids, belong to 88 genera of 34 families (Maheshwari et al, 1983). Meanwhile, chromosome elimination (the bulbosum technique) and *in vitro* culture of unfertilized ovaries and ovules have been proved to be efficient means of haploid induction in some plant species. This progress paved the way for studying and utilizing the haploids in higher plants.

This paper mainly discusses this progress in terms of origin, genetics and application of haploids.

2 ORIGIN OF HAPLOIDS

There are two different generations in the life cycle of higher plants: namely the asexual generation producing spores and the sexual generation producing gametes. The sporophyte in the asexual generation is diploid with two sets of chromosome derived from both parents. Before the formation of the spores through meiosis, the zygotic (diploid) chromosome number is reduced to the gametic (haploid) number, the characteristic for the gamete or the haploid phase of the life cycle. This means, haploids of higher plants are sporophytes with gametic number of chromosomes.

Figure 1 shows the origin of haploid sporophyte of higher plants. In nature the haploids of higher plants are produced via abnormal fertilization and, therefore, it is rarely seen, and the frequency of occurrence of haploids is very low. Since the first report of haploid plants in the early 1920's, many efforts have been made to induce haploids of higher plants, which could be classified roughly into two categories: *in vivo* induction of haploids by various physical, chemical or biological stimulants, and *in vitro* culture which seemed to be more efficient than *in vivo* methods.

From the life cycle point of view, there are three pathways to induce haploid sporophyte of higher plants:

I. Haploid sporophyte originating from meiotic spore by means of pollen(anther) or/and, ovary and ovule culture;

II. Haploid sporophyte originating from zygote due to chromosome elimination;

III. Haploid sporophyte originating from male and female gametes, their precursor cells and gametophytes by the isolation and manipulation of reproductive cells and protoplasts (Yang & Zhou, 1992)

Because it is most difficult to obtain haploids by using the third pathway, this technique has not been applied to cereals so far. Although this paper will not cover it, these techniques of experimental plant reproductive biology show great potential in providing new means for biotechnology and may eventually permit direct reproductive cell engineering. Meanwhile it serves to deepen our knowledge of the control of the reproductive processes.

2.1 Zygotic chromosome elimination

The chromosome elimination technique originally developed for barley (Kasha & Kao, 1970) named bulbosum technique, was extended to wheat by Barcley in 1975. He reported high frequency of haploids through the chromosome elimination process. The developmental stage of immature embryos at the time of excision is critical with respect to plant regeneration; the embryos after 2 weeks of pollination show the highest capacity for

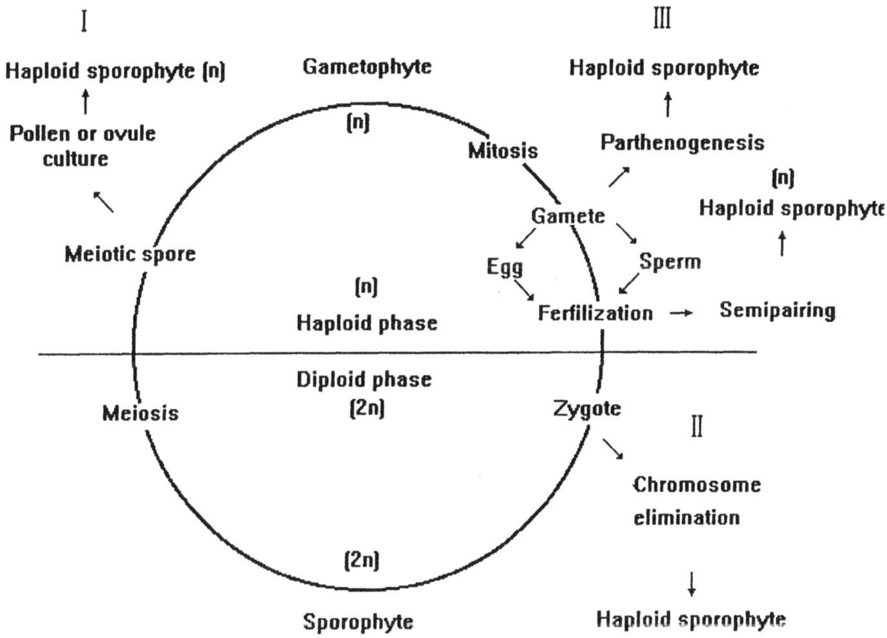

Figure 1. Three Pathways inducing haploid sporophytes of higher plants in the life cycle

development. The development of embryos also varies with the conditions, particularly temperature, and with the genotypes used.The crossability of wheat with tetraploid *H. bulbosum* is restricted to a few genotypes (Falk & Kasha, 1981) and is genetically controlled by the genes *Kr1* and *Kr2* on the chromosome 5B and 5A, respectively which are also responsible for determining the crossability with rye, *Secale cereale* (Falk and Kasha,1983; Sitch et al., 1985; Snape et al., 1979). These studies led to the proposal that this process is under genetic control. Laurie and Bennett (1988a, 1988b) carried out a few wide hybridizations with wheat and found that fertilization occurred between wheat and maize and/or sorghum, and between barley and maize. At the metaphase of zygotes there were 21 wheat chromosomes or 7 barley chromosomes and 10 maize chromosomes. This indicated that fertilization had been taken place, but the karyotype of the hybrid was unstable. In the initial cell cycle, all maize chromosomes are eliminated. Therefore, wheat haploid plants can be recovered via *in vitro* culture of these wheat haploid embryos. Among the 26 crosses between wheat and maize, 25 affected fertilization. The embryological observation was conducted and results are shown in Figure 2. Similar results of cytoembryological investigations were obtained by Wang et al. (1991).

The maize technique has been currently improved as an alternative method for wheat haploid production (Inagaki & Tahir, 1990; Suenage & Nakajima, 1989; Ushiyama et al, 1991). Kisana et al (1993) made a comparison between anther culture and wheat × maize method in a wheat breeding program and found that the wheat × maize method led to a higher frequency of haploids and no variation in progenies. Under field conditions it is difficult to make the cross between wheat and maize simply due to the difference in their flowering time. Recently, Li et al (1994) found that some wheat cultivars, e.g. Gansu No. 4, are highly crossable with *Tipsacum dictloides*, a wild species of maize with the same flowering time as wheat, and gave a frequency of 59% embryos. Therefore, the maize chromosome elimination could be considered to be useful for wheat improvement. In the crosses of wheat with maize and sorghum, 3% and 10% florets were fertilized and set seeds, respectively. It means that DNA fragments of maize or sorghum and maize transposon elements might be transferred into wheat cultivars (Laurie & Bennett, 1988a).

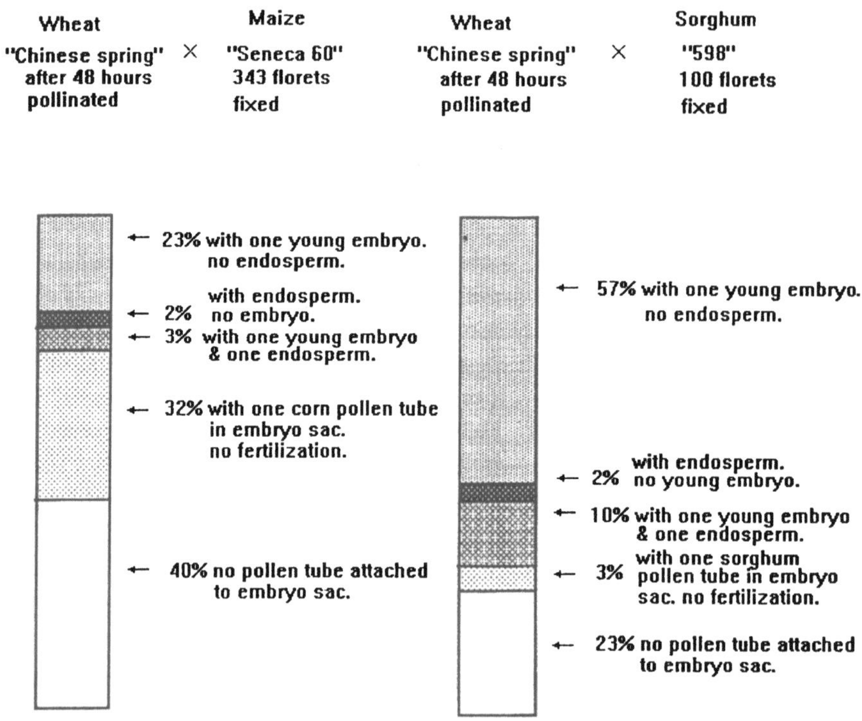

Figure 2. Chromosome elimination in wheat crossed with maize and sorghum

2.2 Haploids in vitro induced by meiotic spores

The pathways of the induction of haploids *in vitro* derived from meiotic spores may be classified into two categories: Anther (pollen) culture: androgenesis, and unpollinated ovary (ovule) culture: gynogenesis.

2.2.1 Androgenesis

In recent years, the frequency of microspore embryogenesis in cereals, mainly barley, wheat and rice has been significantly improved. Now the regeneration rates are measured at green pollen plants per anther compared to previously per 100 anthers. Table 1. demonstrates the maximum frequencies of green plants obtained via anther culture in barley and wheat.

There are two routes of regeneration: Organogenesis via a distinct callus phase, or direct embryogenesis (no callus phase). Since each regenerant is clearly derived from a single celled microspore, the embryogenetic route is preferred in crop improvement programmes using hybrid materials. The organogenetic route is generally longer and, thus, is more appreciated when mutant selection is applied to the *in vitro* tissue cultures and when prolonged culture period is required to maximize the amount of microspore variations induced in the regenerants.

Table 1. Maximum frequencies of green plants reported via anther culture in barley and wheat based on a few selected references

Reference	Yield of green plants per 100 anthers
Barley :	
Hunter (1988)	600 – 2700
Kao (1991)	1300
Olsen (1991)	940
Li W.Z. and Hu Han (1992)	334 – 2795
Wheat:	
Liu C. H. and Hu Han (1989)	117
Chu C. C. (1990)	360
Kasha (1990)	322
Orshinsky B.R. et al. (1990)	200 – 455

The frequency of the induction of green pollen plants has been steadily increased, and the following factors are considered to be important to influence culture response :

1. the effect of genotype,
2. the stage of microspore development,
3. culture media, and
4. culture conditions.

Table 2. Effect of genotypes

Genotypes	Anthers plated	Callus yield	Fre. regeneration (%)		Plant / anther		
			RFGP	RFAP	GPY	APY	G / A
Igri	180	10.57	33.33	16.67	3.52	1.67	2.0
Igri HO	420	12.92	30.64	3.23	3.96	0.42	9.4
Igri H1	180	22.18	30.30	18.18	6.72	4.03	1.7
Atias	60	8.57	7.14	7.14	0.61	0.61	1.0
No. 1 G. N.	120	14.14	4.00	4.00	0.57	0.57	1.0
Doublet	180	11.66	15.00	7.50	1.75	0.87	2.0
Harrington	300	2.12	23.53	5.88	0.50	0.12	4.2
Triumph	240	2.62	66.67	16.67	1.75	0.44	4.0
Goft	180	1.14	16.67	11.11	0.19	0.13	1.5
Sabarlis	300	8.41	8.33	4.17	0.70	0.35	2.0
Wen I	240	1.63	10.00	13.33	0.16	0.22	0.7
Ht / Gm F6	300	2.82	20.00	5.00	0.56	0.14	4.0
No. 3 Z. S/Igri F1	240	10.20	14.29	28.57	1.46	2.91	0.5
G.M.R.T./ Igri F1	60	12.53	11.54	7.69	1.45	0.96	1.5
Harrington/Igri F1	60	8.40	33.33	33.33	2.80	2.80	1.0
Igri/Harry F1	180	38.61	37.50	16.67	14.48	6.44	2.2
Igri/J.Q. F1	120	27.87	56.25	18.75	15.68	5.23	3.0
Average	-	11.55	24.62	12.82	3.34	1.64	-

Note:

1. callus yield: number of callus/per anther plated.
2. RFGP & RFAP—regeneration frequency of green (albino) plantlets: percentage of calli giving green (albino) plantlets.
3. GPY & APY – green (albino) plantlets yield: number of calli giving green (albino) plantlets/ per anther plated.
4. G/A: ratio of the number of calli giving green plantlets to those giving albinos.

2.2.1.1 The effect of genotype

The genotype of donor plants , i.e. genetic factor, has a great influence on the anther culture response. In earlier studies, significant differences in callus formation using varieties or crosses were observed.

There were significant difference between different genotypes of barley, wheat and rice in terms of callus yield per anther plated, regeneration frequency of green plantlets (RFGP) and green plantlet yield per anther plated (GPY). For example, according to the three indices mentioned above, of the 17 barley materials tested, the most responsive genotypes were Igri/Harry F_1, Triumph and Igri/J.Q. F_1, respectively, and the least responding genotypes were Goft and Wen I, respectively(Table 2.).

2.2.1.2 The stage of microspore development

Microspore developmental stage at the time of excision and inoculation is critical for the induction of pollen plants. He and Ouyang (1984) exactly divided the developmental stage into early uninucleate, mid-uninucleate, late-uninucleate and premitotic stage . These stages are characterized by size, shape and position of the nucleus, and by size and presence or absence of the vacuole. The response of anther culture with the change of microspore stage is very sensitive, and the callus yield sharply decreases if the anthers containing microspores at the stage other than mid- or late-uninucleate stage are used.

2.2.1.3 Improvement of media component

The culture media are the main factor inducing green plantlets from microspores. Table 3. showed the components of some basic media mainly used for anther culture in cereals. Modifications of carbon and nitrogen sources have significantly improved the response of anther culture.

(i) *Nitrogen.* In induction media, it has been found that reduced levels of ammonia nitrogen is beneficial. In barley it was decreased from 20 mM to 2 mM (Olsen, 1988). In the meantime, higher levels of nitrate nitrogen and addition of organic nitrogens in the form of amino acids, especially glutamine, were recommended in order to maintain the total nitrogen levels in the medium.

(ii) *Carbon.* Recent studies indicated that carbon sources played a critical role in microspore embryogenesis. In the past, sucrose was the most commenly used carbon source. However, Hunter (1988) revealed that maltose was most suitable for barley Chu et al (1990) obtained high frequency of pollen embryos and regeneration of plants in wheat on monosaccharide containing media. These results suggested that the sugars other than sucrose may be exploited further.

Table 3. Components of anther-pollen culture media of several cereals

Compoenets	Medium (mg/L)					
	MS	FHG	N6	C17	BAC	Potato*
KNO_3	1900	1900	2830	1400	2600	1000
NH_4NO_3	1650	165	...	300	...	...
$(NH_4)_2SO_4$	...	...	463	...	400	100
KH_2PO_4	170	170	460	400	170	200
$CaCl_2.2H_2O$	400	440	166	150	600	...
$MgSO_4.7H_2O$	370	370	185	150	300	125
$NaH_2PO_4.H_2O$	...	...	...	...	150	...
$FeSO_4.7H_2O$	27.8	...	27.8	27.8	...	27.8
$Na_2EDTA.2H_2O$	37.3	40	37.3	37.3	...	37.3
Sesquetrene 330Fe	...	...	...	...	40	...
KCl	...	...	...	...	...	35
$MnSO_4.4H_2O$	22.3	22.3	4.4	11.2	5.0	...
$ZnSO_4.7H_2O$	8.6	8.6	1.5	8.6	2.0	...
H_3BO_3	6.2	6.2	1.6	6.2	5.0	...
KI	0.83	0.83	0.8	0.83	0.8	...
$NaMoO_4.2H_2O$	0.25	0.25	...	...	0.25	...
$CuSO_4.5H_2O$	0.025	0.025	...	0.025	0.025	...
$CoCl_2.6H_2O$	0.025	0.025	...	0.025	0.025	...
Myo-inositol	100	100	...	...	2000	...
Thiamine-HCl	0.4	0.4	1.0	1.0	1.0	1.0
VB_6	0.5	...	0.5	0.5	0.5	...
Nicotinic acid	0.5	...	0.5	0.5	0.5	...
Glycine	2.0	...	2.0	2.0	...	...
Glutamine	...	730	...	...	...	0.5 to 1g l^{-1}
Casein hydrolysate	...	...	500	300	...	...
Sucrose	30,000	...	60,000	90,000	60,000	90,000
Glucose	...	...	...	...	17500	...
Maltose		62,000				
Ficoll-400	...	200,000	...	...	300,000	...
PH	5.7	5.6	5.8	5.8	6.2	5.8

*BAG3 medium also contains (mol/l) $KHCO_3$(50), VC (1.0) Citric acid (10), Pyrunk acid (10), $AgNO_3$(10)

** Potato II medium contains 10% aqueous potato extract

(iii) *2,4-D*. There are contradictory reports on the effects of 2,4-D levels for induction: high levels needed in some species, but low levels (or none) in others. Liu and Hu (1990) found the beneficial effects of high levels(4 mg/l) of 2,4-D in wheat anther culture.

It was known that autoclaving affected the medium composition by degrading some components (Ke-cheng & Bornman, 1991) and caused undesirable reactions with others (Schenk et al, 1991). Filter sterilized media have been shown to lead to higher culture response than autoclaved media (Chu & Hill, 1988). Thus, it is now the preferable procedure for medium preparation.

2.2.1.4 Improvement of culture methods

In his early research work, Kao (1981) observed that by adding Ficoll 400 to increase the density of the culture medium, the frequency of plant formation could be considerably increased. Recently Kao et al. (1991) further investigated and proved the culture conditions for the induction of green plants in anther culture and indicated that cells in callus under anaerobic conditions, e.g. existed in submerged cultures in liquid medium, contained lactate and alcoholdehydrogenase. The accumulation of lactic acid and other organic acids in the tissues may damage the organelles in the cell and result in the formation of albino plants. Thus, for direct embryogenesis from microspores, ample supply of a fresh, well-buffered medium, and proper aeration for microspore-derived embryos are essential for obtaining high frequencies of green plants. High Ficoll and sucrose levels in liquid medium keep the anthers floating and trap the microspores in the anthers for a sufficiently long time under a condition of sugar starvation. This starvation was essential to induce the microspores and eventually leads to androgenesis. (Wei et al., 1986).

Another remarkable improvement has been the multiple step culture procedure successfully developed in cereals. Different stages requires different conditions. This procedure is briefly described as follows:

(i) *Anther starvation*. Anthers dissected from spikes are put into 0.3 M mannitol solution without any media components and incubated for 3-4 days at 25 °C. The length of starvation is more important than osmotic pressure and mannitol solution free of carbon source is essential (Li et al., 1993).

(ii) *Induction culture*. After 3-day starvation, mannitol solution is replaced by liquid induction medium. Various improvements have been conducted in these media concerning nitrogen and carbon source. Lower NH_4NO_3 levels are compensated by raising glutamine concentration, and maltose is used instead of sucrose.

(iii) *Regeneration culture*. After 15-20 days of culture, calli derived from anthers are transferred onto solid regeneration medium containing proline which promotes the regeneration of green pollen plants in wheat, rice and barley. Temperature is another important factor in this step. When four different temperatures (10°C, 15°C, 20°C, 25°C) were tested, the frequency of differentiation and yield of green plantlets was highest at 20°C, followed, in the decreasing order, by 15°C , 10° 3 , °C (Hu et al., 1991).

(iv) *Robust plantlet culture and transplantation.* MET (pp333, Multi-Effect Triazole, a kind of growth retardant) played an important role in this step. MET at certain concentration was found to induce strong root system and hence guarantee the vigorous plantlet growth.

2.2.2 Isolated pollen/microspore culture

Anther culture, in essence, is that microspores are cultured within the anthers. But in isolated microspore culture, microspores are removed out of the anthers and then cultured in the media. Therefore, isolated microspore culture provides more ideal sytem with a large relatively uniform population of haploid single cells for use in mutant selection, genetic transformation and in studies of embryo development. There are two approaches to microspore isolation. One is to induce the immature anthers to dehisce to release the microspores into a medium or a suitable osmoticum. The other method is to mechanically grind or disrupt the anthers to remove the microspores, usually, either by homogenizing or microblending, and then collect and culture them. To date green plants have been regenerated from naturally shed and mechanically-isolated microspores in barley, rice, wheat and maize etc. However, in most of these studies the microspores were isolated from pretreated spikes and anthers or precultured anthers. Wei et al (1986) first reported plantlet regeneration from freshly harvested barley pollen, but the reproducibility and regeneration frequency of this study were not very satisfactory.

By now our laboratory has developed the standard method with cultivation of isolated microspores by microblending. By using this method, the green plants with high regeneration frequency were obtained from the directly-isolated microspores of barley genotypes showing different anther-culture responses (Fig. 3), and also from the microspores of rice and wheat only after cold pretreatment of spikelets for several days but without anther pre-culture. This system offers an attractive model to study the molecular mechanism of switch from the gametophytic pathway to the sporophytic pathway.

Various types of pretreatments (stress treatment), such as sugar starvation in barley (Wei et al., 1986) and tobacco (Imamura et al., 1982), high temperature in *brassica (*Chuong & Beversdorf., 1985*) etc.* can be used as

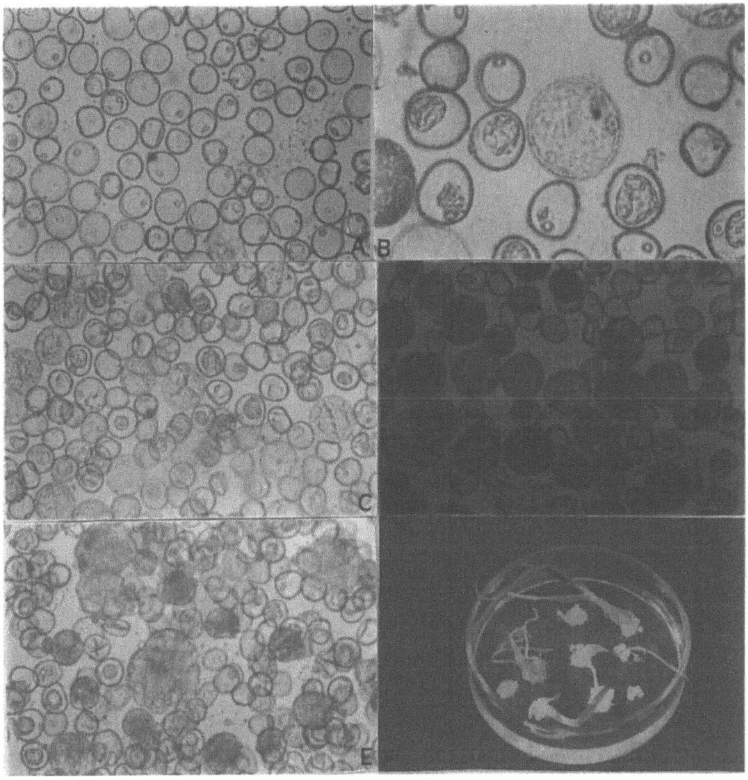

Figure 3. Plant regeneration through direct culture of mechanically isolated microspores of barely A. Freshly isolated microspores from the spikes. B, Embryogenic microspore. C, Dividingmicrospores after 4-5 days of isolation. D, Multicellular pollen grains. E, Cell masses. F, Green regeneration plants.

the external stimualor to trigger the embryogenic process (Pechan & Keller, 1989). It was believed that the initiation of dedifferentiation, or the induction of embryogenesis, takes place during pretreatment period (Huang & Sunderland, 1982, Pechan et al., 1991). In directly-isolated microspore culture of barley, using confocal laser scanning microscope (CLSM) and indirect immuno-fluenocence technique, We made the cytological observations on dedifferentiation process during pretreatment period and embryogenic process during the culture. We found that the dedifferentiation of isolated microspores of barley started within 12 hours of mannitol pretreatment. The main cytological characteristics of initiated microspores were: the cell volume was obviously enlarged, and the volume of nuclei and nucleoli were also greatly increased. Nucleoli were extremely clear and condensed. The ratio of nucleus to cytoplasm was also very high.

2.2.3 In vitro culture of unpollinated ovaries and ovules

The attempt to culture unpollinated and unfertilized ovules or ovaries was made in many materials by the scientists of P. Maheshwari's group, during 1950's and 1960's, without success, while an unexpected breakthrough was made in anther culture.

San Noeum (1976, 1979) was the first to succeeded in regenerating haploid plants of barley by means of ovary culture. Subsequently, haploid plants or plantlets were raised from cultured ovaries of wheat (Zhu and Wu, 1979), tobacco (Zhu and Wu, 1979; Wu and Chen, 1982), rice (Beauville, 1980; Zhou and Yang, 1980,1981; Kuo, 1982), barley (Wang and Kuang, 1981; Huang et al., 1982), maize (Ao et al., 1982; Truong-Andre and Demarly ,1984), lily (Gu and Cheng, 1983) and sunflower (Cai and Zhou, 1984). So much success within such a short period indicated that the induction of haploids via *in vitro* culture was not inaccessible as had been thought earlier.

The procedure for *in vitro* culture of unpollinated ovaries is similar to anther culture which has been discussed in the previous paragraph: pollen (anther culture). *In vitro* culture of unpollinated ovaries should also be regarded as a promising biological technique for crop improvement. To-date it has not been put into practice owing to low frequency of the induction of haploids, and the procedure needs to be further refined and improved. Considering the number of haploid cells per ovary to be much less than the number of pollen per anther, the potential of anther culture for the production and application of haploids is much more than that of *in vitro* culture of unpollinated ovaries.

3 GENETICS OF HAPLOIDS

Pollen-derived haploids are good system for studying the important genetic questions concerning heredity and variation, recombination and expression which are the basis of the mechanism of formation of genetic characteristics.

3.1 Genetic (chromosomal) stability and variability of pollen-derived plants

Stable, homozygous diploid strains can be obtained through chromosome doubling of haploids. For many years, we have been studying karyotypic analysis of somatic cells and pollen mother cells (PMCs) derived from pollen plants. Through the investigation of chromosome configuration of root tip cells, PMCs and genetic analysis of pollen plants, we found that genetic (chromosomal) stability and variability coexisted in anther cultures during the same process (Hu, 1983). This is an important genetic feature of pollen-derived plants.

The genetic analysis and cytological observations were carried out in unselected populations of pollen plants. Results obtained from such populations in wheat (Hu et al., 1978, 1980; Yang et al., 1978), rice (Chen and Li ,1978), maize (Gu et al., 1981) and tobacco (Xu et al., 1980) over a period of several years indicated that about 90% of the diploid lines were genetically uniform. These results suggested that though diploids, heteroploids as well as haploids occur through anther culture, it might be considered that anther culture produces mainly homozygous recombinant lines (90%). The coexisting variants form only about 10% of pollen plants.

3.2 Microsporeclonal variation

Variability of the chromosome number and structure in the plant cells regenerated *in vitro* is a common phenomenon (D'Amato, 1985). The same phenomenon has also been observed in pollen plants. The technology of introducing genetic variation using cell culture has been termed somaclonal and gametoclonal variation (Sharp et al., 1984). But pollen plants derived from microspores were not gametes, it should be termed microsporeclonal variation. We first observed this phenomenon, which was not accepted by classical geneticists before. The recovery of aneuploid plants using tissue culture may appear as an unwanted aspects of instability during *in vitro* culture, but there are many analytical and practical applications of aneuploids (Kudirka et al., 1986). For example, aneuploids can be used for mapping genes on chromosomes or for creating alien substitution lines. Each of these applications is constructive in either expanding our knowledge of the plant genome or introducing diversity into different plant cultivars.

3.2.1 Variation in chromosome number

From 1981 to 1983, aneuploid production in anther culture was conducted in our laboratory using the inbred spring wheat variety Orofen (Hu, 1985). The root-tip cells of 472 pollen plants and PMCs of some plant lines were examined and identified cytologically. According to the basic number of chromosomes, these 472 pollen plants can be classified into four types: euploids (85.%), aneuploids (9.3%), heteroploids (1.7%) and mixoploids (3.2%). It was worthy to note that 18 aneuploids with aberrations in chromosomes were obtained, representing 3.8% of the total plants. These results suggested that genetic variants occur spontaneously in anther culture.

3.2.2 Variation in chromosome structure

Variations in chromosome structure such as dicentrics, telosomis, isochromosomes, deletions and breakages were found in plants regenerated via anther culture (De Buyser et al., 1985; Hu et al., 1982; Hu, 1983). The results of these investigations are as follows.

3.2.2.1 Dicentric chromosome and translocations

In our previous work (Hu, 1983) dicentric chromosomes in pollen plants of Orofen were frequently observed. The somatic chromosome configuration of 453 pollen plants derived from Orofen X Xiaoyen 759 with a ring chromosome indicated that the chromosome translocation occurred in the somatic cell.

3.2.2.2 Deletion chromosomes

A 1B long arm deletion line was acquired from a pure line variety of Orofen via anther culture (Hu et al.,1982). The D 232 pollen plants derived from winter wheat Kedong 58 with the chromosome configuration $2n=41 + 6B^{L}$ which was a mono-long arm deletion often occurred. These deletion lines could be maintained and were fairly stable.

3.2.2.3 Telosomics

Among pollen plants of Orofen the monotelosomic plant line Q31 was found to have the configuration 2n=41+t'. Its spike resembles a speltoid spike, but was significantly different from the parent Orofen. This indicated that monotelosomics gave rise to variation in chromosome number and structure via anther culture. Mackey (1954) confirmed that the chromosome 5A was presumed to carry the spelta gene q. According to their reports, this monotelosomics, which might be 5A monotelosomic. The D19 pollen plant derived from winter wheat Kedong 58, which was a ditelosomic, showing 2n=41+2t' chromosome configuration, was acquired directly via anther culture. The cytological examination of this plant's progenies (10 plants, H_2) showed 5 disomics (2=42), 1 mixoploid, 3 ditelosomics (2n=42+2t') and 1 double ditelosomic 1B.

3.2.2.4 Chromosome breakage

By means of the C-banding technique, the analysis of chromosome configuration of pollen plants was carried out, using the F_1 hybrids of hexaploid triticale, Beagle × Orofen, for anther culture. The haploid monotelosomic with 5R short arm telos and the substitution addition line (2n=46) with chromosome configuration 2n=40W + 6R were obtained. The rye chromosome configuration is 1R", 7R" and $6R'6R^{s}$. These two materials indicated that the chromosome breakage might have occured during the *in vitro* culture process. The breakpoint was located at heterochromatic telomere, which may plays a role in chromosome breakage in tissue culture including anther culture. McCoy et al. (1982) proposed that late-replicating pericentromere heterochromatin may be responsible for chromosome breakage in oat tissue cultures. Lapitan et al. (1984) reported that 12 of 13 breakpoints in chromosomes that involved translocations and deletions were attributed to heterochromatin in tissue culture of wheat-rye hybrids. Based on the high frequency of chromosome breakage accompanied by an increase in the number of single gene recessive variants, Lee and Phillips (1987) suggested that transposable elements may be responsible for some of the observed genetic variation, an activity that remains to be investigated.

3.2.3 Possible mechanisms for variability of pollen plants

It is evident, particularly from the work of McCoy et al. (1982), Ogihara (1981) and Sacristan (1971), that tissue culture can cause chromosome deletions, translocations and other minor rearrangements. Our previous work (Hu et al., 1978) indicated that a number of pollen plants in wheat were mixoploids. We found the evidence in root-tip cells and callus mitosis not only for aneuploid, heteroploid and mixoploid plants but also for chromosome breakage, endomitosis, multipolar mitosis and chromosome decentralization etc. Similar phenomena were observed by Mix et al. (1978) in pollen-derived plants of barley.

De Paepe et al. (1981) proposed that variant tobacco dihaploids derived either from the vegetative nucleus of the pollen or from the fusion of the generative and vegetative nuclei. Zeng and Ouyang (1980) found that in induced wheat microspores *in vitro* , nuclear fusions and endomitosis may double chromosome numbers. Some free chromosomes and chromosomal fragments produced *in vitro* would lead to karyotype variation. Huang (1982) found that in wheat pollen culture, cell wall formation accompanying pollen embryogensis was via either cell plate or wall ingrowth. In the latter case the cell walls were usually incomplete. Nuclear fusion, chromosomal or nuclear movement through gaps in the cell wall were frequently observed. Sun et al. (1983) also found that there was no cell wall, but only a "phragmoplast-plasmalemma complex" and phragmoplast between two equal daughter nuclei, thus these nuclei were free in the pollen. Therefore, anomalous cell-wall formation may be a possible reason for ploidy variation in anther culture. Meanwhile, 3×, 5×, 7 × etc. odd ploidy pollen plants were derived from the endoreduplicated generative nucleus and nuclear fusion. All of these above-mentioned phenomena occurred at the early stage of anther culture.

In the late developmental stage of pollen callus, various mitotic abnormalities occurred frequently. For example, spindle fusion and nuclear endoreduplication may produce doubled euploids or polyploids. On the other hand, multipolar mitosis, chromosome bridge, lagging chromosome, chromsomal fragments and dicentric chromosomes etc. lead to the formation of aneuploids.

3.2.4 Origins of variation

De Buyser et al. (1985) suggested that the chromosomal abnormalities found in doubled haploid (DH) plants originated in the anthers of donor plants or during the process of anther culture. Wang and Hu (1993) proposed that there were two periods in which the chromosomal variation in pollen plants might occur: before anther culture and/or during anther culture.

3.2.4.1 Chromosome variation before anther culture

Nakamura and Keller (1982) and Suarez et al. (1988) demonstrated the instability of secondary hexaploid *triticale* and wheat and where abnormal meiosis might change the gametic chromosome number and induce structural changes. Anther culture allowed these abnormal gametes to be present at the whole plant level. Misdivisions produced isochromosomes, and chromosome non-disjunction made the gametic chromosome number either increase or decrease. The loss of laggard univalents might also reduce the gametic chromosome number. If we suppose that bivalent non-disjunction exists in the meiosis of the donor plant, we can explain why some pollen plants possess an abnormal chromosome number such as 29, 38, or even 74.

3.2.4.2 Chromosome variation during anther culture

If no chromosome variation occurred in culture, then the chromosome number of spontaneously doubled diploids would be even and all chromosomes (including aberrant chromosomes) should appear in pairs. Therefore odd-numbered diploids such as 2n=41, 43, 45, and 47, and those which have unpaired aberrations must have had their chromosome number and structure changed during the culture.

Miao et al (1988), Tao and Hu (1989a), and Wang and Hu (1985) all observed abnormal chromosome constitutions such as 2n=41, 2n=43 as well as chromosome structure variations in pollen plants. B150 was a good example for that its chromosome inversion could only occurred during the culture. Pollen plant B150, derived from a heptaploid hybrid, was a mosaic. Among 50 cells which had been analyzed cytologically, cells with 47 chromosomes were at the highest frequency, i.e. 50%; and cells with 48 chromosomes at 14%. About 25% of the cells had the constitution of 2n=47+a ring chromosome. Comparing two photographs of one cell before and after Giemsa C-banding, Wang and Hu (1993) found a pericentric inversion in one of the two 1R chromosome: one 1R had two strong heterochromatin bands in the long arm and one strong band in the short arm. The other 1R was normal with two heterochromatin bands in the short arm and one in the long arm. Among 23 C-banding cells observed there were 14 cells with 1R inversion.

This indicated that the chromosome pericentric inversions regularly occurred in pollen-derived plants during anther culture. These two abnormal types of cell division, e.g. abnormal meiosis and mitosis were the mechanism of microsporeclonal variation, which produced chromosome fragments due to breakage and chromosome reunion leading to chromosome rearrangement. It opened up the way for studying plant cytogenetics and variation. Meanwhile, it was the genetic basis of obtaining translocation lines, transferring alien chromosomes into donor plants and studying the cytogenetic recombination.

3.3 Gametic genotypes fully expressed at homozygous plant level

Anther culture serves as a tool to solve the questions in genetic studies and plant breeding when conventional methods is not successful. On the other hand, the use of pollen haploids results in rapid achievement of homozygosity and easy expression of recessive characters, thus enhancing the selection efficiency. More importantly, the various types of recombinant gametes can be fully expressed at the homozygous plant level, and the microsporeclonal variation can then recover recombinants and variants, and create new types which are difficult to be obtain using the conventional methods (Hu and Huang, 1987).

Since pollen grains from F_1 hybrids are heterozygous, different gene combinations of both parents of a cross occur in every grain. If pollen grains of F_1 hybrids are induced in pollen plants by anther culture, the plants (H_1) show various phenotypes of both parents and their recombinants. The most common system was to use F_1 hybrids as the parental material for haploid production, thereby fix the new products of recombination between the parental genotypes at the earliest possible opportunity.

In general, it is not easy to test this principle by conventional methods, but there is significant evidence when distant hybrids are used for anther culture.

(A) AABBRR x AABBDD	(B) AABBDDEE x AABBDD
(2n=42) (2n=42)	(2n=56) (2n=42)
↓	↓
F_1 AABBDR	F_1 AABBDDE
↓	↓
Gametes ABD(0-7)R(0-7)	Gametes ABDE(0-7)
15 kinds of gamete genotypes with 14-28 chromosomes	8 kinds of gamete genotypes with 21-28 chromosomes

Figure 4. Two systems that express hybrid gametes at the homozygous plant level

Figure 4 illustrates two systems which produced hybrid gametes at homozygous plant level. In conventional breeding, crosses between hexaploid *triticale* (AABBRR) and hexaploid bread wheat (AABBDD) produced hybrids with the chromosome constitution AABBDR, in which D and R chromosomes were present as two sets of 7 univalents. At meiosis, these two sets were distributed into daughter cells, each with the frquencies ranging from 0 to 7. Fifteen kinds of gametes could be predicated with chromosome numbers ranging from 14 to 28. However, Muntzing (1979) indicated that, due to natural selection, in practice it is difficult to combine the same gamete types into one zygote. Therefore, the various gamete genotypes of hybrids are difficult to be fully expressed in sexual hybridization. From a cross between Rosner (6X) *triticale* and Kedong (6X) bread wheat, a total of 12 aneuploid and haploid plants were identified by Giemsa C-banding technique (Wang and Hu, 1985), in which there were 2-7 rye chromosomes and 18-21 wheat chromosomes. The types of pollen plants were basically the same as the type of pollen chromosome composition formed by meiosis of hybrid. Therefore, all possible recombinant types in the pollen of F_1 hybrid might be fully expressed at plant level by means of anther culture.

Meanwhile, using the same material 10 recombinant types of pollen-derived plants were obtained as against only 4 types of F_2 plants produced by the conventional methods. The frequencies of composition types of H_1 pollen-derived plants and F_2 plants produced by conventional methods were 37.04% (10/27) and 11.76 (4/34), respectively. The difference between these values was significant and indicated that a high number of recombinant types might be obtained in shorter time and from smaller populations by anther culture than by conventional methods.

Using the A system a large variety of diverse pollen-derived plants with different chromosome constitutions were produced. The data of repeated experiments (Table 4.) indicated that from 680 pollen plants there were 15 kinds of gamete types with different chromosome constitutions, including 11 with chromosome numbers ranging from 17 to 27 (the corresponding diploids after spontaneous chromosome doubling), one monosomic, and other groups including heteroploids, mixoploids and chromosome structure variants.

Based on the same principle, anther culture of F_1 hybrids of the octoploid *Agropyron trititrigia* and bread wheat (Zhong 3 × Orofen) was also conducted in the B system (Fig. 4) (Miao et al., 1988). We obtained 112 pollen-derived plants, among which there were 10 kinds of plants with different chromosome constitutions , eight kinds with various chromosome constitutions, one trisomic and 20 mixoploids (Table 5).

These results demonstrated that the possible recombinant types of F_1 hybrid pollen, i.e. recombined gametes, might be fully expressed at the plant level by means of anther culture, whereas some recombined gametes might be lost in sexual hybridization.

Based on the principles of *in vitro* induced haploids mentioned above, genetic studies, including recombination of genomes, chromosome (gene) mapping and creation and identification of alien translocation lines might be carried out.

3.4 Recombination of R and D genomes

The largest proportion of the population of pollen plants were those with 23 chromosomes (haploid) and 46 chromosomes (doubles haploid). These are difficult to obtain by conventional crosses, but in this case, novel gamete can be produced directly. Why were the pollen plants with chromosome constitution of 2n=23/24 in haploids and 2n=46/48 in diploids observed in highest frequency? Our investigation (Tao and Hu ,1989a) showed that using C-banding technique, crosses of *triticale* (6×) × wheat (6×) were examined for wheat and rye chromosome content. X^2 tests indicated that the distribution of chromosomes involved in segregation from the different genomes was quite different. Specifically, the distribution of R genome chromosomes was at random, while the distribution of D genome chromosomes was skewed. Due to the interaction between the two genomes, R and D, gametes having 23 or 24 chromosomes were produced

Table 4. Number of chromosome in pollen-derived plants from different hybrids between triticale (6X) and bread wheat (6X)

F_1 material	Chromosome numbers							
	17 83	18 36	19 38	20 40	41	21 42	22 44	23 46
Rosner × Kedong 58						1	1	7
Beagle × Kedong 58	1	1	2	6		40	40	113
Beagle × Jinghua 1			2	1		3	11	17
Beagle × Orofen			1	2	1	6	17	18
Total No. of plants	1	1	5	9	1	50	69	155

F_1 material	Chromosome numbers							
	24 48	25 50	26 52	27 54	I	II	III	Total
Rosner × Kedong 58	3	4	1	1		9		27
Beagle × Kedong 58	88	32	18	6	5	91	23	466
Beagle × Jinghua 1	16	10	4	2	2	22	5	95
Beagle × Orofen	18	13	3			13		92
Total No. of plants	125	59	26	9	7	135	28	680

*I = Heteroploids, II = Mixoploids, III = Chromosome structure variants
The data of Rosner x Kedong 58, Beaglex Kedong 58, Beaglex Jinghua 1 derived from wang & Hu (1985). The data of Beagle x Orofen derived from Tao & Hu (1989).

Table 5. Number of chromosomes of pollen-derived plants from the F_1 between 8x trititrigia and 6x bread wheat (Zhong 3 X Orofen)

	Chromosome number										
	21 42	43	22 44	23 46	24 48	25 50	26 52	27 54	28 56	Mixoploids	Total
No. plants examined	1	1	6	21	22	23		6	2	20	112
percentage	0.9	0.9	5.4	8.7	19.6	20.5	8.9	5.4	1.8	17.9	100

Data derived from Miao et al., 1988.

predominantly resulting in pollen plants with chromosome constitution of `2n=23/24 in haploids and 2n=46/48 in diploids.

3.5 Chromosome (gene) mapping

3.5.1 Plant height

The alien chromosome can not normally pair with its homoeologous chromosome in hybrid progenies, which confounds analysis and interpretion of the data on agronomic performance. By integrating suitable methods of identification of the alien chromosomes, such as biochemical markers, with a procedure for the phenotypic characterization of the alien chromosome substitution lines with respect to agronomic characters, the effects of the alien chromosome can be distinguished precisely from those of its homoeologous chromosome.

Tao et al. (1991) carried out the genetic analysis of M27, a wheat 1R (1D) substitution line, by backcrossing reciprocal monosomic analysis. A doubled haploid line M27, derived from a hexaploid *triticale* × bread wheat cross and known to be a 1R (1D) whole chromosome substitution line, was characterized for the genetic control of agronomic characters by the use of the backcross reciprocal monosomic analysis. The SDS-PAGE technique was also applied to identify 1R glutelin subunit composition in the monosomic and euploid progenies of the backcross in order to investigate the transmission pattern of the rye chromosome from 1R, 1D double monosomics. The results showed that 1R had the effect on the height.

3.5.2 Disease resistance (powdery mildew)

Zhang et al. (1995) conducted the chromosome (gene) mapping of the resistance to powdery mildew of wheat, using alien substitution line m24 (6R/6D) crossed with 6A deletion line of Apo wheat , and then carrying out anther culture of F_1 hybrids. A number of aneuploids, including 6R addition line, $6R^L$ isochromosome addition, $6R^L$ telomere addition, and $6R^L$ deletion addition line were obtained. If the $6R^L$ (6R long arm) were lost, the plant would be infected with powdery mildew. Therefore, it gave the evidence that the genes resistant to powdery mildew were located on the long arm of $6R^L$ chromosome. We also found that the molecular probe by RFLP, Ps 687 was linked with the characteristic of the resistance to powdery mildew.

3.6 Creation and identification of alien translocation lines

Plant geneticists and breeders pay more attention to the investigation of translocation lines, for using alien translocation as materials, the structure and function of chromosomes, the expression and regulation of alien chromosome fragments or gene transformation could be investigated. Meanwhile the good translocation lines might be utilized directly in breeding programme. Using F_1 seeds of Chinese Spring (CS,) crossed with M27 (1R/1D substitution line) as materials for anther culture, the pollen-derived plants Wer-1, Wer-2, etc were obtained and identified by using biochemical markers, including storage protein, glutelin and gliddin, chromosome C-banding and in situ hybridization (GISH and FISH). One non-Robertosonian translocation line was directly obtained, whose chromosome composition was 1Rd/1D+1AS. 1AL-1RL, namely 1RL long arm deletion, substitution, translocation line (Fig. 5).

4 APPLICATIONS OF *IN VITRO* INDUCED HAPLOIDS

4.1 The possibility of using DH plants in crop improvement

According to the biometrical studies with DH lines, Snape (1989) indicated that DH systems had the unique genetic property of allowing completely homozygous lines to be developed from heterozygous parents in a single generation. In self-pollinating species, such as wheat, barley and rice, this property can be used to increase the efficiency of cultivar production. Time-consuming back crossing can be avoided in getting selected material ready for commercialization. There is also an increase in selection efficiency relative to conventional methods because of an increase in additive genetic variation, absence of dominance variation and family segregation, and a decrease in environmental variation effects through greater replication possibilities.

4.1.1 Time-saving advantage

Time-saving is the most obvious advantage of a DH system because yield and anther evaluation trails can be done much sooner than with conventional lines, particularly with winter materials, just because the requirement for vernalization to initiate flowering sets an upper limit on the speed of generation turn over with winter crops.

It also takes less time to build up pure stocks of a new cultivar. Since DH are completely homozygous, all stocks are identical and no purification system is required other than isolation to avoid outcrossing. But in the conventional system stocks are usually derived from a single plant of an advanced generation, several generations are required to build up sufficient quantities of seed for release.

4.1.2 Increasing selection efficiency

Compared to selection during the early generations of a pedigree programme, the homozygosity obtained from a DH system increases the efficiency of selection for both qualitative, major gene characters, and, in particular, for quantitative characters. Thus, it should be easier to identify the superior genotype in a cross and to produce new cultivars when using DH system.

If the selection of desirable recessive alleles at major gene loci happens in a F_2 population, then only a proportion $(1/4)^n$, where n is the number of segregating loci, will have the desirable allelic combination. However, with a DH population, such genotype will be at a frequency of $(1/2)^n$. Thus, the frequency of fixation in a F_1 derived DH population is the square root of the probability in a F_2 population.

A remarkable advantage of using DH lines is that greater additive genetic variance is expressed between the recombinants produced in a cross than between the relative F_2s and F_3s. Another benefit is that dominance variation is absent. These properties are illustrated genetically by comparing addition and translocation lines. This CETPP method has significant potential as a tool in germplasm enhancement (Hu,1992).

Thirdly, the environmental variance between F_2 individual (VEI) is likely to be greater than that between F_3 or DH plots (VEP), which are made of plots genetically similar individuals. Furthermore, replication can be introduced to reduce VEP so that the individual breeding values of the line can be more accurately assessed.

Conventional early generation plots like F_3 or F_4 will exhibit genetic differences between individuals within the plots unlike DH plots where all individuals are genetically identical. It will make visual selection of desirable lines more difficult in early generations compared to DH plots.

Overall, the different genetic properties of DH populations compared to early conventional generations show that selection efficiency is increased.

Table 6. Expectation of phenotypic variances in different generations derived from a cross between inbred parents (Snape 1989)

Generation	Variance
F_2 (between individual plants)	$V_A+V_A+V_{EI}$
F_3 (between family means)	$V_A+1/2V_d+V_{EP}$
F_1 derived DH family means	$2V_A+ V_{EP}$

V_A = Additive component of variation
V_D = Dominance component of variation
V_{EI} = Environmental variance between F_2 individuals
V_{EP} = Environmental variance within F_3 / dihaploid progenies

Although the DH system allows the greatest time-saving compared to a conventional pedigree programme, it also has two disadvantages. One is that a completely random sample of gametes is fixed, thus, genetic drift will result in a high rejection frequency since only a small proportion of lines will meet desired criteria. For example, if only a single gene, e.g. for disease resistance, is segregating between the parents, then half of the population will be expected to carry the undesirable, susceptible allele. If n major gene loci are segregating, only a proportion of $(1/2)^n$ of the DHs will contain the desirable combinations of alleles. Clearly, even for small number of loci this proportion will be small.

The other disadvantage of an F_1 system is that only one round of recombination is allowed between the parental genomes before fixation. Thus, the DH population will be in linkage disequilibrum if linkage between genes are important components of variation.

The compensation for these disadvantages is to delay haploidization until the F_2 or F_3 generation and to conduct selection prior to haploid production. For example, F_2 inividuals can be classified for major gene characters and also selected visually for quantitative characters of high heritability, such as flowing time and plant height. Only desirable plants are then used for DH production. Such a scheme was illustrated in Fig. 6. This scheme does not save much time compared to an F_1 system, but does increase selection efficiency for both major genes and quantitative characters.

Variety A X Variety B

AAbb ↓ aaBB

F_1: AaBb

↓ selfing

AABB AaBB aaBB
F_2: AABb AaBb aaBb
AAbb Aabb aabb

↓ Field testing and selection

Selected F_3: AABB AABb AaBB

↓ Doubled haploid production

Doubled haploids· AABB AAbb aaBB

↓ Field testing and selection

New variety: AABB

Figure 6. The selected F3 double haploid system for varetal production in self-pollinating crops (Snape 1989)

This should ensure a high frequency of lines with desired levels of performance and, consquently, very large population should not be necessary for genetic advance. Since more rounds of recombination are allowed, there will also be less linkage disequilibrium than in an F_1-derived population. More advanced generations can be used in this way, thereby giving more opportunity for selection prior to haploidization.

In practice, to overcome these disadvantages of the DH system, Zhang et al. (1983) developed a multistep breeding method which is composed of one cycle of anther culture followed by sexual hybridization between different genotypes of pollen plants and another cycle of anther culture of the selected sexual hybrids. By using this method, some polygenic traits have been successfully improved. An example is the release of the cold resistant rice variety named Hua Han Zao. In addition, Hu et al. (1985) developed a method combining anther culture with composite crossing, where the linkage between genes might be broken. By using this method, a winter wheat variety with desirable characters, Jinghua No. 1 was released. Based on the above-mentioned discussion, it can be considered that the advantages of DH lines can be obtained very successfully with self-pollinating crops. In such programmes different filial generations can be chosen as the parental material for haploidization, though the most common system is to use F1 hybrids thus fixing the production of recombination between the two parental genomes at the earliest opportunity. Since each DH produced is a potential cultivar, field selection is practised to identify those lines with the desirable combination of characters.

4.2 Developing and releasing new varieties

In 1979, Ciba-Geigy seeds licensed a barley cultivar, Mingo, the first licensed cultivar by haploid induced *in vitro* based on the chromosome elimination, bulbosum method (Kasha et al., 1980). Since anther culture technique was applied to breeding program, a great progress has been achieved in China. D. F. Hu (1995) described the achievement of pollen haploid breeding. Using anther culture combined with conventional breeding method Chinese scientists have developed and released a number of cultivars, such as wheat, rice , maize, pepper and fruits which have good agronomic characteristics of high yield, well adaptation, disease and draught resistance etc. Meanwhile, by now many Chinese breeding units have established perfect procedure and system of pollen haploid breeding as a routine method of crop improvement. Up to 1994, the main varieties developed from pollen derived plants are as follows;

(i) *Wheat.* More than 10 new varieties have been developed and released into production by using anther culture (Table 7). The area of the land under cultivation of these varieties reached 1,160,000 ha.

(ii) *Rice.* In 1980s 22 new varieties of rice derived from pollen including Zhonghua No.8, No.9, No.10, No.11, No. 12 and Hua Han Zao etc. were developed. The area under cultivation of these varieties was more than 1,000,000 ha. In recent years Chinese scientists have proposed a new way using rice subspecies hybrids, japonica with indica *via* anther culture to increase rice yield more than 20% compared to the control varieties.

Table 7. Wheat varieties released in production via anther culture(Hu 1991)

Name of varieties	Breeding and research institute
Jinghua No. 1	Hu D.F. Laboratory of Plant Cell Engineering , Beijing Academy of Agricultural Sciences
Jinghua No. 3	Ibid
Jinghua No. 5	Ibid.
Huapei 764	Academy of Genesu Agricultural Sciences,
Zheng Chun No. 11	Institute of Agriculture of Gansu Zheng Ye Region
Gan Chun No. 16	Gansu Agricultural University
Yu Mai No. 6	Luo Yang Agricultural Institute
Kui Hua No. 2	Sin Jing Nong Qi Shi Agricultural Institute
Xia Mai No.1	Laboratory of Genetics, Institute of Wheat, Henan Academy of Agriculture
Anther Culture 28	Zhao, Y. L. et al. Laboratory of Genetics, Institute of Wheat, Henan Academy of Agriculture
Hua 555	Wang P. et al. Hebei Academy of Agriculture

(iii) *Maize*. In 1992, the Evaluation Committee for crop variety in Guangxi, China, passed the evaluation of Gui-san No.1. It was the first maize hybrids from pollen in the world to go into production. The area under cultivation of this variety was more than 10,000 ha.

(iv) *Pepper*. The sweet pepper variety Hai Hua No.31 was selected from pollen-derived plants by Z.R. Jiang and C.L. Li in the tissue culture laboratory, Haidian District, Beijing. It has released to production. The area under cultivation reached 5,800 ha.

(v) *Fruits*. The apple and pear pollen-derived plants were developed by K.R. Xue in the institute of Fruits of Chinese Academy of Agricultural Sciences. From pear variety. Jin Feng, Xue et al. developed a new variety derived from pollen which has good agronomic characteristics, with short stature. It is the first success in the world in this field.

4.3 Chromosome engineering in Triticeae using pollen-derived plants (CETPP)

This system has been established by Hu Han and his group. It combines chromosome engineering with anther culture and modified identification methods transferring the desirable chromosome (genes) into cultivars and thus to create new strains of wheat. As chromosome substitution, identified by using biochemical markers, including storage protein, glutelin and gliddin, chromosome C-banding and in situ hybridization (GISH and FISH) . One non-Robertsonian translocation line was directly obtained, whose chromosome composition was $1R^d/1D + 1A^S.1A^L\text{-}1R^L$, namely $1R^L$ long arm deletion, substitution, translocation line (Fig.5)

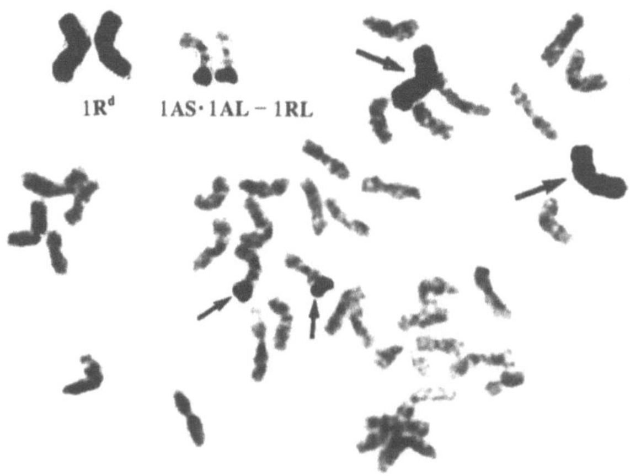

Figure 5. Non-Robertsonian translocation identified with in situ hybridization

4.4 Genetic manipulation

As microspore culture is a single cell system, it makes selection at the single cell level possible and, furthermore, offers new prospects for genetic manipulation, e.g. mutagenesis and transformation. Direct gene transfer by microinjection using isolated multicellular pollen embryoids offers the possibility of transgenic plant formation of all cereals including wheat by using culture of isolated pollen having high regeneration efficiency (Potrykus et al., 1985; Potrykus, 1988). It can be expected that the new and interesting information concerning the genetic manipulation using microspore culture of cereals, including wheat, barley and rice will be obtained in the near future.

4.5 Usage of DH populations

A new field of haploid usage is the molecular genome identification, particularly for QTL analysis. DH populations are an important tool to obtain reproducible DNA polymorphism in barley (Heun et al., 1991) and rice (McMouch and Tanksley, 1992, Xu et al., 1994) etc. One kind of method for RFLP map of these cereals was based on DH populations.

5 CONCLUSION

There are three pathways to produce haploid sporophytes of higher plants:

I. From meiotic spores by anther/pollen and unpollinated ovary (ovule)culture,

II. From reduced zygotes by chromosome elimination,

III. From male and female gametes by isolation and manipulation of reproductive cells and protoplasts.

Among these, approaches I and II have been developed and applied with great success. The progress has paved the way for fundamental studies and practical utilization of DHs.

The frequency of microspore embryogenesis in cereals has been increased significantly. Up to now in several genotypes of barley and wheat one could produce more than one green plantlet per anther. The success is calculated by green pollen plants per anther in comparison with the previous rate of per 100 anthers. But as we know, one anther of cereals contains 2000-3000 pollen grains, so there is a great potential for microspore culture.

The initial period of culture might be a starvation period during which the process of dedifferentiation and the acquisition of embryogenic capacity by cultured pollen occurs (Herberle-Bors, 1989; Li et al., 1993). The direction of pollen development can be influenced during this period, by switching its pathway from gametophyte development to sporophyte development (Harada 1989). So isolated microspore culture can be a model system to study cell differentiation and other fundamental biological problems.

In practice, DHs combined with different breeding methods, such as conventional breeding, chromosome engineering, mutagenesis and wide hybridization will efficiently recover and screen a lot of recombinants and variants. Meanwhile it can also create new types which are usually difficult to obtain by conventional methods for crop improvement.

In addition, isolated microspore culture in connection with genetic transformation may result in an increased research efforts in this field.

6 REFERENCES

Ao, G M, Zhao S X & Li G H, (1982) *In vitro* induction of haploid plantlets from unpollinated ovaries of corn (*Zea mays* L.). Acta Genet. Sin. 9:281-283. (in Chinese with English abstract)

Barcley I R, (1975) High frequencies of haploid production in wheat (*Triticum aestivum*) by chromosome elimination. Nature 256:410-411.

Beauville A M (1980) Obtaintion d'haploides *in vitro* a partir of ovaires nonfecondes de riz, *Oryza sativa* L. C. R. Acad. Sci, Paris 296:489-492.

Blakeslee A F, Belling J, Farnham M E & Begner A D (1922) Science 55:646.

Cai, D.T. & Zhou C (1984) *In vitro* induction of haploid embryoids and plantlets from unpollinated young florets and ovules of *Helianthus annuus* L. Kexue Tongbao 29:680-682.

Chen Y. & Li L (1987) Investigation and utilization of pollen-derived haploid plants in rice and wheat. In: Proc Symp. Plant Tissue Culture. (pp. 199-212) Science Press, Peking.

Chu C C & Hill R D (1988) An improved anther culture method for obtaining higher frequency of pollen embryoids in *Triticum aestivum* L. Plant Sci. 55:175-181.

Chu C C, Hill R D & Brule-Babel A L (1990) High frequency of pollen embryoid formation and plant regeneration *in Triticum aestivum* L. on monosaccharide containing media. Plant Sci. 66:255-262.

Chuong P V & Beversdorf W D (1985) High frequency embryogenesis through isolated microspore culture in *Brassica napus* L. and *B.carinata Brann*. Plant Sci. 39:219-226.

De Buyser J, Henry Y & Table G (1985) Wheat androgenesis: cytogenetical analysis and agronomic performance of doubled haploids. Z. Pflanzenzucht. 95:23-34.

De Paepe R, Bleton E & Gnangbe F (1981) Basis and extent of genetic variability among doubled haploid plants obtained by pollen culture in *Nicotiana sylvestris*. Theor. Appl. Genet. 59:177-184.

D'Amato F (1985) Cytogenetics of plant cell and tissue cultures and their regenerates. CRC Crit. Rev. Plant Sci. 3:73-112.

Falk D E & Kasha K J. (1981) Comparison of the crossability of rye (*Secale cereale*) and Hordeum bulbosum onto wheat (*Triticum aestivum*). Can. J. Gent. Cytol. 23:81-88.

Falk D E & KashaK J (1983) Genetic studies of the crossability of hexaploid wheat with rye and *Hordeum bulbosum*. Theor. Appl. Genet. 64:303-307.

Gu Z P & Cheng K C (1983) In vitro induction of haploid plantlets from unpollinated ovaries of lily and its embryological observations. Acta Bot. Sin. 25:24-28. (in Chinese with English abstracts)

Harada H (1989) Applications of plant cell and tissue culture. (pp. 59-74) Wiley. Chichester Ciba Foundation Symposium 137.

He D G & Ouyang J W (1984) Callus and plantlet formation from cultured wheat anthers at different developmental stages. Plant Sci. Lett. 33:71-79.

Heberle-Bors E (1989) Isolated pollen culture in tobacco: Plant reproductive development in a nutshell. Sex. Plant Reprod. 2:1-10.

Heun M, Kennedy A E, Anderson J A, Lapitan N L V, Sorrelis M E & Tanksley S D (1991) Construction of a restriction fragment length polymorphism map for barley. Genome 34:437-447.

Hu D F (1995) Plant Cell Engineering Breeding Science and China. (in Chinese) 2/95 pp21-24.

Hu D F, Yuan Z D, Tang Y & Liu J P (1985). Jinghua No 1., a winter wheat variety derived from pollen sporophyte. Sci. Sin. 28:733-745.

Hu H & Huang B (1987) Application of pollen-erived plants to crop improvement. Int. Rev. Cytol. 107:293-313.
Hu H (1983) Genetic stability and variability of pollen-erived plants. In: Sen S K & Giles L K L (eds) Plant Cell Culture in Crop Improvement. (pp. 145-157). Plenum, New York.
Hu H (1985) Use of Haploids in Crop Improvement. In: Biotechnology in International Agricultural Research. (pp. 75-84) International Rice Research Institute, Manila.
Hu H, (1992) Germplasm enhancement by anther culture in Triticeae. Hereditas 116:151-154.
Hu H, Hsi T & Chia S, (1978) Chromosome variation of somatic cells of pollen calli and plants in wheat (*Triticum aestivum* L.). Acta Genet. Sin. 5:23-30.
Hu H, Hsi T Y & Ouyang J W (1980) Chromosome variation of pollen mother cells of pollen-derived plants in wheat (*Triticum aestivum* L.). Sci. Sin. 23:905-912.
Hu H, Li W Z & Jing J K (1991) Anther/pollen *in vitro* culture of barley. Barley Genet. 6:203-205.
Hu H, Xi Z Y, Jing J K & Wang X Z (1982). Production of aneuploids and heteroploids of pollen-erived plants. In: Fujiwara A (ed.) Plant Tissue Culture 1982 .(pp. 421-424) Jap. Assoc. Plant Tissue Cult. Tokyo.
Huang B & Sunderland N (1982) Temperature-stress pretreatment in barley anther culture. Ann. Bot. 49:77-88.
Huang Q F, Yang H Y & Zhou C (1982) Embryological observations on ovary culture of unpollinated young flowers in *Hordeum vulgare* L. Acta Bot. Sin. 24:295-300. (in Chinese with English abstract)
Hunter C P (1988) Ph.D. Thesis. Wye College, University of London. Springer, Berlin.
Imamura J, Okabe E, Kyo M & Harada H (1982) Embryogenesis and plantlet formation through direct culture of isolated pollen of *Nicotiana tabacum* cv. Samsum and *Nicotiana rustica* cv. Rustica. Plant Cell Physiol. 23:713-716.
Inagaki M N & Tahir M, (1990) Comparison of haploid production frequencies in wheat varieties crossed with *Hordeum bulbosum* L. and maize. Jpn J. Breed. 40:209-216.
Kao K N (1981) Plant formation from barley anther cultures with Ficoll media. Z Pflanzenphysiol. 103:437-443.
Kao K N, Saleam M, Abrams S, Pedras M, Hom D & Mallard C (1991) Growth conditions for induction of green plants from microspores by anther culture methods. Plant Cell Rep. 9:595-601.
Kasha K J & Kao K N (1970) High frequency of haploid production in barley (*Hordeum vulgare* L.). Nature 225:874-875.
Kasha K J and Reinbergs E (1980) Achievements with Haploids in Barley Research and Breeding. In:Davies DR and Hopwood DA (eds) The Plant Genome (pp.215-230). The John Innes Charity, Norwich.
Ke-cheng H & Bornman C H (1991) An oxygen electrode assay to determine autoclaved-induced toxicity in tissue culture media. Physiol. Plant. 81:55-58.
Kimber G & Riley R (1963). Haploid angiosperms. Bot. Rev. 29:480-531.
Kisana N S, Nkongolo K K, Quick J S & Johnson D L (1993) Production of doubled haploids by anther culture and wheat × maize method in a wheat breeding programme. Plant Breed. 110:96-102.
Kudirka D T, Schaeffer G W & Baenziger P S (1986) Wheat: Genetic variablility through anther culture. In: Bajaj Y P S (ed.) Biotechnology in Agriculture and Forestry 2. Crop I. (pp. 39-54) Springer, Berlin.
Kuo C S (1982) The preliminary studies on culture of unfertilized ovaries of rice *in vitro*. Acta Bot. Sin. 24:33-38. (in Chinese with English abstract)
Laurie D A & Bennett M D (1988a) The production of haploid wheat plants from wheat x maize crosses. Theor. Appl. Genet. 76:393-397.

Laurie D A & Bennett M D (1988b) Cytological evidence for fertilization in hexaploid wheat x sorghum crosses. Plant Breed. 100:73-82.

Lee M & Phillips R L (1987) Genomic rearrangements in maize induced by tissue culture. Genome 229:122-128.

Li D W, Z W Qin , Ouyang P & Yao Q X (1994) Wide Hybridization between *Triticum aestivum* and *Tripsacum dactyloides*. Chinese Journal Genet. 21: 271-275.

Li W Z, Jing J K, Yao G H & Hu H (1993) The effects of genotypes and mannitol pretreatment of high frequency androgenesis in barley. Chinese Sci. Bull. 38 :151-155.

Liu C H & Hu H (1990) High frequency of androgenesis in wheat (*Triticum aestivum*). Genet. Manipul. Plants 5 :24-28

MacKey J (1954) Neutron and X-ray experiments in wheat and a revision of the speltoid problem. Hereditas 40:65-180.

Maheshwari S C, Rachid A & Tyagi A K (1983) Anther pollen culture for production of haploids and their utility. IAPTC Newslett. 41:2-7.

McCoy T J, Phillips R L & Rines H W (1982) Cytogenetic analysis of plants regenerated from oat (*Avena sativa*) tissue culture: High frequency of partial chromosome loss. Can. J. Genet. Cytol. 24: 37-50.

McMouch S R & Tanksley S D (1992) Development and use of RFLP in rice breeding and genetics. Rice Biotechnol. 109-133.

Miao Z H, Zhang J J & Hu H (1988) Expression of various gametic types in pollen plants regenerated from hybrids between *Triticum-Agropyron* and wheat. Theor. Appl. Genet. 75:485-491.

Mix G, Wilson H M & Foroughi-Wehr B (1978) The cytological status of plants of *Hordeum vulgare* L. regenerated from microspore callus. Z. Pflanzenzucht. 80:89-99.

Muntzing A (1979) Triticale: Results and problems. Fortschritte der Pflanzenzuechtung, Adv. Plant Breed. 10. Parey, Berlin.

Nakamura C & Keller W A (1982) Callus proliferation and plant regeneration from immature embryos of hexaploid triticale. Z. Pflanzenzucht. 88:137-160.

Ogihara T (1981) Tissue culture in *Haworthia*, Part 4: Genetic characterization of plants regenerated from callus. Theor. Appl. Genet. 601:353-363.

Olsen F L (1988) Induction of microspore embryogenesis in cultured anthers of *Hordeum vulgare*. The effects of ammonium nitrate, glutamine and asparagine as nitrogen sources. Carlsberg Res. Commun. 52: 393-404.

Orshinsky B R, McGregor L J, Grace I E J, Hucl P & Kartha K K (1990) Improved embryoid induction and green shoot regeneration from wheat anthers cultured in medium with maltose. Plant Cell Rep. 9:365-369.

Pechan P M & Keller W A (1989) Induction of microspore embryogenesis in *Brassica napus* L. by gamma irradiation and ethanol stress. In Vitro 25:1073-1074.

Pechan P M, Bartels D, Brown D O C & Schell J (1991) Messenger RNA and protein changes associated with induction of Brassica microspore embryogenesis. Planta 184:16-165.

Potrykus I (1988) Direct gene transfer to plants. In: Application of Plant Cell Tissue Culture (Ciba Foundation symposium 137). (pp. 144-162) John Wiley, New York.

Potrykus I, Paszkowski J, Saul M, Muller K T, Schocher R, Negrutiu I, Kunzler p & Shillito R (1985) Direct gene transfer to protoplast: an efficient and generally applicable method for stable alteration of plant genomes. In: Feeling M (ed.) Plant genetics. (pp. 181-199) Liss, New York.

Sacristan M D (1971) Karyotypic changes in callus cultures from haploid and diploid *Crepis capillaris* L. Wallr. Chromosoma 33:273-293.

San Noeum L H (1979) *In vitro* induction of gynogenesis in higher plants. In: Proc Conf Broadening Genet Base Crops (pp. 327-329) Pudoc, Wageningen.

San Noeum. L H (1976) Haploids d' *Hordeum vulgare* L. par culture *in vitro* nonfecondes. Ann. Amelior Plant .24:751-754.

Schenk N, Hsiao K C & Bornman C H (1991) Avoidance of precipitation and carbohydrate breakdown in autoclaved plant tissue culture media. Plant Cell Rep. 10:115-119.

Sharp W R, Evans D A & Ammirato P V (1984) Plant genetic engineering: Designing crops to meet food industry specifications. Food Technol. 112-119.

Sitch L A, Snape J W & Firman S T (1985) Intrachromosomal mapping of crossability genes in wheat (*Triticum aestivum*). Theor. Appl. Genet. 70:309-314.

Snape J W, Chapman V, Moss J, Blanchard C E & Miller T E (1979) The crossabilities of wheat varieties with *Hordeum bulbosum*. Z. Pflanzenzucht. 85:200-204.

Suenage K & Nakajima (1989) Efficient production of haploid wheat (*Triticum aestivum*) through crosses between Japanese wheat and maize (*Zea mays*). Plant Cell Rep. 8:263-266.

Sun C S, Chu C C and Li H Q (1983) Electron microscope obervation of microspore division of wheat in vitro. Acta Bot. Sin. 25:295-300. (in Chinese with English abstract).

Tao Y Z & Hu H (1989a) Recombination of R-D chromosomes in pollen plants cultured from hybrid of 6x triticale x common wheat. Theor. Appl. Genet. 77:899-904.

Tao Y Z, Snape J W & Hu H (1991) Genetic analysis of M 27, a wheat 1R(1D) substitution line, by backcross reciprocal monosomic analysis. J. Genet. Breed. 45:189-196.

Truong-Andre I & Demarly Y (1984) Obtaining plants by *in vitro* culture of unpollinated maize ovaries (*Zea mays* L.) and preliminary studies on the progeny of a gynogenetic plant. Z. Pflanzenzucht.92:309-320.

Ushiyama T, Shionizu T & Kuwabara T (1991) High frequency of haploid production of wheat through intergeneric cross with teosinte. Jap. J. Breed. 41:353-357.

Wang C C & Kuang B J (1981) Induction of haploid plants from the female gametophyte of *Hordeum vulgare* L. Acta Bot. Sin. 23:329-330. (in Chinese with English abstract)

Wang J L, Sun C S, Lu T G, Fang A, Cui H R, Cheng S Z & Xiang C (1991) Fertilization and embryological development in wheat x maize crossed. Acta Bot. Sin. 33:674-679.

Wang X Z & Hu H (1985) The chromosome constitution of plants derived from pollen of hexaploid triticale x common wheat F_1 hybrid. Theor. Appl. Genet. 70:92-96.

Wang Y B & Hu H (1993) Gamete composition and chromosome variation in pollen-derived plants from octaploid triticale x common wheat hybrids. Theor. Appl. Genet. 85:681-687.

Wei Z M, Kyo M & Harada H (1986) Callus formation and plant regeneration through direct culture of isolated pollen of *Hordeum vulgare* cv. Sabarlis. Theor. Appl. Genet. 72:252-255.

Wu B J & Chen K C (1982) Cytological and embryological studies on haploid plant production from cultured unpollinated ovaries of *Nicotiana tabacum* L. Acta Bot. Sin. 24:125-129. (in Chinese with English abstract)

Xu H, Ai S, Chen Z & Jia X (1980) Report on heredity and viability of progenies of pollen-derived tobacco. Zhongguo Yancao 1:8-10.

Xu J C, Zhu L H, Chen Y, Lu C F & Cai H W (1994) Construction of a rice molecular linkage map using a doubled haploid population. Acta Genet. Sin. 21:205-214.

Yang D & Gao J (1979) Observation of genetic stability and variability of the progenies of pollen-derived haploid plants in wheat. Ann. Rep. Inst. Agric. Sci. Shandong, Chang Wei 1978:23-32.

Yang H Y & Zhou C (1992) Experimental plant reproductive biology and reproductive cell manipulation in higher plants: Now and the future. Amer. J. Bot. 79:354-363.

Zeng J Z & Ouyang T W (1980) The early androgenesis in *in vitro* wheat anthers under ordinary and low temperature. Acta Genet. Sin. 7:165-172. (in Chinese with English abstract)

Zhang W J, Jing J K & Hu H (1995) The transmission of rye chromosome 6R in wheat background. Chinese J. Genet. 22:113-117.

Zhang Z H, Zheng Z L & Sun W G (1983) Breeding of new rice varieties through multiple recombination between lines of pollen plants. In: Shen J H et al. (eds.) Studies on Anther Culture Breeding in Rice. (pp. 1-6) Agric Press, Beijing. (in Chinese with English abstract)

Zhou C & Yang H Y (1980) *In vitro* induction of haploid plantlets from unpollinated young ovaries of *Oryza sativa* L. Acta Genet. Sin. 7:287-288. (in Chinese)

Zhou C & Yang H Y (1981) Induction of haploid rice plantlets by ovary culture. Plant Sci. Lett. 20:231-237.

Zhu Z C & Wu H S (1979) *In vitro* induction of haploid plantlets from the unpollinated ovaries of *Triticum aestivum* and *Nicotiana tabacum* L. Acta Genet. Sin. 23:499-501. (in Chinese with English abstract)

Chapter 13

SOMATIC HYBRIDIZATION FOR PLANT IMPROVEMENT

Yu-Guang LI [1], Peter A. STOUTJESTIJK [2], and Philip J. LARKIN [2]
[1] Plant Cell Biology Group, RSBS,ANU, GPO box 475, Canberra, ACT, Australia, and
[2] CSIRO Division of Plant Industry, Canberra, ACT 2601, Australia

1 INTRODUCTION

Since the concept of somatic hybridization in plants was first developed in the early 70's there has been great expectation of the impact that it could make to plant breeding. These expectations were generated by the idea of gaining access to nuclear genes that were beyond sexual boundaries and also by the possibility to do breeding of the cytoplasmic genomes, both with phylogenetically near and distant species. This chapter attempts to review the progress to date and to examine the reasons why the expectations have so far been largely unsatisfactory. We will argue that there has in fact been extensive progress and that useful new germplasm is now beginning to emerge. The progress is the consequence of developments in a number of areas: new fusion methods, better heterokaryon selection strategies, better protoplast culture methods and more powerful molecular tools for analysing the hybrids. Hopefully with these first dividends of application will come a renewed interest and investment in this technology.

2 CURRENT STATE OF PROTOPLAST FUSION TECHNOLOGY

2.1 Overview

Somatic hybridization, in contrast to sexual hybridization, refers to the formation of hybrids between somatic, rather than germ (or gametic) cells. In plants, somatic hybrids are produced by protoplast fusion and regeneration.

A serious limitation of conventional sexual hybridization in plant breeding programmes is sexual incompatibility. Diverse reproductive barriers can prevent phylogenetically distant species from yielding any hybrids. Sexual incompatibility includes pre-zygotic incompatibility and post-zygotic incompatibility (Collins et al., 1984).

By combining somatic cells through protoplast fusion, somatic hybridization is expected to bypass sexual incompatibility. In addition, somatic hybridization enables the transfer of both nuclear and cytoplasmic genomes, and it has been suggested that the greatest contribution of protoplast fusion to plant improvement will probably be in the production of cybrids and asymmetric hybrids (Evans, 1983; Rose et al, 1990; Morikawa and Yamada, 1992).

In the context of plant improvement, the following events may be regarded as among the most significant in the development of protoplast fusion technology:

1960 Isolation of protoplasts in large numbers was realised using cell wall degrading enzymes (Cocking, 1960).

1962-1975 Basic culture media and protocols were established for protoplast, cell, and 1975 tissue culture (eg, Murashige and Skoog, 1962; Gamborg et al, 1968; Schenk and Hildebrandt, 1972; Kao and Michayluk, 1975).

1971 First plant regeneration from protoplasts (Takebe et al, 1971).

1972 First somatic hybrid plant production (Carlson et al, 1972).

1974 High-frequency protoplast fusion was obtained using polyethylene glycol as fusogen (Kao and Michayluk, 1974; Wallin et al., 1974).

1978 First intergeneric somatic hybrid plant was produced (Melcher et al., 1978) with introduced chilling tolerance (Smillie et al., 1979).

1978-1980 First asymmetric somatic hybrid/cybrid plants were produced with transferred CMS, using ionising radiation to reduce genomic contribution from the donor (Zelcer et al., 1978; Aviv et al., 1980).

1979-1981 A versatile electrical fusion method, electrofusion, was developed (Senda et al, 1979; Zimmermann and Scheurich, 1981)

1979-1984 A highly efficient method for mass selection of hybrid cells was developed using fluorescence-activated cell sorting (Galbraith and Galbraith, 1979; Harkins and Galbraith, 1984).

1987 Individual selection, culture and manipulation of protoplasts were achieved using a computer-directed system (Schweiger et al, 1987).

1990 First release (to Canadian tobacco industry as quoted by Brown and Thorpe, 1995) of a commercial cultivar derived from protoplast fusion (Douglas et al, 1981a,b,c,d; Pandeya et al, 1986)

1995 First inter-kingdom (*Nicotiana*-mouse) hybrid plant was produced with introduced animal trait (mouse immunoglobulin G) in the leaves (Makankawkeyoon et al., 1995).

Current protoplast fusion technology involves five discrete yet interrelated steps:

1. Isolation and purification of protoplasts.
2. Protoplast fusion.
3. Selection of hybrid cells
4. Protoplast culture and regeneration.
5. Hybridity verification

These five steps will be reviewed in the following subsections.

2.2 Protoplast isolation and purification

In principle protoplast isolation is quite simple: the cell wall is removed by polysaccharide-hydrolysing enzymes, usually a pectinase to dissolve the middle lamella among the walls of adjacent cells, and a cellulase (often combining hemicellulase activities) to degrade the wall of each cell. In practice, however, the optimal conditions for protoplast isolation from any particular donor plant is still not readily predicted and must be decided empirically. The main reason for this is probably variations in wall composition of different source materials. Nevertheless, current protoplast isolation techniques have enabled release of protoplasts from virtually any donor source. Some general guidelines are made from a number of reports (eg Evans and Bravo, 1983; Davies, 1988; Ishii, 1989; Roest and Gilissen, 1989, 1993; Chang and Loescher, 1991):

1. The most widely used and probably best donor sources are the young leaves of in vitro shoot culture and actively dividing suspension cell cultures. Comparatively high protoplast yield and abundance in donor materials are two main reasons for their popularity. In fusion experiments, an additional advantage is that protoplasts from green leaves and suspension cells contrast well and facilitate identification of hybrid cells (see Section 2.4). Nevertheless, a disadvantage of using suspension cell culture is the genetic

instability. Aneuploidy and decrease in regenerability are common phenomena in long term culture of suspension cells. The hypocotyl can be an alternative to suspension cells. They have been reported to be genetically stable and etiolated hypocotyl protoplasts could readily be distinguished from leaf protoplasts (Sundberg and Glimelius, 1986).

2. Commercial enzymes are rarely analytically pure. They may contain various impurities including proteolytic enzymes, phenolics and salts, which in turn exert adverse effects on the yield and viability of released protoplasts. While complete purification may mean prohibitively high cost, partial purification of the enzymes can be done conveniently using activated charcoal. Fine powder of activated charcoal is mixed with the enzyme solution to absorb the impurities and then the enzyme solution is centrifuged to remove the charcoal. In order to minimise the potential deleterious effects of the enzyme preparation, the contact time of the plant materials with the digestion solution should be minimised. The process can be accelerated and exposure time minimised by stripping off the epidermis, cutting leaves to strips prior to enzyme digestion, and vacuum filtration at the beginning of plasmolysis and enzyme digestion to facilitate permeation of the enzyme solution into the intercellular spaces.

3. To obtain high yield of viable protoplasts, two practices may be helpful: (1) Low light or complete dark for a few days to the donor plants prior to sampling. This can result in thinner cell walls with lower pectinate and/or mobilisation of starch deposited in starch granules which otherwise disrupt fragile protoplasts during manipulation (eg, pelleting and resuspension); (2) Plasmolysis prior to the enzyme digestion in a hypertonic solution (usually the non-enzyme version of the digestion solution). This can reduce the amount of enzyme-uptake by endocytosis during the protoplast isolation. The pre-plasmolysis step also washes away cell debris (eg caused by stripping or cutting leaves) and perhaps toxins released from dead cells.

Following enzyme digestion, a raw preparation of protoplasts is obtained, which is contaminated with undigested cells and tissues and debris of overdigested broken cells and must be purified to facilitate subsequent manipulation. To date the most commonly used purification technique is still the filtration-centrifugation as was sixteen years ago (Evans and Bravo, 1983). The protoplast preparation is passed through a set of filters to retain by size (eg, > 50 μm) big clumps and debris, the filtrate is then centrifuged to pellet protoplasts while leaving light debris in the supernatant due to buoyant density. This purification alone is often insufficient because the protoplasts may still be mixed with small yet heavy and so concomitantly pelleted cell debris.

In the past, sucrose or Ficoll solutions were often used in flotation purification of protoplasts, but a suitable density of sugar solution may lead to osmolality harmful to protoplast viability. In recent years, discontinuous density gradients flotation using Percoll® has proved useful and versatile in purification of protoplasts (eg, Li et al, 1993) as well as enrichment of heterokaryons (eg, Kamata and Nagata, 1987). Percoll (non-dialysable polyvinylpyrrolidone coated silica sol) has a very low osmolality and thus can provide a wide range of density gradients which can all be made appropriately iso-osmotic for protoplasts. In addition, Percoll was reported non-toxic to various cells in comparison with Ficoll, metrizamide and sodium metrizoate (Pertoft et al., 1977).

2.3 Protoplast fusion

Based on today's fusion technology, two cells of any kind can be fused successfully. Fusion of protoplasts and recovery of fusion products have so far been best accomplished by using either PEG or electrical stimulation (see 3.2). So this section will focus on these two fusogenic treatments.

PEG is a highly water-soluble, non-ionic weak surfactant. Its ether linkage makes it slightly negative in polarity and capable of forming hydrogen bonds with positively polarised groups of cell membrane components. When the chain of PEG polymer molecule is long enough (MW $\geq$ 1000) it may act as a molecular bridge between the surfaces of adjacent protoplasts, allowing agglutination to occur. PEG can also bind Ca^{++} and other cations in the fusion buffer, and Ca^{++} may further form a bridge between the PEG chain and negatively polarised groups of membrane components and, thus, enhance the agglutination. During subsequent elution, the PEG molecules that are already bound to cell membranes are forced away from the membranes; osmotic shock may also occur on cell membranes by using hypotonic washing solutions. These combined effects can cause disturbance and redistribution of the electric changes on the adhering membranes, leading to fusion (Kao and Michayluk, 1989).

The exact fusion mechanism using PEG is still not fully understood. Numerous factors are known to affect the frequency of PEG-induced fusion, the following are probably among the most important: Concentration, MW, and purity of the PEG used; Concentration of divalent cations (particularly Ca^{++}), pH, osmolality, and methods of application of the solutions involved in fusion and elution; Physiological status of the protoplasts, especially that of membranes, ie the tolerance of protoplasts to the enzymatic; and PEG treatments; and density of the protoplast population in fusion.

The PEG method has been extensively used for protoplast fusion of various plant species. The wide use of PEG has, however, encountered challenges. The impurities present in commercial preparations of PEG have often been associated with cyto-toxicity or reduced viability of fusion products, and ironically, some impurities, particularly lipophilic impurities, were reported to increase the fusion frequency (Smith et al, 1982). PEG was also reported to cause deformation of mitochondria (Rashid, 1988).

In the past ten years, electrofusion has become a competitive alternative to PEG-fusion (see Section 13.3). Electrofusion treatments are relatively non-toxic to protoplasts, convenient, and are efficient and reproducible in the fusion outcome. Electrofusion is also versatile in permitting both mass fusion of millions of protoplasts and individual fusion of defined protoplast pairs.

Electrofusion is also a two-step process, but the fusion mechanism is comparatively clear (Zimmermann, 1986; Bates et al, 1989; Jones, 1991):

(1) Protoplasts are aligned in an alternating current (AC) field which is inhomogeneous and of low voltage and high frequency. The AC causes dielectrophoresis of the protoplasts by inducing a dipole in each protoplast and masking the net surface charge distributed on the membranes. The protoplasts undergoing dielectrophoresis are mutually attracted and consequently bind together to form chains parallel to the field direction. The electrofusion buffer is composed of a non-electrolyte (usually mannitol) to ensure appropriate osmolality and at the same time enable the dielectrophoretic alignment to occur. Calcium ions at greater than micromolar concentrations in the fusion buffer are usually required for better alignment of protoplasts under dielectrophoresis.

(2) The aligned protoplasts are fused by applying a direct current (DC) pulse. The DC pulse is generally of high voltage but short duration, which causes a reversible breakdown at the contact zone of two membranes. Such membrane breakdown occurs at several points in the contact zone simultaneously leading to fusion of cell membranes and subsequent fusion of cell contents.

Optimum electrofusion should produce a large number of viable and binary heterokaryons. Parameters such as voltage, frequency and duration of AC and DC, protoplast density, and cell membrane properties can all affect the alignment and fusion of protoplasts as well as the viability of fusion products. The following phenomena have been observed by a number of researchers (Sale and Hamilton, 1968; Tempelaar and Jones, 1985a; Saunders et al, 1986, 1989; Hibi et al, 1988; Mehrle et al, 1990; Li, 1993) and may be considered as general guidelines in electrofusion experiments:

(1) Protoplast types with different membrane properties have different fusibility, eg mesophyll protoplasts tend to fuse more readily than suspension-cell protoplasts.

(2) Within the same cell type, larger protoplasts tend to fuse more readily than smaller ones; this can be estimated from a simplified equation for a model cell in electric field: $V = 3aE/2$ (where V is the transmembrane potential, a is the cell radius, and E is the electric field intensity).

(3) The rapid alignment in an AC field at low voltage (eg 50-200 V cm^{-1}) has a negligible effect on the viability of protoplasts exposed to a broad range of frequencies (eg 0.1-6 MHz); however, the frequency of binary fusion is usually higher when protoplasts are aligned in shorter chains (eg 2-4 protoplasts) than in longer chains, although the overall fusion frequency might be lower.

(4) The waveform of the DC pulse is preferably square rather than a decay curve, in order to ensure reproducible control as well as higher viability of protoplasts; shorter pulses (eg 10-50 μs) at higher voltages (eg 1,000-2,000 V cm^{-1}) are preferable to longer pulses at lower voltages. Although fusion yield may be comparable between the two combinations, the viability and plating efficiency of protoplasts tend to be higher in the former.

(5) Applying multiple DC pulses at higher voltages may facilitate the overall fusion frequency but may also be at the cost of binary fusions and viability of fusion products.

In recent years, biochemical means (eg, making use of avidin-biotin high affinity) have been attempted to increase frequency of heterokaryons (Tsong and Tomita, 1993), and exogenous proteases (eg, dispase and pronase E) have been used to improve the overall fusion frequency of mammalian cells (Ohno-Shosaku and Okada, 1989).

2.4 Selection of hybrid cells

For a practical breeding programme, it is often necessary to either select (ie, isolate and enrich) hybrid cells immediately after fusion, or inhibit non-hybrid cells with pre- and/or post-fusion treatments. There are a number of factors to be considered. Firstly, there are always more non-hybrid cells than hybrid cells in populations of fusion products. The non-hybrid cells include homokaryons as well as non-fused parental protoplasts. Unless culture conditions are made in favour of the fragile fusant cells they will generally be overgrown by the more numerous non-hybrid cells, though there were some exceptions where hybrid vigour was reported (eg, Grosser et al, 1988a; Cardi et al, 1993a & b; Taguchi et al, 1993). Secondly, cell division from protoplasts requires high culture density (usually 10^4-10^6 ml^{-1}), but the density of the hybrid cells in a fusion population is generally very low and they are often surrounded by toxic dead cells and debris. Thirdly, selection at the cellular level is far more economic than at the plant level.

Today, various techniques are available for hybrid cell selection, which are however based on only two rationales:

(1) Complementation for survival of only the hybrid cells. This is usually achieved through genetic complementation of recessive mutants, physiological complementation of autotrophic deficients, or metabolic complementation of differentially inhibited parental protoplasts.

(2) Physical isolation of the hybrid. This involves separation and enrichment using physical differences between the hybrid and the parental cells, such as buoyant density and visible characteristics in fluorescence, pigmentation, and/or regeneration capability

Genetic or physiological complementations are highly efficient, but the mutants required are not always available. Metabolic complementation relies on irreversible biochemical inhibitors rather than mutants and is more versatile. The parental protoplasts are treated with two different metabolic inhibitors (eg, one with iodoacetate, and the other with rhodamine 6-G; Böttcher et al, 1989). Following fusion, metabolic complementation allows hybrids to survive while non-fused parental protoplasts all died of the irreversible inhibition. It has been suggested that when the two parents are irreversibly inhibited by different agents, the inhibited enzymes or structures are sufficiently non-overlapping, allowing the hybrid cells to achieve a sufficient degree of metabolic functionality for survival and cell division (Wright, 1978).

Instead of using only biochemical inhibitors, the combined application of an inhibitor on one parent (recipient) and ionising radiation on the other (donor) has received far wider acceptance (see Section 3). This is perhaps due to the twofold role played by such a combination: (1) creating metabolic complementation in the fusants, and (2) directing elimination of the donor chromosomes. The latter role is of particular significance in fusion combinations involving phylogenetically distant species. Intergeneric somatic hybrids are usually highly unstable due to spontaneous elimination of parental chromosomes, and the hybrid cells are often non-morphogenic or, once regenerated, infertile (Negrutiu et al, 1989a & b). It is therefore assumed that a directed elimination of the donor chromosomes may help overcome the regeneration and fertility problems in combinations between phylogenetically distant species (Gupta et al, 1984; Bates et al, 1987; Gleba et al, 1988). From a number of recent studies some general conclusions may be made:

(1) Ionising radiation may cause DNA lesions (Hall et al, 1992a & b) and loss of alleles, loci and chromosomes (Wijbrandi et al, 1990 a & b) of the irradiated protoplasts.

(2) This may result in enforced unidirectional elimination of the irradiated nuclear DNA (Wijbrandi et al, 1990b & c; Melzer and O'Connell, 1992) and mitochondrial DNA (Bonnema et al, 1992; Hall et al, 1992b) in the hybrid cells and their progeny.

(3) Chromosome elimination is effected by radiation dosage as well as parental species (Babiychuk et al, 1992; Lefrancois et al, 1993); the extent of asymmetry appears to increase with the phylogenetic distance. To date, however, it appears that the directed elimination of donor genomes using ionising radiation has only limited success in overcoming plant regeneration and fertility problems (see Section 4.2)

Physical isolation for heterokaryon selection is feasible when the two parental protoplast populations differ considerably in physical characters, such as colour, buoyant density, or fluorescence. The hybrid cells are then expected to be a mixture or intermediate between the two parents. For example, fusions between mesophyll and suspension-cell protoplasts yield heterokaryons that contain both green chloroplasts of the mesophyll protoplasts and the translucent cytoplasmic strands of the suspension-cell protoplasts. These heterokaryons can be isolated using a micromanipulator (eg, Fahleson et al, 1988; Mendis et al, 1991), or even with a hand-held micropipettor. Nonetheless, such isolation is too tedious to attract general application. The physical separation can be automated with a FACS (fluorescence-activated cell sorter) as developed by Herzenberg et al. (1976) and applied to protoplast selection (Galbraith and Galbraith, 1979; Harkins and Galbraith, 1984). Once the parental protoplasts are labelled with different fluorochromes (eg, carboxyfluorescein and scopoletin) the FACS is capable of selecting fusant cells that display dual fluorescence. The major drawback with this FACS technology is the high cost of the equipment.

2.5 Protoplast culture and regeneration: a case study of Medicago species

Passage through a successful selection system can provide a heterokaryon population. For breeding purposes, these heterokaryons must undergo cell division, morphogenesis and regeneration into plants. For a general review

Table 1. Plant regeneration from protoplast of *Medicago* species

Species	Cultivar/line	Protoplast source[1]	Recovery medium [2]	Reference
M. arborea		R, L	KM8P	Mariotti et al., 1984
M. arborea		L	KM8P	Arcioni et al., 1985
M. coerulea		L	KM8P	Arcioni et al., 1982
M. difalcata		C	Kao	Gilmour et al., 1987b
M. falcata		C	Kao	Gilmour et al., 1987b
M. falcata	318	L	Modified Kao	Téoulé 1983
M. glutinosa		S	KM8P	Arcioni et al., 1982
M. glutinosa		C	Kao	Gilmour et al., 1987b
M. hemicycla		C	Kao	Gilmour et al., 1987b
M. sativa	Adriana	L, R, S	KM8P/KM8	Pezzotti et al., 1984
M. sativa	Answer	S	Modified Kao	Atanassov & Brown 1984
M. sativa	Canadian #1 & #2	L	Kao	Kao & Michayluk 1980
M. sativa	Citation	S	Modified Kao	Atanassov & Brown 1984
M. sativa	DuPuit	S	Modified KM8P	Niizeki & Saito 1987
M. sativa	Europe	L	KM8P	Dos Santose et al., 1980
M. sativa	Europe	C, R, L	Kao/modified Kao	Lu et al., 1982; 1983
M. sativa	Europe	R	Modified Kao	Xu et al., 1982
M. sativa	Ladak	L	Modified KM8P	Song et al., 1990
M. sativa	MSR 12	L	Modified Kao KM8P	Téoulé 1983
M. sativa	NSVAH	S	Modified B_5	Mezentsev 1981
M. sativa	Rambler	L	Modified Kao	Dijak & Brown 1987
M. sativa	Rambler	S	Modified KM8P	Niizeki & Saito 1987
M. sativa	Rangelander	L	Modified Kao	Dijak & Brown 1987
M. sativa	Rangelander	S	Modified KM8P	Niizeki & Saito 1987
M. sativa	Rangelander	L	Modified KW	Larkin et al., 1988
M. sativa	Regen S	L	Modified KW	Johnson et al., 1981
M. sativa	Regen S	L	Modified Kao	Atanassov & Brown 1984
M. sativa	Regen S	L	Modified Kao	Dijak & Brown 1987
M. sativa	Vernal	S	Modified KM8P	Niizeki & Saito 1987
M. truncatula	Jemalong	L	Modified KW	Rose & Nolan 1995
M. varia		C	Kao	Gilmour et al., 1987 b

[1] R = seedling root; L = leaf; C = seedling cotyledon; S = suspension cell or callus. [2]KM8P and KM8 (Kao and Michayluk, 1975); Kao (Kao, 1977); KW (Kao and Wetter, 1977); B_5 (Gamborg et al., 1968).

of protoplast culture and regeneration the reader is referred to Chapter 2 of this book and several recent review articles (Roest and Gilissen, 1989, 1993). As a case study, the authors wish to present a specific review on an important forage crop, .-3B*Megica o sativa* (lucerne, alfalfa).

Although legume species in general are still regarded as recalcitrant in terms of plant regeneration from protoplasts, protoplast-derived plants have been obtained from a number of *Megica o* species, particularly lucerne. Table 1 is a summary of reports of plant regeneration from protoplasts of *Megica o* species. There have been at least 20 independent reports on plant regeneration from the protoplasts of 9 *Megica o* species, involving 14 reports on 13 lucerne cultivars. In addition, a much larger number of lucerne cultivars, breeding lines and proprietary clones have been identified capable of regeneration from callus tissue (Mitten et al, 1984; Brown and Atanassov, 1985; Bingham et al, 1988). The protoplasts have been isolated and regenerated into plants from leaves, cotyledons, roots and suspension cells, but more frequently from leaves (Table 1). With one exception (Mezentsev, 1981) all these successful recoveries of protoplasts of *Megica o* species have been based on culture in Kao's media (Table 1), either KM8P (Kao and Michayluk, 1975), KW (Kao and Wetter, 1977), Kao (Kao, 1977), or a modified version of Kao's media. Some general observations are listed below:

(1) Regenerability is genotype-specific, heritable and probably a dominant trait fitting a two-gene model (Reisch and Bingham, 1980; Hernández-Fernández and Christie, 1989; Crea et al, 1995). Most lucerne stocks contain about 10% of genotypes capable of regeneration (Bingham et al, 1988).

(2) Creeping-rooted cultivars (ie, those form adventitious shoots from roots) are often better regenerators than non-creepers (Brown and Atanassov, 1985). The cultivar Rangelander is the best regenerator currently known in lucerne from which almost all genotypes can regenerate (Atanassov and Brown, 1984; Bingham et al, 1988).

(3) Somatic embryogenesis is the principal pathway of regeneration in lucerne (Stuart et al, 1985), but in protoplast culture it occurs more often with an intervening callus stage. The presence of 2,4-D in culture medium is usually required for induction of somatic embryogenesis (Saunders and Bingham, 1972; Davies, 1988).

(4) Kao's media are particularly suitable for recovery of *Megica o* protoplasts (Table 1), and the basic culture protocol developed by Atanassov and Brown (1984) appears to be widely applicable in plant regeneration of different lucerne cultivars.

Using KWM, a modified KW medium, Davies et al (1989) demonstrated a significant role of ultrafiltration of the medium in recovery of lucerne protoplasts. The divided cells were increased by 50-400% in ultrafiltered KWM over those in the non-ultrafiltered. This effect was interpreted as due to the elimination of high MW inhibitors in the highly enriched KW medium. Using ultrafiltered KWM and calcium-alginate beads for nurse culture, the regeneration efficiency was further improved. Colony formation at 4% relative to the initial protoplast population was recorded (Larkin et al, 1988), which represented a 30 fold improvement over the previous best recoveries of colonies. Embedding protoplasts in calcium alginate matrix has two advantages over that in agarose: (1) absence of an elevated temperature treatment as required by the latter; and (2) easy release of the entrapped cells in sodium citrate.

In an assessment of nurse methods using feeder layers, conditioned media or membrane chamber, Gilmour et al (1987a) concluded that the only system capable of sustaining division of low numbers of forage legume protoplasts was that employing membrane chambers. In our laboratory, three nurse systems have been used for lucerne protoplast culture, calcium alginate beads (Larkin et al, 1989), membrane chambers using Millipore™-insert (Li et al, 1993) and agarose embedding (Stoutjesdijk, 1996). All three were effective in improving protoplast recovery and regeneration. Millipore™-insert membrane chambers were the simplest to use but were the least effective. Alginate beads were more effective than agarose embedding . These three nurse systems have a common advantage in allowing protoplasts to be subcultured rapidly and conveniently. This is of particular importance for sustained division in the first month of culture, during which medium changes are essential to achieve a gradual drop of osmotic pressure, a steady supply of nutrients and necessary alterations of growth regulators.

2.6 Hybridity verification

The recovery of protoplasts through a selection system is not a conclusive proof for hybridity, since selection systems sometimes permit parental escapes. Too stringent selection is often at the cost of recovery rate of hybrid cells. In addition, not all daughter cells derived from a fusant cell will be genetically identical due to changes such as unbalanced translocations and chromosome assortment; consequently the colonies formed may be of chimerical origin and the plants regenerated vary in their genetic composition. It is, therefore, necessary to verify the hybridity at the plant level. Verification may come from a genotypic assay such as karyotype and DNA analyses, or from a phenotypic assay such as morphological and isoenzyme analyses. For plant breeding purpose, the transfer of the desired trait to the target plant is the most significant verification.

While gross genomic changes (eg, at ploidy level) may be judged by chromosomal and/or morphological observations , molecular techniques are required to detect subtle genomic changes in asymmetric hybrids or cybrids.

A powerful technique for DNA analysis in hybrid verification is Southern analysis (eg, Yarrow et al., 1990; Gavrilenko et al.., 192). As little as 0.1 pg blotted DNA complimentary to the probe is detectable under optimised conditions (Sambrook et al., 1989). In conventional Southern analysis, species-specific DNA probes are required to verify the presence of donor DNA in hybrids. However, species-specific DNA probes are not always available, and this is particularly true in breeding programmes aiming at transfer of a trait from a wild relative of the crop species. Hunting for species-specific probes, which are ideally highly repetitive and dispersed over the whole donor genome, can be difficult and laborious, especially when large homologies in DNA sequences exist between the two parental species (Anamthawat-Jónsson et al., 1990).

Total genomic probing offers an alternative to conventional Southern analysis. It was initially used in the identification of breeding lines and sexual hybrids in cereal crops (Schwarzacher et al. 1989, 1992; Anamthawat-Jónsson et al., 1990). Li et al. (1993) adapted this technique for identification of somatic hybrids. Unlabelled total genomic DNA of the recipient parent is used as a blocking agent, while the donor total DNA is labelled and used as a total genomic probe. The blocking agent is applied in excess and the hybridization conditions optimised to block the DNA sequences that are common between the two parental species. This leaves only donor-specific sequences in the blotted putative hybrid DNA available for binding the radioactive complementary sequences in the probe DNA. Using this technique highly asymmetric somatic hybrids (about 0.4% donor DNA in the recipient genome) are detectable as quantitated using a PhosphorImager (Li et al, 1993). Compared with hunting for species-specific probes, the total genomic probing is more economical in both time and labour. Moreover, a total genomic probe involves all the species-specific sequences, and, thus, more versatile than any single cloned probe if detection of all types of hybrids/cybrids is desired.

Restriction fragment length polymorphisms (RFLP) have been used to idendify somatic hybrids as well as cybrids. For example, the electrophoretic patterns of cytoplasmic DNA after digestion with an appropriate restriction endonuclease are characteristic of the cytoplasmic genomes, which can be detected by staining with ethidium bromide, or by Southern hybridization with a repetitive sequence of either mtDNA, cpDNA, or cDNA (eg D'Hont et al., 1987; Kao et al., 1991; Narasimhulu et al., 1992). The resolution power of RFLPs with Southern hybridisation has been reported as 0.1-0.5% donor DNA in the recipient genome, in contrast to about 5% with direct ethidium-bromide staining (Scowcroft and Larkin, 1981; Clark et al., 1986).

Recently, random amplified polymorphic DNA (RAPD) technology has been used for verification of somatic hybrids (eg, Nakano and Mii, 1993; Craig et al., 1994; Hansen and Earle, 1995; Stoutjesdijk, 1996). When identifying hybridity, random primers are first screened for their ability to generate specific polymorphisms for each fusion parent using PCR (polymerase chain reaction). Once satisfactory primers are identified, they are then used to detect parental polymorphisms in putative somatic hybrids. Under optimal conditions, this PCR-based method is capable of detecting single copy genes within total genomic DNA in a comparatively short time span and without use of radioactive labelling (Yu et al., 1993; Guidet, 1994).

3 PROGRESS USING SOMATIC HYBRIDIZATION FOR PLANT IMPROVEMENT

3.1 Tabulation of somatic hybrid/cybrid plants produced in breeding programs

Since Carlson et al. (1972) reported the first protoplast-fusion-derived plant, a large number of somatic hybrids/cybrids have been produced through protoplast fusion. To provide a big picture of the current application of protoplast fusion technology for plant improvement, we have tabulated the somatic hybrid/cybrid plants produced so far where there was a specified plant breeding purpose (Table 2), ie, they are to improve an existing target species or breeding line (recipient) by introducing a specified trait from another species or breeding line (donor). This table does not include those hybrids/cybrids which (1) failed to regenerate plants, (2) were produced primarily for basic research purposes or (3) for creating genetic diversity without specifying the desired trait(s). In a plant breeding context, we believe the table can serve as a general guide without losing significant information by excluding the three groups of hybrids/cybrids as mentioned above.

Table 2. Somatic hybrid plants produced for breeding purpose

Recipient species (R)	Donor species (D)	Desired donor trait (DDT)	Fusion method	Selection and verification methods	DDT transfer	Best fertility	Genome symmetry	Reference
Caryophyllaceae								
Dianthus carophyllus	*D. chinensis*	Floral traits	PEG	R: non-regenerating D: IOA MA, IZA, CC, RAPD	+	ND	S	Nakano & Mii, 1993
Dianthus barbatus	*Gypsophila paniculata*	Floral traits	EF	R: IOA D: non-regenerating CC, dot blot (rDNA)	+	ND	S	Nakano *et* al., 1996
Compositeae								
Cichorium intybus	*Helianthus annuus*	CMS	PEG	D: non-regenerating MM, RFLP (mtDNA)	+	FF	S	Rambaud *et al.*, 1993
Helianthus annuus	*H. giganteus*	Disease and insect resistance	PEG	R: non-regenerating D: IOA, MA, IZA, CC	ND	FF	S	Krasnyanski & Men czel, 1995
Cruciferae								

Recipient species (R)	Donor species (D)	Desired donor trait (DDT)	Fusion method	Selection and verification methods	DDT trans-fer	Best fer-tility	Genome symme-try	Reference
Brassica Napus	1. *B. napus* CMS line 2. *B. campestris*	CMS TZ tolerance	PEG	R: green, erucic acid (EA) negative D: yellow at low temperature, EA positive (D1). RFLP (cpDNA, mtDNA), CC, thylakoid protein assay, Southern (mtDNA)	+	FF	S & Cy	Pelletier *et al.*, 1983; Chetrit *et al.*, 1985
B. napus\ CMS line	*B. napus* TZ-tolerant line	TZ tolerance	PEG	R: IA-treated D: γ-irradiated RFLP (cpDNA, mtDNA), CC	+	F	A	Barsby *et al.*, 1987a
B. napus winter line	*B. napus* CMS spring line	CMS	PEG	R: IA-treated D: γ-irradiated RFLP (cpDNA, mtDNA), CC	+	F	A	Barsby *et al.*, 1987b
B. oleracea CMS line	*B. campestris* TZ tolerant Line	TZ tolerance	PEG	R: CMS D: etiolated, no protoplast recovery RFLP (rDNA, cpDNA, mtDNA), MA	+	F	S	Robertson *et al.*, 1987
B. napus	*Eruca sativa*	Tolerance to drought and aphids	PEG	R: Etiolated, carboxyfluorescein-stained D: Green Isolation by MM or FACS, IZA, FC, cpDNA RFLP gel stain	-	F	S	Fahleson *et al.*, 1988

Recipient species (R)	Donor species (D)	Desired donor trait (DDT)	Fusion method	Selection and verification methods	DDT transfer	Best fertility	Genome symmetry	Reference
B. napus	*B. hirta*	Tolerance to drought and *Alternaria brassicae*	PEG	R: high regenerability D: no plant regeneration MA, RFLP (cpDNA, mtDNA),CC, RFLP (rDNA)	ND	F	S	Primard *et al.*, 1988
B. napus winter line	*B. napus TZ* – tolerant spring line	TZ tolerance	PEGm	RFLP (cpDNA)	+	F	Cy	Thomzik & Hain, 1988
B. oleracea	*B. campestris*	TZ tolerance	PEG	R: CMS Southern (mtDNA), RFLP (rDNA, cpDNA), MA, IZA, CC	+	F	S	Jourdan *et al.*, 1989a
B. oleracea	*B. napus*	TZ tolerance	PEG	R: CMS,etiolated D: green MA,IZA,CC,RFLP (mtDNA,cpDNA)	+	F	S & Cy	Jourdan *et al.*, 1989b

Recipient species (R)	Donor species (D)	Desired donor trait (DDT)	Fusion method	Selection and verification methods	DDT transfer	Best fertility	Genome symmetry	Reference
B. napus	*B. nigra*	Resistance to *Phoma lingam* & *Plasmodiophora brassicae*	PEGm	D:Hygro^{r}, x-irradiated CC,IZA	-	F	A	Sacristán *et al.*, 1989
B. napus	*B. juncea* *B. nigra* *B. carinata*	Resistance to *Phoma lingam*	PEG	D:X-irradiated RFLP(rDNA, nDNA), IZA	+	F	A	Sjödin & Glimelius, 1989
B. napus	*Raphanus sativus*	CMS	PEG	R: IOA-treated D: cytoplasts used MA,CC,Southern (mtDNA)	+	ND	Cy	Sakai & Imamura, 1990
B. juncea	*Eruca sativa*	Tolerance to drought and aphids	PEGm	D: γ-irradiated MA,IZA	ND	F	A	Sikdar *et al.*, 1990
B. oleracea	*B. napus*	CMS	PEG/ high pH	R: IA-treated D: γ-irradiated Southern, RFLP(cpDNA,mtDNA)	+	FF	Cy	Yarrow *et al.*, 1990
B. oleracea	*B. campestris*	TZ tolerance	PEG	R: CMS MA,IZA,Southern(mtDNA), RFLP(cpDNA)	+	FF	ND	Christey *et al.*, 1991

Recipient species (R)	Donor species (D)	Desired donor trait (DDT)	Fusion method	Selection and verification methods	DDT trans-fer	Best fer-tility	Genome symme-try	Reference
B. napus cv. Westar	*B. napus* CMS line	CMS	PEG/ high pH	R:TZ tolerant MA,RFLP(mtDNA, cpDNA),CC	+	FF	Cy	Kao *et al.*, 1991
B. juncea	*B. spinescens*	High photosyn-Thetic deficiency, white rust resistance, salt tolerance	PEG	MA,CC,IZA	ND	FF	S	Kirti *et al.*, 1991
Lactuca sativa	*L. virosa*	Bacterial rot resistance	EF	R:IOA-treated D:no protoplast recovery CC,IZA,MA	ND	St	S	Matsumoto, 1991
B. oleracea	*Raphanus sativus*	Clubroot disease resistance	EF	R:IOA-treated D:no plant regeneration MA,IZA,RFLP(cpDNA, mtDNA,rDNA),CC	+	F	S	Hagimori *et al.*, 1992
B. oleracea	*B. napus* CMS line	CMS, novel organelles	PEG/ high pH/ high Ca^{++}	R: IOA-treated D: γ-irradiated RFLP(mtDNA,cpDNA),CC	+	FF	S	Kao *et al.*, 1992
B. juncea	*Moricandia arvensis*	Water use efficiency, resistance to	PEG	RFLP(rDNA,mtDNA), CC, MA	-	FF	S	Kirti *et al.*, 1992a

Recipient species (R)	Donor species (D)	Desired donor trait (DDT)	Fusion method	Selection and verification methods	DDT transfer	Best fertility	Genome symmetry	Reference
		fungal diseases						
B. juncea	*Trachystoma ballii*	Indehiscent pods, alternaria blight resistance	PEGm	CC,MA,RFLP(rDNA)	+	FF	S	Kirti *et al.*, 1992b
B. napus	*Raphanus Sativus*	Resistance to beet cyst nematode	PEG	CC, FC, RFLP (rDNA, cpDNA, mtDNA)	+	MF	S	Lelivelt & Krens, 1992
B. oleracea ssp. *Botrytis*	*B. oleracea* coldtolerant line	Cold tolerance	PEGm	R: CMS,IA-treated,green D: γ-irradiated,etiolated and FDA-stained MM or FACS, RFLP (mtDNA,cpDNA)	+	ND	Cy	Walters *et al.*, 1992
B. oleracea	*Raphanus sativus*	Resistance to fungal diseases	PEGm	R: high regenerability, IOA-treated MA,IZA,CC,RFLP (rDNA)	ND	F	S	Yamanaka *et al.*, 1992
B. napus	*Arabidopsis thaliana*	Herbicide resistance	PEG	D: γ-irradiated, Chlorsulfuron resistant, MA, IZA, RFLP (rDNA, cpDNA)	+	St	A	Bauer-Weston *et al.*, 1993
B. napus	*Sinapis alba*	Beet cyst nematode resistance	PEG	R: chlorophyll autoflourescence D: FDA stained MM, MA, CC, RFLP(rDNA,cpDNA,	+	St	A & S	Lelivelt *et al.*, 1993

Recipient species (R)	Donor species (D)	Desired donor trait (DDT)	Fusion method	Selection and verification methods	DDT transfer	Best fertility	Genome symmetry	Reference
				mtDNA)				
B. napus	*Thlaspi perfoliatum*	Long-chain fatty acids	PEG	R:carboxyfluoresce-Indiacetate stained D:chlorophyll autoflourescence, x-irradiated FACS, IZA, dot blot	+	FF	A & S	Fahleson *et al.*, 1994
B. carinata	*Camelina sativa*	*Alternaria* blight resistance	PEG	CC, MA, RFLP (rDNA, MtDNA,cpDNA)	ND	ND	A & S	Narasimhulu *et al.*, 1994
B. napus	*B. tournefortii*	*Phoma lingam resistance*	PEG	R:carboxyfluoresce in diacetate stained D: chlorophyll autoflourescence FACS, IZA, MA, Southern,RFLP(cpDNA)	+	FF	S	Liu *et al.*, 1995
B. juncea	*Diplotaxis harra*	Drought tolerance	PEG	D: non-dividing protoplasts MA, IZA, CC	ND	FF	A & S	Begum *et al.*, 1995
B. napus	*B. nigra*	*Plasmodiaphora brassicae* resistance	PEGm	R: $Hygro^r$ D: x-irradiated MA,slot blot(ssDNA)	+	ND	A & S	Gerdemann-Knork et al., 1995
B. oleracea	*B. napus*	Resistance of black rot caused by	EF	R:IOA D:no division MA, CC, FACS, RAPD	+	FF	S	Hansen & Earle, 1995

Recipient species (R)	Donor species (D)	Desired donor trait (DDT)	Fusion method	Selection and verification methods	DDT transfer	Best fertility	Genome symmetry	Reference
		Xanthomonas campestris pv campestris						
B. oleracea	*B. rapa*	Cold tolerance	PEG	R: IOA MA, FACS, IZA, RFLP(mtDNA)	+	FF	S	Heath & Earle, 1996
Cucurbitaceae								
Cucumis melo	Pumpkin hybrid(*Cucurbita maxima* × *C. moschata)*	Unspecified fruit characteristics	EF	Selection: colony hybrid vigour MA, IZA, CC	ND	F	A	Yamaguchi & Shiga, 1993
Gramineae								
Oryza sativa	*Echinochloa oryzicola*	C4 Photosynthesis capability	EF	R : no protoplast recovery D: high regenerability, IOA-treated MA, IZA, CC	ND	ND	S	Terada *et al.*, 1987
O. sativa cv. Nipponbare	*O. sativa* CMS Line	CMS	EF	R: IOA-treated D : x-irradiated CC , IZA, Southern(mtDNA), mtDNA RFLP gel stain	+	FF	Cy	Kyozuka *et al.*, 1989
O. sativa cv. Fujiminori	*O. sativa* CMS line	CMS	EF	R: IOA-treated D : γ-irradiated	+	ND	Cy	Yang *et al.*, 1989

Recipient species (R)	Donor species (D)	Desired donor trait (DDT)	Fusion method	Selection and verification methods	DDT transfer	Best fertility	Genome symmetry	Reference
				MA, IZA, CC, mtDNA RFLP gel stain				
Lolium multiflorum	*Festuca arundinacea*	Agronomic traits such as persistance, CMS	EF	R: x-irradiated or non-irradiated D: IOA, CC,RFLP(cpDNA, mtDNA), dot blot	ND	ND	A	Spangenberg *et al.*, 1994; 1995
Labiatae								
Mentha piperita	*Mentha gentilis*	Novel volatile compounds	EF	MA, CC, RAPD, volatile oil analysis	+	ND	S	Sato *et al.*, 1996
Leguminosa (Fabaceae)								
Medicago sativa	*M. falcata*	Winter hardiness	High pH/High Ca^{+}	R: purple flower D: yellow flower MA, IZA, CC, RFLP (mtDNA, cpDNA)	ND	ND	S	Téoulé1983; D'Hont *et al.*, 1987
Lotus corniculatus	*L. Conimbricensis*	Indehiscent pods	PEGm	R: etiolated, IA-treated D: no protoplast recovery IZA, MA	ND	St	S	Wright *et al.*, 1987
L. corniculatus	*L. tenuis*	Rapid recovery from defoliation	EF	R: geen, CT positive D: colourless SC, CT negative, Kan^{r}, IA-treated, MA,IZA, CC, CT assay	ND	ND	S	Aziz *et la.*, 1990
Medicago sativa	*M. intertexta*	Resistance to pests, downy	PEG	R: Kan^{r} D: Hygro^{r}, no plant	ND	ND	S	Thomas *et al.*, 1990

Recipient species (R)	Donor species (D)	Desired donor trait (DDT)	Fusion method	Selection and verification methods	DDT transfer	Best fertility	Genome symmetry	Reference
		mildew and anthracnose		regeneration RFLP(rDNA), CC				
M. sativa	*M. falcata*	Winter hardiness	PEG	R: purple flower, FDA-labelled D: yellow flower, rhodamine isothiocyanate-labelled Isolation by MM, MA, IZA, CC	ND	F	ND	Mendis *et al.*, 1991
M. sativa	*Onobrychis viciifolia*	Foliar condensed tannins (CT)	EF	R: IOA-treated mesophyll D: no plant regeneration, γ-irradiated SC MA,CC, Southern (total genomic probe), CT assay	+ but transient	MF	A	Li *et al.*, 1993;1996
M. sativa	*M. coerulea*	Hexaploid breeding lines	EF	R: green mesophyll D: colourless SC MM, MA, CC, IZA, Southern	+	F	S	Pupilli *et al.*, 1992; 1995
M. sativa	*M. arborea*	*Coryne bacterium insidiosum* resistance	EF	R: Green D: fluorescein isothiocyanate fluorescence MM,MA, CC, IZA, Southern	ND	ND	S	Nenz *et al.*, 1996
Lilliaceae								
Asparagus officinalis	*A. macowanii*	*Fusarium oxysporum,*	EF	R: IOA D: no protoplast recovery	ND	ND	S	Kunitake *et al.*, 1996

Recipient species (R)	Donor species (D)	Desired donor trait (DDT)	Fusion method	Selection and verification methods	DDT transfer	Best fertility	Genome symmetry	Reference
		F. moniliforme, *Phomopsis* asparagi resistance		MA, CC, RFLP (RdNA), RAPD				
Passifloaceae								
Passiflorae dulis	*P. alata*, *P. amethystina*, *P. cincinnuata*, *P. giberti*, *P. coccinea*	Pest and disease resistance	PEG	MA, CC, IZA	ND	ND	S	Dornelas *et al.*, 1995
Rutaceae								
Citrus sinensis rootstock	*P. trifoliata*	Cold hardiness, resistance to tristeza virus, foot rot, and citrus nematode	PEG	R: no plant regeneration, unifoliate leaves D: no protoplast recovery, trifoliate leaves MA, IZA, CC	ND	NA	S	Grosser *et al.*, 1988a
C. sinensis rootstock	*Severina disticha*	Cold hardiness, salt and boron tolerance,	PEG	R: no plant regeneration, embryogenic D: non- embryogenic MA, IZA, CC	ND	NA	S	Grosser *et al.*, 1988b

Recipient species (R)	Donor species (D)	Desired donor trait (DDT)	Fusion method	Selection and verification methods	DDT transfer	Best fertility	Genome symmetry	Reference
		resistance to *Phythophthora* and nematodes						
Rudbeckia hirta	*R. laciniata*	Floral traits	EF	R: plant via shoot regeneration, pigmented root D: plant via rhizogenesis MA, IZA, cpDNA RFLP gel stain	+	MF	S	Al-Atabee *et al.*, 1990
Citrus reticulata rootstock	*Citropsis gillatiana*	Resistance to burrowing nematode & phytophthora foot rot	PEG	MA, CC, IZA	ND	NA	S	Grosser *et al.*, 1990
C. limon	*Citrus sinensis* *C. jambhiri* hybrid line	Mal secco tolerance, cold hardiness, seedless triploid	PEG	MA, IZA, CC	-	ND	S	Tusa *et al.*, 1990; 1992
C. sinensis *rootstock*	*Fortunella crassifolia*	Cold hardiness, dwarfing	PEG	CC, IZA	ND	NA	S	Deng *et al.*, 1992

Recipient species (R)	Donor species (D)	Desired donor trait (DDT)	Fusion method	Selection and verification methods	DDT transfer	Best fertility	Genome symmetry	Reference
		growth habit						
Citrus hybrid "Tangelo"	*Somatic hybrid of Citrus. sinensis and C. reticulata*	Autotetraploid lines for seedless cultivar breeding	PEG	D: non-regenerating MA, IZA, CC	ND	MF	S	Grosser *et al.*, 1992
Citrus reticulata, C. sinensis, C. aurantium, C. limonia	*Poncirus trifoliata, C. reticulata*	Rootstock characteristics	PEG	R: leaf meosphyll, non-regenerating D: non-dividing MA,IZA	ND	NA	S	Grosser *et al.*, 1994
Citrus paradisi, C. reticulata, C. sinensis, Fortunella crassifolia	*Altantia ceylanic* *C. retiulata, C. ichangensis, Citropsis gilletiana Feronia limonia, F. crassifolia,* *Microcitrus papuana, Poncirus trifoliata, Severinia buxifolia,*	New varities, multiple agronomic traits (increased disease resistance of rootstock), new citrus varities	EF	R: non-dividing D: leaf meosphyll, non-regenerating MA, CC, IZA, RAPD	ND	ND	S	Grosser *et al.*, 1996

Recipient species (R)	Donor species (D)	Desired donor trait (DDT)	Fusion method	Selection and verification methods	DDT transfer	Best fertility	Genome symmetry	Reference
	S. disticha							
Solanaceae								
Lycopersicon esculenum	*Solanum tuberosum*	Chilling tolerance	PEG/ high pH	R: yellowish when grown unshaded CC, RuBPCase IEF, MA	+	ND	S	Melchers *et al.*, 1978; Smillie *et al.*, 1979
Nicotiana sylvestris	*N. tabacum*	CMS	PEG	R: no plant regeneration D: x-irradiated CC, MA	+	ND	A&Cy	Zelcer *et al.*, 1978; Aviv *et al.*, 1980
N. tabacum	*N. rustica*	Resistance to blue mole & black root rot, elevated nicotine content	PEGm	R: albino line D: chlorophyll deficient chlorophyll complementation, MA, CC, IZA, RUBPCase IEF	+	F	S	Dounglas *et al.*, 1981 a,b, c, d; Pandeya *et al*, 1986
N. tabacum	*N. nesophila* *N. stocktonii*	Resistance to TMV & other diseases	PEG	R: albino mutant D: normal green CC, IZA, MA	+ TMV	F	S	Evans *et al.*, 1981
N. tabacum	*N. glutinosa*	Resistance to TMV	PEG	R: CMS RuBPCase IEF, MA, CC	+	FF	S	Uchimiya 1982
Solanum tuberosum	*S. nigrum*	TZ tolerance	PEG/ high pH	MA, CC	+	F	S	Binding *et al.*, 1982
S. tuberosum	*S. chacoense*	Potato virus Y resistance	PEG(?)	R: no plant regeneration D: no protoplast recovery MA, CC, RuBPCase IEF,	+	F	S	Butenko *et al.*, 1982

Recipient species (R)	Donor species (D)	Desired donor trait (DDT)	Fusion method	Selection and verification methods	DDT transfer	Best fertility	Genome symmetry	Reference
				RFLP (cpDNA, mtDNA)				
S. tuberosum	*S. brevidens*	Disease & physiological resistance	PEG/ high pH	MA, CC, cpDNA RFLP gel stain, RuBPCase IEF	ND	St	S	Barsby *et al.*, 1984
S. tuberosum diploid line	*S. brevidens*	PLRV resistance	PEG	MA, hybrid vigour, CC	+	F	S	Austin *et al.*, 1985
S. tuberosum tetraploid line	*S. brevidens*	PLRV resistance	PEG	R: late blight resistant MA, hybrid vigour, CC	+	F	S	Austin *et al.*, 1986, Helgeson *et al.*, 1986
S. melongena	*S. sisym-briifolium*	Resistance to root knot nematode & carmine spider mite	PEG	R: 6-azauracil resistant D: purple callus with anthocyanins MA, CC, IZA, RuBPCase IEF	+	St	S	Gleddie *et al.*, 1985, 1986
Lycopersicone sculentum	*S. lycopersi-coides*	Chiling tolerance	PEGm	R: protoplast inhibited by fusogen D: no protoplast recovery Hybrid vigour, MA, IZA, CC, RuBPCase IEF	ND	F	S	Handley *et al.*, 1986
L.esculentum	*L. peruvianum*	Insect & disease resistance, high carotene & vitamin C	PEG	R: no plant regeneration MA, RuBPCase IEF, CC	ND	F	S	Kinsara *et al.*, 1986
Nicotiana	*N. plumbaginifol-ia*	TZ tolerance	PEGm	R: albino mutant	+	F	Cy	Menczel *et al.*,

Recipient species (R)	Donor species (D)	Desired donor trait (DDT)	Fusion method	Selection and verification methods	DDT transfer	Best fertility	Genome symmetry	Reference
				fluorescence transient, x-irradiated MA, Southern(cpDNA)				
Solanum tuberosum	*S. phureja*	Unspecified disease resistance	EF	D: bleached, FDA-stained Isolation by MM, CC,FC, MA, Giemsa C-banding	ND	ND	S	Puite *et al.*, 1986
S. tuberosum dihaploid line	*S. brevidens* diploid line	Resistance to PLRV & potato viruses X & Y	EF PEG	MA, IZA, RFLP(cDNA), CC, cpDNA RFLP gel stain	+	FF	S	Fish *et al.*, 1987, 1988a,b; Pehu *et al.*, 1989, 1990
Solanum tuberosum	*S. pinnatisectum*	Unspecified disease resistance	PEGm	R: albino mutant D: normal green,γ-irradiated MA, IZA, cpDNA RFLP gel stain	ND	ND	S	Sidorov *et al.*, 1987
S. tuberosum dihaploid or mono haploid lines	*S. tuberosum* fungusresistant lines	Resistance to *Phytoph-Thora* or *Fusarium* in heterozygous hybrid	PEGm	CC, IZA, MA	ND	ND	S	Deimling *et al.*, 1988
N. tabacum	*N. africana*	New source of CMS	PEG/ high pH /high Ca^{++}	D: x-irradiated, CC, MA, RFLP (mt DNA), RuBPCase IEF, RFLP (cpDNA, mtDNA)	+	NA	Cy	Kumashiro *et al.*, 1988

Recipient species (R)	Donor species (D)	Desired donor trait (DDT)	Fusion method	Selection and verification methods	DDT transfer	Best fertility	Genome symmetry	Reference
S. tuberosum	*Lycopersicon Pimpinelli-folium*	Resistance to bacterial wilt & soft rot, tolerance to high temperature (THT)	EF	MA, IZA, CC	+ but ND (THT)	F	S	Okamura 1988; Okamura & Momose 1990
S. melongena	*S. knasianum*	Resistance to *Verticillium dahliae*, shoot and fruit borer & root knot nematode	EF	MA, CC, IZA	ND	MF	S	Sihachakr *et al.*, 1988
N. tabacum	*N. repanda*	Unspecified disease resistance	PEG	R: tentoxin resistant, CMS when combined with cytoplasm of D D: x-irradiated, CC, MA, RFLP (mtDNA), RuBPCase IEF	ND	MF	A	Kumashiro *et al.*, 1989
S. melongena	*S. torvum*	Resistance to Verticillium wilt & root knot nematode	EF	D: no plant regeneration MA, IZA, CC	ND	ND	S	Sihachakr *et al.*, 1989

Recipient species (R)	Donor species (D)	Desired donor trait (DDT)	Fusion method	Selection and verification methods	DDT trans-fer	Best fer-tility	Genome symme-try	Reference
N. tabacum	*N. repanda*	TMV resistance	EF	D: γ-irradiated, Kan^r MA, IZA, CC	+	FF	A	Bates 1990
S. tuberosum cv. BF15	*S. tuberosum* dihaploid clones	Resistance to potato viruses × & Y & nematodes	EF	D: no protoplast recovery MA, CC, IZA, hybrid vigour	ND	ND	S	Chaput *et al.*, 1990
S. melongena	*S. integrifolium*	Resistance Pseudomo-nas solanacea-rum	Dex-tran	R: no plant regeneration D: IOA-treated MA, IZA, CC	+	F	S	Kameya *et al.*, 1990
S. tuberosum cv Atzimba & cv Atlantic	*S. tuberosum* CMS line	CMS	PEG/ high Ca^{++}	R: IA or rhodamine-6G-treated D: γ-irradiated MA,CC, mtDNA RFLP gel stain	+	FF	Cy	Perl *et al.*, 1990
S. tuberosum	*S. brevidens*	Frost toleance	PEGm	MA, IZA, CC, hybrid vigour	+	ND	S	Preiszner *et al.*, 1991
L. esculentum	*S. muricatum*	Fruit appearance & flavour	EF	R: no protoplast recovery Hybrid vigour, MA, IZA, CC, RuBPCase IEF	ND	F	S	Sakomoto & Taguchi 1991
S. tuberosum	*S. berthaultic*	Glandular trichomes i.e.	EF	MA, IZA, CC, hybrid vigour	+	F	S	Serraf *et al.*, 1991

Recipient species (R)	Donor species (D)	Desired donor trait (DDT)	Fusion method	Selection and verification methods	DDT trans-fer	Best fer-tility	Genome symme-try	Reference
		resistance to insects						
N. tabacum	*N. debneyi*	Resistance to black root rot	PEG/ high Ca^{++}	R: kan^r D: metrotrexate resistant MA, CC , IZA, RFLP (rDNA, cpDNA, mtDNA)	ND	F	S	Sproule *et al.*, 1991
L. esculentum	*S.etuberosum; S. brevidens* × *S. etuberosum*	Resistance to viruses & bacterial disease, frost tolerance	PEGm	Hybrid vigour, IZA, CC, MA, Southern	ND	F	S	Gavrilenko *et al.*, 1992
L. esculentum	*S. etuberosum and S. brevidens*	Frost, viral and bacterial resistance	PEGm	hybrid vigour and callus morphology MA, IZA, CC, RAPD	ND	MF	S	Gavrilenko *et al.*, 1994
S. tuberosum	*S.phureja*	Agronomic traits e.g higher tuber yield	EF	R(orD) : green D(or R) : bleached, FDA-stained Isolation by MM, CC, FC, RFLP (cDNA)	+	ND	S	Mattheij & Puite 1992
S. tuberosum	*S. circaeifolium*	Resistance to diseases e.g *Phytophthora infestans & Globodera pallida*	EF	R: bleached D: green CC, FC, MA, RFLP(cDNA)	+	FF	S	Mattheij *et al.*, 1992

Recipient species (R)	Donor species (D)	Desired donor trait (DDT)	Fusion method	Selection and verification methods	DDT transfer	Best fertility	Genome symmetry	Reference
L. esculentum	*L. chilense*	Drought tolerance	PEG	R: IOA CC, Southern, RFLP (cpDNA, mtDNA)	ND	St	S	Bonnema & O'Connell 1992
S. tuberosum	*S. torvum*	Verticillium resistance	EF	R: late regeneration D: early regneration, colony morphology MA,CC, IZA	+	ND	S	Jadari *et al.*, 1992
S. tuberosum	*S. commersonii*	Frost tolerance	EF PEG/EF	Colony hybrid vigour MA,CC, IZA	+	FF	S	Cardi *et al.*, 1993 a, b
N. tabacum	*N. benthamiana*	Aphid resistance	EF	R: IOA D: Kan^r MA, RFLP(rDNA, cpDNA),IZA	+	F	S	Hagimori *et al.*, 1993
Petunia hybrida	*P. variabilis*	Floral traits and growth habit	EF	Colony hybrid vigour MA, CC, IZA, RFLP(rDNA, cpDNA, mtDNA)	+	F	S	Taguchi *et al.*, 1993
S. tuberosum	*S. brevidens*	Virus resistance	EF	D: γ-irradiated MA, CC, dot blot, RFLP (mtDNA , cpDNA, chromosome specific markers), RAPD	ND	ND	A	Xu *et al.*, 1993a, b, c; Xu & Pehu, 1993, 1994
S. tuberosum susceptible dihaploid with good tuber shape and high yields	*S. tuberosum* resistant dihaploid	Nematode (*Globodera pallida*) resistance in resynthised tetraploid	EF	D: non-regenerating MA, CC, Patatin isoforms	+	ND	S	Cooper-Bland *et al.*, 1994

Recipient species (R)	Donor species (D)	Desired donor trait (DDT)	Fusion method	Selection and verification methods	DDT transfer	Best fertility	Genome symmetry	Reference
S. tuberosum	*S. phureja*	Reducing sugar accumulation	EF	D: non-regenerating MA, reducing sugars, RAPD	+	ND	A	Craig *et al.*, 1994
N. tabacum	*N. glutinosa*	Improved breeding lines	PEG	R: methotrexate resistance D: Kan^r MA, IZA, RFLP (rDNA, cpDNA, mtDNA)	ND	F	S	Donaldson *et al.*, 1994
Hyoscyamus muticus	*H. albus*	Improved medicinal and resistance traits	PEG	R: albino, regenerating D: green, non-regenerating MA, CC, IZA,	ND	FF	S	Rahman *et al.*, 1994
S. tuberosum	*S. bulbocastanum* *S. pinnatisectum*	Cybrid lines for crop improvement	PEG	R: Streptomycin resistance D: γ-irradiated, nitromethlyurea(plastome mutation) IZA, CC, RFLP(cpDNA, mtDNA)	+	NA	A & Cy	Siderov *et al.*, 1994
S. tuberosum	*S. pinnatisectum*	*Phytophthora* resistance	EF	R: no protoplast recovery D: no protoplast recovery MA, CC, IZA, FC, RFLP (rDNA)	ND	ND	S	Ward *et al.*, 1994
S. tuberosum	*S. chacoense*	Colorada	EF	MA, IZA,	+	MF	S	Cheng *et al.*,

Recipient species (R)	Donor species (D)	Desired donor trait (DDT)	Fusion method	Selection and verification methods	DDT transfer	Best fertility	Genome symmetry	Reference
		potato beetle resistance		Glycoalkaloid analysis				1995
N. tabatum	*N. megalosiphon*	Nuclear encoded disease resistance	PEG	R: methotrexate resistance D: Kan^r MA, IZA, RFLP (nDNA, rDNA, mtDNA)	ND	FF	S	Donaldson *et al.*, 1995
L. esculentum	*S. ochranthum*	Unspecifeid pest resistance	PEG	D: non-regenerating IZA, RAPD	ND	ND	S	Kobayashi *et al.*, 1996
Umbelliferae								
Daucus carota variety K5	*D. carota* CMS line	CMS	PEG	R: IOA-treated D: x-irradiated CC, mtDNA RFLP gel stain, MA	+	FF	Cy	Tanno-Suenaga *et al.*, 1988, 1991

Abbrevations:
A: asymmetric, CC: chromosome counting, cDNA: complementary DNA, CMS:cytoplasmic male sterile / sterility, cpDNA: chloroplast DNA, Cy: cybrid, D: donor species, cultivar or breeding line, DDT: desired donor trait, DNA: deoxyribonucleic acid, EF: electrofusion, F: fertile, FACS: fluorescence-activated cell sorter, FC: flowcytometry to measure DNA content and thus estimate the ploidy level, FDA: fluorescein diacetate, FF: female fertile, $Hygro^r$: hygromycin resistance, IA: iodoacetic acid, IOA: iodocaetamide, IZA: isozyme analysis, Kan^r: kanamycin resistant, MA: morphological analysis, MF: male fertile, MM: Micromanipulator, mtDNA: mitochondrial DNA, NA: not applicable, ND: not determined, nDNA: nuclear DNA, PEG: polyethylene glycol, PEGm: modified PEG methods using also dimethyl sulfoxide based on Menczel et al., (1981) and Menczel & Wolfe(1984), PLRV: potato leaf roll virus, R: recipient species, cultivar or breeding line, RAPD: random amplified polymorphic DNA, rDNA: ribosomal DNA, RFLP: restriction fragment length polymorphisims (probe type in brackets), RuBPase IEF: isoelectric focusing of small &/or large subunit(s) of ribulose-1,5-bisphosphate carboxylase/oxygenase (EC 4.1.1.39), S: symmetric, SC: suspension cells, St: sterile, TMV: tobacco mosaic virus, TZ: triazine.

3.2 Implications and trends

Some general or statistical observations can be drawn from Table 2:

To date 33 species in 22 genera of 12 families have been targeted for plant improvement using protoplast fusion technology. Approximately 65% of the fusion combinations have been interspecific, 25% intergeneric and 10% intraspecific, with species from some 40 genera used as the donor of the desired traits.

Successful hybrid/cybrid plants have mostly been produced in the families *Solanaceae,* and *Cruciferae*, particularly in the genera *Solanum, Nicotiana, Brassica* and *Lycopersicon* of these two families, and *Citrus* of *Rutaceae* family and *Medicago* of *Leguminosae* family. Some successful hybrid/cybrid plants have also been produced in the families *Carophyllaceae, Compositeae, Gramineae, Cucurbitaceae, Labiatae, Liliaceae, Passifloraceae,* and *Umbelliferae*. However, no intergeneric somatic hybrid plants for breeding purposes have been reported in the last five families yet. Disease resistance is the most common desired trait, followed by CMS and pest resistance. CMS is also among the most successful in the transfer of desired traits.

All fusions, except two (Téoulé, 1983; Kameya et al, 1990), have used PEG-based and/or electrofusion methods. Earlier fusions were almost exclusively achieved using PEG or modified PEG methods; in the last four years, however, electrofusion has become popular. More electrofusion has been employed in *Solanaceae* and *Leguminosae* than in *Cruciferae* and *Rutaceae*. Differential inactivation (eg, IOA to one parent and γ-ray to the other) and/or differential regenerability (eg, recipient is protoplast-regenerable while donor is not) have been most frequently used in hybrid selection. All the cases of using FACS for hybrid selection have been in *Cruciferae*. Hybrid vigour has been used mostly in *Solanaceae* and *Rutaceae*. Various techniques have been used in hybrid/cybrid verification: morphological comparison, isoenzyme analysis and chromosome counting have been most frequently reported. Of the DNA-based techniques, RFLP analysis is the most popular. Depending on what probes are used, RFLP's can readily distinguish between nuclear hybrids and cybrids.

The introduced desired traits have included both qualitative (eg, CMS, virus resistance, herbicide tolerance etc) and quantitative characters (eg, yield, drought tolerance etc). There was no apparent correlation between the symmetry of nuclear genome and the best fertility of the plants. Nonetheless, the proportion of fertile versus infertile plants may differ between symmetric and asymmetric hybrids (yet no data were given in many reports), since the best fertility in Table 2 refers to any fertile plant(s) in hybrid populations.

About 70% of the fusion experiments produced symmetric hybrids. Since the late 1980s, however, more asymmetric hybrids and cybrids have been produced, particularly in *Cruciferae*. All the reported hybrids in *Rutaceae* and most hybrids in *Leguminosae* have been symmetric. Where the donor protoplasts were irradiated, the resultant hybrids were generally found to be asymmetric; more or less symmetric hybrids were produced otherwise.

The first somatic hybrid plants with useful transferred traits were with introduced chilling tolerance (Melchers et al, 1978; Smillie et al, 1979), CMS (Zelzer et al, 1978; Aviv et al, 1980), and disease resistance (Evans et al, 1981; Douglas et al, 1981a, b, c, d; Pandeya et al, 1986). The first somatic hybrid plant produced with a higher yield (a polygenetic trait) than the best commercial cultivar (recipient) was in potato (Mattheij and Puite, 1992). The first commercial cultivar derived from protoplast fusion (Douglas et al, 1981a, b, c, d; Pandeya et al, 1986) was released in 1990 to the Canadian tobacco industry (Brown and Thorpe, 1995). Nevertheless, very little germplasm from somatic hybridization has yet contributed to new released cultivars so far, perhaps highligting the fact that it often takes more than ten years to bring initial research results to the stage of a recognised cultivar.

4 FUTURE PERSPECTIVE

4.1 Attractions and limitations of employing somatic hybridization in plant improvement

There are three genomes co-existing in plant cells: nuclear, plastid and mitochondrial genome. In sexual hybridization, the latter two genomes of the paternal parent are virtually excluded from the hybrids due to maternal inheritance. In addition, sexual hybrids between two pure-breeding lines are genetically uniform obeying Mendelian principle of uniformity in F1. Somatic hybridization brings together all the three genomes of both parents in the hybrid cells and the regenerated plants vary dramatically regardless of homozygosity of the parents, thus, yielding far more variability than sexual hybridization:

In terms of the nuclear genome, a somatic hybrid can be broadly classified as (1) symmetric, or more frequently, (2) asymmetric (ie, most or majority of the nuclear genome is from one parent). It may be that absolute exclusion of nuclear genome of one parent is very unlikely and pure cybrids perhaps never exist, being instead those having a nuclear genome mostly from one parent and at least some elements of the cytoplasmic genomes from the other (Galun, 1993). In fact, recurrent backcrossing in conventional breeding has long been producing such “cybrids”, such as the application of alloplasmic male sterility in hybrid maize production. Nevertheless, recurrent backcrossing is very time-consuming requiring many generations to substitute the recipient nucleus. Using an inhibitor to inactivate cytoplasmic genomes of one parent and ionising irradiation to inactivate the nuclear genome of the other, somatic hybridization offers opportunity to obtain cybrids by one step. Somatic hybridization is the only breeding approach that can produce cybrids by one step.

In terms of the plastid genome, a somatic hybrid can also be broadly classified as (1) recipient type , or (2) donor type , neglecting the rare events of heteroplasmy (Fitter and Rose, 1993). There appears to be processes operating in chloroplasts and mitochondria which favour the assortment to one type or the other (Scowcroft and Larkin, 1981; Rose et al, 1990; Walters and Earle, 1993). Nevertheless, especially in mitochondria, considerable genetic recombination can occur between the parental genomes (Chetrit et al, 1985; Rose et al, 1990; Landgren and Glimelius, 1994). So in terms of the mitochondrial genome a somatic hybrid can be broadly classified as (1) donor type, (2) recipient type, and (3) recombinant type.

There are, therefore, a rich diversity of genetic combinations that can be brought about in a somatic hybrid population through protoplast fusion. In fact somatic hybridization offers the opportunity to create variability in all three genomes which is unsurpassed by any other breeding techniques known today.

Another attraction of employing somatic hybridization is that polygenetic traits can be transferred without knowing the molecular basis of the genes coding the traits. This is of immediate significance because many agronomic traits are polygenically controlled while current molecular breeding technology is still at a stage of developing and refining its single-gene cloning and transformation strategies. The successful transfer of high-tuber-yield genes in potato (Mattheij and Puite, 1992) is an encouraging example. In addition, successful transfer of many putative single-gene characters of unknown molecular details (eg, insect and disease resistance) in somatic hybrids (eg, Lelivelt et al, 1993; Liu et al, 1995) may prove valuable in molecular studies of the genes because the hybrids may serve as “mutants” from the “wild type” (the recipient parent) facilitating cloning of the desired gene.

The fourth attraction of employing somatic hybridization is in improvement of the plants wherein (1) sexual reproduction is either difficult or absent such as banana, cassava, potato, sweet potato, sugarcane and yam (Bhojwani and Razdan, 1996). An additional advantage to these cases is that the fusion-derived plants need not be fertile, thereby circumventing the infertility problem (see Section 4.2); (2) novelty is the goal of breeding, such as ornamental flowers. This is probably a very promising field to be further explored; so far there are only a few reports aiming at improvement of floral characters in somatic hybrid plants (Table 2).

There are several limitations of the current protoplast fusion technology. Firstly, the low chance integration of the desired gene(s) into the target genomes due to the still uncontrollable mitotic or meiotic recombination events between the target and the donor genomes. This is arguably a main reason why the desired traits failed to express in many of the somatic hybrid/cybrid plants produced so far (Table 2). Secondly, there is the probability of concomitant integration of unwanted or deleterious gene(s). Unfortunately, that which offered the potential of creating greatest variability in plant breeding materials is in turn causing the limitations of this technology. In current application of somatic hybridization in plant breeding, there are several practical questions to be answered.

4.2 Questions to be answered

Infertility in somatic hybrids is perhaps the primary question to be answered. The majority of the intergeneric plants produced so far have suffered some degree of infertility. This is particularly true in combinations between phylogenetically distant species, despite the directed elimination of donor chromosomes using ionising irradiation (eg, Bauer-Weston et al, 1993). Unless the plants are vegetatively propagated in field production, an infertile plant is of little practical value in breeding programmes. Normally the hybrid plant is expected to be backcrossed to the target (recipient) species to maximise the recipient genomes while retaining the fusion-introduced desired gene(s).

The pollen viability of somatic hybrids is generally low. An additional complication is that the pollen viability can be unstable even in vegetative clones of the hybrid plants and/or the ability of viable hybrid pollen to cross to the target plants can be restricted. For example, normal pollen germination was observed in 24 out of 36 tested somatic hybrid plants between lucerne (*M. sativa*) and sainfoin (*Onobrychis viciifolia*) (Li et al, 1993), but in 72 rooted cuttings of these male-fertile plants only 7 produced viable pollen (Y-G Li, unpubl.). The cause is still unclear, however, chromosomal instability could be an important factor. Chromosome elimination continued in a few hybrid plants over a 7-month period of observation and the chromosome number was different between the clones and the original hybrid plants (Y-G Li, unpubl.).

The infertility in somatic hybrids, particularly intergeneric, has been ascribed to: (1) the structural and dynamic instability of chromosomes in hybrid cells, and (2) the nuclear-cytoplasmic incompatibility caused by interactions between nuclear genome and plasmon (Harms, 1983).

Chromosome number has been rather variable in most somatic hybrids reported to date. Based on results from transposon research in maize (Walbot et al, 1985), the authors also suspect the likely involvement of transposon bursts and hybrid dysgenesis in somatic hybrid infertility. Various types of abnormalities (including infertility) in somatic hybrids may be analogous to the post-zygotic incompatibility that is already known in sexual hybrids between phylogenetically distant species. In this sense, although somatic hybridization has circumvented the pre-zygotic incompatibility, it has not bypassed the post-zygotic incompatibility yet. A study of the causes of post-zygotic incompatibility is expected to help solve the infertility problem.

Closely related to the above question is the transient gene expression in somatic hybrids. Four asymmetric hybrids out of about 4230 plants derived from lucerne (recipient) fused to sainfoin or *Lotus pedunculatus* (donors), were observed to have detectable levels of foliar condensed tannins even though these plants were lucerne-like in appearence. However in each of the four cases rooted cutting had no foliar tannins (Li et al, 1996). Chromosomal instability and gene silencing due to inherent and/or fusion-induced genetic regulation could be two contributing factors to this transient gene expression. Similar phenomena have already been reported in a number of transgenic plants produced through direct DNA transfer. The instability in expression of transgenic DNA is frequently observed, and this is usually not due to deletion or mutation of the introduced DNA but rather to the inactivation of the transgene (Finneagan and McElroy, 1994). Methylation (Hamill and Rhodes, 1993), and homology-dependent gene silencing including co-suppression and epigene conversion (Jorgensen, RA, 1995) have been suggested.

The genetic instability due to post-zygotic incompatibility as expressed by infertility and transient gene expression in the somatic hybrids will prove to be the greatest obstacle in employing somatic hybridization in plant breeding. A greater understanding of this problem may benefit not only somatic hybridization but direct DNA transfer as well.

Another obstacle in employing somatic hybridization in plant breeding is the difficulty associated with plant regeneration from protoplasts. Although by 1993 there have been 320 species in which plant regeneration from protoplasts was reported, many important plant species still remain recalcitrant or not yet been investigated for their regenerability, particularly woody plants (eg, grape, oil palm, pineapple, rubber, tea, banana, datepalm, coconut, mango, etc) (Li, 1989; Roest and Gilissen, 1993). Among the regenerated species a single family *Solanaceae* accounts for about 1/4 (ie, 76 out of 320); regeneration was reported in 29, 23, 9, 9, 8, and 6 species belonging to the genera *Solanum, Nicotiana, Medicago, Citrus, Brassica,* and *Lycopersicon*, respectively (Roest and Gilissen, 1989, 1993; Table 1).

Most somatic hybrids/cybrids for breeding purposes achieved so far also belong to these 6 genera (Table 2), indicating a strong correlation between a genus' regeneration performance and its employment frequency in somatic hybridization breeding.

To be of practical value for plant improvement, not only the hybrid cells must be regenerated into plants but also the regeneration frequency must be reasonably high to allow effective selection. For example, starting with a population of 1,000,000 protoplasts for the fusion experiment, 2% frequency of heterokaryon binary fusions and 0.1% plant regeneration frequency, then only 20 plants will be recovered. This will give a low probability of finding the desired gene transfer. Currently a general problem in numerous protoplast regeneration protocols seems to be that they are either less efficient and hardly reproducible in others' hands, or they are efficient but only to specific genotype(s) (D. He, pers. comm.). To extend application of protoplast fusion technology to more plant species will require a deeper understanding of the mechanisms underlying plant cell development such as differentiation, dedifferentiation and redifferentiation.

4.3 CONCLUDING REMARKS

Twenty-four years have passed since the first protoplast-fusion derived plant was obtanined. A steady accumulation in both our basic knowledge and technical capabilities is obvious especially in the past decade. The world's first protoplast-fusion-derived commercial product (a tobacco cultivar) has been released since 1990, creating an estimated value of US$ 199,000,000 in 1995 for Canadian tobacco industry (Brown and Thorpe, 1995). The world's first transgenic plant product (interestingly also a tobacco cultivar) entered commerce around 1992-1993 in China, and since then transgenic tomatoes, cotton, canola, and squash have been commercialised (Redenbaugh et al, 1995).

Despite these exciting achievements, a significant proportion of literature on either protoplast fusion technology or DNA transfer technology remains concentrated on empirical approaches, and there is a considerable gap between our basic and applied studies. As a result, the developmental biology inherent to morphogenesis in vitro and molecular genetics inherent to mitotic recombination and genetic regulation are still largely unknown. As techniques for isolation and fusion of protoplasts and selection of hybrid cells come of age, problems associated with regeneration difficulties may seriously impede the wider application of somatic hybridization in plant improvement. As more and more somatic hybrids and transgenic plants leave the test tube and become subject to natural environment, problems associated with genetic instability may surface with increasing regularity. To solve these problems, far more basic research is required. Somatic hybridization in plant breeding has not yet realised its great potential. Further investigation in basic research and development of practical means to overcome the obstacles of genetic instability and plant regenerability will pay more dividends to those hoping to realise the full potential of this technology in the 21st century.

5 REFERENCES

Al-Atabee JS, Mulligan BJ & Power JB (1990) Interspecific somatic hybrids of *Rudbeckia hirta* and *R. laciniata* (*Compositae*). Plant Cell Rep. 8: 517-520

Anamthawat-Jónsson K, Schwarzacher T, Leitch AR, Bennett MD & Heslop-Harrison JS (1990) Discrimination between closely related *Triticeae* species using genomic DNA as a probe. Theor. Appl.Genet. 79: 721-728.

Arcioni S, Davey MR, dos Santos AVP & Cocking EC (1982) Somatic embryogenesis in tissues from mesophyll and cell suspension protoplasts of *Medicago coerulea* and *M. glutinosa.* Z Pflanzenphysiol. 106: 105-110.

Arcioni S, Mariotti D & Pezzotti M (1985) Plant regeneration of *Medicago arborea* from protoplasts culture and preliminary experiments of somatic hybridization with *Medicago sativa.* Genet. Agraria. 39: 307

Atanassov A & Brown DCW (1984) Plant regeneration from suspension culture and mesophyll protoplasts of *Medicago sativa* L. Plant Cell Tissue Organ Cult. 3: 149-162

Austin S, HK Ehlenfeldt, MA Baer & JP Helgeson (1986) Somatic hybrids produced by protoplast fusion between *Solanum tuberosum* and *S. brevidens*: phenotypic variation under field conditions. Theor. Appl. Genet. 71: 682-690.

Austin S, MA Baer & JP Helgeson (1985) Transfer of resistance to potato leaf roll virus from *Solanum brevidens* into *Solanum tuberosum* by somatic fusion. Plant Sci. 39: 75-82.

Aviv D, Fluhr R, Edelman M & Galun E (1980) Progeny analysis of the interspecific somatic hybrids: *Nicotiana tabacum* (cms) + *Nicotiana sylvestris* with respect to nuclear and chloroplast markers. Theor. Appl. Genet. 56: 145-150.

Aziz MA, Chand PK, Power JB & Davey MR (1990) Somatic hybrids between the forage legumes *Lotus corniculatus* L and *L. tenuis* Waldst et Kit. J. Exp. Bot. 41:471-479.

Babiychuk E, Kushnir S & Gleba YY (1992) Spontaneous extensive chromosome elimination in somatic hybrids between somatically congruent species *Nicotiana tabacum* L. and *Atropa belladonna* L. Theor. Appl. Genet. 84: 87-91.

Barsby TL, Shepard JF, Kemble RJ & Wong R (1984) Somatic hybridization in the genus Solanum: S. tuberosum and S. brevidens. Plant Cell Rep. 3: 165-167.

Barsby TL, Yarrow SA, Kemble RJ & Grant I (1987b) The transfer of cytoplasmic male sterility to winter-type oilseed rape (Brassica napus L.) by protoplast fusion. Plant Sci. 53: 243-248.

Barsby TL, Yarrow SA, Wu S-C, Coumans M, Kemble RJ, Powell AD, Chuong PV, Beversdorf WD & Pauls KP (1987a) The combination of polima CMS and cytoplasmic triazine resistance in *Brassica napus*. Theor. Appl. Genet. 73:809-814.

Bates GW (1990) Asymmetric hybridization between *Nicotiana tabacum* and *N. repanda* by donor recipient protoplast fusion: transfer of TMV resistance. Theor. Appl. Genet. 80: 481-487.

Bates GW, Hasenkampf CA, Contolini CL & Piastuch WC (1987) Asymmetric hybridization in *Nicotiana* by fusion of irradiated protoplasts. Theor. Appl. Genet. 74: 718-726.
Bates GW, Saunders JA & Sowers AE (1989) Electrofusion: principles and application. In: Neumann E et al.(eds) Electroporation and electrofusion in cell biology. (pp 387-395) Plenum Press, NY.
Bauer-Weston B, Keller W, Webb J & Gleddie S (1993) Production and characterization of asymmetric somatic hybrids between *Arabidopsis thaliana* and B*rassica napus*. Theor. Appl. Genet. 86: 150-158.
Begum F, Paul S, Bag N,. Sikdar SR & Sen SK (1995) Somatic hybrids between *Brassica juncea* (l) czern and *Diplotaxis harra* (forsk) boiss and the generation of backcross progenies. Theor. Appl. Genet. 91:1167-72.
Bhojwani SS & Razdan MK (1996) Plant Tissue Culture: Theory and Practice, A sevised edition. Chapter 13 (pp. 371-405) Elsevier, Amsterdam.
Binding H, Jain SM, Finger J, Mordhorst G, Nehls R & Gressel J (1982) Somatic hybridization of an atrazine resistant biotype of *Solanum nigrum* with *Solanum tuberosum*: Part 1. Clonal variation in morphology and in atrazine sensitivity. Theor. Appl. Genet. 63:273-277.
Bingham ET, McCoy TJ & Walker KA (1988) Alfalfa tissue culture. In: Hanson AA (ed.) Alfalfa and alfalfa improvement. Agronomy monograph, Vol 29. (pp.809-826) ASA-CSSA-SSSA, Madison, USA.
Bonnema AB & O'Connell MA (1992) Molecular analysis of the nuclear organellar genotype of somatic hybrid plants between tomato (*Lycopersicon esculentum*) and *Lycopersicon chilense*. Plant Cell Rep. 10: 629-632.
Bonnema AB, Melzer JM, Murray LW & O'Connell MA (1992) Non-random inheritance of organellar genomes in symmetric and asymmetric somatic hybrids between *Lycopersicon esculentum* and *L. pennellii*. Theor. Appl. Genet. 84: 435-442.
Brown DCW & Thorpe TA (1995) Crop improvement through tissue culture. World J. Microbiol Biotechnol. 11: 409-415.
Brown DCW & Atanassov A (1985) Role of genetic background in somatic embryogenesis in *Medicago*. Plant Cell Tissue Org Cult. 4: 111-122.
Butenko R, Kuchko A & Komarnitsky I (1982) Some features of somatic hybrids between *Solanum tuberosum* and *S. chacoense* and its F1 sexual progeny. In: Fujimara A (ed.) Plant Tissue Culture 1982. (pp. 643-644) Marizen co., Tokyo.
Böttcher UF, Aviv D& Galun E (1989) Complementation between protoplasts treated with either of two metabolic inhibitors results in somatic hybrid plants. Plant Sci. 63:67-77.
Cardi T, D'Ambrosio F, Consoli D, Puite KJ & Ramulu KS (1993b) Production of somatic hybrids between frost-tolerant *Solanum commersonii* and *S. tuberosum*: characterization of hybrid plants. Theor. Appl.Genet. 87: 193-200.
Cardi T, Puite K J, Ramulu K S, D'Ambrosio F & Frusciante L (1993a) Production of somatic hybrids between frost-tolerant *Solanum commersonii* and *S. tuberosum*: Protoplast fusion, regeneration and isozyme analysis. Am. Potato J. 70:753-764.
Carlson PS, Smith HH & Dearing RD (1972) Parasexual interspecific plant hybridization. Proc. Nat. Acad. Sci. USA 69: 2292-2294.
Carr DH & Walker JE (1961) Carbol fuchsin as a stain for human chromosomes. Stain Technol. 36:233-236.
Carruthers VR (1991) Pasture composition and grazing management on dairy farms differing in the incidence of bloat. In: Ellinbank DRI (ed.) Bloat. DRDC Bloat Workshop, Melboune, Aug 1991, pp. 1-3.

Chang PK & Loescher WH (1991) Effects of preconditioning and isolation conditions on potato *Solanum tuberosum* L. cv. Russet Burbank) protoplast yield for shoot regeneration and electroporation. Plant Sci. 72: 103-109.

Chaput MH, Sihachakr D, Ducreux G, Marie D & Barghi N (1990) Somatic hybrid plants produced by electrofusion between dihaploid potatoes: BF15 (H1), Aminca (H6) and Cardinal (H3). Plant Cell Rep. 9: 411-414.

Cheng J, Saunders JA & Sinden SL (1995) Colarado potato beetle resisitant somatic hybrid potato plants produced via protoplast electrofusion. In Vitro Cell. Dev. Biol. 31P: 90-95.

Chetrit P, Mathieu C, Vedel F, Pelletier G & Primard C (1985) Mitochondrial DNA polymorphism induced by protoplast fusion in Cruciferae. Theor. Appl. Genet. 69:361-366.

Christey MC, Makaroff CA & Earle ED (1991) Atrazine-resistant cytoplasmic male-sterile-nigra broccoli obtained by protoplast fusion between cytoplasmic male-sterile *Brassica oleracea* and atrazine-resistant *Brassica campestris*. Theor. Appl. Genet. 83: 201-208 .

Clark E, Schnabelrauch L, Hanson MR & Sink KC (1986) Differential fate of plastid and mitochondrial genomes in Petunia somatic hybrids. Theor. Appl. Genet. 72: 748-755.

Cocking EC (1960) A method for the isolation of plant protoplasts and vacuoles. Nature 187: 962-963.

Collins GB, Taylor NL & DeVerna JW (1984) In vitro approaches to interspecific hybridization and chromosome manipulation in crop plants. In: Gustafson JP (ed.) Gene manipulation in Plant improvement. (pp. 323-383) Plenum Press, NY.

Cooper CS, Duke JA, Forde MB, Reed CF & Weder JKP. (1981) *Onobrichis viciifolia* Scop. In: Duke JA (ed.) Handbook of Leumes of World Economic Importance. (pp. 157-161) Plenum Press, NY.

Cooper JP (1973) Genetic variation in herbage constituents. In: Butler GW & Bailey RW (eds) Chemistry and Biochemistry of Herbage. Vol. 2. (pp. 379-417) Academic Press, NY.

Cooper-Bland S, DeMaine M J, Fleming M L M H, Phillips M S, Powell W & Kumar A (1994) Synthesis of intraspecific somatic hybrids of *Solanum tuberosum*: assessments of morphological, biochemical and nematode (*Globodera pallida*) resistance characteristics. J. Exp. Bot. 45: 1319-1325.

Craig A L, Morrison I, Baird E, Waugh R, Coleman M, Davie P & Powell W (1994). Expression of reducing sugar accumulation in interspecific somatic hybrids of potato. Plant Cell Rep. 13: 410-405.

Crea F, Bellucci M, Damiani F & Arcioni S (1995) Genetic control of somatic embryogenesis in alfalfa (*Medicago sativa* L cv Adriana). Euphytica 81: 151-155.

D'Hont A, Quetier F, Teoule E & Dattee Y (1987) Mitochondrial and chloroplast DNA analysis of interspecific somatic hybrids of a Leguminosae: *Medicago* (alfalfa). Plant Sci. 53: 237-242.

Davies PA (1988) Chapter 3: Development of techniques for protoplast culture. In: Cell culture of forage legumes for the production of nutritionally improved, bloat-safe lucerne. PhD thesis, (pp. 78-180) ANU, Canberra, Australia

Davies PA, Larkin PJ & Tanner GJ (1989) Enhanced protoplast division by media ultrafiltration. Plant Sci. 60: 237-244.

Deimling S, Zitzlsperger J & Wenzel G (1988) Somatic fusion for breeding of tetraploid potatoes. Plant Breed. 101: 181-189.

Deng XX, Grosser JW & Gmitter FG (Jr.) (1992) Intergeneric somatic hybrid plants from protoplast fusion of *Fortunella crassifolia* cultivar 'Meiwa' with *Citrus sinensis* cultivar 'Valencia'. Sci.Hortic. 49: 55-62.

Dijak M & DCW Brown (1987) Patterns of direct and indirect embryogenesis from mesophyll protoplasts of *Medicago sativa*. Plant Cell Tissue Organ Cult. 9: 121-130.

Donaldson PA, Bevis EE, Pandeya RS & Gleddie SC (1994). Random chloroplast segregation and frequent mtDNA rearrangements in fertile somatic hybrids between *Nicotiana tabacum* L. and *N. glutinosa* L. Theor. Appl. Genet. 87: 900-908.

Donaldson, PA., Bevis E, Pandeya R &Gleddie S (1995) Rare symmetric and asymmetric *Nicotiana tabacum* (plus) N-megalosiphon somatic hybrids recovered by selection for nuclear-encoded resistance genes and in the absence of genome inactivation. Theor. Appl. Genet. 91: 747-55.

Dornelas, MC, Tavares FCA., Deoliviera JC & Vieira MLC (1995) Plant regeneration from protoplast fusion in *Passiflora* spp. Plant Cell Rep. 15: 106-10.

Dos Santos AVP, Outka DE, Cocking CE & Davey MR (1980) Organogenesis and somatic embryogenesis in tissue derived from leaf protoplasts and leaf explants of *Medicago sativa*. Z. Planzenphysiol. 99: 261-270.

Douglas G C, Keller W A & Setterfield G (1981a). Somatic hybridisation between *Nicotiana rustica* and *N. tabacum*. I. Isolation and culture of protoplasts and regeneration of plants from cell cultures of wild-type and chlorophyll deficient strains. Can. J. Bot. 59 :208-219.

Douglas G C, Keller W A & Setterfield G (1981b). Somatic hybridisation between *Nicotiana rustica* and *N. tabacum*. II. Protoplast fusion and selection and regeneration of hybrid plants. Can. J. Bot. 59: 220-227.

Douglas G C, Wetter L R, Keller W A & Setterfield G (1981d) Somatic hybridisation between *Nicotiana rustica* and *N. tabacum*. IV. Analysis of nuclear and chloroplast genome expression in somatic hybrids. Can. J. Bot. 59: 1509-1513.

Douglas G C, Wetter L R, Nakamura C, Keller W A & Setterfield G (1981c) Somatic hybridization between *Nicotiana rustica* and *N. tabacum*. III. Biochemical, morphological, and cytological analysis of somatic hybrids. Can. J. Bot. 59: 228-237

Evans DA (1983) Protoplast fusion. In: Evans DA et al.(eds) Handbook of plant cell culture. Vol 1. (pp. 291-321) Macmillan, NY.

Evans DA & Bravo JE (1983) Protoplast isolation and culture. In: Evans DA et al.(eds) Handbook of plant cell culture. vol 1. (pp 124-176). Macmillan, NY.

Evans DA, CE Flick & RA Jensen (1981) Disease resistance: incorporation into sexually incompatible hybrids of the genus *Nicotiana*. Science 213: 907-909.

Fahleson J, Eriksson I, Landgren M, Stymne S & Glimelius K (1994). Intertribal somatic hybrids between *Brassica napus* and *Thlaspi perfoliatum* with high content of the *T. perfoliatum*-specific nervonic acid. Theor. Appl. Genet. 87: 795-804.

Fahleson J, Råhlén L & Glimelius K (1988) Analysis of plants regenerated from protoplast fusions between *Brassica napus* and *Eruca sativa*. Theor. Appl. Genet. 76: 507-512.

Finnegan J & McElroy D (1990) Transgene inactivation: plants fight back! Bio/Technology 12: 883-888.

Fish N, Karp A & Jones MGK (1987) Improved isolation of dihaploid *S. tuberosum* protoplasts and the production of somatic hybrids between dihaploid *S. tuberosum* and *S. brevidens*. In Vitro 23: 575-580.

Fish N, Karp A & Jones MGK (1988a) Production of somatic hybrids by electrofusion in *Solanum*. Theor. Appl. Genet. 76: 260-266.

Fish N, Steele SH & Jones MGK (1988b) Field assessment of dihaploid *Solanum tuberosum* and *S. brevidens* somatic hybrids. Theor. Appl. Genet. 76: 880-886.

Fitter JT & Rose RJ (1993) Investigation of chloroplast DNA heteroplasmy in *Medicago sativa* L. using cultured tissue. Theor. Appl. Genet. 86: 65-70.

Galbraith DW & Galbraith JEC (1979) A method for identification of fusion of plant protoplasts derived from tissue cultures. Z. Pflanzenphysiol. 93: 149-158.

Galun E (1993) Cybrids-an introspective overview. Int. Assoc. Plant Tissue Cult. Newslett. 70: 2-10.

Gamborg OL, RA Miller & K Ojima (1968) Nutrient requirements of suspension cultures of soybean root cells. Exp. Cell. Res. 50:151-158.
Gavrilenko TA, Barbakar NI & Pavlov AV (1992) Somatic hybridization between *Lycopersicon esculentum* and non-tuberous *Solanum* species of the Etuberosa series. Plant Sci. 86: 203-214.
Gavrilenko TA, Dorokhov DB, Nikulenkova TV (1994) Characteristics of intergeneric somatic hybrids of *Lysopersicon esculentum* Mill. and nontuberous potato species of the *Etuberosa* series. Russian Journal of Geneties 30(12) :1388-1396.
Gerdemannknorck, M, Nielen S, Tzscheetzsch C, Iglisch J & Schieder O (1995) Transfer of disease resistance within the genus *Brassica* through asymmetric somatic hybridization Euphytica 85: 247-53.
Gilmour DM, Davey MR & Cocking EC (1987b) Plant regeneration from cotyledon protoplasts of wild *Medicago* species. Plant Sci. 48: 107-112.
Gilmour DM, Davey MR, Cocking EC & D Pental (1987a) Culture of low numbers of forage legume protoplasts in membrane chambers. J. Plant Physiol. 126: 457-465.
Gleba YY, Hinnisdaels S, Sidorov VA, Kaleda VA, Parakonny AS, Boryshuk NV, Cherep NN, Negrutiu I & Jacobs M (1988) Intergeneric asymmetric hybrids between *Nicotiana plumbaginifolia* and *Atropa belladonna* obtained by "gamma-fusion". Theor. Appl. Genet. 76: 760-766.
Gleddie S, Fassuliotis G, Keller WA & Setterfield G (1985) Somatic hybridization as a potential method of transferring nematode and mite resistance into eggplant. Z Pflanzenzücht 94: 348-351.
Gleddie S, WA Keller & G Setterfield (1986) Production and characterization of somatic hybrids between *Solanum melongena* L. and *S. sisymbriifolium* Lam. Theor. Appl. Genet. 71: 613-621.
Grosser J W, Gmitter F G, Louzada E S & Chandler J L (1992). Production of somatic hybrid and autotetraploid breeding parents for seedless Citrus development. Hortscience 27: 1125-1127.
Grosser J W, Louzada E S, Gmitter F G & Chandler J L (1994). Somatic hybridization of complementary Citrus rootstocks: Five new hybrids. Hortscience 29: 812-813.
Grosser JW, Gmitter FG (Jr.) & Chandler JL (1988a) Intergeneric somatic hybrid plants of *Citrus sinensis* cv. Hamlin and *Poncirus trifoliata* cv. Flying Dragon. Plant Cell Rep. 7: 5-8.
Grosser JW, Gmitter FG (Jr.) & Chandler JL (1988b) Intergeneric somatic hybrid plants from sexually incompatible woody species: *Citrus sinensis* and *Severinia disticha*. Theor. Appl. Genet. 75: 397-410.
Grosser JW, Gmitter FG (Jr.), Tusa N & Chandler JL (1990) Somatic hybrid plants from sexually incompatible woody species: *Citrus reticulata* and *Citropsis gilletiana*. Plant Cell Rep. 8: 656-659.
Grosser JW, Mouraofo FAA, Gmitter FG, Louzada ES, Jiang J, Baergen K, Quiros A, Cabasson C, Schell JL & Chandler JL. (1996) Allotetraploid hybrids between citrus and seven related genera produced by somatic hybridization. Theor. App. Genet. 92: 577-82.
Guidet F (1994) A powerful new technique to quickly prepare hundreds of plant extracts for PCR and RAPD analyses. Nucleic Acids Res. 22: 1772-1773.
Gupta PP, Schieder O & Gupta M (1984) Intergeneric gene transfer between somatically and sexually incompatible plants through asymmetric protoplast fusion. Mol. Gen. Genet. 197: 30-35.
Hagimori M, Matsui M, Matsuzaki T, Shinozaki Y, Shinoda T & Harada H (1993) Production of somatic hybrids between *Nicotiana benthamiana* and *N. tabacum* and their resistance to aphids. Plant Sci. 91: 213-222.

Hagimori M, Nagaoka M, Kato N & Yoshikawa H (1992) Production and characterization of somatic hybrids between the Japanese radish and cauliflower. Theor. Appl. Genet. 84:.891-824.

Hall RD, Rouwendal GJA & Krens FA (1992a) Asymmetric somatic cell hybridization in plants I. The early effects of (sub)lethal doses of UV and gamma radiation on the cell physiology and DNA integrity of cultured sugarbeet (*Beta vulgaris* L.) protoplasts. Mol. Gen. Genet. 234:.306-314.

Hall RD, Rouwendal GJA & Krens FA (1992b) Asymmetric somatic cell hybridization in plants II. Electrophoretic analysis of radiation-induced DNA damage and repair following the exposure of sugarbeet (*Beta vulgaris* L.) protoplasts to UV and gamma rays. Mol. Gen. Genet. 234: 315-324.

Handley LW, Nickels RL, Cameron MW, Moore PP & Sink KC (1986) Somatic hybrid plants between *Lycopersicon esculentum* and *Solanum lycopersicoides*. Theor. Appl. Genet. 71: 691-697.

Hansen LN & Earle ED (1995) Transfer of resistance to *Xanthomonas campestris* pv campestris into *Brassica oleracea* l by protoplast fusion. Theor. Appl. Genet. 91: 1293-300.

Harkins KR & Galbraith DW (1984) Flow sorting and culture of plant protoplasts. Physiol. Plant.60: 43-52.

Harms CT (1983) Somatic incompatibility in the development of higher plant somatic hybrids. Q. Rev. Biol. 58: 325-353.

Heath D W & Earle ED (1996) Synthesis of ogura male sterile rapeseed (*Brassica napus*) with cold tolerance by protoplast fusion and effects of atrazine resistance on seed yield. Plant Cell Rep. 15: 939-44.

Helgeson JP, Hunt GJ, Haberlach GT & Austin S (1986) Somatic hybrids between *Solanum brevidens* and *Solanum tuberosum*: expression of a late blight resistance gene and potato leaf roll resistance. Plant Cell Rep. 3: 212-214.

Hernández-Fernández MM & Christie BR (1989) Inheritance of somatic embryogenesis in alfalfa (*Medicago sativa* L.). Genome 32: 318-321.

Herzenberg LA, Sweet RG & Herzenberg LA (1976) Fluorescence-activated cell sorting. Sci. Am. 234: 108-117.

Heslop-Harrison JS, Leitch AR, Schwarzacher T & Anamthawat-Jónsson K (1990) Detection and characterization of 1B/1R translocations in hexaploid wheat. Heredity 65: 385-392.

Hibi T, Kano H, Sugiura M, Kazami T & Kimura S (1988) High-speed electro-fusion and electro-transfection of plant protoplasts by a continuous flow electro-manipulator. Plant Cell Rep. 7: 153-157.

Jadari R, Sihachakr D, Rossignol L & Ducreux G (1992). Transfer of resisitance to *Verticillium dahliae* Kleb. from *Solanum torvum* S.W. into potato (*Solanum tuberosum* L.) by protoplast electrofusion. Euphytica 64: 39-47.

Johnson LB, Stuteville DL, Higgins RK & Skinner DZ (1981) Regeneration of alfalfa plants from protoplasts of selected Regen-S clones. Plant Sci. Lett. 20: 297-304.

Jones MGK (1991) Electrical fusion of protoplasts. In: Lindsey K (ed.) Plant Tissue Culture Manual. (pp. D3:1-11) Kluwer, Netherlands.

Jourdan PS, Earle ED & Mutschler MA (1989a) Synthesis of male sterile, triazine-resistant *Brassica napus* by somatic hybridization between cytoplasmic male sterile *B. oleracea* and atrazine-resistant *B. campestris*. Theor. Appl. Genet. 78: 445-455.

Jourdan PS, Earle ED & Mutschler MA (1989b) Atrazine-resistant cauliflower obtained by somatic hybridization between *Brassica oleracea* and atrazine-resistant *B. napus*. Theor. Appl. Genet. 78: 271-279.

Kamata Y & Nagata T (1987) Enrichment of heterokaryocytes between mesophyll and epidermis protoplasts by density gradient centrifugation after electric fusion. Theor. Appl. Genet. 75: 26-29.

Kameya T, Miyazawa N & Toki S (1990) Production of somatic hybrids between *Solanum melongena* L. and *S. integrifolium* Poir. Jap. J. Breed. 40: 429-434.

Kao HM, Brown GG, Scoles G & Seguinswartz G (1991) Ogura cytoplasmic male sterility and triazine tolerant *Brassica-napus* cv. Westar produced by protoplast fusion. Plant Sci. 75: 63-72.

Kao HM, Keller WA, Gleddie S & Brown GG (1992) Synthesis of *Brassica oleracea-Brassica napus* somatic hybrid plants with novel organelle DNA compositions. Theor. Appl. Genet. 83: 313-320.

Kao KN & Michayluk MR (1974) A method for high-frequency intergeneric fusion of plant protoplasts. Planta 115: 355-367.

Kao KN & Michayluk MR (1975) Nutritional requirements for growth of *Vicia hajastana* cells and protoplasts at a very low population density in liquid media. Planta 126: 105-110.

Kao KN & Michayluk MR (1980) Plant regeneration from mesophyll protoplasts of alfalfa. Z. Pflanzenphysiol. 96: 135-141.

Kao KN & Michayluk MR (1989) Fusion of plant protoplasts - techniques. In: Bajaj YPS (ed.) Biotechnology in Agriculture and Forestry. Vol 8. (pp. 277-288) Springer Berlin.

Kao KN & Wetter LR (1977) Advances in techniques of plant protoplast fusion and culture of heterokaryocytes. In: Brinkley BR & Porter K (Eds) International cell biology. (p 216) Rockefeller Univ Press, NY.

Kao KN (1977) Chromosomal behaviour in somatic hybrids of soybean-*Nicotiana glauca*. Mol. Gen. Genet. 150: 225-230.

Kinsara A, Patnaik SN, Cocking EC & Power JB (1986) Somatic hybrid plants of *Lycopersicon esculentum* Mill. and *Lycopersicon peruvianum* Mill. J. Plant. Physiol. 125: 225-234.

Kirti PB, Narasimhulu SB, Prakash S & Chopra VL (1992a) Somatic hybridization between *Brassica juncea* and *Moricandia arvensis* by protoplast fusion. Plant Cell Rep. 11: 318-321.

Kirti PB, Narasimhulu SB, Prakash S & Chopra VL (1992b) Production and characterization of intergeneric somatic hybrids of *Trachystoma ballii* and *Brassica juncea*. Plant Cell Rep. 11: 90-92.

Kirti PB, Prakash S & Chopra VL (1991) Interspecific hybridization between *Brassica juncea* and *B. spinescens* through protoplast fusion. Plant Cell Rep. 9: 639-642.

Kobayashi RS, Stommel JR & Sinden SL. (1995) Somatic hybridization between *Solanum ochranthum* and *Lycopersicon esculentum*. Plant Cell Tissue Organ. Cult. 45: 73-78.

Krasnyanski S & Menczel L (1995) Production of fertile somatic hybrid plants of sunflower and *Helianthus giganteus* L. by protoplast fusion. Plant Cell Rep. 14: 232-235.

Kumashiro T, Asahi T & Komari T (1988) A new source of cytoplasmic male sterile tobacco obtained by fusion between *Nicotiana tabacum* and x-irradiated *N. africana* protoplasts. Plant Sci. 55: 247-254.

Kumashiro T, Asahi T & Nakakido F (1989) Transfer of cytoplasmic factors by asymmetric fusion from a cross-incompatible species, *Nicotiana repanda* to *N. tabacum* and characterization of cytoplasmic genomes. Plant Sci. 61:137-144.

Kunitake H, Nakashima T, Mori K, Tanaka M, Saito A & Mii M (1996) Production of interspecific somatic hybrid plants between *Asparagus officinalis* and *A. macowanii* through electrofusion. Plant Sci. 116: 213-22.

Kyozuka J, Kaneda T & Shimamoto K (1989) Production of cytoplasmic male sterile rice (*Oryza sativa* L.) by cell fusion. Bio/Technology 7: 1171-1174.
Landgren M & Glimelius K (1994) A high frequency of intergenomic mitochondrial recombination and an overall biased segregation of *B. campestris* or recombined *B. campestris* mitochondria were found in somatic hybrids made within Brassicaceae. Theor. Appl. Genet. 87: 854-862.
Larkin PJ, Davies PA & Tanner GJ (1988) Nurse culture of low numbers of Medicago and Nicotiana protoplasts using calcium alginate beads. Plant Sci. 58: 203-210.
Lefrancois C, Chupeau Y & Bourgin J P (1993) Sexual and somatic hybridization in the genus *Lycopersicon*. Theor. Appl. Genet. 86: 533-546.
Lelivelt CLC & Krens FA (1992) Transfer of resistance to the beet cyst nematode (*Heterodera schachtii* Schm.) into the *Brassica napus* L. gene pool through intergeneric somatic hybridization with *Raphanus sativus* L. Theor. Appl. Genet. 83: 87-894
Lelivelt CLC, Leunissen EHM, Frederiks HJ, Helsper JPFG & Krens FA (1993). Transfer of resistance to the beet cyst nematode (*Heterodera schachtii* Schm.) from *Sinapis alba* L. (white mustard) to the *Brassica napus* L. gene pool by means of sexual and somatic hybridization. Theor. Appl. Genet. 85: 688-696.
Li Y-G (1989) Grape tissue culture: review and perspective. Arid Zone Res. 6(2): 35-53
Li Y-G (1993) Protoplast fusion for bloat-safe and nutritionally improved lucerne. PhD thesis, Australian National University, Canberra.
Li Y-G, Tanner G J, Delves A C & Larkin P J (1993) Asymmetric somatic hybrid plants between *Medicago sativa* L. (alfalfa, lucerne) and *Onobrychis viciifolia* Scop. (sainfoin). Theor. Appl. Genet. 87: 55-463.
Li Y-G, Tanner K, Stoutjesdijk P & Larkin P (1996) A highly sensitive method for screening of condensed tannins and its application in plants. In: Vercauteren J et al. (eds) Polyphenols Communications 96. Vol. 1 (pp. 195-196) Groupe Polyphenols, Universite Bordeaux 2, Bordeaux Cedex, France
Liu J-H, Dixeluis C, Eriksson I & Glemelius K (1995). *Brassica napus* (+) *B. tournefortii*, a somatic hybrid containing traits of agronomic importance for rapeseed breeding. Plant Sci. 109: 75-86.
Lu DY, Davey MR & Cocking EC (1983) A comparison of the cultural behaviour of protoplasts from leaves, cotyledons and roots of *Medicago sativa*. Plant Sci. Lett. 31: 87-99.
Lu DY, Davey MR, Pental D & Cocking EC (1982) Forage legume protoplasts: somatic embryogenesis from protoplasts of seedling cotyledons and roots of *Medicago sativa*. In: Fujiwara A (ed.) Plant Tissue Culture 1982. (pp. 597-598) Marizen Co, Tokyo.
Makonkawkeyoon S, Smitamana P, Hirunpetcharat C & Maneekarn N (1995) Production of mouse immunoglobulin G by a hybrid plant derived from tobacco-mouse cell fusions. Experientia 51: 19-25
Mariotti D, Arcioni S & Pezzotti M (1984) Regeneration of *Medicago arborea* plants from tissue and protoplast cultures of different organ origin. Plant Sci. Lett. 37: 149-156.
Matsumoto E (1991) Interspecific somatic hybridization between lettuce (*Lactuca sativa*) and wild species *L. virosa*. Plant Cell Rep. 9: 531-534.
Mattheij WM & Puite KJ (1992) Tetraploid potato hybrids through protoplast fusions and analysis of their performance in the field. Theor. Appl. Genet. 83: 807-812.
Mattheij WM, Eijlander R, de Koning JRA & Louwes KM (1992) Interspecific hybridization between the cultivated potato *Solanum tuberosum* subspecies *tuberosum* L. and the wild species *S. circaeifolium* subsp. circaeifolium Bitter exhibiting resistance to *Phytophthora infestans* (Mont.) de Bary and *Globodera pallida* (Stone) Behrens 1.Somatic hybrid. Theor. Appl. Genet. 83: 459-466.

Mehrle W, Naton B & Hampp R (1990) Determination of physical membrane properties of plant cell protoplasts via the electrofusion technique: prediction of optimal fusion yields and protoplast viability. Plant Cell. Rep. 8: 687-691.

Melchers G, Sacristan MD & Holder AA (1978) Somatic hybrid plants of potato and tomato regenerated from fused protoplasts. Carlsberg Res. Comm. 43: 203-218.

Melzer JM & O'Connell MA (1992) Effect of radiation dose on the production of and the extent of asymmetry in tomato asymmetric somatic hybrids. Theor. Appl. Genet. 83: 337-344.

Menczel L & Wolfe K (1984) High frequency of fusion induced in freely suspended protoplast mixtures by polyethylene glycol and dimethyl sulphoxide at high pH. Plant Cell Rep. 3: 196-198.

Menczel L, Nagy F, Kiss ZS & Maliga P (1981) Streptomycin resistant and sensitive somatic hybrids of *Nicotiana tabacum* + *Nicotiana knightiana*: correlation of resistance to *N. tabacum plastids*. Theor. Appl. Genet. 59: 191-195.

Menczel L, Polsby LS, Steinback KE & Maliga P (1986) Fusion-mediated transfer of triazine-resistant chloroplasts: characterization of *Nicotiana tabacum* cybrid plants. Mol. Gen. Genet. 205: 201-205.

Mendis MH, Power JB & Davey MR (1991) Somatic hybrids of the forage legumes *Medicago sativa* L. and *M. falcata* L.. J. Exp. Bot. 42:1565-1573.

Mezentsev AV (1981) Massive regeneration of alfalfa plants from cells and protoplasts. Soviet Agric.Sci. 4: 31-34.

Mitten DH, Sato SJ & Skokut TA (1984) In vitro regenerative potential of alfalfa germplasm sources. Crop Sci. 24: 943-945.

Mol JNM, Stuitje TR, Gerats AGM & Koes RE (1988) Cloned genes of phenylpropanoid metabolism in plants. Plant Mol. Biol. Rep. 6: 274-279.

Morikawa H & Yamada Y (1992) Protoplast fusion. In: Fowler M & W Warren GS (eds) Plant Biotechnology: Comprehensive Biotechnology. 2nd Suppl. (pp. 199-222) Pergamon Press, Oxford.

Murashige T & Skoog F (1962) A revised medium for rapid growth and bioassays with tobacco tissue cultures. Physiol. Plant. 15: 473-497.

Nakano M & Mii M (1993) Interspecific somatic hybridization in Dianthus: selection of hybrids by the use of iodoacetamide inactivation and regeneration ability. Plant Sci. 88: 203-208.

Nakano M, Hoshino Y & Mii M (1996) Intergeneric somatic hybrid plantlets between *Dianthus barbatus* and *Gypsophila paniculata* obtained by electrofusion. Theor. Appl. Genet. 92:170-172.

Narasimhulu SB, Kirti PB, Prakash S & Chopra VL (1992) Resynthesis of *Brassica carinata* by protoplast fusion and recovery of a novel cytoplasmic hybrid. Plant Cell Rep. 11: 428-432.

Negrutiu I, Hinnisdaels S, Mouras A, Gill BS, Gharti-Chhetri GB, Davey MR, Gleba YY, Sidorov V & Jacobs M (1989a) Somatic versus sexual hybridization: features, facts and future. Acta. Bot. Neerl. 38: 253-272.

Negrutiu I, Mouras A, Gleba YY, Sidorov V, Hinnisdaels S, Famelaer Y & Jacobs M (1989b) Symmetric versus asymmetric fusion combinations in higher plants. In: Bajaj YPS (ed) Biotechnology in agriculture and forestry. Vol 8. (pp.304-319) Springer, Berlin.

Nenz E, Pupilli F, Damiani F & Arcioni S (1996) Somatic hybrid plants between the forage legumes *Medicago sativa* L. and *Medicago arborea* L. Theor. Appl. Genet. 93: 183-189.

Niizeki M & Saito K (1987) Genotypic variation in plant regeneration from calli and protoplasts of alfalfa, *Medicago sativa* L.. Plant Tissue Cult. Lett. 4: 27-31.

Ohno-Shosaku T & Okada Y (1989) Role of proteases in electrofusion of Mammalian cells. In: Neumann E et al. (eds) Electroporation and Clectrofusion in Cell Biology. (pp. 193-202) Plenum, NY.

Okamura M & Momose M (1990) Somatic hybridization, back-fusion and the production of asymmetric hybrid plants in potato. In: Abstracts 7th Intl Cong Plant Tissue Cell Cult. Amsterdam, p 217.

Okamura M (1988) Regeneration and evaluation of somatic hybrid plants between *Solanum tuberosum* and *Lycopersicon pimpinellifolium*. In: Puite KJ et al.(eds) Progress in plant protoplast research. (pp. 213-214) Proc Intl Protoplast Symp, Wageningen.

Pandeya RS, Douglas GC, Keller WA, Setterfield G & Patrick ZA (1986) Somatic hybridisation between *Nicotiana rustica* and *N. tabacum*: development of tobacco breeding strains with disease resistance and elevated nicotine content. Z. Pflanzenzuch. 96: 346-352.

Pehu E, Gibson RW, Jones MGK & Karp A (1990) Studies on the genetic basis of resistance to potato leaf roll virus, potato virus-Y and potato virus-X in *Solanum* brevidens using somatic hybrids of *Solanum brevidens* and *Solanum tuberosum*. Plant Sci. 69: 95-101.

Pehu E, Karp A, Moore K, Steele S, Dunckley R & Jones MGK (1989) Molecular, cytogenetic and morphological characterization of somatic hybrids of dihaploid *Solanum tuberosum* and diploid *S. brevidens*. Theor. Appl. Genet. 78: 696-704.

Pelletier G, Primard C, Vedel F, Chetrit P, Reny R, Rouselle P & Renard M (1983) Intergeneric cytoplasmic hybridization in Cruciferae by protoplast fusion. Mol.Gen. Genet. 191: 244-250.

Perl A, Aviv D & Galun E (1990) Protoplast-fusion-derived CMS potato cybrids: potential seed-parents for hybrid, true-potato-seeds. J. Hered. 81: 38-442.

Pertoft H, Rubin K, Kjellen L, Laurent TC & Klingeborn B (1977) The Viability of cells grown or centrifuged in a new density gradient medium, Percoll($^+$). Exp. Cell Res.110: 449-457.

Pezzotti M, Arcioni S & Mariotti D (1984) Plant regeneration from mesophyll, root and cell suspension protoplasts of *Medicago sativa* cv. Adriana. Genet. Agraria. 38: 195-208.

Preiszner J, Fehér A, Veisz O, Sutka J & Dudits D (1991) Characterization of morphological variation and cold resistance in interspecific somatic hybrids between potato (*Solanum tuberosum* L.) and S. brevidens Phil. Euphytica 57: 37-49.

Primard C, Vedel F, Mathieu C, Pelletier G & Chèvre AM (1988) Interspecific somatic hybridization between *Brassica napus* and *Brassica hirta* (*Sinapis alba* L.). Theor. Appl. Genet. 75: 546-552.

Puite KJ, Roest S & Pijnacker LP (1986) Somatic hybrid plants after electrofusion of diploid *Solanum tuberosum* and *Solanum phureja*. Plant Cell Rep. 5: 262-265.

Pupilli F, Businelli S, Caceres M E, Damiani F & Arcioni S (1995) Molecular, cytological and morpho-agronomical characterization of hexaploid somatic hybrids in *Medicago*. Theor. Appl. Genet. 90: 347-355.

Pupilli F, Scarpa GM, Damiani F & Arcioni S (1992) Production of interspecific somatic hybrid plants in the genus *Medicago* through protoplast fusion. Theor. Appl. Genet. 84: 792-797.

Rahman L, Ahuja P S & Banerjee S (1994) Fertile somatic hybrid between sexually incompatible *Hyoscyamus muticus* and *Hyoscyamus albus*. Plant Cell Rep. 13: 537-540.

Rambaud C, Dubois J & Vassuer J (1993) Male-sterile chicory cybrids obtained by intergeneric protoplast fusion. Theor. Appl. Genet. 87: 347-352.

Rashid A (1988) Cell Physiology and Genetics of Higher Plants. Vol 2. pp. 85-117, CRC Press, Florida.

Redenbaugh K, Malyj LD, Emlay D & Lindemann J (1995) Commercialisation of Plant Biotechnology Products. Contents Calgene.
Reis PJ & Schinckel PG (1963) Some effects of sulfur amino acids on the growth and composition of wool. Aust. J. Biol. Sci. 16: 218-230.
Reisch B & Bingham ET (1980) The genetic control of bud formation from callus cultures of diploid alfalfa. Plant Sci. Lett. 20: 71-77.
Robertson D, Palmer JD, Earle ED & Mutschler MA (1987) Analysis of organelle genomes in a somatic hybrid derived from cytoplasmic male-sterile Brassica oleracea and atrazine-resistant *B. campestris*. Theor. Appl. Genet. 74: 303-309.
Roest R & Gilissen JW (1989) Plant regeneration from protoplasts: a literature review. Acta Bot. Neerl. 38: 1-23.
Roest R & Gilissen JW (1993) Regeneration from protoplasts: a supplementary literature review. Acta Bot. Neerl. 42: 1-23.
Rose RJ & Nolan KE (1995) Regeneration of *Medicago truncatula* from protoplasts isolated from kanamycin-sensitive and kanamycin-resistant plants. Plant Cell Rep. 14: 349-353.
Rose RJ, Thomas MR & Fitter JT (1990) The transfer of cytoplasmic and nuclear genomes by somatic hybridisation. Aust. J. Plant Physiol. 17: 303-321.
Sacristán MD, Gerdemann-Knörck M & Schieder O (1989) Incorporation of hygromycin resistance in *Brassica nigra* and its transfer to *B. napus* through asymmetric protoplast fusion. Theor. Appl. Genet. 78: 194-200.
Sakai T & Imamura J (1990) Intergeneric transfer of cytoplasmic male sterility between *Raphanus sativus* (CMS line) and *Brassica napus* through cytoplast-protoplast fusion. Theor. Appl. Genet. 80: 421-427.
Sakomoto K & Taguchi T (1991) Regeneration of intergeneric somatic hybrid plants between *Lycopersicon esculentum* and *Solanum muricatum*. Theor. Appl. Genet. 81: 509-513.
Sale AJH & Hamilton WA (1968) Effects of high electric fields on micro-organisms III. Lysis of erythrocytes and protoplasts. Biochem. Biophys. Acta 163: 37-43.
Sambrook J, Fritsch EF & Maniatis T (1989) Analysis of genomic DNA by Southern hybridization. In: Molecular Cloning - A Laboratory Manual (2nd edn). vol 2. (pp. 9.31-9.62) Cold Spring Harbor Lab. Press, USA.
Sato H, Yamada K, Mii M, Hosomi K, Okuyama S, Uzawa M, Ishikawa H & Ito Y (1996) Production of an interspecific somatic hybrid between peppermint and gingermint. Plant Sci. 115:101-7.
Saunders JA, Matthews BF & Miller PD (1989) Plant gene transfer using electrofusion and electroporation. In: Neumann E et al. (eds) Electroporation and Electrofusion in Cell Biology. (pp. 343-354) Plenum Press, NY.
Saunders JA, Roskos LA, Mischke S, Aly MA & Owens MLD (1986) Behaviour and viability of tobacco protoplasts in response to electrofusion parameters. Plant Physiol. 80: 117-121.
Saunders JW & Bingham ET (1972) Production of alfalfa plants from callus tissue. Crop Sci. 12: 804-808
Schenk RU & Hildebrandt AC (1972) Medium and techniques for induction and growth of monocotyledonous and dicotyledonous plant cell cultures. Can. J. Bot. 50: 199-204.
Schoenmakers H C H, Wolters A M A, de Hann A, Saiedi A K & Koornneef M (1994) Asymmetric somatic hybridization between tomato (*Lycopersicon esculentum* Mill) and gamma-irradiated potato (*Solanum tuberosum* L.): a quantitative analysis. Theor. Appl. Genet. 87: 713-720.
Schwarzacher T, Anamthawat-Jónsson K, Harrison GE, Islam AKMR, Jia JZ, King IP, Leitch AR, Miller TE, Reader SM, Rogers WJ, Shi M & Heslop-Harrison JS (1992) Genomic in

situ hybridization to identify alien chromosomes and chromosome segments in wheat. Theor. Appl. Genet. 84: 778-786.

Schweiger HG, Dirk J, Koop HU, Kranz E, Neuhaus G, Spangenberg G & Wolff D (1987) Individual selection, culture and manipulation of higher plant cells. Theor. Appl. Genet. 73: 769-783.

Scowcroft WR & Larkin PJ (1981) Chloroplast DNA assorts randomly in intraspecific somatic hybrids of *Nicotiana debneyi*. Theor. Appl. Genet. 60: 179-184.

Senda M, Takeda J, Abe S & Nakamura T (1979) Induction of cell fusion of plant protoplasts by electrical stimulation. Plant Cell Physiol. 20:1441-1443.

Serraf I, Sihachakr D, Ducreux G, Brown SC, Allot M, Barghi N & Rossignol L (1991) Interspecific somatic hybridization in potato by protoplast electrofusion. Plant Sci. 76: 115-126.

Shwarzacher T, Leith AR, Bennett MD & Heslop-Harrison JS (1989) In situ localization of parental genomes in a wide hybrid. Ann. Bot. 64:.315-324.

Sidorov V A, Yevtushenko D P, Shakhovsky A M & Gleba Y Y (1994) Cybrid production based on mutagenic inactivation of protoplasts and rescueing of mutant plastids in fusion products: potato with a plastome from *S. bulbocastanum* and *S. pinnatisectum*. Theor. Appl. Genet. 88: 525-529.

Sidorov VA, Zubko MK, Kuchko AA, Komanitsky IK & Gleba YY (1987) Somatic hybridization in potato: Use of gamma-irradiated protoplasts of *Solanum pinnatisectum* in genetic reconstruction. Theor. Appl. Genet. 74:364-368.

Sihachakr D, Haicour R, Chaput M-H, Barrientos E, Ducreux G & Rossignol L (1989) Somatic hybrid plants produced by electrofusion between *Solanum melongena* L. and *Solanum torvum* Sw. Theor. Appl. Genet. 77:1-6.

Sikdar SR, Chatterjee G, Das S & Sen SK (1990) "Erussica", the intergeneric fertile somatic hybrid developed through protoplast fusion between *Eruca sativa* Lam and *Brassica juncea* (L.) Czern. Theor. Appl. Genet. 79: 561-567.

Sjödin C & Glimelius K (1989) Transfer of resistance *Phoma lingam* to *Brassica napus* by asymmetric somatic hybridization combined with toxin selection. Theor. Appl. Genet. 78: 513-520

Smillie RM, Melchers G & von Wettstein D (1979) Chilling resistance of somatic hybrids of tomato and potato. Carlsberg Res. Commun. 44:127-132.

Smith CL, Ahkong QF, Fisher D & Lucy JA (1982) Is purified poly(ethylene glycol) able to induce cell fusion? Biochem. Biophys. Acta 692: 109-114.

Song J, Sorensen EL & Liang GH (1990) Direct embryogenesis from single mesophyll protoplasts in alfalfa (*Medicago sativa* L.). Plant Cell Rep. 9: 21-25.

Spangenberg G, Valles M P, Wang Z Y, Montavon P, Nagel J & Potrykus I (1994). Asymmetric somatic hybridization between tall fescue (*Festuca arundinacea* Schreb.) and irradiated Italian ryegrass (*Lolium multiflorum* Lam.) protoplasts. Theor. Appl. Genet. 88: 509-519.

Spangenberg G, Wang ZY, Legris G, Montavon P, Takamizo T, Perezvicente R, Valles MP, Nagel J & Potrykus I (1995) Intergeneric symmetric and asymmetric somatic hybridization in *Festuca* and *Lolium*. Euphytica 85: 235-245.

Sproule A, Donaldson P, Dijak M, Bevis E, Pandeya R, Keller WA & Gleddie S (1991) Fertile somatic hybrids between transgenic *Nicotiana tabacum* and transgenic *N. debneyi* selected by dual-antibiotic resistance. Theor. Appl. Genet. 82: 450-456.

Stoutjesdijk PA, (1996) Improving the acid-soil tolerance of *Medicago sativa* by asymmetric somatic hybridisation. PhD thesis, LA Trobe University, Victoria, Australia.

Stuart DA, Nelson J, Strickland SG & Nichol JW (1985) Factors affecting developmental processes in alfalfa cell cultures. In: Henke RR et al. (eds) Tissue Culture in Forestry and Agriculture. (pp. 59-73) Plenum , NY.

Sundberg E & Glimelius K (1986) A method for production of interspecific hybrids within *Brassiceae* via somatic hybridization, using resynthesis of *Brassica napus* as a model. Plant Sci. 43: 155-162.

Taguchi T, Sakamoto K & Terada M (1993) Fertile somatic hybrids between *Petunia hybrida* and a wild species, Petunia variabilis. Theor. Appl. Genet. 87: 75-80.

Takebe I, Labib G & Melchers G (1971) Regeneration of whole plants from isolated mesophyll protoplasts of tobacco. Naturwissenschaften 58: 318-320.

Tanno-Suenaga L, Ichikawa H & Imamura J (1988) Transfer of the CMS trait in *Daucus carota* L. by donor-recipient protoplast fusion. Theor. Appl. Genet. 76: 855-860.

Tanno-Suenaga L, Nagao E & Imamura J (1991) Transfer of the petaloid-type CMS in carrot by donor-recipient protoplast fusion. Jap. J. Breed. 41: 25-33.

Tempelaar MJ & Jones MGK (1985a) Fusion characteristics of plant protoplasts in electric fields. Planta 165: 205-216.

Tempelaar MJ & Jones MGK (1985b) Directed electrofusion between protoplasts with different responses in a mass fusion system. Plant Cell Rep. 4: 92-95.

Terada R, Kyozuka J, Nishibayashi S & Shimamoto K (1987) Plantlet regeneration from somatic hybrids of rice (*Oryza sativa* L.) and barnyard grass (*Echinochloa oryzicola* Vasing). Mol. Gen. Genet. 210: 39-43.

Thomas MR, Johnson LB & White FF (1990) Selection of interspecific somatic hybrids of *Medicago* by using *Agrobacterium*-transformed tissues. Plant Sci. 69:189-198.

Thomzik JE & Hain R (1988) Transfer and segregation of triazine tolerant chloroplasts in *Brassica napus* L. Theor. Appl. Genet. 76:165-171.

Tsong TY & Tomita M (1993) Selective B lymphocyte-myeloma cell fusion. Methods Enzymol 220: 238-246.

Tusa N, Grosser JW & Gmitter FG (Jr.) (1990) Plant regeneration of 'Valencia' sweet orange, 'Femminello' lemon, and the interspecific somatic hybrid following protoplast fusion. J Am. Soc. Hortic. Sci. 115: 1043-1046.

Tusa N, Grosser JW, Gmitter FG (Jr.) & Louzada ES (1992) Production of tetraploid somatic hybrid breeding parents for use in lemon cultivar improvement. HortScience 27: 445-447.

Téoulé E (1983) Hybridation somatique entre *Medicago sativa* L. et *Medicago falcata* L. C R Acad Sc Paris 297:Série III: 13-16.

Uchimiya H (1982) Somatic hybridisation between male sterile *Nicotiana tabacum* and *N. glutinosa* through protoplast fusion. Theor. Appl. Genet. 61: 69-72.

Walbot V & Cullis CA (1985) Rapid genomic change in higher plants. Ann. Rev. Plant Physiol. 36: 367-396.

Wallin A, Glimelius K & Eriksson T (1974) The induction of aggregation and fusion of *Daucus carota* protoplasts by polyethylene glycol. Z. Pflanzenphysiol. 74: 64-80.

Walters TW & Earle ED (1993) Organellar segregation, rearrangement and recombination in protoplast fusion-derived *Brassica oleracea* calli. Theor. Appl. Genet. 85: 761-769.

Walters TW, Mutschler MA & Earle ED (1992) Protoplast fusion-derived Ogura male sterile cauliflower with cold tolerance. Plant Cell Rep. 10: 624-628.

Ward A C, Phelpstead J S-J, Gleadle A E, Blackhall N W, Cooper-Bland S, Kumar A, Powell W, Power J B & Davey M R (1994) Interspecific somatic hybrids between dihaploid *Solanum tuberosum* L. and the wild species *S. pinnatisectum* Dun. J. Exp. Bot 45:1433-1440.

Wijbrandi J, Wolters AMA & Koornneef M (1990a) Asymmetric somatic hybrids between *Lycopersicon esculentum* and irradiated *Lycopersicon . peruvianum*. 2. Analysis with marker genes. Theor. Appl. Genet. 80: 665-672.

Wijbrandi J, Zabel P & Koornneef M (1990b) Restriction fragment length polymorphism analysis of somatic hybrids between *Lycopersicon esculentum* and irradiated *L. peruvianum:* evidence for limited donor genome elimination and extensive chromosome rearrangements. Mol .Gen. Genet. 222: 270-277.

Wright RL, Somers DA & McGraw RL (1987) Somatic hybridization between birdsfoot trefoil (*Lotus corniculatus* L.) and *L. conimbricensis* Willd. Theor. Appl. Genet .75: 151-156.

Wright WE (1978) The isolation of heterokaryons and hybrids by a selective system using irreversible biochemical inhibitors. Exp. Cell Res. 112: 395-407.

Xu Y S & Pehu E (1993) RFLP analysis of asymmetric somatic hybrids between *Solanum tuberosum* and irradiated *S. brevidens*. Theor. Appl. Genet. 85: 754-760.

Xu Y S & Pehu E (1994). Analysis of chloroplast and mitochondrial DNA in asymmetric somatic hybrids between *Solanum tuberosum* and irradiated *S. brevidens*. Transgenic Res. 3: 256-259.

Xu Y S, Jones M G K, Karp A & Pehu E (1993b) Analysis of the mitochondrial DNA of the somatic hybrids of *Solanum brevidens* and *S. tuberosum* using non-radioactive digoxigenin-labelled DNA probes. Theor. Appl. Genet. 85: 1017-1022.

Xu Y S, Murto M, Dunckley R, Jones M G K & Pehu E (1993a) Production of asymmetric hybrids between *Solanum tuberosum* and irradiated *S. brevidens*. Theor. Appl. Genet. 85: 729-734.

Xu Y-S, Clark MS & Pehu E (1993) Use of RAPD markers to screen somatic hybrids between *Solanum tuberosum* and *S. brevidens*. Plant Cell Rep. 12: 107-109.

Xu Z-H, Davey MR & Cocking EC (1982) Organogenesis from root protoplasts of the forage legumes *Medicago sativa* and *Trigonella foenum-graceum*. Z. Pflanzenphysiol. 107: 231-235.

Yamaguchi J & Shiga T (1993) Characteristics of regenerated plants via electrofusion between melon (*Cucumis melo*) and pumpkin (interspecific hybrid, *Cucurbita maxima* x *C. moschata*). Jap.J. Breed. 43: 173-182.

Yamanaka H, Kuginuki Y, Kanno T & Nishio T (1992) Efficient production of somatic hybrids between *Raphanus sativus* and *Brassica oleracea*. Jap. J. Breed. 42: 329-339.

Yang ZQ, Shikanai T, Mori K & Yamada Y (1989) Plant regeneration from cytoplasmic hybrids of rice (*Oryza sativa* L.). Theor. Appl. Genet. 77: 305-310.

Yarrow SA, Burnett LA, Wildeman RP & Kemble RJ (1990) The transfer of 'Polima' cytoplasmic male sterility from oilseed rape (*Brassica napus*) to broccoli (*B. oleracea*) by protoplast fusion. Plant Cell Rep. 9: 185-188.

Yu KF, Van Deynze A & Pauls KP (1993) Random amplified polymorphiv DNA (RAPD) analysis. Meth. Plant Mol. Biol. Biotech. 15: 287-301.

Zelcer A, Aviv D & Galun E (1978) Interspecific Transfer of cytoplasmic male sterility by fusion between protoplasts of normal *Nicotiana sylvestris* and x-ray irradiated protoplasts of male-sterile *N. tabacum*. Z. Pflanzenphysiol. 90: 397-407.

Zimmermann U & Scheurich P (1981) High frequency fusion of plant protoplasts by electric fields. Planta 151: 26-32.

Zimmermann U (1986) Electrical breakdown, electropermeabilization and electrofusion. Rev. Physiol. Biochem. Pharmacol. 105: 175-256.

Chapter 14

GERMINATION OF SYNTHETIC SEEDS :

NEW SUBSTRATES AND PHYTOPROTECTION FOR CARROT SOMATIC EMBRYO CONVERSION TO PLANT.

Jean-Marc DUPUIS[1], Christine MILLOT[1], Eberhard TEUFEL[2], Jean-Louis ARNAULT[1] and Georges FREYSSINET[3].

[1]Rhône-Poulenc Agro, Département de Biotechnologies,14-20 rue Pierre Baizet, 69009 Lyon, France, [2]Rhône-Poulenc Rhodia AG, Forshung und Entwicklung, Engesserstrasse 8, 79108 Freiburg, Germany and [3]Rhône-Poulenc S.A., Direction Scientifique, 25 Quai Paul Doumer, 92408 Courbevoie cedex, France.

1 INTRODUCTION

Despite the fact that somatic embryos develop from sporophytic plant cells as opposed to the natural zygotic origin, the somatic and zygotic embryos are identical in appearance and possess the same morphogenetic potential linked to numerous physiological similarities (Raghavan, 1986). Furthermore, somatic embryos can develop in many instances to a whole fertile plant as do the natural mature seeds. Thus, the production of a plant from a somatic embryo appears to be comparable to that from a true seed. It is therefore, hoped that production of plants may occur through schemes involving the sowing of a coated somatic embryo as an artificial seed, its germination and acclimatization in a non-sterile environment such as a greenhouse or even directly in the field.

In fact, monitoring germination of a somatic embryo and its subsequent development into a plant means monitoring two basically different physiological stages (Fig. 1): (i) Unlike a zygotic embryo inside a true seed, the somatic embryo is heterotrophic and may lack nutritional reserves. This is especially true for embryos of species such as carrot in which reserves are located in the endosperm. For this reason, the embryo culture medium has to contain carbohydrates which, under non-sterile conditions will favour development of micro-organisms. Hence, it is necessary to maintain an aseptic environment around the embryo (Molle et al., 1993). (ii) During development, the young plant turns green and becomes photo-autotrophic, no longer requiring exogenous carbohydrates. Therefore, one can envisage a direct or progressive weaning from aseptic conditions.

In reality, the difficulties in controlling the asepsis during the heterotrophic phase of the embryo to plant development make the plant production very close to vegetative multiplication systems such as micropropagation.

In order to monitor the conversion of somatic embryos to plants, we chose to use culture techniques developed for various horticultural species. Thus, we were able to focus our efforts on two major roadblocks : (i) the coating of the somatic embryo which represents the first substrate for the culture of the young plantlet, and (ii) the sanitary protection of both the embryo and the developing plantlet.

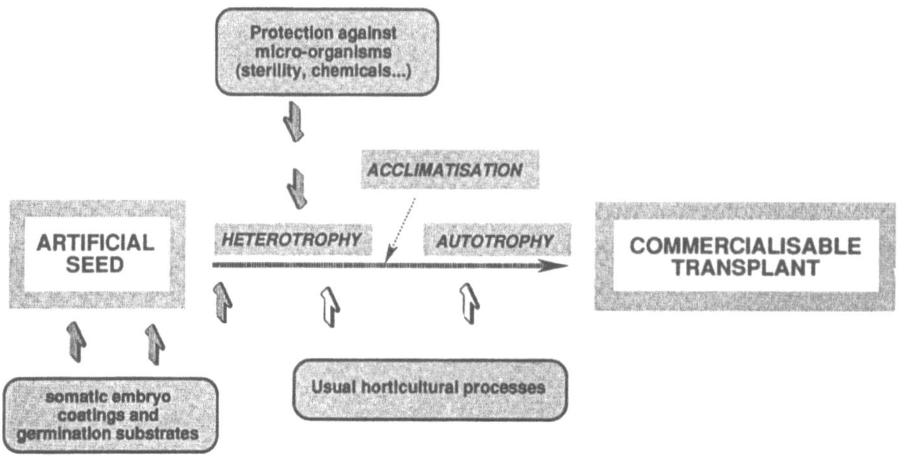

Figure 1. Flowchart of the main technical points (filled narrows) of the embryo to plant conversion process that are described in the text.

2 ARTIFICIAL SEEDS AND THE DELIVERY PROCESS

All the procedures described for conversion of somatic embryos to plants, and especially those involving species with albuminous seed reserves, require an exogenous carbon supply in order to support the primary and heterotrophic phases of somatic embryo germination and plantlet development. For this purpose , sugar is the most widely used carbon source and asepsis is required to protect the somatic embryo and the germination medium against micro-organisms. Instead of breaking sterility, technical evolution was recently characterized by :

- the bulk conditioning, in jars, of the aseptic embryos complemented by the controlled release of sugar;
- the control of sterility limited in space at the embryo or plant scale.

Thus, the techniques developed by Sakamoto et al. (1995), Dupuis et al. (1994), Carlson et al. (1992), may illustrate these technical advances. We will not give a full description of all the techniques, but will only try to point out some specific germination techniques that can easily be integrated in a whole germination and plant production process.

2.1 Alginate beads

Using celery and carrot somatic embryos as models, Onishi et al. (1994) developed a multiple step system which integrates :

- "In jar" somatic embryo culture, maturation and desiccation to obtain encapsulable units capable of growing photo-autotrophically after sowing (Onishi et al., 1994). In practice, the growth of encapsulated unit is promoted by the use of a very small sucrose supply.
- The use of micro-capsules that progressively release sucrose inside alginate beads. The feasibility of such a process had already been demonstrated by Redenbaugh et al. (1987). According to the process developed by Kirin Brewery (Japan), the sucrose micro-capsules, ca. 0.5 mm in diameter, were granulated by a centrifugal tumbling, granulating, and coating apparatus (Sakamoto et al., 1995). The core sucrose granule was coated by spraying a mixture of an interpolymer of ethylene-vinyl acetate-organic acid (Elvax 4260) and bees wax followed by drying. When this micro-capsule was introduced into the gel beads, it released sucrose gradually. This process reduced the daily sucrose supply and, consecutively, a lowering the risk for the development of micro-organisms. A slight phytoprotection by incorporation of the fungicide 0.1% Topsin M (Thiophanate methyl ester; Nippon Soda Co, LTD) in the micro-capsules allowed further control of the risk of infection.
- The use of self-breaking beads. The feasibility of such a process had been demonstrated by Redenbaugh et al. (1987). In the methods developed by Onishi et al. (1994), alginate beads obtained after hardening are rinsed thoroughly with tap water and then immersed in a monovalent cation solution (e.g. potassium nitrate), followed by another rinsing with running tap water. After sowing in moist conditions, such gel beads gradually swell

and become brittle, and finally split spontaneously. This limits the risks linked to oxygendeficiency within the gel, or those linked to unsuitable elasticity or strength of the gel bead (Onishi et al., 1994).

The entire scheme described by these authors led to a 50% conversion rate of embryo to plant under non-sterile conditions. It is evident that encapsulable units are not desiccable, becoming unsuitable for conservation and subsequent commercialization. In such a case the sowing place has to be close to the capsule manufacturing place. Despite this drawback, the technology has demonstrated its ability to solve some of the main problems posed by the alginate bead techniques, by controlling both the weaning stage of the encapsulable units and the micro-capsule technology.

It is well known that oxygen deficiency occurs in alginate beads (Barbotin et al., 1993). The second advantage of the whole technique was to place the germinating embryo in contact with the atmospheric oxygen by the use of self-breaking beads. In a same manner, the use of an aerated gel or the use of perfluorocarbon could be another powerful technique for all types of artificial seeds.

To conclude this topic, Kawaguchi (1992) has devised a different gel breaking system for splitting the gel capsules. The coating contains a high molecular weight compound having a solution-gel transition temperature. The carrier reversibly exhibits liquid state at lower temperature than the solution-gel transition one. By changing the temperature, the method allows the gel to be removed and dissolved before transplantation. The researcher has suggested the use of this technique for controlling germination. Germination-inhibiting substances may be included into the gel. Their removal would allow embryo to germinate.

2.2 Pharmaceutical type capsules

The pharmaceutical type capsules appear to be an efficient coating system for the production of synthetic seeds, allowing germination and subsequent development of plantlets. Such a capsule was presented in Figure. 2. The capsule body is covered on its inner surface by a watertight film. The capsule is filled with 500 µl of germination medium containing Gelrite after the application of this film. Cotton or vermiculite may replace the gel. Finally, a torpedo shape carrot somatic embryo is placed on the internal medium without special orientation. The capsule is closed with a cap (Dupuis et al., 1994). The usefulness of a coating consisting of gelatin containing glycerol depends on the mechanical and physical properties of the inner film that covers the inner surface of the capsule. The most critical property of the coating is that the inner film must be impermeable to water,

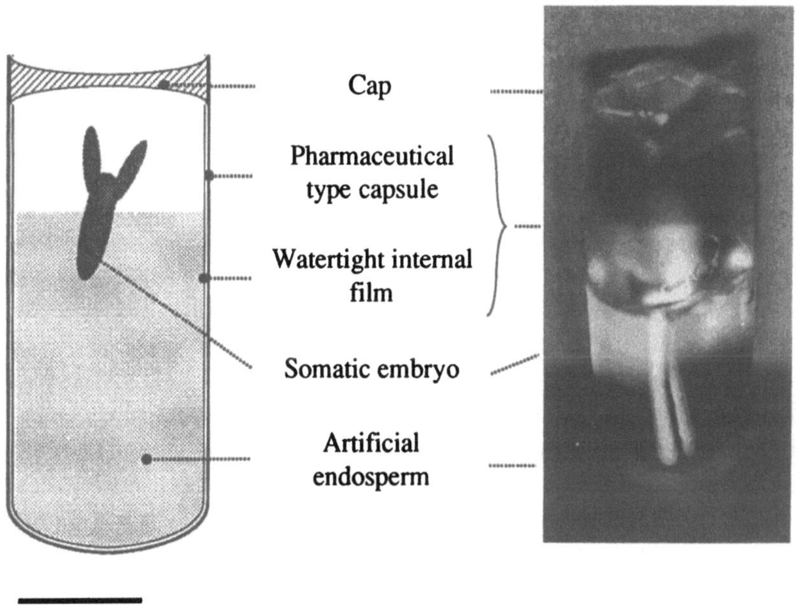

Figure 2. Representation of an artificial seed consisting of a carrot somatic embryo placed on an artificial endosperm, in a pharmaceutical capsule. The top of the capsule was cut to show the embryo. Bar : 5 mm. (From Dupuis et al., 1994 , with the permission of Bio/Technology).

yet it must be sufficiently flexible to allow root growth through the capsule and should not be toxic for the plant.

The film mixture we have described meets these requirements, as well as forms an easily manipulatable and homogeneous covering of the inner capsule surface. For instance, 90% conversion rate and 65% root emergence rate were achieved with films made of 3.33% polyvinyl chloride, 3.33% polyvinyl acetate and 3.33% bentone as a thickener, in trichloroethane (Dupuis et al., 1994). Motoyama et al. (1988) described other types of watertight films for the same use, such as nitro-cellulose (8% in an ethyl acetate solution), ethyl cellulose (10% in ethanol), PVA (20% in methanol) and even types of acrylic paints. Unfortunately, the data concerning rates of germination, conversion and root emergence are imprecise making comparisons between the two types of capsules difficult.

Pharmaceutical type capsules described by Dupuis et al. (1994) provide the nutritional requirements of germinating somatic embryos by limiting sugar loss during planting and germination. They also ensure the mechanical protection of the embryos until after seed sowing without interfering with embryo conversion. On the other hand, the gelatin nature of the capsule may favour microbial development. It also impedes cold conservation of the artificial seed, the capsule becoming smooth even under very low relative humidity (data not shown). By replacing the gel with a solid carrier, made of cellulose acetate, and then replacing the coated capsule with only a film, it was possible to develop a new type of artificial seed with a very wide range of possible applications.

2.3 Sterile plugs and trays

Some authors (Gupta et al., 1993) considered that applications of somatic embryogenesis in forestry will use modifications of existing container systems. Even if these are supplanted by direct sowing of artificial seeds into nurseries, the authors have improved the use of usual horticulture plugs and trays for germinating somatic embryos.

The somatic embryos are embedded in a pre-sterilized plug made of a mass of soil like medium such as vermiculite, perlite or sand. The plugs are placed in trays and moistened with an aqueous solution of a source of carbon and energy for the embryo. The plugs containing the embryos are maintained in an aseptic moist environment. Furthermore, the trays are exposed to atmospheric gas and light necessary for the germinating embryos. When the plantlets have developed several true leaves, the plants are exposed to the non-sterile ambient environment and transplanted into soil by removing the sealing film. This method was utilized for culturing somatic embryos of yellow poplar (*Liriodendron tulipifera* L.). Out of 48 embryos sown, 80% survived in autotrophic state for 42 days after removal of the film (Carlson et al., 1992).

3 CELLULOSE ACETATE MINI-PLUGS

To us, doing away with the requirement for aseptic conditions during the embryo germination and plant development phases appears to be the future of the protection against micro-organisms. Indeed, this would make the techniques for the embryo-to-plant culture to be very close to the traditional horticultural plant production systems. We have attempted to achieve this goal by defining mixtures of chemicals which are able to control micro-organism development. These mixtures have to be non-toxic to both embryo and plantlet. Such a control would allow us to adapt the somatic embryo culture techniques to the current horticultural or field culture practices. For this purpose we defined a universal synthetic endosperm technology : the cellulose acetate mini-plug (CAMP) system.

3.1 Manufacture of the mini-plugs

We have significantly improved the use of novel substrates made of cellulose acetate fibers which we called mini-plugs. The mini-plugs are 4 to 8 mm high cylinders with a 4 to 8 mm diameter (Fig. 3). Except when indicated otherwise, the mini-plugs used during these studies were made of cellulose acetate fibers. These fibers, manufactured by Rhône-Poulenc, are gathered together to form a cable similar to the one for cigarette filter tips. The cable was cut out in sections to create the mini-plugs. These can be used unmodified if we control the ambient relative humidity, the nutrient supply and the development of micro-organisms. During some experiments, the mini-plugs were partially coated on their outer surface by a film to control water and nutrient supplies.

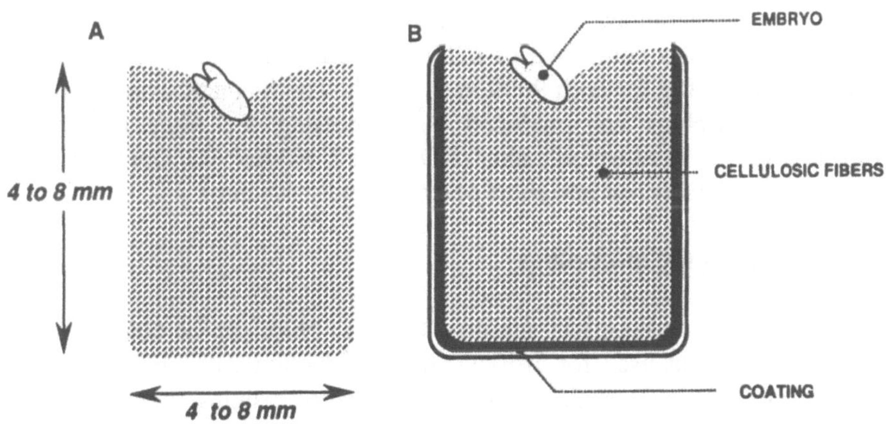

Figure 3. Schematic representation of an artificial seed consisting of a somatic embryo placed on a cellulose acetate mini-plug. The mini-plug may be devoid of any supplementary coating (A) or coated on its outer surface by a film to control water and nutrient supplies (B).

3.1.1 Manufacture of cellulose acetate tow

According to the process described by Serad and Sanders (1978), acetate flakes are prepared by esterification of high purity chemical cellulose with acetic anhydrate. Polymer solutions are converted into fibers by extrusion.

The extrusion process consists of four main operations : (1) dissolution of the cellulose acetate flakes in a volatile solvent, (2) filtration of the solution to remove insoluble matter, (3) extrusion of the solution to form fibers and (4) lubrication and take-up of the yarn on a suitable package. Acetone is the universal solvent for cellulose acetate fiber production in dry extrusion.

Many different acetate continuous tows are manufactured. The different parameters are "tex" [weight (g) of a 1000 m filament] or "denier" [weight (g) of a 9000 m filament], cross-sectional shape, and number of filaments. Individual filament denier (denier per filament or dpf) is usually in the 2-4 dpf range (0.2-0.4 tex per filament).

Varying the cross sectional shape of the spinneret hole directly influences the cross-sectional shape of the fiber. Circular holes produce filaments with an approximately circular cross section, but with crenulated edges; triangular holes produce filaments in the shape of a Y. The same basic extrusion technology that produces continuous filament yarn also produces tow. The filaments from a number of spinnerets are gathered together into a ribbon-like strand or tow. A mechanical device uniformly crimps the tow which is later wrapped into paper to give a finished filter rod. In filter tow processing there are several denier terms to consider : denier per filament and total denier of the tow. A tow described as 8.0 Y 50 may be interpreted, in our studies, as : an uncrimped tow band that weighs 50,000 g for each 9000 m of length and which is composed of 6250 individual filaments (50,000 / 8.0) each weighing 8.0 g for each 9000 m in length. The filament is in the shape of a Y. Mini-plugs are produced by processing the filter tow on a conventional model filter rod maker (KDF2 AF2, Körber AG, Germany) and then cutting the filter rod into short peaces (usually 4-8 mm).

3.1.2 Coating the mini-plugs

To control water and nutrient supplies, the mini-plugs can be coated with a film similar to that used for artificial seed production with pharmaceutical type capsules (Dupuis et al., 1994). In the case of cellulose acetate, a first layer was used as plastering material prior to the subsequent film application. The second layer, controlling water and nutrient leakage, was a film made of Polyvinyl chloride (PVC, Sigma, France), Polyvinyl acetate (PVA, Mowilith M70^{r}, Hoechst, Germany) and bentone SD1^{r} (Eurindis, France). We have already described the properties of such a film in a previous work (Dupuis et al., 1994). The most critical parameters of this film are : (1) it must control water leakage, (2) it must be non-toxic to the plant and (3) it must be flexible enough to allow root growth through the film. The film mixtures we have described above meet these requirements, as well as

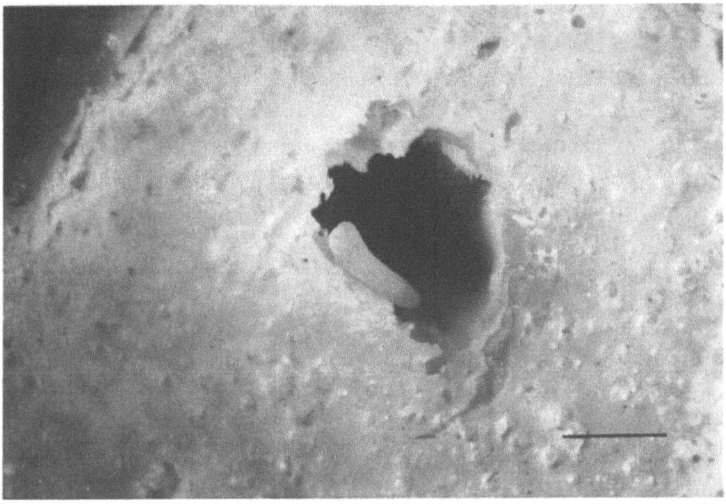

Figure 4. Artificial seed (as described in Fig. 3B) consisting of a carrot somatic embryo placed in a cellulose acetate mini-plug coated on its outer surface by a film. Bar : 1 mm

forming an easily manipulated and homogeneous covering of the outer surface of the mini-plug.

A hole was also made to allow a better root orientation and penetration through the mini-plugs during germination. Fig. 4 shows a carrot somatic embryo within a cellulose acetate mini-plug.

3.2 Carrot somatic embryo germination on naked cellulose acetate mini-plugs (CAMP)

Germination tests have allowed us to specify the structure of the non-coated mini-plugs with respect to the conversion rate of embryos into plants. After collection, torpedo shape (0.5 to 1.5 mm) somatic embryos were placed on non-coated mini-plugs containing 300 ml of a germination medium composed of the inorganic macronutrients of Heller (1953) and the micronutrients of Murashige and Skoog (1962) with 15 g l^{-1} sucrose. The pH was adjusted to 5.6 before autoclaving. The mini-plugs bearing the somatic embryos were sown aseptically by placing them into glass test tubes containing 10 ml of germination medium gelled with 0.6 % Phytagel^{r} (Sigma, France). These tubes were placed in a 16 hour daylight cycle at 55 $\mu E\ m^{-2}\ s^{-1}$ (400-700 nm) and 25 °C. The germination rate was defined as the percentage of germinating embryos showing both shoot and root growth divided by the total number of embryos sown. Root emergence was defined as the percentage of synthetic seeds showing roots emerging from the plug divided by the total number of embryos sown. In a same manner, conversion rate was defined as the percentage of embryos giving rise to normal plantlets divided by the total number of embryos sown. The plantlets were characterized by the appearance of at least 2 leaves and the growth of the root system.

In the first experiment we have compared different CAMPs to Gelrite as the control. The CAMPs were characterized by the section of fibers (the Y

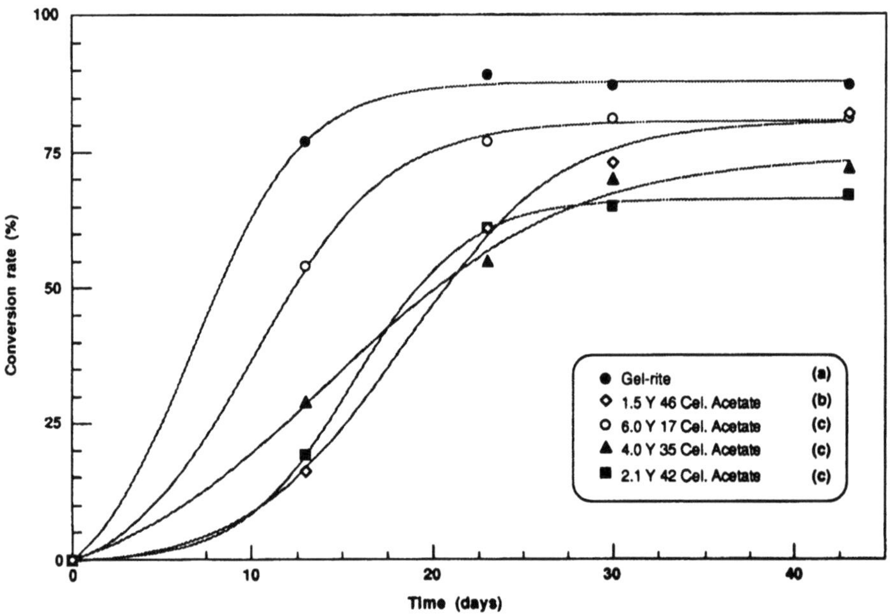

Figure5. Conversion rates of carrot somatic embryo on various cellulose acetate miniugs which differ in number of filaments per cable and filament tex. Analysis of variance gave a F test value F4, 40 = 27.9, significant at the 0.01 level. The mean effects of treatments were compared using the Neumann and Keuls test. N=45 embryos per treatment. a, b, c, : for F to be significant, the same letters indicate that means are not statistically different at the 0.05 level.

form) and their denier. The denier per filament were comprised between 1.5 and 6.0 and the total denier was between 17,000 and 40,000. This lead to different mini-plug densities. Forty-three days after sowing, conversion rates ranged from 65 % to 90%. The cables 6,0 Y 17 gave conversion rates very close to the control (81% versus 87% for Gelrite; Fig. 5). Moreover, the kinetics of conversion were significantly different depending on the substrates. This was particularly obvious 11 days after sowing, when only the control and the 6.0 Y 17 CAMP gave a conversion rate higher than 50%. The maximum conversion rates were reached after 3 weeks for all mini-plugs except for the 6.0 Y 17 ones which showed the same conversion pattern as the control. These observations were confirmed by fresh and dry weight data which reflect whole plant development. Only cables with a weak total denier gave similar or higher results than the control. In the case of Gelrite, the average fresh weight was about 125 mg after 6 weeks of sowing, versus 150 mg in the case of the best cellulose acetate mini-plugs (Fig. 6).

3.3 Variations in mini-plug shape

Another way to vary the mini-plug density and structure was to use 6.0 Y 17 cables differing in diameters. In such a case, denier per filament and total denier are similar, but the number of fibers per cm^2 of cable section varies. We used mini-plugs with 5 and 8 mm diameters.

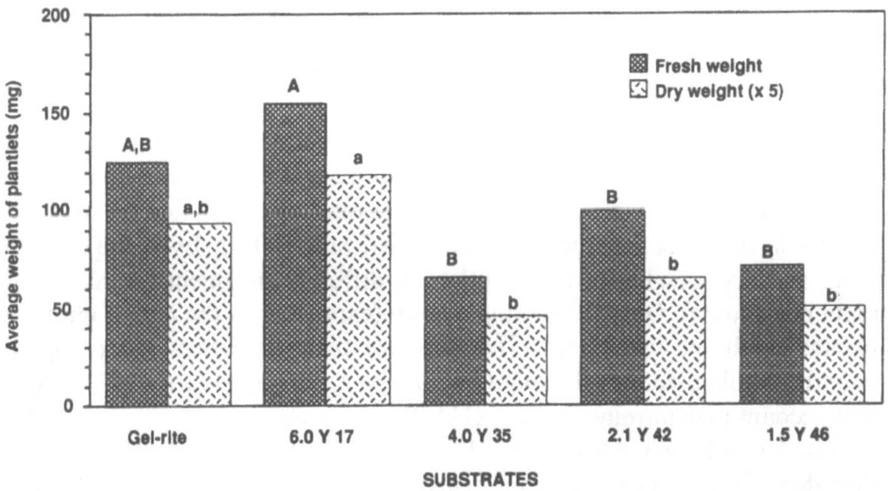

Figure 6. Effects of cellulose acetate mini-plug structures on the growth of carrot plantlets derived from somatic embryos. Measurements were taken 6 weeks after sowing the embryos. Analysis of variance made for fresh weight measurements gave a F test value $F_{4,60} = 5.56$, significant at the 1‰ level. Results of the Neumann and Keuls test are symbolized by small arabic letters (a, b). The same letters indicate that means are not statistically different at the 1% level. Analysis of variance made for dry weight measurements gave a F test value $F_{4,60} =$ 6.3, significant at the 1‰ level. Results of the Neumann and Keuls test are symbolized by capital letters (A, B). The same letters indicate that means are not statistically different at the 1% level.

Table 1 presents the mini-plug density, the conversion rate and the uprooted plant rate corresponding to the different plug types. The emergence rate is defined as the number of plants with roots emerging under the plug divided by the total number of embryos sown. The uprooted plant rate is defined as the number of uprooted plants divided by the total number of embryos sown. In our case, an prooted plantlet is a plant with a root not entirely penetrating the plug. At that time, the upper part of the root appeared over the plug and the apical shoot appeared uprooted (Fig. 7). On the other hand, a well rooted plantlet was defined as a plant whose root system was totally rooted inside the plug.

Table 1. Effects of plug structure on plant development and rooting. Measurements were taken 5 weeks after sowing. n = 45 mini-plugs per treatment.

	Conversion rate (%)	Emergence rate (%)	Uprooted plant rate (%)
Control	91	100	0
Pure cellulose	98	100	56
Cellulose acetate 8 mm	87	96	20
Cellulose acetate 8 mm	89	96	60
χ_2	3.84	2.05	17.42
Signification level (%)	ns	ns	Ns

In the experiment, mini-plugs made of pure cellulose were also compared to the cellulose acetate ones. Gelrite was chosen as a control. All these substrates gave statistically indifferent rates of conversion (between 87 and 98%) and root emergence (between 96 and 100%). Even the average fresh and dry weights of the plantlets in different treatments were similar (data non shown). Only the rooting quality (expressed as the rate of unrooted plants) exhibited differences. This suggests that parameters related to not only the whole plant development but also to rooting quality should be taken into account to determine the best CAMP to be used.

The 5 mm large CAMP have a 137 kg m-3 density. This density is higher than that of the plugs used in the first experiment. This suggests that density is not the only physical property of the plug influencing plant rooting and growth. Other parameters relating to volume occupancy such as number of filaments and filament size and shape should be taken into account. Variations these parameters may affect variations in nutrient supply, oxygen to water ratio, pF1 or other chemical or physiological factors. Such parameters are usually taken into consideration to characterize the soilless substrates in greenhouse substrates (Gras, 1987). Further studies on the physical and chemical properties of the CAMP would allow us to offer products suitable for each plant species and all somatic embryo requirements.

CAMP may also be manufactured with size and shape of the plugs usually used for growing plants in soilless culture systems. For example, 2 cm wide and 2,5 cm high cellulose acetate plugs were used for sustaining the

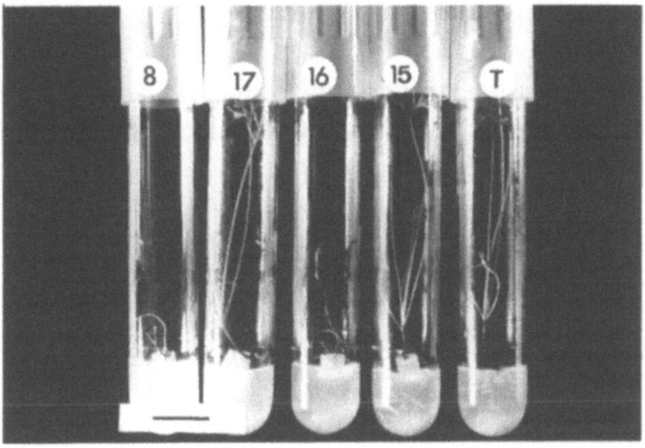

Figure 7. Typical carrot young plantlets growing from somatic embryos on cellulose acetate mini-plugs. The numbers on test tube caps represent different treatments: (8): up-rooted plantlet growing on a 1.5 Y 46 cellulose acetate mini-plug; (16 and 17) : up-rooted plantlet growing on a 4.0 Y 35 cellulose acetate mini-plug; (15) : plantlet growing on 6.0 Y 17 cellulose acetate mini-plug; (T) : plantlet growing on Gelrite commonly used as control. Only plantlets figuring with references 15, 17 and T are considered as converted. Bar : 2 cm.

growth of the germinating embryos. Different plug densities were tested (Table 2). In this experiment, the test tubes containing the plugs were not filled with germination medium. The plugs fill up the test tube section so that only 10 ml of the liquid germination medium were added per plug to assure embryo germination. This volume of medium was enough to reach the "container capacity". As can be seen from the Table 2, neither the germination and conversion rates, nor the fresh and dry weights of the plants differred from the control. This shows, once again, that not only density but also the plug structure are critical for embryo conversion.

Table 2. Effects of plug density on plant development and rooting. Measurements were taken 6 weeks after sowing.

	Gelrite	100 kg· m^{-3} plugs	150 kg·m^{-3} plugs	Statistical variable
N	48	48	48	
Germination rate (%)	94	92	96	$\chi_2 = 0.71$ ns
Conversion rate (%)	90	94	81	$\chi_2 = 5.79$ ns
Fresh weight (mg $plant^{-1}$)	121	128	119	$F_{2,81} = 0.30$ ns
Dry weight (mg $plant^{-1}$)	18	21	19	$F_{2,80} = 1.02$ ns

3.4 Coated mini-plugs

In practice, it is possible that the mini-plugs will need to be physically isolated from the outside environment. We have already seen that films composed of PVC, PVA and bentone as a thickener fully meet this requirement without impeding germination and subsequent plant development (Dupuis et al., 1994). Such films were used to improve the mini-plug coating technology. Toward this goal different film compositions were tested :

films made with 3% PVC, 3% PVA and 3% bentone in a single layer (called C3%);

films made with 5% PVC, 5% PVA and 5% bentone in a single layer (called C5%);

(3) - films made with 3% PVC, 3% PVA and 3% bentone in a first layer and with 5% PVC, 5% PVA and 5% bentone in a second layer (called C3*5%);

Mixtures (2) and (3) were defined to strengthen the basic composition of mixture (1). In this experiment the test tubes containing the mini-plugs were filled with 10 ml of germination medium devoided of any sugar, before artificial seed positioning. Fig. 8 presents coated mini-plugs similar to those involved in the experiment.

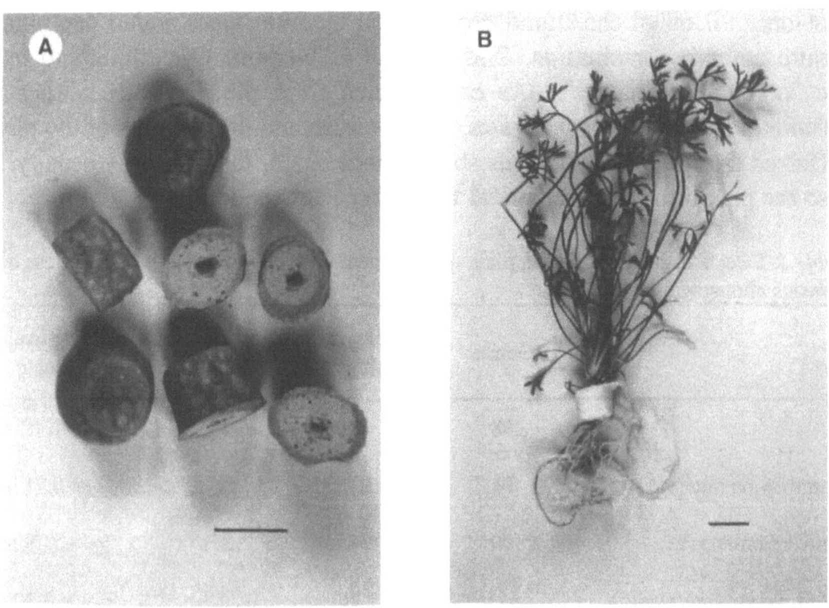

Figure 8. An outer view of coated cellulose acetate mini-plugs (A) as schematized in Fig. 3B and a typical plantlet obtained from such an artificial seed (B). Bar : 8 mm

As can be seen in Figure 9, the embryo-to-plant conversion rates, three weeks after sowning, appeared to be similar in all the treatments, ($X2 = 0.2$ with 4 df, 33 days after sowing). However, these rates were reached more slowly in the case of coated mini-plugs than in the case of the controls. Intermediate values, after 9 days, were less than 10% for coated mini-plugs as against 30-40% for the controls. In a same manner, the fresh and dry weights of plants were significantly lower for treated mini-plugs than for the non-treated ones (data not shown). Finally, root emergence outside the coating was not reduced by the presence of the film, and the emergence rates were identical to conversion rates (between 85 and 95% depending on the treatment).

These results were consistent with those published on pharmaceutical type capsule (PCT) by Dupuis et al. (1994). Moreover, the mini-plug coating technology is easier to use than the PCT one; the film covers the outer surface of the plug instead of the inner surface in the case of a capsule.In addition, the film does not need to maintain the rigidity of the artificial seed as was the case with the capsules. As we used the PCT film technique, one can imagine improvements in the mini-plug coating film properties and composition. An example would be to decrease the film rigidity and

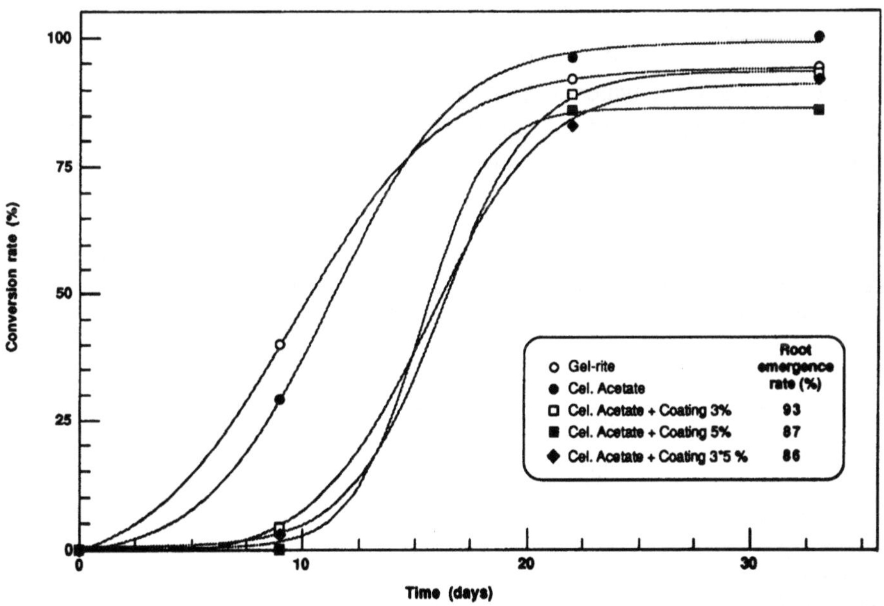

Figure 9. Conversion rates of carrot somatic embryos on coated cellulose acetate mini-plugs differing in coating composition. Measurements were taken 5 weeks after sowing. n = 48 embryos per treatment. Khi-square values were $\chi 2 = 24$; 9 days after sowing, significant at the 0.01 level and $c\chi 2 = 0.2$; 33 days after sowing, non significant. Note: Curves were not closely related to models. They were represented as the result of an interpolation.

increase the fluidity of the film-forming mixture (film components in a solvent).

3.5 Desiccation of the mini-plugs

The kinetics of desiccation were then determined for mini-plugs made of cellulose or cellulose acetate. Non-coated mini-plugs were placed in desiccators at 45% relative humidity (RH). As can be seen in Fig. 10, the results were similar between cellulose acetate and cellulose, suggesting that water retention may be related more to the plug structure than to the polymer itself. For mini-plugs, the minimum water content was reached after 12 days. The kinetics of desiccation of carrot somatic embryos, under these conditions as established by Lecouteux et al. (1991). They demonstrated that only a few hours were necessary to reach a quasi-stationary state of embryo dehydration under 45% RH. Therefore, the use of mini-plugs appears to be a potential method to gently decrease the water content of somatic embryos to a defined RH value. Such a water control may be useful for desiccation of recalcitrant somatic embryos.

Preliminary viability tests are presented in table 3. The embryo germination and conversion rates were similar after desiccation with or without mini-plugs.

Water status of the desiccated plugs was similar to that described above. The real value of the water content of embryo for coated embryos is missing. Nevertheless, this experiment reveals the potential for somatic embryos to be dehydrated inside a plug and, subsequently, the potential for commercialization of such an artificial seed.

Table 3. Effects of desiccation on coated carrot somatic embryo survival and subsequent plant development. Measurements were taken 5 weeks after sowing. The initial water content was 280 µl per mini-plug, approximatively 8 g $H_2O \cdot g^{-1}$ DW.

	Desiccated non coated embryos	Desiccated coated embryos	Non desiccated non coated embryos	Plugs
N	49	42	36	18
Germination rate (%)	100	98	100	
Conversion rate (%)	98	95	92	
Plug water content (0.01 g $H_2O \cdot g^{-1}$ DW)				1.6 ± 0.7
Statistical variable (conversion)	$\chi_2 = 1.83$ ns			

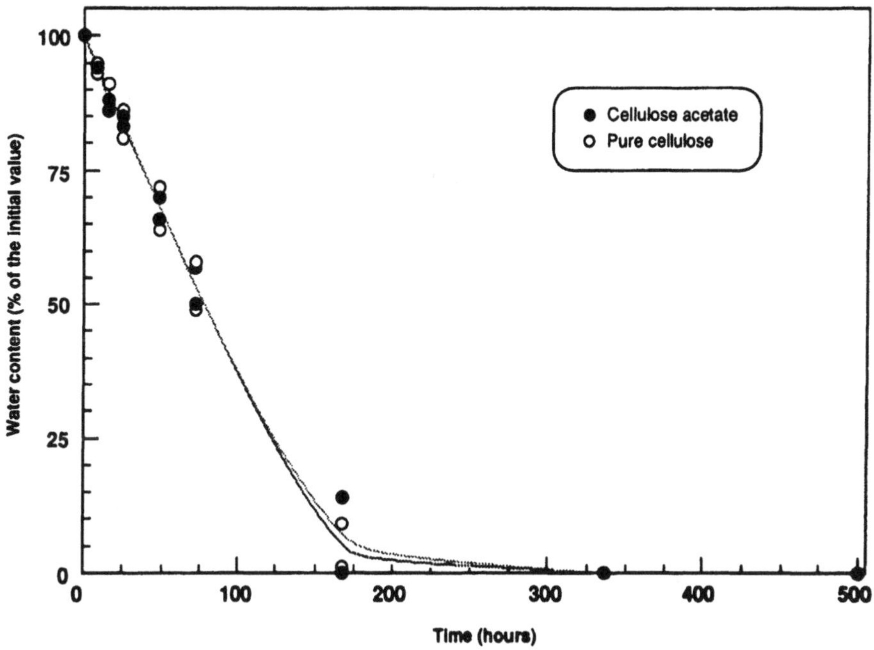

Figure 10. Change in the water content of mini-plug with time during storage under 45% relative humidity at 4°C. The initial water content was approximatively 8.5g H_2O. g^{-1} DW. The experiment consisted in two repetitions. Each point represents the average of 20 individual measurements in a repetition.

4 PHYTOPROTECTION OF CARROT SOMATIC EMBRYOS

There are relatively few papers on the chemical protection of somatic embryos against micro-organisms. Some of them are related to genetic engineering studies. Thus, Tsang et al. (1989) studied the toxicity of antibiotics like kanamycin, hygromycin B, geneticin, methotrexate and cefotaxime on Picea glauca somatic embryos. Mathias and Boyd (1986) have also shown that carbenicillin and cefotaxime were able to improve significantly the performance of wheat in culture by stimulating regeneration and inducing embryogenesis. Hama (1986) has described the use of copper hydroxyquinoline solutions containing penicillin and streptomycin, or of zineb solutions containing chloramphenicol.

For the protection of the germination medium, the phytoprotection mixture has to control all micro-organisms, pathogenic or otherwise, that could develop on a sucrose containing medium.

According to Molle et al. (1993), the chemical agents with significant protection of synthetic seeds, can be classified into three large families : biocides, bactericides, and fungicides. Biocides are compounds which act on a broad spectrum of bacteria and fungi and are used to disinfect floors. Bactericides, most of which have been developed in the medical field, control bacteria but usually have a relatively narrow spectrum of activity. Fungicides have been developed in both crop protection and medical fields.

Falkiner (1988) listed the desirable features of an anti-microbial chemical for plant tissue culturs. We can retain three main points that are very critical towards the synthetic seed technology: (i) possess a broad spectrum of activity; (ii) be suitable for use in combination with other antimicrobial agents; and (iii) have minimal side effects (i.e. no phytotoxicity and no impact on the environment).

Among the molecules that we have tested, fungicides and biocides were too toxic for somatic embryos in their range of efficiency on micro-organism growth. However, some antibiotics such as cefotaxime or ampicilline promoted embryo conversion, but were not able to control all types of micro-organisms, especially the fungi. The first results obtained were disappoint-

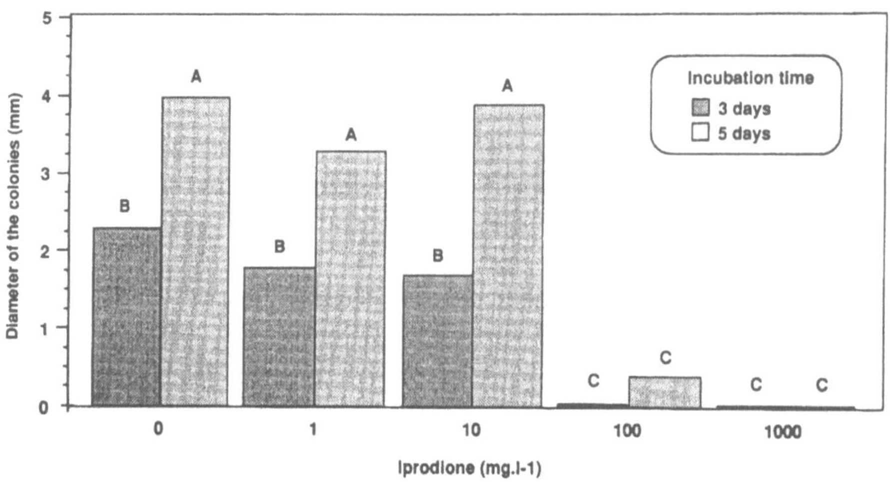

Figure 11. Effects of iprodione on micro-organism development. Measurements were taken 3 or 5 days after contamination. n = 10 dishes per treatment. Analysis of variance gave a F test value $F_{4,90}$ = 187, significant at the 0.001 level. The mean effects of treatments were compared with the Neumann and Keuls test. The same capital letters (A,B) indicate that means were not statistically different at the 0.05 level.

ing, leading us to conclude that mixtures of several compounds will have to be considered to effectively control a broad range of micro-organisms (Molle et al., 1993). The results obtained with iprodione reflet perfectly our progress achieved since 1993 and the actual state of art. Iprodione is one of the more widely used fungicides in horticulture. It is characterized by its wide spectrum of activity covering numerous diseases as *Botrytis*, *Sclerotinia* or *Helminthosporium*. The screening of fungicides involved injection of a peat filtrate into the germination medium in the presence of an antibiotic, cefotaxime and the fungicide. Among the chemicals tested, iprodione controlled fungal growth at 100 mg l^{-1} (Fig. 11).

Table 4. Effects of iprodione on carrot somatic embryo conversion to plant. Measurements were taken 4 weeks after sowing. n = 100 embryos per treatment.

Iprodione (mg l^{-1})	Culture time on the iprodione Containing media		Statistical Variable\
	1 week	4 weeks	
0	97	97	
1	98	89	χ_2 = 15.2
10	89	95	
100	93	92	
1000	0	0	

This value was absolutely compatible with the selectivity data. The toxic effects of iprodione on carrot somatic embryo germination and subsequent plant conversion are described in Table 4 and Figure. 11. As can be seen, iprodione did not impede the conversion rate, nor the plant development expressed by fresh and dry weight data at concentrations less than 100 mgl^{-1}. Figure. 13 shows the nature of the plantlets obtained in the presence of the fungicide.

The next step was to conduct phytoprotection tests, incorporating this data, in a production area (glasshouse). Cellulose mini-plugs were saturated with germination medium containing sucrose, bactericide and iprodione. Eleven days after inoculation, the contaminated surfaces (expressed as % of the contaminated plugs) were : control-67.5 %, iprodione (100 mg l^{-1}) alone - 70.0%, bactericide alone - 80.0%, and iprodione (100 mg l^{-1}) + bactericide - 47.5%. n = 40 plugs per treatment. The $\chi 2$ value was $\chi 2 = 9.5$, significant at the 0.01 level. In this preliminary experiment, most of the infectious colonies were of fungi. Thus, the iprodione alone is unable to control the development of the micro-organisms. But in association with a bacteriostatic compound the contamination level was significantly reduced as compared to the control. Even if this result was in disagreement with the screening test, the level of efficiency expressed some improvement over the published data. It also shows that data from the laboratory have to be adapted to production culture conditions. With this in mind, we are screening new mixtures and accumulating data and experience in this field.

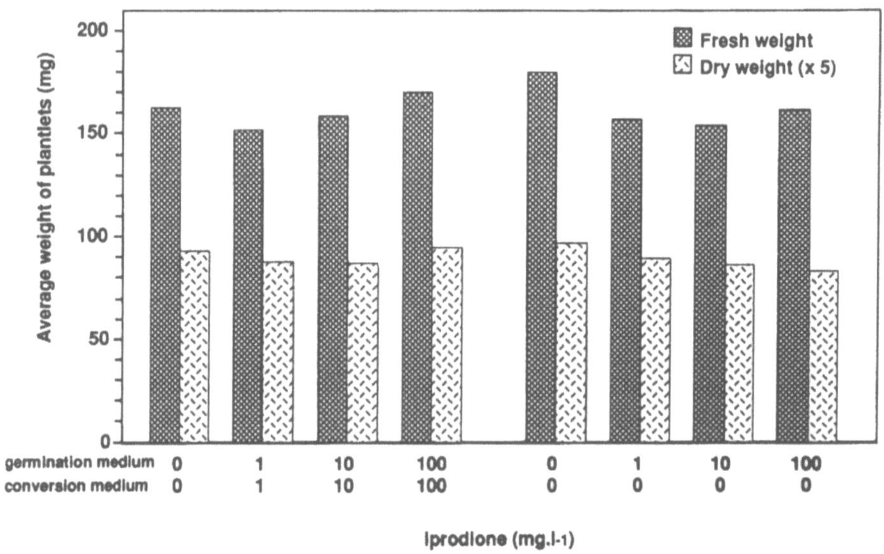

Figure 12. Effects of different iprodione concentrations in germination or conversion media on carrot plantlet development from somatic embryos. Measurements were taken 5 weeks after sowing. n = 40 plants per treatment in 4 repetitions. Analysis of variance gave a F test value $F_{3,24} = 1.34$, non significant, for fresh weight measurements. Analysis of variance gave a F test value $F_{3,23} = 1.58$, non significant, for dry weight measurements.

At this time we can consider that our experiment with iprodione reveals the potential of phytoprotection, demonstrating that the future in this area is to optimize the mixture of non phytotoxic compounds to reduce the microbial growth. Todate no solution giving less than 45% of uncontaminated area (in case of using gelified germination medium) or uncontaminated plugs after more than 10 days in culture has been reported.

One can imagine combining our present data and methods with other ways, such as the use of CO_2, a reduction in the sucrose content or a progressive release of sucrose and chemicals. At this time, no data in these directions have been published. The use of encapsulable units may be a way,

Figure13.:Plantlets growing on germination medium containing (right) 100 $mg.l^{-1}$ of iprodione or devoided (left) of any phytoprotectant. Bar : 2 cm.

as reported by Onishi et al. (1994) who were able to protect somatic embryos with thiophanate methyl ester at a very low concentration.

5 CONCLUSION

We have significantly improved the use of cellulose acetate as a substrate for embryo to plant development. Used as mini-plugs, coated or otherwise, this substrate sustained conversion rates of somatic embryos to plantlets higher than 90%. Otherwise, high conversion rates were also reached after dry storage of the artificial seeds.

In addition, we have attempted to define a mixture of compounds with anti-microbial properties to ensure that problems do not arise during the sowing of the capsule or its delivery in non-sterile conditions. This way, our immediate objective is to define a mixture that controls at least partially the growth of micro-organisms, but which is sufficiently non-phytotoxic to promote plant growth and development. If the phytoprotection mixture has only a partial effect, we will have to control the micro-organism growth by the use of alternative methods such as: (i) controlled sucrose release during the heterotrophic germination phase; (ii) carbon supply as CO2, as soon as possible during the acclimatization process; and (iii) the use of "encapsulable units" instead of somatic embryos. Presently we are not in such a situation, and our other targets are to adapt the acclimation process to the common horticultural techniques, to integrate all these results in an industrial pilot plant for carrot artificial seed delivery and to adapt our solutions to species other than carrot.

6 ACKNOWLEDGMENTS

This work was supported by the BioAvenir programme financed by Rhône-Poulenc, the Ministry in charge of Research and by the Ministry in charge of Agriculture. The authors thank Anne Gayraud, Juliette Barre and Myriam Rossi for their excellent and efficient technical assistance. Special acknowledgments are granted to Marie-Hélène Liegeois and to Denis Chenevotot for their contribution and support. The authors also wish to express their gratitude to Martine Freyssinet, Dominique Job and Jullian Little for help in editing the manuscript and several expert and helpfull discussions.

7 REFERENCES

Barbotin JN, Nava Saucedo JE, Bazinet C, Kersulec A, Thomasset B & Thomas D (1993) Immobilization of whole cells and somatic embryos : coating process and cell-matrix interaction. In Redenbaugh K (ed.) Synseeds: Applications of Synthetic Seeds to Crop Improvement. (pp. 65-104) CRC Press, Boca-Raton.

Carlson WC, Hartle JE & Bower BK (1992) World patent number 92/07457.

Dupuis JM, Roffat C, DeRose RT & Molle F (1994) Pharmaceutical capsules as a coating system for artificial seeds. Bio/Technology 12 : 385 - 389.

Falkiner FK (1988) Strategy for the selection of antibiotics for use against common bacterial pathogens and endophytes of plants. Acta Hortic. 225 : 53 - 56.

Gupta PK, Pullman G, Timmis R, Kreitinger M, Carlson WC, Grob J & Welty E (1993) Forestry in the 21st century. The biotechnology of somatic embryogenesis. Bio/Technology 11 : 454-459.

Hama I (1986) Artificial seeds. Japanese patent application no 4078/1986.

Heller R (1953) Research on the mineral nutrition of plant tissues, Ann. Sci. Nat. Bot. Biol. Veg. 14:1-223.

Kawaguchi M (1992) Carrier for artificial seed and production of artificial seed. Japanese Patent 06125671-A.

Lecouteux C, Florin B, Tessereau H, Bollon H & Pétiard V (1991) Cryopreservation of carrot somatic embryos using a simplified freezing process. Cryo-Letters 12 : 319 - 328.

Mathias RJ & Boyd LA (1986) Cefotaxime stimulates callus growth, embryogenesis and regeneration in hexaploid bread wheat (Triticum aestivum L. Em. Thell). Plant Sci. 46 : 217 - 223.

Molle F, Dupuis JM, Ducos JP, Anselm A, Crolus-Savidan I, Pétiard V & Freyssinet G (1993) Carrot somatic embryogenesis and its application to synthetic seeds. In : Redenbaugh K (ed.) Synseeds; Applications of Synthetic Seeds to Crop Improvement. (pp. 257-288) CRC Press, Boca-Raton.

Motoyama S, Umeda S, Ogishima H & Motegi S (1988) US Patent number 4,769,945.
Murashige T & Skoog F (1962) A revised medium for rapid growth and bioassays with tobacco tissue culture. Physiol. Plant. 15 : 473-497.
Onishi N, Sakamoto Y & Hirosawa T (1994) Synthetic seeds as an application of mass production of somatic embryos. Plant Cell Tissue Organ Cult. 39 : 137 - 145.
Raghavan V (1986) Embryogenesis in Angiosperms. Cambridge Univ. Press, Cambridge.
Redenbaugh K, Slade D, Viss P & Fujii JAA. (1987) Encapsulation of somatic embryos in synthetic seed coats. HortScience 22 : 803 - 809.
Sakamoto Y, Onishi N & Hirosawa T (1995) Delivery systems for tissue culture by encapsulation. In : Aitken-Christie J et al. (eds) Automation and Environmental Control in Plant Tissue Culture. (pp. 215 - 244) Kluwer, Dordrecht.
Serad GA & Sanders JR (1978) Cellulose acetate and triacetate fibers,. In : Grayson M & Eckroth D (eds) Encyclopedia of Chemical Technology. Vol. 5. (pp. 89-117) John Wiley, New-York.
Tsang EWT, David H, David A & Dunstan DI (1989) Toxicity of antibiotics on zygotic embryos of white spruce (Picea glauca) cultured in vitro. Plant Cell Rep. 8 : 214 - 216.

Chapter 15

MORPHOGENESIS IN MICROPROPAGATION

Abel PIQUERAS[1] and Pierre C. DEBERGH[2]
[1]Department of Plant Nutrition and Physiology, CEBAS(CSIC), P O Box 4195, 30080 Murcia, Spain, and [2]Department of Plant Production – Horticulture, Faculty of Agriculture and Applied Biological Sciences, University Gent, Coupure links 653, 9000 Gent, Belgium

1 INTRODUCTION

Both nouns in the title of this article are very often used in many different ways, and are, therefore, ambiguous. In writing this chapter we clearly wanted to delimit its content and, therefore, definitions are required. Micropropagation will be considered as the true-to-type *in vitro* propagation of an initial explant; and morphogenesis as changes in the morphology of the micropropagated material while still *in vitro,* and, to a certain extent, the carry-over effect from *in vitro* to *in vivo.*

Many factors are responsible for changes in the morphology (forms and structure) of micropropagated plants; while some are very obvious (plant growth regulators), others are more surprising and often not given due consideration or even ignored (gas phase, container type, place on a shelf, etc.).

Control of the morphology of a plant or of one of its organs is of considerable importance in micropropagation. Micropropagation techniques rely basically on the use and manipulation of plant morphogenesis *in vitro,* and it can be said that it uses and some times abuses the plasticity in morphogenesis exhibited by plant tissues. This is controlled by three groups of factors that can be related to the plant material, to the culture medium and to the in vitro environment during micropropagation. All these factors need to be balanced during the micropropagation stages.

An appropriate equilibrium between the aforementioned groups of factors is required to control or to manipulate the efficiency of a system: e.g., the number of propagules, the ease to cut and to transplant the material, meeting the desiderata of a customer (e.g. a cluster with a given number of shoots as final product or a single rooted shoot).

At the outset, it is necessary to emphasize that there is not a unique answer to any kind of problem related to micropropagation and/or morphogenesis, because each situation is unique. For each situation interpretation of phenomena is a *conditio sine qua non*, and this requires proper understanding of the factors involved.

Therefore, in this chapter we will try to approach systematically the different factors we consider to have an influence on the morphology of micropropagated material. However, one should keep in mind that care is required in interpretation of results obtained during any one cycle of a culture. Both the medium and the plant are in a state of flux, within cycles and from one subculture to the next, and they influence the third main phase, the head space. Culture conditions change and plant requirements change. Recognising these changes is essential to a better understanding of the whole process (Williams, 1995).

2 PLANT FACTORS

2.1 Inoculum - explant

Inoculum is defined as the plant material taken from the mother plant to initiate a culture (= original explant). Explant is the plant material used to subculture.

For micropropagation purposes most often systems are favoured which are based on the development of existing meristems (axillary or apical buds). This practice prevents many of the negative effects associated with *de novo* formation of buds (caulogenesis) and embryos (embryogenesis). Mutations and somaclonal variations may be usefull in other applications of plant tissue culture but not for commercial plant production as a general principle, and also not in genetic engineering (the expectation is to study the engineered character; complications caused by modifications of the genetic background in which the gene is inserted need to be avoided).

2.1.1 Genotype

Genotype is the most important factor that controls morphogenic responses in plant tissue cultures for micropropagation. These morpho-organogenic responses have their origin in the interaction of genotype with the environment and have been related to genetic factors that were grouped by George (1993) in three categories, considering the location of the genes involved, as nuclear, cytoplasmic and gene interaction products.

For most tissue culture techniques, for micropropagation in particular, it is well documented that not all the genotypes belonging to a genera, or even species behave identically. Some cultivars or varieties are easy to micropropagate, while others can be recalcitrant for one or different aspects of the micropropagation process (e.g. initiation, elongation, rootability) (Pierik, 1990). Different cultivars of *Gerbera jamesonii* can have a completely different behaviour when micropropagated, e.g. one has a tendency for bushiness, while others not; or their growth habit on the same medium is completely different.

2.1.2 Physiology

The physiological status of the source of inocula can have a dramatic effect on its behaviour in culture. The most striking examples are related to juvenility (Franclet 1987). Especially in woody species, lack of juvenility is often a barrier to easy rooting (Vieitez et al. 1994). Juvenility in a plant is quite often associated with absence of flowering; however, this is not always the case.

Juvenility (whatever it means) does not only interfere when a meristem (or an axillary or apical bud) is taken as inoculum, but also when adventitious buds are initiated to start a micropropagation scheme. The latter case is very well illustrated in many cereals; regeneration is sometimes limited to very specific organs and even to a particular state of their development (embryos with endosperm in the milky stage, innermost leaves) (Wernicke *et al.*, 1982).

This can be related to gene expression and function, which can be modulated by external environmental stimuli or by a developmental state of the plant material which is used as inoculum or explant source. The concept of "long term memory", as described by Swartz (1991) for post-*in vitro* stages, is also applicable to the inoculum and the explant. Indeed, cells can retain their memory, and this can be transmitted to successive cell generations and expressed as epigenetic changes in form and function that have been explained as stable and heritable changes linked to mitotic divisions but unable to pass on to sexual progeny.

2.1.3 Growing conditions of the mother plant

It is well documented that the growing conditions of the mother plant can have a considerable effect on the success rate in haploid production. In micropropagation schemes too, their impact is notable. Field grown material can cause serious problems during the initiation stage, mainly because of contaminations. Therefore, more stringent disinfection procedures are required, and this can have far reaching consequences, especially because of their interference in the metabolism of the phenolics, emphasizing hypersensitivity reactions (Rhodes and Woolthorton, 1978; Debergh and Maene, 1981).

More appropriate growing conditions of the mother plants can be the key to the initiation of a successful micropropagation procedure. Many reports have mentioned the impact of the environment in which mother plants are grown, and many factors have been investigated (lighting, temperature, growth regulator pre-treatments, forcing of sources of inocula, ...) (George, 1993), but the why is still not understood.

Figure 1. Eighteen month old plants of *Daphne odora* 'Aureomarginata' derived from apical (left) and lateral (right) shoot types displaying divergent growth habits. (courtesy of T.R. Marks; published by permission of the Journal of Horticultural Science).

2.1.4 Explant type

Morphogenesis in micropropagation can be affected by the choice of the explant in each subculture. This situation has been defined as physiological variability by Marks and Myers (1994). Working with *Daphne* and *Kalmia*, they have identified some factors wich increased morphogenic variability. Two distinct shoot types were identified: central apical- types- formed directly by elongation of the cultured shoots, and lateral types- produced from axillary buds on the central apical shoots. These two shoot types differed in their capacities for growth and rooting in vitro. They also showed a different behaviour when transplanted in the greenhouse (Fig. 1). Eighteen month old plants of *Daphne odora* exhibited a bushy growth in the greenhouse when derived from an apical shoot *in vitro* and an elongated growth when they originated froma lateral shoot. Within these shoot types variability was further increased by the selection of shoots according to length. This divergence in growth potential between shoot types has also been observed by Vanderschaeghe and Debergh (1988) in *Fagus sylvatica*. The explanations for this phenomena are related to alterations in apical dominance patterns caused by topophysically induced changes in the concentrations of growth regulators in the selected explants which predetermined their future morphogenic potential.

An often neglected aspect is the way an explant is cut. On different occasions we have witnessed that manager of commercial tissue culture operations, especially those working with *Gerbera jamesonii*, could identify the person who operated (transplanted) the cultures, just by looking at the container (difference in morphology of the cultures) (among others, J Uvin, pers. comm.). This observation emphasizes the importance of the interpretation of acts. Can a robot do a better job?

2.2 Final product wanted

The final goal of any micropropagation process (s.l.) is to yield rooted plants established in a greenhouse. However, there are many variations on this subject, even for a same cultivar, because the market demand is not always the same. The customer's requirement can be, e.g.: a rooted shoot, a rooted cluster of shoots, a non-rooted shoot, or the required size can be different from one customer to another. For orchids the deal can be for protocorms or plantlets. In the future encapsulated embryos may be the required final product (Reuther, 1990).

When a particular micropropagation protocol needs to be developed on the request of a customer, commercial considerations have to be made in terms of productivity and cost price per unit. One type of micropropagation can be more expensive than the other, because the input required and /or the difficulties to produce a unit with certain morphogenic characteristics can be more than one with different specifications.

As clusters of meristemoids and somatic embryos are in a growing demand, there is an urgent need for future large scale micropropagation methods to use automated and bioreactors systems (Aitken-Christie *et al.* 1995), which requires the use of liquid culture media, as well as specific plant growth regulators and environmental conditions (Ziv 1991; Takayama and Akita 1995). The morphogenic demands in liquid cultures are completely different from those on solid media.

Especially for micropropagation purposes the development of cheap types of bioreactors is required. The system developed by Alvard *et al.* (1993), using plastic filter units, and its recent improvement R.I.T.A. (Récipient d'Immersion Temporaire Adapté) (C. Teisson, pers. comm.) are certainly consistent steps in this direction. New standards of morphogenesis will be required once bioreactor types of culture have been adapted.

3 THE CULTURE MEDIUM

3.1 Gelling agents

The impact of agar concentration and type on the morphogenesis of tissue cultured plants is well documented (Debergh *et al.*, 1981; Debergh, 1983; Romberger and Tabor, 1971). Increasing the agar concentration can reduce hyperhydricity (Debergh *et al.*, 1992). Symptoms characterizing hyperhydricity are not identical in all plants. Development of the morphological symptoms of hyperhydricity depends upon multiple factors expressed over time. Morphological symptoms can be: shorter internodes and rozetting; the stem diameter can be larger or smaller; leaves can be thick and elongated, wrinkled and or curled, brittle, translucent, with a reduced or hypertrophied surface; the color can be abnormal, and in conifers the needles may stick together; initiation of roots on shoots is usually difficult (Debergh *et al.*, 1992). The major factor responsible for the dramatic influence of agar is the matric potential. Beruto *et al.* (1995) have proved that although the contribution of the matric potential to the water potential is small, its effect is considerable.

As the concentration, the type of agar also has a serious influence on many physicochemical characteristics of the gelled substrate. Parameters which are influenced are among others: hardness, compressibility, availability of growth regulators to the plant material, and input of mineral impurities (Debergh, 1983). Impurities present in the gelling agents have been found to inhibit morphogenic responses in shoot cultures of *Pyrus communis* (Singha, 1984) and sugar cane (Anders *et al*., 1988) when several brands of agar were compared. All these factors have been shown to have considerable effect on the propagation ratio and/or the morphogenesis of the micropropagated material. Moreover, the scene is complicated by the fact that many of the aforementioned physicochemical parameters are also influenced by the factors involved in the preparation of the agarized media (e.g. stiring, cooking time), and by the ingredients (e.g. kind and amount of carbohydrates, concentration of mineral elements) (Matthijs, 1991; Debergh *et al*., 1994).

The impact of the medium constituents on the physicochemical parameters of a culture medium, and thereby on the morphogenesis of the cultured material is confirmed by the work of Huang *et al.*(1995) with gelrite. Rigidity of Gelrite media depended on combined levels of MS macrosalts; basal nutrient formulations, sucrose concentration, pH and Gelrite concentration.

3.2 Nutrtional factors (minerals and carbohydrates)

Morphological variability in micropropagated cultures has not often been related to nutritional components of the culure medium. Although most often all the media used for micropropagation are well defined (exceptions are quite often media for orchids) in their composition, abiotic factors or interactions can determine the availability of the components, causing alterations in their expected morphogenic capacity (Debergh *et al*., 1994). As an example of these interactions, one can cite the precipitation of phosphate and iron in MS-medium, which changes its composition and after 7 days up to 50% of the iron and 13% of the phosphate from the medium may not be readily available (Dalton *et al*., 1983). Also, when EDTA is used as an iron chelating agent, formaldehyde can be formed during illumination of the cultures with fluorescent light, creating iron deficiency (Hangarter and Stasinopoulos, 1991 a,b). The mineral concentration can also affect the morphology of micropropagated plants by changes in the osmotic pressure which mainly affects the development of roots *in vitro*. Indeed it is well documented that for rhizogenesis media with low salt concentration are prefered. It is said that these ion-induced changes, target properties of the cell wall and appear to be specific for each species (Pritchard 1994). More information on the impact of the chemical microenvironment is presented by Williams (1995).

Carbohydrates, incorporated in the culture medium as a carbon source, have been found to play an important role, not only in regulating the morphogenic behaviour of cultured plant tissues, but also in changing the propagation rate (Welander et al. 1989) and the rooting behaviour (Romano et al., 1995) of several micropropagated plant species. Carbohydrates control organogenesis by acting as energy source and also by altering the osmotic potential of the culture medium, which alters cell wall properties as extension, hardening and compostion with subsequent modifications in the morphogenic pattern (Pritchard et al., 1991).

3.3 Bioregulators and products with hormonal activity

Among the components of a culture medium, bio-regulators (we favour this term over plant growth regulators, because everything does in fact regulate plant growth, and also this terminology is open to criticism) are the best documented, although probably the least understood, elements controling the morphogenic response in micropropagation. Notwithstanding, it can be said that the classical paper by Skoog and Miller (1957) is the only valuable central dogma in tissue culture: the type of genesis (neoformation) in a tissue culture system is controlled by the ratio between bio-regulators, endogenous and exogenous. The aforementioned paper emphasised the possibities to manipulate caulogenesis, callogenesis and rhizogenesis. However, experience with a multitude of observations *in vitro* and *in vivo* (pruning, induction of branching by hormonal sprays) also taught us that the habit of a plant or a culture (e.g. apical dominance vs. branching habit) is to a certain extent also controled by bio-regulators.

The classical concept of morphogenic effects of bio-regulators is well documented in different review papers and books (among others: George, 1993), and will, therefore, not be elaborated in this document. In many systems bio-regulators are used at fix concentrations during the complete culture period. Some developmental responses can be obtained by exposing the cultured tissues to bio-regulators over shorter periods, as in the case of somatic embryogenesis induction in *Rosa* (S Gillis, pers. comm.) and rooting of micropropagated shoots (Morte *et al.*, 1991). The study of bio-regulators is complex. In the recent past many suggestions have been made to include new types of bio-regulators, such as polyamines, brassinosteroids, jasmonic acid and turgorins (Salisbury and Ross, 1992); this just reflects our tremendous ignorance. In this chapter we mainly want to consider the cytokinins, because this group of bio-regulators underwent the most drastic changes and fundamental research efforts, as well as products with cytokinin-like activity, growth retardants, and non conventional regulators.

The discovery that several substituted ureas have cytokinin activity is important for the tissue culture technology in general and for micropropagation in particular. Indeed, they are active at much lower concentrations than the N^6-substituted adenines. Another advantage is that they avoid or minimize browning reactions (Nitsch & Strain, 1969), which can be responsible for considerable changes in the vigour of the cultures and in the appearance of shoots and other organs. The most active substituted ureas are thidiazuron (TDZ) (Mok *et al.*, 1980) and N-(2-chloro-4-pyridyl)-N'-phenylurea (CPPU) (Fellman *et al.*, 1987). TDZ is said to be most powerful (Mok *et al.*, 1982), and its main morphogenic effect in micropropagation is its strong promotion of shoot proliferation (superior to any conventional cytokinin), offerring an alternative to the propagation of woody species with poor response to the conventional cytokinins (Huetteman and Preece 1993). However, despite of its beneficial effect TDZ can stimulate callus induction and shoot organogenesis, which eventually may threaten the uniformity of the micropropagated plants by generating somaclonal variation. The physiological basis for such strong morphogenic effects may be the persistance of TDZ in plant tissues where it is less susceptible to degrading enzymes than are endogenous cytokinins (Mok *et al.*,1987). However, it is also possible that phenylureas act by means other than pure cytokinin activity *per se;* they may also act as inhibitors of cytokinin oxidase (Horgan, 1987).

The recent use of growth retardants in micropropagation has been introduced to prevent the abnormal morphogenic patterns observed in micropropagation systems in liquid media using bioreactors (Ziv *et al.*, 1994). Inhibitors of giberrellin biosynthesis have been shown to prevent abnomal morphogenic development of leaves and other organs in plants and tissue cultures (Graebe, 1987; Ziv, 1990). Paclobutrazol (PAC) and ancymidol (ANC), inhibitors of gibberellin biosynthesis, have been tested in micropropagation systems using liquid media. PAC showed a pronunced effect on bud proliferation with low proliferation of leaves and an extended effect on regenerated plants developing *ex vitro*; ANC caused a significant inhibition of shoot growth (Ziv and Ariel, 1991). Reduced leaf surface might help to overcome hyperhydricity problems.

It has been demonstrated that some derivatives from the metabolism of plant growth regulators, generated in cultured tissues, can cause morphogenic abnormalities and interfere with the acclimatization of micropropagated plants (Blakesley, 1991). In a recent study, Werbrouck *et al.* (1995) observed tremendous accumulation of glucose conjugate in the 9-position of benzyladenine (9G-BA) in the basal parts of the *in vitro* multiplied shoots of *Spathiphyllum*; it might be associated with inhibited rooting and poor acclimatization. For other plants, other types of conjugates have been identified (Van Staden and Drewes , 1994).

Different pesticides have a cytokinin-like effect. When carbendazim (Debergh *et al.*, 1993.) or imidazol pesticides (Werbrouck and Debergh, 1995) is added to a culture media, containing a cytokinin (urea derivates or

N^6-substituted adenines) a tremendous increase in number of buds occurred (e.g. 150 vs. 5; Fig. 2), especially with monocots (*Cordyline terminalis, Anthurium scherzerianum, Spathiphyllum floribundum*). This phenomenon is not observed when media are supplemented with only pesticides, indicating that they are not cytokinins *per se* (Fig. 2).

Figure 2. Shoot inducing effect of BA is enhanced by imazalil in *Spathiphyllum floribundum* 'Petite'. Left: subcultured on a medium containing $5 mgL^{-1}$ BA; right: subcultured on a medium containing $5 mgL^{-1}$ of BA + 16 mgL^{-1} imazalil (from Werbrouck & Debergh, 1995).

Recent studies have shown that non-ionic co-polymer surfactants can act as stimulators of tissue cultures in several plant species (Khatun *et al.*, 1993). So far, the most frequently employed surfactant is Pluronic F-68 which stimulated shoot production from cultured cotyledons of flax (Khatun *et al.*, 1993) and cell aggregates of *Arabidopsis* (Ribeiro *et al.*, 1992). The possible mechanisms of action proposed for pluronics have been the stimulation of nutrient uptake through formation of short-lived transmembrane pores, enhancement of MDH (malate dehydrogenase) and Apase (acid phophatase) activities, increased oxidation of carbohydrates via tricarboxylic acid cycle accelerating substrate accumulation and cellular phosphate utilization, both processes contributing to increased metabolism and growth (Kumar *et al.*, 1992).

3.4 pH of the medium

Changes in the pH of a medium during plant tissue culture can be caused by autocatalytic changes (Dalton *et al.*, 1983) and by the shoots growing in the medium, which can alter its pH depending on the species (the ability to raise the pH has been related to increased multiplicaton and rooting rates) (Leifert *et al.*, 1992). The physiological basis for this effect may be nutrient deficiency and an increase of free aluminium ions produced at pH values below 4.5 and related toxicity as observed in plants cultured *in vivo* (Mengel, 1984). Another notable effect of pH change is slowing down of bacterial growth, thereby altering the morphogenic effects (Leifert *et al.*, 1992).

4 THE CULTURE ENVIRONMENT

4.1 Gas phase

To avoid contaminations, tissue cultures are carried out in closed vessels, which restricts the exchange of gases between the internal atmosphere of the culture vials and the surrounding gaseous environment. It has been clearly demonstrated that the gaseous composition in the headspace of culture vials can change, and it is affected by many factors (Matthijs *et al.*, 1995). These changes can affect morphogenesis (growth, development and quality) in tissue cultures. An effective control of this gaseous environment may offer a way to manipulate the growth and differentiation of micropropagated plants.

Indeed, accumulation of gases can be responsible for losses during micropropagation (De Proft *et al.*, 1985a,b), and probably also during other applications of the tissue culture technology. They may also have positive effects: faster, better, more, ... growth. The body and closure device of a tissue culture container are of paramount importance in determining the headspace composition and, consequently, in plant growth and development. In fact, the container determines gas exchange with the environment quantitatively and qualitatively (Demeester *et al.*, 1995).

It is not easy to give a simple interpretation of results, because most often the various parameters, interfering in the system, have not been scored or were not considered in the published information. The composition of the headspace is, among others, influenced by:

- the elements of the container, whereby the possibilities and the way of gaseous exchange with the environment (convection, permeability, diffusion) are of basic importance (Matthijs *et al.*, 1995);
- the plant material: the stage of *in vitro* culture (Matthijs *et al.*, 1995); the cultivar (De Riek *et al.*, 1991); the stress situation of the explant; days in

culture (Demeester *et al.*, 1995) or sampling time in relation to the photo- and nyctoperiod (Debergh *et al.*, 1992)

- the environment: place on the shelve, type of air conditioning, ... ;
- the culture medium: see Section 3.

Another complication is the interaction between the various volatiles in the headspace of the culture vial and, thereby, the effect of one on the concentration of the other gassed (De Proft *et al.*, 1985 a,b). Hereafter, oversimplified examples of the influence of particular gaseous components will be presented.

4.1.1 CO_2

The morphogenic effects engendered by the CO_2-concentration in the head space of a tissue culture container are most often not very obvious, although they are sometimes very dramatic. Indeed, excessive accumulation of CO_2 (up to 20%) (De Proft *et al.*, 1985 a,b) can be responsible for extreme dwarfing or the appearance of albinism in the shoots (e.g. *Cordyline terminalis, Ficus benjamina*).

Cultured plant tissues generate CO_2 during respiration, and it is only taken up in significant quantities by photosynthetically active cells or tissues. Sucrose added to the mediun is most often considered to be the primary substrate for respiration in cultured plant tissues. However, this is not always the case. Under certain conditions the ratio of C-incorporation originating from the CO_2 released is as important as the C taken up from sucrose in the culture medium (De Riek *et al.*, 1991). Often a switch in the mixotrophic balance from more heterotrophic to more autotrophic can be accompanied by morphogenic alterations: larger leaves, more anthocyanins in the leaves and/or petioles and stems.

4.1.2 Ethylene

Ethylene influences different morphogenic responses in plant tissue cultures: callogenesis, adventitious bud development, embryogenesis and rhizogenesis. Each of these processes can be influenced in a stimulatory or inhibitory way, depending on the species considered, in a very complex relationship between the cultured tissues, vessel environment and physical properties (Matthijs *et al.* 1995). As under *in vivo* conditions, ethylene can inhibit growth and accelerate senescence *in vitro*. For example, potato tuberization in vitro (Hussey & Stacey 1984), growth of carnation explants (Melé *et al.* 1982) and shoot elongation in roses (Kevers *et al.* 1992) are inhibited by ethylene. At the same time several positive effects of ethylene have been described. These include suppresion of apical dominance and increase in axillary shoot production in bromeliads (De Proft *et al.* 1985a,b), enhancement of bulb primordia development in regenerated shoots of tulips (Taeb & Alderson 1990), stimulation of shoot formation in lavandin (Panizza *et al.* 1988), growth of *Cymbidium* protocorms, *Dendranthema* and potato minicuttings and *Gerbera* shoots (Cachita *et al.* 1987).

The use of inhibitors of ethylene biosynthesis and action (Matthijs *et al.*, 1995) have often proved to be useful in the development of reliable micropropagation systems (mainly to avoid leaf drop).

4.1.3 Oxygen

The oxygen concentration in the headspace of a culture vial influences the metabolism of cultured tissues by regulating the amount of energy available for growth and development, and thus determins organogesesis. Each morphogenic response may require an optimal oxygen concentration for a given organ and species (Norton 1988; Miyagawa *et al.* 1986). Most often the O_2-concentration is not monitored and even not measured. For micropropagation, the level of oxygen present in vessels at a given time depends, among other aspects, on the possibilities of gaseous exchange between the container atmosphere with the environment, regulated by the type of closure. Tightly sealed vessels will favour an increase in CO_2-concentration with time. There are some reports on stimulation of shoot proliferation by hypoxia (Imamura and Harada, 1981; Norton, 1988), that may be caused by the inhibition of ethylene biosynthesis under the prevailing environmental conditions.

With certain types of plastic films, used to close tissue culture vials, almost anaerobic conditons can be created (less than 4.5% of oxygen) (Matthijs *et al.*, 1995) within a few days after subculture. It is still not clear to which extend it influences morphogenesis.

4.2 Light conditions

Light has a strong influence on the developmental morphology of plants and can modify it in two ways, via photosynthesis and via photomorphogenesis.

Normally, photosynthesis provides green plants with a sufficient supply of energy by capturing light energy and carbon incorporation, which is used for the synthesis of cellular structures and components required to sustain growth of green plants. In tissue cultures the conditions are most often not sufficient to support autotrophic development; mixotrophy is predominant (De Riek *et al.*, 1991). By increasing light and providing more CO_2, autotrophy can be achieved (Kozai, 1991).

Plant growth is also controlled by the photomorphogenic pigments, which include phytochrome, blue light and UV absorbing receptors. They act as detectors of light-dependent environmental conditions (such as quality, quantity and photoperiod), modulating growth in response to them. Photomorphogenic pigments regulate the rate and direction of growth of the various plant tissues and organs. At the same time, complex interactions take place between photosynthesis and photomorphogenesis (Cosgrove, 1994).

4.2.1 Light intensity

The light intensity influences the morphology of tissue cultured plants. However, from the published information it is most often difficult to distinguish pure light quantum effects from temperature effects. Indeed, due to the greenhouse effect more light will also engender a higher temperature in the container, and also of the objects (e.g., leaves) the photons hit.

Light intensity requirements for morphogenesis are different for each explant type, and among different stages of micropropagation. Light intensity influences shoot growth and proliferation as well as rhizogenesis (Murashige,1974). Rooting of microshoots can be affected directly by the light intensity under which the mother plant or *in vitro* cultures were raised.

Lateral lighting of micropropagated potato plants gives rise to larger leaves than when the same light sources are used for top lighting (Y. Desjardins, pers. comm.). Very high light intensities (up to 200 μmol m^{-2} s^{-1} compared to 30 μmol m^{-2} s^{-1}) can avoid symptoms of hyperhydricity in micropropagated *Spathyphyllum* and *Ficus benjamina* (L. Maene, pers. comm.). Plants grown in the dark (skotomorphogenesis) are etiolated.

4.2.2 Light quality

In micropropagation, daylight is completely replaced by artificial light sources, with a different spectrum which conditions photomorphogenesis *in vitro*. Tubular fluorescent lamps are the most frequently used light source in plant tissue culture laboratories. Although several types of "white" fluorescent lamps have been considered interchangeable for micropropagation, they have different morphogenic effects (small leaves, short stems and compact appearance) and are not equivalent in terms of growth and differentiation, because of changes in R:FR (Smith,1982) and in the UV-A range radiation (Mohr, 1986). It can be considered that the spectral balance of the lamp rather than its light (photons) output is determinant for photomorphogenic effects. Spectral changes can induce morphogenic responses by direct interaction with photosensory systems. UV, near UV and blue light inhibit *in vitro* growth of plant cells and tissues

(Seibert *et al.*, 1975; Chee and Pool, 1989) and the control of light spectral quality has been proposed as an alternative to growth regulators or other environmental regulations (Wilson *et al.*, 1993).

The majority of photomorphogenic responses are induced by either red or blue wavelengths. Blue light is often responsible for dwarf plants, while red light induces elongation [e.g. Appelgren (1991) on *Pelargonium* × *hortorum*]. Letouzé (1970) described how axillary buds of decapitated *Salix babylonica* remained dormant in blue light. The growth inhibition could be overcome by red light (Letouzé 1974). The light effect can be suppressed by bioregulators (George, 1993, Fig. 79). From *in vivo* studies it is evident that the effects are not due to the color of the light *per se*, but due to the ratios R (red):FR (far red), B (blue):R and B:FR (e.g. Rajapakse, 1992).

Earlier (see Section 3.2) it has been reported that the light quality can interfere with the availability of medium ingredients to the plant material. Formaldehyde is formed from EDTA and the medium becomes deficient in iron as it becomes unchelated. The use of appropriately filtered light (eliminate wavelengths lower than 450nm) when culturing plant material can eliminate unnecessary variability by stabilizing the culture media composition (Hangarter and Stasinopoulos, 1991a,b; Stasinopoulos & Hangarter, 1990).

Photodegradation was also used as a tool to induce rooting in *Carica papaya:* rooting by a combination of IBA and riboflavin could be controlled by manipulation of the photoperiod (Drew *et al.*, 1991).

4.2.3 Photoperiod

The photoperiod has a considerable effect on both shoot and root development *in vitro*. During micropropagation, plants are usually grown under fluorescent lamps for 18 to 16 hours/day. This photoperiod covers the average needs of white light for shoot induction in many species, while shorter photoperiods are required to stimulate rooting (Economou and Read, 1987).

The most spectacular effect of the photoperiod on morphogenesis is flowering. This has been very well documented by Nitsch (1968) with *Plumbago indica*, a short day plant. Also, the subsequent behaviour in the field of micropropagated plants can be conditioned by modications in photoperiod. For example, in strawberry stolon production by the plants can be promoted by exposing them to long photoperiods (Jones *et al.*, 1986).

4.3 Humidity (water retention capacity)

Hyperhydricity, formerly termed vitrification (Debergh *et al.*, 1993), is caused by the lack of a gradient in water retention capacity between the container environment and the stomatal cavity, and can be responsible for significant or negligible modifications in morphology, mainly of the leaves. Obvious symptoms on the leaves can be translucency, prominant veins and reduced lamina. Normal looking leaves can also be hyperhydric. In this case the symptoms are more subtle, such as non- or inadequately-functioning stomata, poor or altered cuticle and epicuticular wax.

Container type, closure device, culture room environment, the quality of the gelled medium are factors which can be responsible for this altered morphology. The type and concentration of cytokinin can also induce this phenomena (Debergh, 1983).

5 ACKNOWLEDGEMENTS

A. Piqueras is gratefull to the E.C. for a postdoctoral research fellowship in the framework of the FLAIR programme. The authors would like to thank the Journal of Horticultural Science and Dr. T R. Marks for the permission granted to publish Fig. 1. Some of the studies described in this chapter have been sponsored by the IWONL.

6 REFERENCES

Aitken-Christie J, Kozai T & Smith MAL (1995) (Eds) Automation and Environmental Control in Plant Tissue Culture. Kluwer, Dordrecht.

Alvard D, Cote F & Teisson C (1993) Comparison of methods of liquid medium culture for banana micropropagation. Plant Cell Tissue Organ Cult. 32:55-60.

Anders J, Larrabee PL & Fahey JW (1988) Evaluation of gelrite and numerous agar sources for in vitro regeneration of sugar cane. HortScience 23: 755.

Appelgren M (1991) Effect of light quality on stem elongation of *Pelargonium in vitro*. Sci. Hortic. 45: 345-351.

Beruto D, Beruto M, Cicarelli C & Debergh PC (1995) Matric potential evaluations and measurements of gelled substrates. Physiol. Plant. 94: 151-157.

Blakesley D (1991) Uptake and metabolism of 6-benzyladenine in shoot cultures of Musa and Rhododendron. Plant Cell Tissue Organ Cult. 25:69-74.

Cachita CD, Achim F, Cristea V & Gerley K (1987) 2-chloroethylphosphonic acid and morphogenesis. Symposium Florizel 87 Arlon-Belgium. Plant micropropagation in horticultural industries. pp 234-237.

Chee R & Pool RM (1989) Morphogenic responses to propagule trimming, spectral irradiance, and photoperiod in grapevine shoots recultured in vitro. J. Amer. Soc. Hortic. Sci. 114:350-354.

Cosgrove DJ (1994) Photomodulation of growth. In: Kendrick RE & Kronenberg GHM (eds), Photomorphogenesis in Plants- 2nd Edition (pp. 631-658) Kluwer, Dordrecht.

Dalton CC, Iqbal K & Turner DA (1983) Iron phosphate precipitation in Murashige and Skoog medium. Physiol. Plant. 57: 472-476.

De Proft MP, Maene LJ & Debergh PC (1985 a) Carbon dioxide and ethylene evolution in the culture atmosphere of *Magnolia* cultured *in vitro*. Physiol. Plant. 65: 375-379.

De Proft MP, Van Den Broek G & Van Dijck R (1985 b) Implications of the container-atmosphere during micropropagation of plants. Med. Fac. Landbow. R.U.G., 50: 129-132.

De Riek J, Van Cleemput O & Debergh PC (1991) Carbon metabolism of micropropagated *Rosa multiflora* L.. In Vitro Cell. Dev. Biol. 27P: 57-63.

Debergh PC & Maene LJ (1981) A scheme for commercial propagation of ornamental plants by tissue culture. Sci. Hortic.14: 335-345.

Debergh PC (1983) Effects of agar brand and concentration on the tissue culture medium. Physiol. Plant. 59:270-276.

Debergh PC, Aitken-Christie J, Cohen D, von Arnold S, Zimmerman R & Ziv M (1992) Reconsideration of the term 'vitrification' as used in micropropagation. Plant Cell Tissue Organ Cult. 30: 135-140.

Debergh PC, De Coster G & Steurbaut W (1993) Carbendazim as an alternative plant growth regulator in tissue culture systems. In Vitro Cell. Dev. Biol. 29P: 89-91.

Debergh PC, De Riek J & Capellades M (1992) Responses and mechanisms of adaptation of plants under *in vitro* culture conditions. Transactions Malaysian Soc. Plant. Physiol .3: 140-142.

Debergh PC, De Riek J, & Matthys D (1994) Nutrient supply and growth of plants in Culture.In: Lumsden PJ et al. (eds) Physiology, Growth and Development of Plants in culture. (pp. 58 - 68) Kluwer, Dordrecht.

Debergh PC, Harbaoui Y & Lemeur R (1981) Mass propagation of globe artichoke (*Cynara* scolymus): Evaluations of different hypotheses to overcome vitrification with special reference to water potential. Physiol. Plant. 53: 181-187.

Demeester J, Matthijs D, Pascat B & Debergh P (1995) Toward a controllable headspace composition - growth, development and headspace of a micropropagated *Prunus* rootstock in different containers. In Vitro Cell. Dev. Biol. 31P: 105-112.

Drew RA, Simpson BW & Osborne WJ (1991) Degradation of exogenous indole-3-butyric acid and riboflavin and their influence on rooting response of papaya in vitro. Plant Cell Tissue Organ Cult. 26: 29-34.

Economou AS & Read PE (1987) Light treatments to improve efficiency of in vitro propagation systems. Hortscience 22: 751-754.

Fellman CD, Read PE & Hosier MA (1987). Effects of thidiazuron and CPPU on meristem formation and shoot proliferation. HortScience 22: 1197-1200.

Franclet A (1987) Exposé introductif. In: Ducaté G et al.(Eds) Plant Micropropagation in Horticultural Industries - Preparation, Hardening and Acclimatization Processes. Belgian Plant Tissue Culture Symposium, Florizel, Arlon, Belgium.

George EF (Ed.) (1993) Plant Propagation by Tissue Culture - Part I The Technology. Exegetics, England.

Graebe J.E. (1987). Gibberellin biosynthesis and control. Ann. Rev. Plant Physiol. 38: 197-201.

Hangarter RP & Stasinopoulos TC (1991a) Effect of Fe-catalyzed photooxidation of EDTA on root growth in plant culture media. Plant. Physiol. 96: 843-847.

Hangarter RP & Stasinopoulos TC (1991b) Repression of plant tissue culture growth by light is caused by photochemical change in the tissue culture medium. Plant Sci. 79: 253-257.

Horgan R (1987) Plant growth regulators and the control of growth and differentiation in plant tissue cultures. In: Green CE et al. (eds) Plant Tissue and Cell Culture. (pp. 153-149) Alan R Liss, New York.

Huang LC, Kohashi C, Vangundy R & Murashige T (1995) Effects of common components on hardness of culture media prepared with Gelrite^{TM}. In Vitro Cell. Dev. Biol. 31P: 84-89.

Huetteman CA & Preece JE (1993). Thidiazuron : a potent cytokinin for woody plant tissue culture. Plant Cell Tissue Organ Cult. 33: 105-119.

Hussey G & Stacey NJ (1984) Factors affecting the formation of in vitro tubers of potato (*Solanum tuberosum* L.) Ann. Bot. 53: 565-578.

Imamura J. & Harada H. (1981). Stimulation of tobacco pollen embryogenesis by anaerobic treatments. Z. Pflanzenphysiol. 103: 259-263.

Jones OP, Waller BJ, Hadlow WC & Beech MG (1986) Modification of micropropagation porcedures to control field performance. Rep. E. Malling Res. Sta. 1985, p 95.

Kevers C, Boyer N, Courduroux JC & Gaspar T (1992) The influence of ethylene on proliferation and growth of rose shoot cultures. Plant Cell Tissue Organ Cult. 28: 175-181.

Khatun A, Laouar L, Davey MR, Power JB, Mulligan BJ & Lowe KC (1993) Effects of Pluronic F-68 on shoot regeneration from cultured jute cotyledons and on growth of transformed roots. Plant Cell Tissue Organ Cult. 34: 133-140

Kozai T (1991) Micropropagation under photoautotrophic conditions. In: Debergh PC & Zimmerman RH (eds) Micropropagation: Technology and Application (pp. 447-470) Kluwer, Dordrecht

Kumar V, Laouar L, Davey MR, Mulligan BJ & Lowe KC (1992) Pluronic F-68 stimulates growth of solanum dulcamara in culture. J. Exp. Bot. 43: 487-940.

Leifert C, Pryce S, Lumsden PJ & Waites WM (1992) Effect of medium acidity on growth and rooting of different plant species growing in vitro. Plant Cell Tissue Organ Cult. 30: 171-179.

Letouzé R (1970) Maintien de la dominance apicale après décapitation d'une bouture de saule (*Salix babylonica)* en culture *in vitro*. Influence de la lumière bleue-violette. C. R. Acad. Sci. Paris 271: 2309-2312.

Letouzé R (1974) The growth of the axillary bud of a willow cutting culured *in vitro*: apical dominance and light quality. Physiol. Vég. 12:397-412.

Marks TR & Myers PE (1994) Physiological variability arising from in vitro culture is induced by shoot selection and manipulation strategies. J. Hortic. Sci. 69:1-9.

Matthijs D (1991) Fysicochemische aspecten van agarmedia.Thesis, University of Gent.

Matthijs D, Gielis J & Debergh P (1995) Ethylene. In: Aitken-Christie J et al (eds) Automation and Environmental Control in Plant Tissue Culture. (pp. 473-491) Kluwer, Dordrecht

Matthijs DG, Pascat B, Demeester J, Christiaens K & Debergh P (1994) Factors controlling the evolution of the gaseous atmosphere during *in vitro* culture. Proc. Int. Symp. Ecopyhysiology *in vitro*, Aix en Provence (in press).

Melé E, Messeguer J & Camprubi P (1982) Effect of ethylene on carnation explants grown in sealed vessels. Proc. 5th Intl. Cong. Plant Tissue & Cell Culture.(pp. 69-70) Jpn. Assoc. Plant Tissue Cult., Japan.

Mengel K. (1984) Ernährung und Stoffwechsel der Pflanzen. (pp. 134-137). Gustav Fischer, Stuttgart.

Miyagawa H, Fujikoa N, Kohda H, Yamasaki K, Taniguchi K & Tanaka R (1986) Studies on the tissue culture of Stevia rebaudiana and its componenets; II induction of shoot primordia. Planta Med. 52:321-323.

Mohr H. (1986) Coaction between pigment systems. In: Kendrick RE & Kronenberg GMH (Eds.). Photomorphogenesis in Plants. (pp 547-564) Martinus Nijhoff, Dordrecht.
Mok MC, Mok DWS & Armstrong DJ (1980). Cytokinin activity of N-phenyl-N'-1,2,3-thidiazol-5-yl urea and its effect on cytokinin autonomy in callus cultures of Phaseolus. Plant. Physiol. 65 (suppl): 24 (abstr).
Mok MC, Mok DWS, Armstrong DJ, Shudo K, Isogai Y & Okamoto T (1982) Cytokinin activity of N-phenyl-N'-1,2,3-thidiazol-5-urea (thidiazuron). Phytochemistry 21:1509-1511.
Mok MC, Mok DWS, Turner JE & Mujer CV (1987) Biological and biochemical effects of cytokinin-active phenylurea derivatives in tissue culture systems. HortScience 22: 1194-1196.
Morte MA, Olmos E, Hellin E & Piqueras A (1991) Micropropagation of Holly (*Ilex Aquifolium*). Acta Hortic. 289:139-141.
Murashige T (1974) Plant propagation through tissue cultures. Ann. Rev. Plant Physiol. 25:135-166.
Nitsch C (1968) Induction *in vitro* de la floraison chez une plante de jours courts: *Plumbago indica* L.. Ann. Sci. Nat. Bot. Paris 12th series, 9: 1-91.
Nitsch JP & Strain GC (1969) Effet de diverses cytokinines sur le brunissement d'explants de canne à sucre. C. R. Ac. Sci. Paris 268: 806 - 809.
Norton C.R. (1988) Metabolic and non-metabolic gas treatments to induce shoot proliferation in woody ornamental plants in vitro. Acta Hortic. 227: 302-204.
Panizza M, Mensuali-Soldi A & Tognoni F (1988) "In vitro" propagation of lavandin: ethylene production during plant development. Acta Hortic. 227: 302-304.
Pierik RLM (1990) Rejuvenation and micropropagation. In: Nijkamp HJJ et al.(eds) Progress in Plant Cellular and Molecular Biology (pp 91-101). Kluwer, Dordrecht.
Pritchard J (1994) Factors affecting cell expansion; hydroponic roots as a model system. In: Lumsden PJ et al.(eds) Physiology, Growth and Development of Plants in Culture. (pp. 104-119) Kluwer, Dordrecht
Pritchard J, Wyn-jones RG & Tomos AD (1991) Turgor, growth and rheological gradients of wheat roots following osmotic stress. J. Exp. Bot. 42: 1043-1049.
Rajapakse NC, Pollock RK, McMahon MJ, Kelly JW & Young RE (1992) Interpretation of light quality measurements and plant response in spectral filter research. HortScience 27: 1208-1211.
Reuther G (1990) Current status and future aspects of large scale micropropagation in commercial plant production. Food Biotechnol. 4: 445 - 459.
Rhodes JM & Wooltorton LSC (1978) The biochemistry of phenolic compounds in wounded plant storage tissues. In: Kahl G (ed.) Biochemistry of Wounded Plant Tissues. (pp. 243-308) W De Gruyter, Berlin.
Ribeiro RCS, Laouar L, Kumar V, Davey MR, Power JB, Mulligan BJ & Lowe KC (1992) Effects of Pluronic F-68 on the growth of *Arabidopsis thaliana* in culture. Soc. Exp. Biol. Lancaster, Abstract A3.20.
Romano A, Noronha C & Martins-Loucao MA (1995) Role of carbohydrates in micropropagation of cork oak. Plant Cell Tissue Organ Cult. 40: 59-167.
Romberger JA & Tabor CA (1971) The *Picea abies* shoot apical meristem in culture. I. Agar and autoclaving effects. Am. J. Bot. 58: 131- 140.
Salisbury FB & Ross CW (1992) (Eds) Plant Physiology. Wadsworth, Belmont, USA.
Seibert M, Wetherbee PJ & Job DD (1975) The effects of light intensity and spectral quality on growth and shoot initiation of tobacco callus. Plant. Physiol. 56:130-139.
Singha S (1984) Influence of two commercial agars on in vitro shoot proliferation of "Almey " crabapple and "Seckel" pear. Hortscience 19:227-228.

Skoog F & Miller CO (1957) Chemical regulation of growth and organ formation in plant tissues cultured in vitro. Symp. Soc. Exp. Biol.11: 118-131.
Smith H (1982) Light quality, photoperception , and plant strategy. Ann. Rev. Plant Physiol. 33: 481-458.
Stasinopoulos TC & Hangarter RP (1990) Preventing photochemistry in culture media by long-pass light filters alters growth of cultured tissues. Plant. Physiol. 93: 1365-1369.
Swartz HJ (1991) Post culture behavior: genetic and epigenetic effects and related problems. In: Debergh PC & Zimmerman RH (eds) Micropropagation: Technology and Application. (pp. 95-121) Kluwer, Dordrecht.
Taeb AG & Alderson PG (1990) Shoot production and bulbing of tulip *in vitro* related to ethylene. HortScience 65: 199-204.
Takayama S & Akita M (1995) The types of bioreactors used for shoots and embryos. Plant Cell Tissue Organ Cult. 39: 109-115.
Van Staden J & Drewes FE (1994) The effect of benzyladenine and its glucosides on adventitious bud formation on lachenalia leaf sections. South African J. Bot. (in press).
Vanderschaege AM & Debergh PC (1988) Influence of explant type on micropropagation of woody species. Med. Fac. Landb. Univ. Gent.53: 1763-1768.
Vieitez AM, Sanchez MC, Amo-Marco JA & Ballester A (1994) Forced flushing of branch segments as a method for obtaining reactive explants of mature *Quercus robur* trees for micropropagation. Plant Cell Tissue Organ Cult. 37: 287-295.
Welander M, Welander NT & Brackman AS (1989) Regulation of in vitro shoot multiplication in Syringia, Alnus and malus by different carbon sources. J. Hortic. Sci. 64: 361-366.
Werbrouck S & Debergh PC (1995) Imazalil enhances the shoot inducing effect of benzyladenine in *Spathiphyllum floribundum* Schott. J. Plant Growth Reg (in press).
Werbrouck SPO, Van der Jeugt B, Dewitte W, Prinsen E, Van Onckelen HA & Debergh PC (1995) Ba metabolism in *Spathipyllum* floribundum Schoot 'Petite' in relation to acclimatization problems. Plant Cell Rep (in press).
Wernicke W, Potrykus I & Thomas E (1982) Morphogenesis from cultured leaf tissue of *Sorghum bicolor* - The morphogenic pathways. Protoplasma 111: 53-62
Williams RR (1995) The chemical microenvironment. In: J Aitken-Christie J (eds.) Automation and Environmental Control in Plant Tissue Culture. (pp. 405- 440) Kluwer, Dordrecht.
Wilson DA, Weigel RC, Wheeler RM & Sager JC (1993) Light spectral quality effects on the growth of potato (Solanum tuberosum L.) nodal cuttings in vitro. In Vitro Cell. Dev. Biol. 29P:5-8.
Ziv M & Ariel T (1991) Bud proliferation and plant regeneration in liquid-cultured Philodendron treated with ancymidol and aclobutrazol. J. Plant Growth Regul. 10:53-57.
Ziv M (1991) Morphogenetic patterns of plants micropropagated in liquid medium in shaken flasks or large scale bioreactor cultures. Israel J. Bot. 40:145-133
Ziv M (1990) The effects of growth retardants on shoot proliferation and morphogenesis in liquid cultured Gladiolus plants. Acta Hortic. 280: 207-214.
Ziv M, Kahany S & Lilien-Kipnis H (1994) Scaled-up proliferation and regeneration of *Nerine* in liquid cultures Part I. The induction and maintenance of proliferating meristematic clusters by paclobutrazol in bioreractors. Plant Cell Tissue Organ Cult. 39: 109-115.

Chapter 16

CELL DIFFERENTIATION AND SECONDARY METABOLITE PRODUCTION

F. CONSTABEL[1] and W.G.W. KURZ[2]
[1]Courtenay/BC, Canada, RR6, Site 662, C95, V9N 8H9 and [2]514 Copland Crescent, Saskatoon/SK, Canada, S7H 2Z5

1 THE CHALLENGE

Cell or cytodifferentiation marks the structural and physiological development of cells sequestering from mitotic cycles and acquiring specific functions within a plant. This process is sometimes also referred to as cell maturation. Cytodifferentiation has been the subject of general botany for generations, but recently seems to fade from textbooks (cf. Esau, 1965, Salisbury and Ross 1985, Strasburger et al., 1991, Eschrich, 1995), since it has triggered little experimentation to analyze its nature. Rather, a modern molecular approach is needed to substantiate the significance of cytodifferentiation or maturation.

The product of cytodifferentiation, i.e. parenchyma and sclerenchyma of various functions, i.e. photosynthesis, uptake of water and nutrients or the storage of primary and secondary metabolites, has conventionally been covered by histology or physiological or functional plant anatomy (Haberlandt, 1914; Eschrich, 1995). In this field of well established disciplines, the production of secondary metabolites more than others continues to intrigue and attract interest, out of curiosity and social concern. For one, the biology and chemistry of secondary metabolites is inspiring due to their structural diversity, their species specificity, and their function. Labelled as by-products until a quarter of a century ago (Paech, 1950), secondary metabolites are now seen as mediators in the interaction of plants and their environment. Depending on their nature, they may attract pollinators or defend against herbivores, pathogens, and encrochement by other plants (Mothes, 1972; Swain, 1977). They may actually be referred to as survival kits. On the other hand, plant secondary metabolites, i.e. primarily alkaloids, terpenoids, flavonoids and a host of glycosides have been critical for human well-being since the dawn of history. Today more than before, they are being used as food additives and health care ingredients. They, therefore, are sought after, refined, and marketed by the processing and pharmaceutical industries. Missing in all these descriptions are the developmental cues which trigger cytodifferentiation and the formation of secondary metabolites in plants.

Plant cell culture is technology introduced half a century ago and aimed at growing cells, tissues, and organs *in vitro,* in pursuit of better understanding the physiology of development. It has been the method of choice when analyzing cytodifferentiation. The demonstration of totipotency of plant cells accomplished by using *in vitro* culture techniques (Reinert 1958, Steward et al., 1958) underscores the point. And, using a variety of parameters, cytodifferentiation has been quantified to a great extent. Physiological experimentation applied to single cells or tissues has even allowed to control the degree of cytodifferentiation, here the production of secondary metabolites. The quality or specific nature of cytodifferentiation, however, has remained elusive until very recently, until molecular approaches have been added (DeLuca et al., 1989, Hashimoto & Yamada, 1994).

When reviewing plant cell culture, the occurrence of secondary metabolites has been observed in callus in early experimentation, sporadically. The scope of products was limited to pigments and fragrances and to simple reaction products like tannins and lignin (Gautheret, 1959). Efforts to elucidate the physiology and chemistry of these products took shape when, for example, tanniferous cells in explants were observed to undergo unequal divisions, to appear in a non-random manner, to consist of hydrolyzable and non-hydrolyzable compounds (Constabel, 1963 -1969). Such efforts were accelerated when experimental results signalled chances for employment of plant cell cultures as sources for desirable products through industrial fermentation (Tulecke and Nickell, 1959). Since then the culture of plant cells has been extended to an ever increasing number of species and resulted in an ever growing list of metabolites found. The last comprehensive description of secondary metabolites in plant cell cultures (Constabel and Vasil, 1988) should benefit from an update and presentation in a digitized mode in the near future. The biology of cytodifferentiation has not substantially benefitted from this activity, cytodifferentiation has not been included in any title on secondary metabolites for an entire decade (cf. Luckner et al., 1977; Kutchan et al., 1983; Rush et al., 1985; Constabel

1988) and, in recent years, may rather have been regarded by some as a "red herring", an irrelevant diversion.

2 THE GOALS

Cytodifferentiation or maturation has been deemed a principle of secondary metabolite production when viewing the correlation between differentiation and production. Increasing yields by applying preference for root cultures over cell suspension cultures and, thus, for material with a higher degree of differentiation or a greater percentage of differentiated cells has become practice in many laboratories. A description of such correlation, therefore, is a desirable first goal and be it for practical purposes. Recent observations with pigmented cells cultured *in vitro*, however, would force issue with correlation as the only interpretation of differentiation and product formation of interest. Observations to be presented later in this review may even cancel the claim of cell maturation as the condition for metabolite production. Or, linking metabolite formation and cytodifferentiation may demand an alternate interpretation and, therefore, lead to a second goal: metabolite production seen as expression of cytodifferentiation. This view still leaves unanswered, how metabolite production is subject to differential developmental regulation and what the cues might be. Recent investigations provide more and more insights into this, the very nature of cytodifferentiation, a third and most important goal. This review will attempt to approach all three goals.

3 THE APPROACH

The correlation of cytodifferentiation and production of secondary metabolites has three components. A first body of evidence is contained in observations made with material differing in development or maturity. Following is physiological experimentation which effects the formation of secondary metabolites. Finally, the focus is directed at genetic manipulations which effect secondary metabolite production. All three approaches have greatly gained by employment of plant cell cultures. Increasingly, though, intact plants are required to demonstrate the advantage of genetic manipulation for an elucidation of secondary metabolism.

In the following investigation preference will be given to methodology which used pigments, indole, and isoquinoline alkaloids as markers to demonstrate diversity and yields or rate of production. For general reviews on secondary metabolites in plant cell cultures see: Verpoorte et al. (1993),

Endress (1994) or Kurz and Constabel (1995) and the classic: Production of secondary metabolites by plant tissue and cell cultures, present aspects and prospects (Fontanel and Tabata, 1987).

Table 1. Cell suspension cultures with prominent capacity for production of secondary metabolites (for references cf. Constabel and Vasil, 1988, Constabel & Tyler, 1994)

Species	Products	Species	Products
Coleus blumei	Rosmarinic acid	*Duboisia leichhardtii*	Tropane alkaloids
Ruta graveolens	Coumarins	*Eschscholtzia californica*	Isoquinoline alkaloids
Daucus carota	Anthocyanins	*Coptis japonica*	Isoquinoline alkaloids
Vitis vinifera	Anthocyanins	*Papaver somniferum*	Isoquinoline alkaloids
Lithospermum spec.	Naphthoquinones	*Peganum harmala*	β-carbolines
Cinchona spec.	Anthraquinones	*Catharanthus roseus*	Monoterpene indole alkaloids
Morinda citrifolia	Anthraquinones	*Ruta graveolens*	Acridone alkaloids
Tripterygium wilfordii	Diterpenes	*Phytolacca americana*	Betalains
Panax ginseng	Saponins	*Beta vulgaris*	Betalains

4 DEVELOPMENTAL PHYSIOLOGY AND PRODUCT FORMATION

4.1 Tissue organization

4.1.1 Callus and cell suspension cultures

Callus initiation by explanting a given plant tissue onto nutrient medium and subsequent culture *in vitro* (cf. Wetter & Constabel, 1982; Gamborg and Phillips, 1995) will demonstrate the correlation of cytodifferentiation and production of secondary metabolites, immediately and almost invariably. Cell division in the explant leading to neo-formation of tissue, i.e. meristems and callus, results in loss of capacity for photosynthesis and storage of metabolites like pigments. Closer inspection may show cell lineages which have sprung from unequal divisions, leaving the pigmented cells in the explant, ageing, and diluted by non-pigmented daughter cells. As callus grows over time, parenchyma cells will re-occur and with them resumption of metabolite, i.e. pigment synthesis and accumulation, often at a lower rate than shown by the source plant. Subculture of callus on fresh medium or in liquid medium, now accompanied by formation of a cell suspension, will repeat the process. Pigmentation or tannin formation preferentially occurs in vacuolated, i.e. mature, differentiated cells (Gautheret, 1959; Constabel, 1968). Since cell suspension cultures are amenable to physiological experimentation and industrial exploitation, occurrence of secondary metabolites in such material has impacted the biology and technology of product formation substantially and widely (Kurz and Constabel, 1995). The most prominent cell suspension cultures and secondary metabolites produced by them are listed in Table 1.

Intriguingly, the presentation in Table 1 also reveals the difficulty of demonstrating various secondary metabolites in callus or cell suspension cultures sought after in academia and industry, i.e. phytochemicals with a sizeable market, by not being listed. Noteworthy are cell suspensions which have been found ambivalent regarding product formation, i.e. they may or may not show sizeable yields when under enhanced culture conditions, i.e. vanillin, cardiac glycosides, taxanes (Table 2). On the other hand, there have always been cases where despite variation of culture conditions cells lacked specific product formation as observed for vinblastine, morphine, or menthol (Table 2).

Table 2. Cell suspension cultures with traces only of desirable secondary metabolites

Species	Products	Reference
Vanilla planifolia	Vanillin	Westcott et al., 1994
Digitalis lanata	Cardiac glycosides	Kreis et al., 1993
Taxus cuspidata	Taxanes	Fett-Neto et al., 1993
Taxus brevifolia	Taxanes	Durzan & Ventimiglia, 1994
T. baccata	Taxanes	Ma et al., 1994a
Taxus × media cv. Hicksii	Taxanes	Wickremesinhe & Arteca, 1994
Papaver somniferum	Morphine	Rush et al., 1985
Mentha piperita	Menthol	Spencer et al., 1993

Failure to develop specialized cells, idioblasts, can be cited. In plants like *Mentha piperita*, for example, only the peltate glandular trichomes accumulate monoterpenes in an extracellular, subcuticular space (McCaskill et al., 1992). Callus and cell suspension cultures do not generate gland cells and storage compartments, thus lacking capacity for accumulation of essential oil, i.e. mono- and sesquiterpenes (Spencer et al., 1993). Besides, the presence of such metabolites could adversely affect growth and stability of such cultures. Cytodifferentiation, thus, must be recognized as a condition for the accumulation of some secondary metabolites.

Callus and cell suspension cultures may show structurally different compounds in different cells. For example, callus of red beet (*Beta vulgaris*) may be more or less rich in cells with either red or yellow pigments, betalains. Consequently, subcultures and selection may allow to generate white, red, yellow, and orange callus, all growing similar amounts of parenchyma under the same culture conditions at about equal rate (Constabel and Nassif-Makki, 1971). The observation raises the question whether the occurrence of different pigments in callus or cells growing under the same culture condition in the same fashion demands a *caveat* regarding the correlation of cytodifferentiation and product formation. Or, differences in betalain pigmentation may simply be interpreted by gradients of precursors available.

Differences in the spectrum of secondary metabolites in cell cultures and source plants have been highlighted, frequently. Most recently, in *Maclura pomifera,* simple and complex flavonoids have been isolated from the wood, the fruits, and the root bark. Cells cultured *in vitro* preferentially accumulated flavones and flavanones with phenyl substituents exclusively in ring A, while isoflavanones did not show typical 3',4'-dihydroxyl substitution as in fruits. Xanthones and stilbenes as in source material were not found at all (delle Monache et al., 1995). Moreover, the *Maclura pomifera* cell suspension culture selected for experimentation showed a greater level of metabolite accumulation (0.91% DW) than the stem (0.26% DW), leaves (0.32 %), and fruits (0.08% DW) of the source plant (Pasqua et al., 1991).

Differences in the ability to produce metabolites between cell cultures and source plants can result in the formation of entirely new compounds. The search for bioactive compounds in *Taxus* spp., for example, has led to the detection of yunnanxane and its homologous esters in cell cultures of *Taxus chinensis* var. *mairei* (Ma et al., 1994 a/b). Or, six triterpenoids were isolated from the callus tissues of *Paeonia* species. Four of the compounds were new from a natural source and demand chemotaxonomic and phylogenetic clarification (Ikuta et al., 1995). Or simply, a new xanthone has been isolated from callus cultures of *Hypericum patulum* (Ishiguro et al., 1995). Such observations increasingly challenge the importance of cytodifferentiation as governing principle of secondary metabolite production. Still, a case can be made for correlation of cytodifferentiation and product formation when a broader spectrum and higher yields of secondary metabolites are the result of employing material more highly organized than callus and cell suspension cultures, i.e. organ cultures.

4.1.2 Root cultures

Since White (1934) demonstrated potentially unlimited growth of tomato roots grown *in vitro*, initiation and maintenance of such cultures was found to be relatively facile. Analyses of secondary metabolites in roots cultured *in vitro* showed fair yields (Table 3), comparable to plant roots.

Table 3. Production of secondary metabolites in root cultures (1994-95)

Species	Product	Yield (root/callus)	Reference
Phytolacca acinosa	Triterpene saponins	1.5 g l^{-1}	Strauss et al., 1995
Alkanna tinctoria	Naphthoquinones	300 µg g^{-1} FW	Mita et al., 1994
Catharanthus roseus	Catharanthine Ajmalicine	0.36% DW 0.57% DW	Vazquez-Flota et al., 1994
Duboisia myoporoides	Scopolamine	1.3 g l^{-1}	Yukimune et al., 1994a
Menispermum dauricum	Isoquinoline alkaloids	0.5% DW	Sugimoto et al., 1994
Vanilla planifolia	Vanillin	0.4 mg g^{-1} DW	Westcott et al., 1994

For example, root cultures of *Phytolacca acinosa* employed for an analysis of triterpene saponins showed the occurrence of twelve such saponins, seven of which were isolated at 0.5 to 1.0 g/l^{-1} quantities, similar to source material (Strauss et al., 1995). In cultures of *Atropa belladonna* even a new group of tropane alkaloids has been detected: calystegines, first identified in extracts of *Calystegia sepium*. They may have a role as nutritional mediator, i.e. they are excreted into the soil and attract bacteria that are of advantage in the rhizosphere of the plant (Draeger et al., 1994).

Experimentation to achieve high yields of Valepotriates with *Centranthus ruber* (Granicher et al., 1995) and more so of ginseng saponins with *Panax ginseng* (Ushiyama et al., 1986) proved the capacity of root cultures. And, more to the point, nimbin, a tetranortriterpenoid similar to azadirachtin, was obtained with callus cultures of *Azadirachta indica* only when regenerating roots (Sanyal et al., 1981). Results regarding diversity and yields, thus, confirm the thesis that a higher level of tissue and cell organization allows for a higher level of performance in secondary metabolite production. Root cultures, however, are slow growing and, therefore, are being phased out quickly by faster growing hairy root cultures.

A strangely morphogenetic, but potentially profitable system of root culture for production of tropane alkaloids has been devised by subjecting *Duboisia myoporoides* roots first to lateral root induction followed by enhancement of root elongation and product formation. A total of 2.5 g/dm^{-3} of scopolamine was produced when this two-stage culture was combined with a high-density culture method (Yukimune et al., 1994 b).

4.1.3 Hairy root cultures

Technology to generate *Agrobacterium rhizogenes*-induced hairy roots, i.e. genetically transformed pluri-branched and fast growing roots as described by Chilton et al. (1982) and the detection of alkaloids in such roots (Flores and Filner, 1985) changed prospects for metabolite production by root cultures dramatically (cf. Ciau-Uitz et al., 1994). A recent review (Loyola-Vargas and Marinda-Ham, 1996) lists 78 species and demonstrates the extent to which hairy roots can be produced and exploited for production of metabolites (Table 4). When looking at industrial application, hairy roots offer a number of advantages over plants and cell suspension and root cultures. For example, in experimentation using *Catharanthus roseus* doubling times for roots and hairy root cultures were 19.5 and 2.8 days, respectively, and alkaloid contents were 2-3 fold higher. Hairy root cultures of *Papaver somniferum* were shown to accumulate morphinane alkaloids, i.e. thebaine and codeine (only non-transformed root cultures accumulated morphine proper). Cell suspension cultures of *Papaver,* on the other hand, showed at best traces of morphinan alkaloids (Yoshimatsu and Shimomura, 1992). Remarkably, hairy root cultures have been demonstrated to perform stably, qualitatively and quantitatively, for over several years of subculturing (Ciau-Uitz et al., 1994). Differentiation of chloroplasts and obvious greening of some hairy root clones has brought along even "shoot metabolites" such as indole alkaloids in *Amsonia elliptica* (Sauerwein et al., 1991a) and scopolamine in *Datura stramonium* (Maldonado-Mendoza and Loyola-Vargas, 1995), polyacetylenes in *Lobelia sessilifolia* (Ishimaru et al., 1994),

Table 4. Production of secondary metabolites in hairy root cultures (1994-95)

Species	Product	Yield	Reference
Leontopodium alpinum	Volatile oils	Traces	Hook, 1994
Astragalus membranaceus	Triglycosidic triterpenes	n.d.	Zhou et al., 1995
Astragalus membranaceus	Cyclortane triterpenes	Traces	Hirotani et al., 1994
Nicotiana tabacum	Sesquiterpenoids	20 $\mu g\ g^{-1}$ FW	Wibberley et al., 1994
Valeriana officinalis	Iridoid diester	Traces	Gränicher et al., 1995
Centranthus ruber	Valepotriates	3% DW	Gränicher et al., 1995
Artemisia annua	Artemisinin equivalent	0.0004% DW	Jaziri et al., 1995
Catharanthus roseus	Catharanthine	0.36 mg l^{-1} DW	Vazquez et al., 1994
Catharanthus roseus	Catharanthine	60 mg l^{-1}	Jung et al., 1994
Catharanthus roseus	Serpentine	16 mg g^{-1} DW	Ciau-Uitz et al., 1994
Datura stramonium	Tropane alkaloids	0.8-1.1% DW	Hilton & Rhodes, 1994
Datura stramonium	Scopolamine	0.2 mg l^{-1}	Maldonado & Loyola, 1995
Hyoscyamus × gyoerffyi	Scopolamine	0.004-0.19 % DW	Ionkova et al., 1994
Hyoscyamus albus	Scopolamine	0.01% DW	Doerk-Schmitz etl., 1994
Solanum aviculare	Steroidal alkaloids	32 mg g^{-1} DW	Subroto and Doran, 1994
Solanum acualetissimum	Steroidal saponin	0.04% DW	Ikenaga et al., 1995
Lobelia sessiliflora	Polyacetylenes	< 0.2% DW	Ishimaru et al., 1994
Platycodon grandiflorum	Polyacetylenes	0.64% DW	Tada et al., 1995
Tagetes patula	Thiophenes	5 $\mu mol\ g^{-1}$ FW	Arroo et al., 1995

n.d.= not determined

and several terpenes in addition to hernandulcin in *Lippia dulcis* (Sauerwein et al., 1991b). In hairy root cultures, thus, cytodifferentiation strongly holds as principle of secondary metabolite production. Interestingly, textbooks on plant anatomy do not even mention the phenomenon of hairy roots (Esau; 1965, Salisbury and Ross, 1985; Strasburger et al., 1991; Eschrich, 1995).

Tissue organization as a precondition for product synthesis and accumulation can be a formidable principle as demonstrated for the production of forskolin, a diterpene. Since this compound is found in tuberous roots of *Coleus forskohlii, in vitro* methodology for its production would suggest hairy root and root, rather than cell suspension and shoot cultures. Actual experimentation revealed that forskolin occurred in cell suspensions at 0.03 % dw only under conditions of a medium designed for root formation (Mersinger et al., 1988), in root and hairy root cultures in concentrations up to 0.12 and 0.3 % dw, respectively (Krombholz et al., 1992). Other results, however, showed forskolin in shoot cultures only or in root (0.0 4% dw) and tumorous rhizogenic callus (0.01 % dw), but again not in shoot (teratoma) cultures (Mukherjee et al., 1996). It would appear that *Coleus forskohlii* cultures would benefit from interchangeable modes of tissue organization to be productive in scale-up bioreactors as demonstrated here: cells of *Catharanthus roseus* have been selected which will grow hairy roots upon transfer to1/3 strength SH medium without growth hormones and within 3 weeks, and will return to become cell suspensions upon return to full strength SH medium with hormones. Such change in tissue organization allows the management of hairy root material, in particular the inoculation with hairy roots of bioreactors of any size. Catharanthine analyses of cells as suspensions and hairy roots revealed 0.7 and 1.7 mg g^{-1} dry weight, respectively (Jung et al., 1994, 1996).

4.1.4 Shoot cultures

Culture of shoots in liquid medium for production of secondary metabolites (Table 5) may have received some publicity to attract investment in , for example, vanilla, but never really seriously challenged the culture of cells and hairy roots due to obstacles in scale-up. Shoot cultures of *Catharanthus roseus* provided for the occurrence of vindoline and vinblastine, a demonstration of the chemosynthetic capacity of green tissues and, thus, the positive correlation between cytodifferentiation and secondary metabolite production. (Miura et al., 1988). And, *Papaver somniferum* callus tissues which produced tracheary elements, shoot buds and/or embryos were seen to produce morphinane alkaloids, i.e. thebaine and/or codeine (Kutchan et al., 1983; Schuchmann and Wellmann, 1983; Yoshikawa and Furuya, 1985). A striking effect of shoot culture, moreover, was found in the production of essential oils. In *Pimpinella anisum*, for example, callus and cell suspension cultures appeared not to accumulate any essential oils, contrary to shoot cultures which produced measurable amounts (Reichling et al., 1985).

Table 5. Production of secondary metabolites in shoot culture

Species	Product	Yield	Reference
Hyacinthus orientalis	essential oils	n.d.	Hosokawa & Fukunaga, 1995
Lavandula latifolia	camphor et al.	12 $\mu g\ g^{-1}$ FW	Calvo & Sanchez-Gras, 1993
Leonurus cardiaca	furanic labdane diterpenes	30-72 $\mu g\ g^{-1}$ FW	Knöss, 1994, 1995
Artemisia annua	atemisinin	38 mg g^{-1} oil	Brown 1994
Artemisia annua	terpeniids, camphor	3.41mg%FW	Fulzele et al., 1995
Digitalis purpurea	carednolides	940 nmol g^{-1} DW	Stuhlemmer et al., 1993
Digitalis purpurea	cardiac aglycons	200 $\mu g\ g^{-1}$ DW	Seitz& Gärtner, 1994
Vinca minor	vincamine	0.4% DW	Tanka et al., 1995
Scoparia dulcis	benzoxazolinone	1.1 mg g^{-1} DW	Hayashi et al., 1995

Finally, transformed *Pimpinella* shoot cultures obtained by treatment with *Agrobacterium tumefaciens*, allowed the accumulation of geraniol and beta-bisabolene amongst various phenylpropanoids at levels which were lower (18%) than those in non-transformed shoot cultures and much lower (89%) than in intact plants, but thoroughly demonstrable (Reichling et al., 1988). Culture of transformed shoots (Table 6) has made some inroads towards increases in yields, but has not reached the level of proficiency as with transformed roots.

Table 6. Production of seceondary metabolites in tumorous shoot cultures

Species	A. tumefaciens	Product	Yield	Reference
Pimpinella anisum	T37	essential oil	18% lower than non-tum. culture	Salem & Charlwood, 1995
Coleus forskohlii	C58	forskolin	0.03% DW	Mukherjee et al., 1996

In *Stevia rebaudiana* a sweet herb of Paraguay known for its ent-kaurene glycosides, 300 times sweeter than 0.4% sucrose solution, among *in vitro* cultures of callus, roots, shoots, and rooted-shoots only the latter were capable of *de-novo* synthesis of the aglykone moiety of stevioside, steviol (Swanson et al., 1992). Here, two explanations present themselves: (a) a stevioside precursor or derivative is produced in the roots and transported to the leaves for further metabolism, or (b) the roots produce a factor required to elicit the biosynthesis of stevioside in the leaves. The combined use of cell-free enzyme systems and of individual tissues as described may be used to resolve the problem. The case, certainly, underpins the importance of cytodifferentiation for product formation.

And finally, microspore-derived embryos of *Brassica napus* cultured *in vitro* fail to accumulate detectable amounts of glucosinolates until they are placed in a germination medium that supports true shoot and root formation. The resulting induction of glucosinolate accumulation is restricted initially to indole derivatives, but the pattern then becomes very similar to that observed in zygotic seedlings. Results are interpreted by saying that microspore-derived embryos are functionally capable of indole glucosinolate synthesis prior to induction of germination, but they are not able to store significant levels of the thioglucosides in their tissues (McClellan et al., 1993). In shoot and embryo culture the correlation of cytodifferentiation and metabolite production certainly reaches its ultimate expression and confirmation. But, industrial application of shoot and embryo culture for production of secondary metabolites appears non-profitable, so far.

4.1.5 Immobilized cell cultures

In plants cell differentiation is effected by chemical gradients, hormones for instance. The idea of providing for such gradients in *in vitro* culture through encapsulation, i.e. immobilization of cells (Table 7) in a gel has subseqently been realized (Brodelius et al., 1979; Brodelius, 1983).

Table 7. Production of secondary metabolites in immobilized cell cultures (1994-95) (for previous record see Constabel & Tyler, 1994)

Species	Matrix	Product	Reference
Datura innoxia	Ca-alginate	scopolamine hyoscyamine	Gontier et al., 1994
Papaver somniferum	Glutaraldehyde	codeine/cideine biotransformation	Stano et al., 1995

Today, cells are being cultured as aggregates or para-tissues in a variety of matrices, i.e. calcium alginate, carrageenan, polyacrylamide, and agarose, as well as on a number of inert substrates like ceramic webs and surrounded by liquid medium (Constabel and Tyler, 1994; Gontier et al., 1994). Physiological experimentation and biochemical analysis may be hampered by the matrices surrounding the cells, immobilized cells have the advantage, however, of allowing nutrient medium replenishment or exchange without loss of biomass as may be required for industrial application of plant cell cultures. When, for example, high yields of sanguinarine had been a target, immobilized cultures of *Papaver somniferum* or *Sanguinaria canadensis* cells have been established. In *Sanguinaria* alkaloid levels of immobilized cells were comparable to suspension cultures, i.e. ca. 1% dw, in *Papaver* alkaloid yields were higher than in suspension cultures (2.9 -12.3% dw) (Rho et al., 1992). And technical details are still being explored and improved. Recently, for example, cells of *Papaver somniferum* have been cross-linked by first permeabilizing them in 5% Tween 80 dissolved in 0.15 M NaCl for 3 h and then washing and suspending them in 0.15 M NaCl and 25% glutaraldehyde. Cross-linking provided for cultures showing substantial tyrosine decarboxylase and DOPA decarboxylase activity over a period of several months (Stano et al., 1995).

In the context of cytodifferentiation and metabolite production, the paradigm of a tissue as presented by cells immobilized in a matrix is most intriguing. If a gradient is at play and if such gradient effects differentiation accompanied by metabolite production, then cells associated in immobilized systems should present a chance to analyse the chemical nature of the gradient. Only recently, confirmation of gradient induction of cytodifferentiation and product formation has been provided: cells of *Chenopodium album* and whole plants of duckweed *Wolffia arrhiza* were co-cultivated. As a result, synthesis of betalain was induced in several cells of *Chenopodium album* callus (Rudat and Ehwald, 1994). Analysis of the inducing agent for betacyanin formation and cytodifferentiation or vice versa would appear most enticing.

In conclusion, in *in vitro* cell cultures level of tissue organization strongly affects the occurrence and yield of secondary metabolites. As a result, hairy root cultures are favoured candidates for industrial production of such metabolites, phytochemicals.

4.2 Cellular organization

4.2.1 Vacuoles, vesicles, endoplasmic reticulum

There probably has never been a callus or cell suspension culture composed of true meristmatic cells only or rather, callus with vacuolated cells is the most common configuration. Since vacuoles generally are the site of flavonoid, alkaloid and glycoside storage, metabolite accumulation in callus is a fairly common occurrence, the more so the older the callus.

Subcellular compartmentalization leading to formation of vacuoles is of particular concern where target compounds, whether accumulated in high concentrations or not, are cytotoxic. The capacity of *Coleus blumei* cell cultures to accumulate rosmarinic acid, an ester of caffeic acid and 3,4-dihydroxyphenyllactic acid, at levels up to 20% of dry weight, is a case in point. Methodology which included rapid isolation of protoplasts and vacuoles resulted in the demonstration and confirmation of rosmarinic acid as an anion, not able to pass a membrane, requiring a carrier system for transport into the vacuole (Hausler et al., 1993). The formation of vacuoles as sink and depository for secondary metabolites is a common feature of plant cells. The alternative would be excretion into the medium.

Vacuolation enables cells to also absorb secondary metabolites from the medium. Enigmatic remains the specificity expressed by the cells regarding the quality of metabolites or the specificity of differentiation allowing for uptake of certain compounds only. Cultured cells of *Coptis japonica* were shown to take up exogenously added quarternary benzylisoquinoline alkaloids indigenous to the species in high amounts, other quarternary benzylisoquinolines were barely absorbed by the cells (Sato et al., 1993a). Specificity of metabolite uptake had previously been shown for indole alkaloids in *Catharanthus roseus* (Deus-Neumann and Zenk, 1984).

While different in development, structure and function, vesicles like vacuoles characterize cytodifferentiation and condition the appearance of metabolites. In *Berberis wilsoniae* at least two out of eight enzymes involved in the biosynthesis of isoquinoline alkaloids like berberine, were found to be strictly associated with vesicles of a specific gravity. These vesicles were the first highly specific und unique compartment serving only alkaloid biosynthesis; they were found in members of four different plant families, and in cell cultures as well as in differentiated tissue (Amann et al., 1986).

Specificity even of the endoplasmic reticulum (ER) may be related to occurrence of some secondary metabolites. As shown for *Lithospermum erythrorhizon* cell suspension cultures, in shikonin biosynthesis, geranylpyrophosphate (GPP) is needed as a substrate for the p-hydroxybenzoate geranyltransferase localized in the ER. GPP synthase, on the other hand, may be observed in the microsomes, in contrast to *Vitis vinifera* where GPP synthase was found in the plastids (Sommer et al., 1995). The rather complex intracellular organization of enzymes catalyzing secondary metabolism would underscore the need for a level of structural and functional differentiation presented by mature cells of older callus or tissues only.

4.2.2 Photoautotrophy

Mixotrophic and photoautotrophic cultures have been obtained both with cells and with roots and hairy roots. This in itself is a technological achievement and was thought to simplify culture technology, albeit with a heavy dependency on energy and costs. But in such cultures significant amounts of secondary metabolites have not been shown to accumulate. More specifically, a comparison of photoautotrophic, photomixotrophic, and heterotrophic cell suspension cultures of *Nicotiana tabacum* showed that patterns of secondary metabolites were different for each kind of culture, and yields did not measure up to the source plants (Ikemeyer and Barz, 1989). Only recently did photosynthesis in hairy root cultures of *Datura stramonium* result in an increase in levels of hyoscyamine and scopolamine production, up to 0.64% dw and 0.03 % dw, respectively, which is higher than in plant roots, but not as high as in selected root cultures. Photosynthesis may redirect secondary metabolism and affect the level of product formation by the production of reducing agents, or providing enzymatic cofactors (Flores et al., 1993; Maldonado-Mendoza and Loyola-Vargas, 1995). The generally inconclusive or negative response by cultures having arrived at photoautotrophy may not surprise, since nature itself does not show an intimate relationship between photosynthesis and secondary product formation except the obvios, i.e. primary products must cascade into secondary metabolism for products to occur.

Even fungi, like ergot or the fly agaric, accumulate alkaloids related to those in green plants. Observations regarding photoautotrophic cell cultures, thus, do not significantly contribute to better understanding the correlation of differentiation and metabolite production. A different situation arises with the development of chloroplasts as the site of secondary metabolism as will be revealed later under molecular biology of product formation.

4.2.3 Laticifers

The development of specialized cells as occur in plants, i.e. idioblasts like gland cells, resin canals, laticifers has rarely been witnessed in cell suspension cultures, has not been found with immobilized cells, but naturally, was seen in root and specifically embryo and shoot cultures, as mentioned above. Or, structural deficiencies in *Papaver* cell suspension cultures negatively impact the accumulation of morphine. Most noteworthy, differentiation of laticifers did not affect the accumultion of benzophenanthridine alkaloids, sanguinarine and related compounds, nor their mutual precursor, dopamine (Rush et al., 1985). Recently, callus cultures of *Calotropis procera* permitted investigations of the hormonal regulation of laticifer formation and resulted in the oberservation that maximum laticifer differentiation occurred when maintained in medium after Murashige and Skoog (1962) supplemented with a certain combination of kinetin and indole-3-acetic acid (IAA) (Suri and Ramawat, 1995). So far, no information regarding the correlation between status of laticifer differentiation and accumulation of secondary metabolites has been added. One might be tempted, for example, to probe for the occurrence of a major latex protein to test the functionality of such material as demonstrated by Nessler and von der Haar (1990) and Nessler(1994).

4.2.4 Lignification

A most convincing demonstration of a correlation between cytodifferentiation and secondary metabolite formation in plants and cell cultures certainly derives from analyzing lignin formation and deposition. Lignin is the most widely occurring secondary metabolite, it is associated with sclerenchyma, and basically characterizes all plants adapted to terrestrial life. When stained with phloroglucinol and HCl, lignin can be shown to accompany the formation of tracheids in callus cultures. A more intimate picture of lignin formation has been obtained by a brilliant demonstration of the power of *in vitro* culture. Mesophyll cells of the first true leaves of 14-day-old seedlings of *Zinnia elegans* have been isolated (Fukuda and Komamine, 1980), shown to respond to auxin and cytokinin in the medium by differentiation of tracheary elements (Fukuda and Komamine (1982), and allowed to elucidate the relationship between wall-bound peroxidases and lignification. Specifically, 2 out of 5 peroxidases identified appeared to be specific for tracheary element differentiation and linked to peroxidase bound to secondary cell walls (Sato et al., 1993b). Further insights into lignification of cell walls and its compartmentalization can be expected from this unique model system. Also, *Pinus taeda* cell suspension cultures grown in medium containing 2,4-dichlorophenoxyacetic acid (2,4-D) showed only primary cell wall formation and essentially no lignification. Once the cultures had been transferred to a succession of fresh media with 1-naphthaleneacetic acid (NAA) instead of 2,4-D, cells showed thickening of the cell wall and concomitant lignification (Eberhardt et al., 1993).

4.2.5 Excretion

Cytodifferentiation by acquisition of a capability for excretion of secondary metabolites has been observed first with *Thalictrum minus* and *Lithospermum erythrorhizon* cell suspension cultures (Nakagawa et al., 1984; Yamamoto et al., 1987; Shimomura et al., 1991). Subsequently, hairy root cultures of *Nicotiana rustica* were observed to release a fair amount of nicotine into the medium (Hamill et al., 1986). And, spent medium of *Lupinus polyphyllus* was shown to contain ethanol, organic acids, amino acids, alkaloids, polysaccharides, and enzymes (Wink, 1994). Excretion of products would be a reaction of utmost importance to improving the application of plant cell cultures, in particular with immobilized cell as well as root and even shoot cultures. Consequently, the use of adsorbents in culture media as sink for secondary metabolites has gained favour. Polymeric adsorbant resins such as Amberlite XAD-2, XAD-4, XAD-7 or XAD-16 (Rohm & Haas) have led to an increase in the extracellular accumulation of sanguinarine in *Papaver somniferum* (Williams et al., 1992), nicotine in *Nicotiana glauca* (Green et al., 1992), of shikonin in cultures of *Lithospermum erythrorhizon* (Shimomura et al., 1991).

The biology of excretion, on the other hand, still leaves much speculation regarding its mechanism and function. Production of furanic labdane diterpenes in shoot cultures of *Leonurus cardiaca* was improved by changing the medium or subculture of the shoots every 5 days. This procedure was required since substances released inhibited growth and diterpene production (Knoss, 1994).

4.3 Growth

No other reaction by plant cells to *in vitro* culture has received more attention than their response to growth and growth factors. Accordingly, the production of secondary metabolites as a function of growth *in vitro* has been scrutinized thoroughly, and the general observation still holds: cells suspended in fresh medium will grow at a rate which follows a sigmoidal pattern. The rate of synthesis and accumulation of secondary metabolites, on the other hand, follows a curve which shows a maximum at the time of decelerated growth and early stationary or idiophase. The classic example shows the formation and accumulation of serpentine in cell suspension cultures of *Catharanthus roseus* (Zenk et al., 1977). The observation is a powerful demonstration that subsequent to cell division only a certain degree of cell maturation predisposes cells to form and accumulate metabolites. Fujita et al. (1981) were the first to derive industrial technology and application from such observations in processes to scale-up *Lithospermum erythrorhizon* cell suspension cultures for production of shikonin.

Maximal production of metabolites coinciding with signifiant depletion of nutrients corroborates the observation made before. *Coleus blumei* cell suspensions, for example, top performers in producing secondary metabolites, i.e. rosmarinic acid at levels of 20% of dry weight or 2 mg ml^{-1}, started product formation only, when growth was reduced due to depletion of phosphate. And, the amount of rosmarinic acid formed depended on the concentration of carbohydrates left in the medium (Razzaque and Ellis, 1977, Gertlowski and Petersen, 1993).

In hairy root cultures a different response has been observed. Alkaloid levels in hairy roots of *Catharanthus roseus* and *Datura stramonium,* for example, remained high over the entire gowth period and peaked at the stationary or idiophase with about 0.2% dw, over 9 and 16 subcultures of 4 weeks each (Loyola-Vargas and Miranda-Ham, 1996). Such data would indicate that fast growth of hairy roots did not repress metabolite synthesis and accumulation to levels of only traces as in cell suspension cultures. With cytodifferentiation intact, biosynthetic capacity appears to have remained intact. Such statement does not detract from the observation that a stimulus to increase the growth rate in hairy roots may be correlated with a respective diminuation in metabolite accumulation. For example, when the treatment of hairy roots of *Chaenactis douglasii* with NAA increased growth, the level of thiarubrine decreased to near zero (Constabel and Towers, 1988). On the other hand, stimulation of growth in hairy root cultures of *Rubia tinctorum* through application of 5 μM IAA resulted in equally increased anthraquinone production (Sato et al., 1991). Obviously, with hairy root cultures the correlation of growth and product formation appears to be ambivalent.

In stark contrast to all of the above, *Phytolacca americana* cells responded to *in vitro* culture by a remarkable accumulation of betacyanin during the log-phase of growth and feeding of phosphate, while product accumulation ceased upon phosphate removal or inhibition of cell division due to aphidicolin (Hirose et al., 1990). Subsequent experimentation confirmed inhibition of betacyanin accumulation in *Phytolacca* cells by aphidicolin and a cycle-specific inhibitor, i.e. propyzamine. The results are interpreted by relating betacyanin inhibition to suppression of the reaction converting tyrosine to DOPA, the initial step in betacyanin formation, which may be closely related to cell division (Hirano and Komamine, 1994). Here, the correlation of cytodifferentiation and production of secondary metabolite accumulation bursts, and a new answer by molecular analyses will have to be found.

4.4 Growth regulators

Topping of tobacco plants, a common practice among tobacco growers, temporarily eliminates apical dominance and increases the level of nicotine content in tobacco leaves. With hydroponic plants the practice has shown to increase the p-methyltransferase activity in the root more than 10-fold in one day. But the presence of 50 mM indolylacetic acid in the nutrient solution blocked the increase completely (Mizusaki et al., 1973). Regarding cells cultured *in vitro*, similar observations hold true. Carrot cells do not synthesize anthocyanin when cultured in medium with 2,4-D. Upon transfer to medium without 2,4-D, anthocyanin formation is being induced within 4-5 days, and together with embryogenesis. Here, investigations suggest a close correlation between growth regulator activity and the induction of embryogenesis, morphological differentiation, anthocyanin synthesis and metabolic differentiation (Ozeki and Komamine, 1986). Results like these heighten expectations to arrive at a negative correlation of growth regulator (phytohormone) treatment and secondary metabolite production in plant tisssue cultures. A closer examination of the effect of phytohormones on metabolite production in cell cultures, however, is warranted.

4.4.1 Growth promoting hormones

Auxins, cytokinins, and gibberellic acids effect elongation and division of plant cells and, thus, are indispensible nutrient complements for the initiation and maintenance of regular, i.e. non-tumorous cell cultures. Given the correlaton of cytodifferentiation and production of secondary metabolites, the presence of growth promoting hormones should hinder any and all productivity in such cultures. And, indeed, many experiments will attest to the reduction, if not suppression, of metabolite synthesis and accumulation in callus and cells suspensions, even hairy root cultures, as shown for the effect of cytokinin on pigmentation in *Haplopappus gracilis* cells years ago (Constabel et al., 1971). While generally valid, such observations do not reflect all experimentation. As a result, the concept of a correlation of differentation and product formation cannnot be accepted unequivocally. For example, cell suspensions of *Morinda citrifolia* were cultivated in media after Gamborg et al. (1968) containing 1 mg l^{-1} NAA or 2,4-D. Both auxins were able to support growth, but anthraquinone formation was observed in the presence of NAA only. Apparently, NAA was able to allow the expression of genes responsible for anthraquinone production. Adding 2,4-D to a culture while in the mode of anthraquinone production led to an immediate cessation of this process while not affecting growth of the culture. 2,4-D inhibited anthraquinone synthesis rather than lacked the capacity to induce it (Haagendoorn et al., 1994). Or, *Thalictrum minus* var. *hypoleucum* cell suspension cultures produce berberine and excrete it into the medium. Although berberine biosynthesis was completely suppressed by 2,4-D, synthesis was restored by adding 6-benzylamino purine (BAP) to the 2,4-D medium . Tracer experiments using tyrosine showed that the suppression of berberine synthesis by 2,4-D was a result of inhibition of 2 enzymes, norcoclaurine 6-O-methyltransferase, a crucial enzyme controlling the formation of benzylisoquinolines, and (S)-tetrahydroberberine oxidase (Hara et al., 1993). Furthermore, berberine synthesis was restored upon transfer of cells to medium with BAP due to activation of three (S)-tetrahydroberberine oxidases, one of them non-stereospecific (Hara et al., 1995). Furthermore, promotion of anthocyanin production by cytokinins was confirmed by obervation of callus cultures derived from *Oxalis linearis*. A concentration of 8 μM thidiazuron (N-phenyl-N'-1,2,3-thidiazol-5-ylurea) and zeatin, respectively, while inhibiting callus growth stimulated pigment production (Meyer and van Staden, 1995). And finally, exogenously supplied gibberellic acid (GA3) enhanced branching in two transformed root clones of the tropane alkaloid producing species *Brugmansia candida* and so enhanced their typical hairy root phenotype. This growth substance also had the effect of reducing the overall alkaloid accumulation. In one case, however, it significantly altered the relative concentrations of different tropine esters (Rhodes et al., 1994). Addition of gibberellic acid enhanced growth but reduced the specific steroidal alkaloid level in hairy roots of *Solanum aviculare*. But taking into account both, growth and alkaloid accumulation, the latter was improved by 40% at gibberellic acid concentrations of 10 and 100 $\mu g\ l^{-1}$ (Subroto and Doran, 1994).

Results from experiments in which metabolite production is linked to growth as affected by growth promoters like auxins, cytokinins, and gibberellic acids are confusing often enough to warrant a different approach. It would focus on the effect of growth regulators on gene activation and transcription/translation of enzymes catalyzing specific steps in the synthesis of metabolites, and disregard growth. Now, some clarification may result. For example, berberine synthesis was restored upon exposure of *Thalictrum minus* cells to kinetin due to activation of three enzymes in its biosynthetic pathway, as referred to earlier (Hara et al., 1995). Even more to the point, auxin rapidly down-regulated transcription of the tryptophan decarboxylase gene from *Catharanthus roseus*, thus reduced the production of tryptamin, the first intermediate towards synthesis of monoterpene indole alkaloids (Goddijn et al., 1992).

4.4.2 Abscission and senescence hormones

Remarkably, experimentation to elucidate the effect of abscissic acid (ABA) on metabolite formation in cell cultures is scarce. This finding is surprising since many observations point to its activity in enhancing maturation processes, in somatic embryos for instance, in a variety of cultures. As a plant growth regulator, ABA is most often associated with the inhibition of growth processes. When added to cell suspension cultures of *Catharanthus roseus* ABA stimulated intracellular accumulation of the indole alkaloids catharanthine and ajmalicine. The response varied depending on the concentration and source of ABA, and the growth phase at which the cells were treated. Precise timing of the ABA addition to the medium resulted in a catharanthine yield of 85 mg l^{-1} after 10 days of cultivation (Smith et al., 1987).

The effect of ethylene on the production of secondary metabolites in plant cell cultures depends on metabolite and species. For example, an ethephon-mediated ethylene release promoted alkaloid production in *Coffea arabica* and *Thalictrum rugosum* cell suspension cultures (Cho et al., 1988; Schulthess and Baumann, 1995). On the other hand, in *Papaver somniferum* cell cultures ethylene did not affect the production of sanguinarine (Songstad et al., 1989). And furthermore, treatment of *Eschscholtzia californica* cultures with ACC (1-aminocyclopropane carboxylic acid), thus, stimulating ethylene formation, did not affect the production of alkaloid (sanguinarine) and its accumulation (Piatti et al., 1991). On the other hand, cardenolide accumulation in tissue cultures of *Digitalis lanata* was increased by an ACC inhibitor and decreased by ethephon (Berglund and Ohlson, 1992). Again, a molecular rather than a conventional physiological approach of ABA and ethylene effects on metabolite production would be required to interprete the broad spectrum of observations made, so far.

5 STRESS PHYSIOLOGY AND PRODUCT FORMATION

5.1 Elicitation of stress products

While observations of cell cultures treated with growth hormones did not always render confirmation of a correlation between cytodifferentiation and production of secondary metabolites, the exposure of cell cultures to stress appears to discount such correlation altogether. Or, can the principle in question here be saved by viewing production of metabolites as an expression of cytodifferentiation?

Exposure of cells cultured *in vitro* to physical, chemical or infectional stress generally results in the formation of secondary metabolites, stress products, in the case of biotic elictors called phytoalexins (Eilert, 1987). Classic experimentation featured the irradiation of cell suspension cultures of *Petroselinum crispum* with ultraviolet light. Its analysis has broken through physiological and biochemical barriers to molecular biological elucidation of elicitation and resulted in details of regulatory mechanisms related to genes encoding enzymes which catalyze the formation of phytoalexins, flavonoids (Zimmermann and Hahlbrock, 1975; Hahlbrock and Scheel, 1989). Following a suggestion by Wolters and Eilert (1983) and due to ease of application biotic elicitors have become prime material for stimulating the production of secondary metabolites in plant cell cultures.

The first experiments aimed at employing elicitation for improved recovery of target metabolites led to ambiguous results: cell cultures of *Papaver somniferum* expected to respond to treatment with a homogenate of

Botrychium spec. failed to produce morphinane alkaloids as had been expected, but accumulated substantial amounts of sanguinarine, later to become a much sought after chemical (Eilert et al., 1985); cell cultures derived from *Catharanthus roseus* treated with homogenates of *Pythium aphanidermatum* produced catharanthine, and this result was a significant advance towards demonstration of monoterpene indole alkaloids, but they failed to synthesize vindoline as well and, thus, precluded hopes for a vinblastine production (Eilert et al., 1986). Both morphine and vindoline synthesis obviously are linked to levels of cytodifferentiation not achieved under conditions of stress. Or, cells exposed to stress differ fundamentally from cells unaffected by stress, in secondary metabolism.

The list of cell species employed in elicitation and products found is growing monthly (Table 8). In all cases, elicitation has led to a transient induction of enzymes which catalyze the synthesis of secondary products as shown for tryptophan decarboxylase and strictosidine synthase in cell suspension cultures of *Catharanthus roseus* (Roewer et al., 1992). Critical were the sensitivity of the cell strain to the elicitor, its specificity, concentration, and time and duration of application.

Additional experimentation and questions regarding the implication of morphogenesis may be helpful. Cell cultures of *Pinus banksiana* treated with an elicitor derived from the ectomycorrhizal fungus *Thelephora terrestris* rapidly increased activities of several enzymes involved in phenylpropanoid metabolism and accumulated true gymnosperm lignin (Campbell and Ellis, 1992). Cells of *Ruta graveolens* responded to the presence of autoclaved homogenates of yeast, *Rhodotorula rubra*, with rapid accumulation of acridone epoxides, furoquinolines, and furanocoumarines (Bohlmann and Eilert, 1994). In *Sanguinaria canadensis* cell suspension cultures, quercetin at 100 μM increased the sanguinarine production by more than 300% (Mahady and Beecher, 1994).

Table 8. Production of secondary metabolites by plant cell cultures in response to elicitation (1994-1995)

Species	Elicitor	Product	Reference
Daucus carota	yeast extract	6-methoxymellein	Guo & Ohta, 1994
Capsicum frutescens	*Gliocladium deliquescens*	phenolics	Holden & Yoeman,1994
Cicer arietinum	*Ascochyta rabiei*	isoflavones, pterocarpan	Barz & Mackenbrock, 1994
Daucus carota	*Rhodotorula rubra*	anthocyanin	Suvarnalatha et al., 1994

Species	Elicitor	Product	Reference
Ruta graveolens	*Rhodotorula rubra*	acridone epoxides, furoquinoline furanocoumalins	Bohlmann & Eilert, 1994
Ephedra distachya	Mannan glycopeptide	p-coumaroylamino acids	Song et al., 1995
Pueraria libata	yeast extract	isoflavonoids	Park et al., 1995
Thalictrum minus	Pectate lyases	isoquinolines	Smolko & Peretti, 1994
Lycopersicon esculentum	yeast extract	rishitin	D'Harlingue et al., 1995
Capsicum annuum	Arachidonic acid	rishitin, capsidiol	Hoshino et al., 1994
Nicotiana tabacum, HR	yeast extract	capsidiol	Wibberley et al., 1994

None of the experiments referred to point to a relationship with morphogenesis except those described by Eilert et al. (1989), where differentiated material, i.e. hydroponically grown *Ruta graveolens* plants, exposed to elicitor maintained a constitutive level of metabolites, while old and new cell suspensions and even shoot teratomata responded by substantial increases in acridone, furoquinoline, and furanocoumarine levels. On the other hand, analyses with *Cicer arietinum* (Barz and Mackenbrock, 1994) have shown that elicitation increases constitutive (isoflavones) as well as de-novo synthesis of secondary metabolites (pterocarpans), thus achieving a rapid infection-induced accumulation of phytoalexins. Both types of products were found in cell suspension cultures, thus the same (low) level of cytodifferentiation. However, there is a limit in the reaction to elicitation. When cell suspension cultures of *Catharanthus roseus* were fed terpenoid precursors like secologanin, loganin, and loganic acid, an increase in the level of ajmalicine and strictosidine was observed. Elicitation did not increase alkaloid accumulation in precursor-fed cells (Moreno et al., 1993). Returning to *Ruta* material, could it be that there existed an upper limit of (constitutive) metabolites in differentiated tissues as presented by hydroponically grown plants and prior to treatment with an elicitor?

While many more metabolites and ever higher yields have been described as products of elicitation, elucidation of the biological significance of elicitation demands a return to the whole plant level and inclusion of developmental regulation (Paiva et al., 1994). Furthermore, the signal transduction mechanism may have to be considered as well. A variety of factors have been postulated to be involved in the transduction of elicitor signals, for example phosphorylation of proteins, changes in ion permeability of the plasma membrane, active oxygen species, ethylene, and jasmonic acid and its methyl ester. Generation of superoxide is one of the earlierst reactions, i.e. a rapid burst of hydrogen peroxide formation in soybean cells has been oberserved. And indeed, treatment of carrot cells with yeast extract as elicitor and buthionine sulfoximine, inhibiting the synthesis of glutathione, led to increased production of 6-methoxymelleine, a response to the synergistic effect of an elicitor and an active oxygen species (Guo and Ohta, 1993). In as much as *in planta* signals follow the phloem, a degree of cytodifferentiation may be required to effect a response in cell cultures as well.

5.2 Allelopathy

The Mexican *Piqueria trinerva* (*Compositae*) as sunflower (*Helianthus annuus*), shows the effect of allelopathy when grown in association with other plants. *Piqueria* callus produces piquerol, a monoterpenoid diasterisomer known for its allelopathic function (Rubluo et al., 1994). It should be possible to use the plant tissue culture system to learn more about allelopathy *per se* and allelopathy as a function of morphogenesis using this material as the producer and co-cultured material as the sensor.

6 STABILITY/VARIABILITY OF PRODUCT FORMATION

Demonstration of stability of metabolite production in plant cell cultures has been of great concern whenever such cultures were meant to be employed by industry. But rather than stability, variation in performance has been the rule, at least in initial phases of cell culture. For example, cell clones obtained with protoplasts of one and the same leaf of a *Catharanthus roseus* plant differed greatly in spectrum and amount of alkaloid production (Constabel et al., 1981). Or, over time, twenty clones from single cells of a suspension culture of *Capsicum frutescens* exhibited marked differences in growth, chlorophyll content, and response to treatment with a fungal elicitor (Holden and Yoeman, 1994). Finally and not uncommon, some *Papaver* and *Catharanthus* cell lines or clones have been found to lose capacity for product formation altogether (Constabel, unpubl.). The question here is, whether variability/stability is paralleled by changes or stability in levels of cytodifferentiation.

Poppy (*Papaver somniferum*) cell cultures which maintained a capacity for sanguinarine synthesis and accumulation over more than a decade did show a certain percentage of single, large cells over the entire length of that period. A more recent observation appears to be more pertinent: here, the relationship between the morphology and indole alkaloid production of *Catharanthus roseus* cells was investigated (Kim et al., 1994). In each protoplast derived clone most of the cells maintained only one of two shapes, either spherical or cylindrical. The cell aspect ratio (cell length/width) for most isolates was stable for more than two years of subculture. The pattern of division remained stable in each phenotype and were not considerably affected by auxin or cytokinin in the culture media. Observations showed that cell morphology of the isolates was stable and probably internally determined. Production of the indole alkaloids, ajmalicine and catharanthine was significantly greater when the cell aspect ratio was more than 2.8. This observation indicated that there was a threshold of cell aspect ratio relevant to metabolite productivity. Intriguingly, major components of cell walls of spherical cells in suspensions of *Catharanthus roseus* were determined as galactans and glucans, those of cylindrical cells as arabinans (Suzuki et al., 1990). The latter might have been secreted into the medium and acted as an elicitor to stimulate the production of indole alkaloids. The morphology of cells, thus, would appear not to have been the primary factor in metabolite production of respective cultures.

The need for stability of product formation or avoidance of variation over long periods of time did prompt the development for industry of hairy root cultures, organized tissues. And progress has been remarkable. For example, in an effort to select for the most productive material to be cultured *in vitro* 500 hairy root cultures of *Datura stramonium* were screened for growth patterns, biomass accumulation, and tropane alkaloid production over a period of 2 years. In the end, some cultures were identified which had minimal capacity, others showed a hyoscyamine and scopolamine level 2 orders of magnitude higher than the source plants (0.08 to 0.33 % dw, Maldonado-Mendoza et al., 1992). Here, cytodifferentiation has clearly become a mere co-factor and secondary to the genetic make-up of material under investigation.

7 MOLECULAR BIOLOGY OF PRODUCT FORMATION

Expansion of methodology by enzymology, gene manipulation, and genetic cell or plant transformation has substantially widened the scope of investigations regarding the production of secondary metabolites and its relation to cytodifferentiation. Sense and antisense expression of genes encoding enzymes catalyzing the synthesis of, for instance, phenylpropanoids and alkaloids have allowed to critically trace biosynthetic pathways and identify branch points (Hahlbrock and Scheel, 1989, Hashimoto and Yamada, 1994). As a consequence, higher rates of metabolite production have been obtained and genetic manipulation has subsequently been proposed as adjunct technique to boost tissue culture performance to levels acceptable to industry. On occasion, reporter genes have been involved to locate reaction sites and, thus, allowing for measured correlation with cytodifferentiation.

Regarding plant cell cultures, several experiments have gained a degree of prominence. Recently, a full length cDNA clone coding for tryptophan decarboxylase (tdc) of *Catharanthus roseus* was inserted into the binary vector pBin19. *Agrobacterium tumefaciens* clones containing the plasmid pTDC were then co-cultivated with cell and root cultures of *Peganum harmala*. As a result, the tdc gene was expressed in a number of transgenic cell suspension and root cultures. Product levels of serotonin were increased up to 10-fold in material now expressing the cDNA encoding tryptophan decarboxylase of *Catharanthus roseus*. And the decarboxylase system was found to be rate limiting (Berlin et al., 1993, 1994). The level of cytodifferentiation appears not to have been a factor. Or, direction of tryptophan to increase the level of tryptamine and its derivatives in transgenic canola (*Brassica napus*) plants by expression of the tdc gene isolated from *Catharanthus roseus* (DeLuca et al., 1989) led to drastic reduction of the level of indole glucosinolate (Chavadej et al., 1994). The lack of the glucosinolates was caused by putative depletion of the available tryptophan pool through direct competition for that pool. The expression of TDC in transgenic canola appeared to redirect tryptophan into tryptamine rather than into tryptophan-derived indole glucosinolates. Finally, when hyoscyamine 6β-hydroxylase cDNA of *Hyoscyamus niger* was inserted in *Atropa belladonna*, hairy roots of this species, naturally void of scopolamine, did accumulate this alkaloid (Hashimoto et al., 1993). Insertion of the foreign DNA into *Belladonna* cells and suspension cultures would not have been followed by scopolamine production, since the site of synthesis was found to be the root pericycle (Kanegae et al., 1994).

So far, cytodifferentiation as co-factor of product formation has not received substantial new evidence by employing molecular biological methodology, except for the following: cell suspension cultures of *Catharanthus roseus* have been developed which accumulate sizeable amounts of monoterpene indole alkaloids, none, however, produced vindoline, vinblastine and vincristine, as did shoot cultures (Miura et al.,

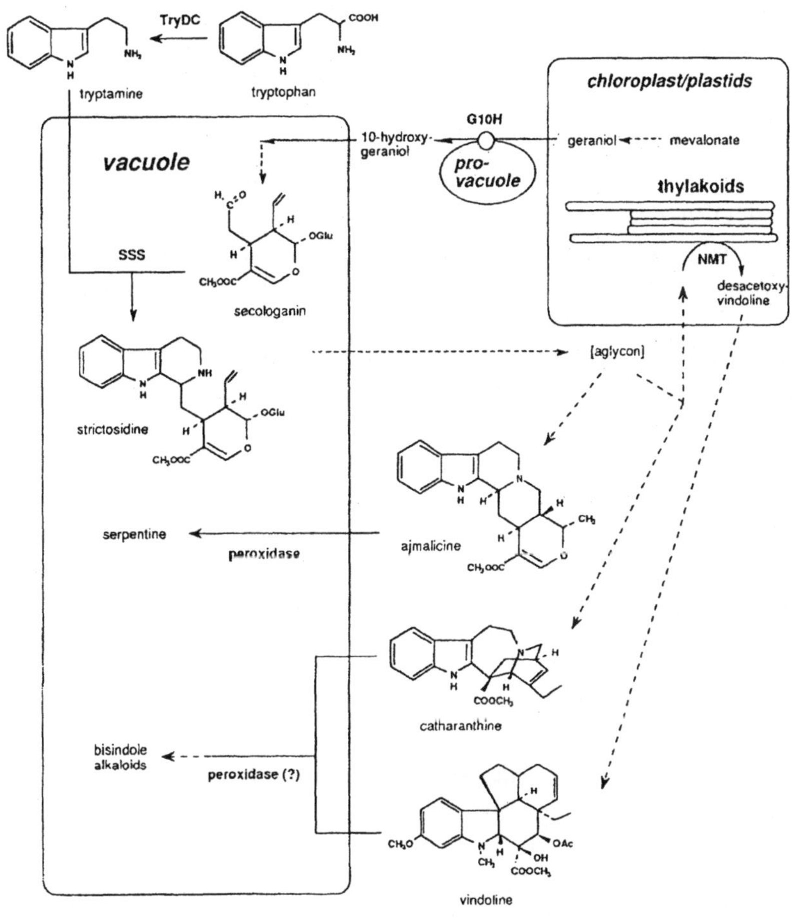

Figure 1. Biosynthetic pathways of monoterpenoid indol alkaloids. TryDC, tryptophan decarboxylase; G10H, geraniol 10-hydroxylase; SSS, strictosidine synthase; NMT, 11-methoxy-2, 16-hydroxytabersonine N-methyltransferase (Hashimoto & Yamada, 1994).

1988). When each of the cell lines was essayed for tryptophan decarboxylase (TDC), strictosidine synthase (SS), *N*-methyltransferase(NMT), and *O*-acetyl transferase (DAT) enzyme activities, all enzymes intimately involved in vindoline synthesis, constant levels of TDC and SS were found, none of the cell lines expressed NMT and DAT activities, catalysts of the first and last step of vindoline, and thus vinblastine synthesis. All these enzymes, however, were readily demonstrated for developing seedlings. In addition,

DAT activity was increased approximately 10-fold after light treatment of dark grown seedlings (DeLuca et al., 1987, 1989). In conclusion, cell suspension cultures of *Catharanthus roseus* were found incapable of vindoline production due to developmental deficiencies. A recent review demonstrates the level of cytodifferentiation required to allow vindoline synthesis to occur (Fig.1).

Also, new light was focussed on the potential of poppy (*Papaver somniferum*) cell cultures for morphine and sanguinarine production. Both these compounds are derived from tyrosine and DOPA and require as a first step towards their synthesis a decarboxylation. With morphine being synthesized in shoots and sanguinarine in roots, shoot and root tyrosine/DOPA decarboxylase enzymes with 75% identity were found and characterized, while their catalytic activity was similar. *In situ* hybridization now revealed that one group of TYDC genes is expressed in root , the other in shoot phloem, specifically. One may be coordinately expressed with enzymes leading to sanguinarine synthesis, the other similarly to morphine synthesis (Facchini and DeLuca, 1994; 1995). It would appear that further experimentation will reveal cell suspension cultures having the cytodifferentiation level of roots, thus allowing for sanguinarine formation, that on the other hand, only embryo or shoot regeneration in cell cultures will permit the occurrence of morphine. Indeed, morphinan alkaloids occurred in cell suspension cultures rarely (Tam et al., 1980), generally however in embryogenic callus cultures (Schuchmann and Wellmann, 1983).

8 CONCLUDING REMARKS

An array of experimental results has been scrutinized and analyzed regarding a correlation between cytodifferentiation and production of secondary metabolites and it may well be that this correlation turns out to be a "red herring", inasmuch as the nature of cytodifferentiation remains somewhat obscure while gains in improving metabolite yields continue to be made, daily.

Undeniably, in most cell cultures accumulation of secondary metabolites occurs maximally when (a) the growth rate of the culture decreases or is being decreased and/or (b) the cultures exhibit some structural differentiation (cytodifferentiation or the organization of cells into structures such as roots, shoots and embryos). This tenet has become the "leitmotif" of much experimentation, held by Lindsey and Yoeman (1983) and is rightly still being promoted today (Petersen and Alfermann, 1993). It is fully appropriate, in particular when accompanied by experimentation which identifies and rectifies limiting factors such as quality of plant source, lack of precursors or low activity of enzymes. Present activity to arrive at taxol-producing cell cultures of Taxus sp. may suffice as an example (cf. Kurz and Constabel, 1995). The striking exception to the rule, betalain production in growing cell cultures derived from *Phytolacca* referred to earlier (Hirano and Komamine, 1994) deserves attention, cannot invalidate, however, the observation that *Phytolacca americana* callus was observed to fade in pigmentation after each subculture and over many years of culture, fully in accordance with the "leitmotif" (Constabel unpubl.).

Non-accordance, it would appear, is the signal given by cell cultures which have been subjected to elicitation. The sudden occurrence of stress products like isoflavonoids, of catharanthine and ajmalicine or of sanguinarine treatment with fungal material would preclude cytological differentiation, expected to develop over time. The problem, however, lies with the observer and his/her interpretation of cytodifferentiation. The latter phenomenon is still too often regarded as a structural, anatomical or histological change, too little a change in function. Substantial and significant increase in the activity of phenylalanine ammonia-lyase over a few minutes as observed in response to elicitation, for instance, is cytodifferentiation. Stress response, therefore, must not interfere with claiming a correlation between cytodifferentiation and product formation. A bridge may be built by accepting stress response in form of product formation as an expression of cytodifferentiation. A molecular biological analysis should help to clarify the situation.

A final argument to test the correlation under investigation may be derived from the existence of cell cultures derived from the same plant source and showing the same level of differentiation, but differing in quality, as shown for *Beta vulgaris* earlier (Constabel and Nassif-Makki, 1971, but see also Girod and Zryd, 1991), or quantity of product, as reported repeatedly and recently for *Coptis japonica* (Sato et al., 1994). The difference may easily have arisen with somaclonal variation. The phenomenon itself, however, does not contribute to clarification of a correlation to cytodifferentiation. Rather, it entices to analyze the molecular biology of product formation and experimentation in an attempt to succeed with synthesis and accumulation of secondary products and with a new view of cytodifferentiation.

The molecular approach of secondary metabolite formation not only presents state-of-the-art methodology, but the only methodology to shed further light on the behaviour of cells cultured *in vitro*. And, in reproducing the reaction scheme for vindoline synthesis in cells of *Catharanthus roseus* as presented by DeLuca (in Hashimoto and Yamada, 1994), a precise formulation of respective cytodifferentiation as condition for synthesis and accumulation will result.

9 REFERENCES

Amann M, Wanner G & Zenk MH (1986) Intracellular compartmentation of two enzymes of berberine biosynthesis in plant cell cultures. Planta 167: 310-320.

Arroo RRJ, Develi A, Meijers H, van de Westerlo E, Kemp AK, Croes AF & Wullems GJ (1995) Effect of exogenous auxin on root morphology and secondary metabolism in *Tagetes patula* hairy root cultures. Physiol. Plant. 93: 233-240.

Barz W & Mackenbrock U (1994) Constitutive and elicitation-induced metabolism of isoflavones and pterocarpans in chickpea (*Cicer arietinum*) cell suspension cultures. Plant Cell Tissue Organ Cult. 38: 199-211.

Berglund T & Ohlsson BA (1992) Effects of ethylene and aminoethoxyvinylglycine on cardenolide accumulation in tissue cultures of *Digitalis lanata*. J. Plant Physiol. 140: 395-398

Berlin J, Ruegenhagen C, Greidziak N, Kuzovkina IN, Witte L & Wray V (1993) Biosynthesis of serotonin and β-carboline alkaloids in hairy root cultures of *Peganum harmala*. Phytochemistry. 33: 593-597.

Berlin J, Ruegenhagen C, Kuzovkina IN, Fecker LF & Sasse F (1994) Are tissue cultures of *Peganum harmala* a useful model system for studying how to manipulate the formation of secondary metabolites. Plant Cell Tissue Organ Cult. 38: 289-298.

Bohlmann J & Eilert U (1994) Elicitor-induced secondary metabolism in *Ruta graveolens* L. Plant Cell Tissue Organ Cult. 38: 189-198.

Brodelius P, Deus B, Mosbach K & Zenk MH (1979) Immobilized plant cells for the production and transformation of natural products. FEBS Lett. 103: 93-97.

Brodelius P (1983) Plant cell cultures. In: Mattiasson B (ed.), Immobilized Cells and Organelles. (pp. 27-55) CRC Press, Boca Raton.

Brown GD (1994) Secondary metabolism in tissue cultures of *Artemisia annua*. J. Nat. Prod. 57: 975-977.

Calvo MC, Sanchez-Gras MC (1993) Accumulation of monoterpenes in shoot-proliferation cultures of *Lavandula latifolia*. Plant Sci. 91: 207-212.

Campbell B & Ellis BE (1992) Fungal elicitor-mediated responses in pine cell cultures: cell wall bound phenolics. Phytochemistry. 31: 737-742.

Chavadej S, Brisson N, McNeil JN & DeLuca V (1994) Redirection of tryptophan leads to production of low indole glucosinolate canola. Proc. Natl. Acad. Sci. USA 91: 2166-2170.

Chilton, MD, Tepfer DA, Petit A, David C, Casse-Delbart F & Tempe J (1982) *Agrobacterium rhizogenes* inserts T-DNA into the genomes of the host plant root cells. Nature 295: 432-434.

Cho GH, Kim DI, Pedersen H & Chin CK (1988): Ethephon enhancement of secondary metabolite synthesis in plant cell cultures. Biotech. Prog: 4, 184-188.

Ciau-Uitz R, Miranda-Ham ML, Coello-Coello J, Chi B, Pacheco LM & Loyola-Vargas VM (1994) Indole alkaloid production by transformed and non-transformed root cultures of Catharanthus roseus. In vitro Cell Dev. Biol. 30P: 84-88.

Constabel F & Nassif-Makki H (1971) Betalainbildung in *Beta* Calluskulturen. Ber. Dtsch. Bot. Ges. 67: 201-206.

Constabel F & Tyler RT (1994) Cell culture for the production of secondary metabolites. In: Vasil IK & Thorpe TA (eds), Plant Cell and Tissue Culture (pp. 271-292), Kluwer Dordrecht.

Constabel F & Vasil IK (eds) (1988) Cell Culture and Somatic Cell Genetics of Plants, Vol. 5: Phytochemicals in Plant Cell Cultures. Academic Press, New York.

Constabel F (1963) Über die Gerbstoffe in Gewebekulturen von *Juniperus communis* L. Planta Med. 11: 417-421.

Constabel F (1968) Dedifferenzierung von Gerbstoffzellen. Planta Med. 16: 241-247.
Constabel F (1968) Gerbstoffproduction der Calluskulturen von *Juniperus communis* L. Planta 79: 58-64
Constabel F (1969) Über die Entwicklung von Gerbstoffzellen in Calluskulturen von *Juniperus communis* L. Planta Med. 17: 101-115.
Constabel F (1988) Principles underlying the use of plant cell fermentation for secondary metabolite production. Biochem. Cell Biol. 66: 658-664
Constabel F, Rambold S, Chatson KB, Kurz WGW & Kutney JP (1981) Alkaloid production in *Catharanthus roseus* (L.) Don. VI. Variation in alkaloid spectra of cell lines derived from one single leaf. Plant Cell Rep. 1: 3-5.
Constabel F, Shyluk JP & Gamborg OL (1971) The effect of hormones on anthocyanin accumulation in cell cultures of *Haplopappus gracilis*. Planta 96: 306-316.
Constabel CP & Towers GHN (1988) Thiarubrine accumulation in hairy root cultures of *Chaenactis douglasii*. J. Plant Physiol. 133: 67-72.
D'Harlingue A, Mamdough AM, Malfatti P, Soulie MC & Bompeix G (1995) Evidence for rishitin biosynthesis in tomato cultures. Phytochemistry. 39: 69-70.
delle Monache G , de Rosa MC, Scurria R, Vitali A, Cuteri A, Monacelli B, Pasqua G & Botta B (1995) Comparison between metabolite production in cell culture and in whole plant of *Maclura pomifera*. Phytochemistry. 39: 575-580
DeLuca V, Balsevich J, Tyler RT & Kurz WGW (1987) Characterization of a novel N-methyltransferase from *Catharanthus roseus* plants. Plant Cell Rep. 6: 458-461.
DeLuca V, Brisson N, Balsevich J & Kurz WGW (1989) Regulation of vindoline biosynthesis in *Catharanthus roseus* Molecular cloning of the first and last steps in biosynthesis. In: Kurz WGW (ed.) Primary and Secondary Metabolism of Plant Cell Cultures. (pp. 154-161). Springer, Berlin.
Deus-Neumann B & Zenk MH (1984) A highly selective alkaloid uptake system in vacuoles of higher plants. Planta 162: 250-260.
Doerk-Schmitz K, Witte L & Alfermann AW (1994) Tropane alkaloid patterns in plants and hairy roots of *Hyoscyamus albus*. Phytochemistry. 35: 107-110.
Draeger B, Funck B, Höler A, Mrachatz G, Nahrstedt A, Portsteffen A, Schaal A & Schmidt R (1994) Calystegines a new group of tropane alkaloids in *Solanaceae*. Plant Cell Tissue Organ Cult. 38: 235-240.
Durzan DJ & Ventimiglia F (1994) Free taxanes and the release of bound compounds having taxane antibody reactivity by xylanase in female, haploid-derived cell suspension cultures of *Taxus brevifolia*. *In vitro* Cell. Dev. Bio. 30P: 219-227.
Eberhardt TL, Bernards MA, He LF, Davin, LB, Wooten JB & Lewis NG (1993) Lignification in cell suspension cultures of *Pinus taeda* J. Biol. Chem. 268: 21088-21096.
Eilert U (1987) Elicitation: Methodology and aspects of application. In: Constabel F & Vasil IK (eds) Cell Culture and Somatic Cell Genetics of Plants, Vol. 4. (pp. 153-196) Academic Press, New York.
Eilert U (1989) Elicitor induction of secondary metabolism in dedifferentiated and differentiated in vitro systems of *Ruta graveolens* In: Kurz WGW (ed.), Primary and Secondary Metabolism of Plant Cell Cultures (pp. 219-228), Springer, Berlin.
Eilert U, Constabel F & Kurz WGW (1986) Elicitor-stimulation of monoterpene indole alkaloid formation in suspension cultures of *Catharanthus roseus* J. Plant Physiol. 126: 11-22.
Eilert U, Kurz WGW & Constabel F (1985) Stimulation of sanguinarine accumulation in *Papaver somniferum* cell cultures by fungal elicitors. J. Plant Physiol. 119: 65 - 76.
Endress R (1994) Plant Cell Biotechnology. Springer, Berlin.
Esau K (1965) Plant Anatomy. John Wiley, New York.

Eschrich W (1995) Funktionelle Pflanzenanatomie. Springer, Berlin.

Facchini PJ & DeLuca V (1994) Differential and tissue specific expression of a gene family for tyrosine/DOPA decarboxylase in opium poppy. J. Biol. Chem. 269: 26684 - 26690.

Facchini PJ & DeLuca V (1995) Phloem-specific expression of tyrosin/DOPA decarboxylase genes and the biosynthesis of isoquinoline alkaloids in opium poppy. Plant Cell 7:

Fett-Neto AG, Melanson SJ, Sakata K & DiCosmo F (1993) Improved growth and taxol yield in developing callus of *Taxus cuspidata* by medium composition modification. Bio/Technology 11: 731-734.

Flores HE & Filner P (1985) Hairy roots of *Solanaceae* as a source of alkaloids. Plant Physiol. 77: 12s.

Flores HE, Yao-Rem D, Cuello-Cuello JL, Maldonado-Mendoza IE & Loyola-Vargas VM (1993) Green roots: photosynthesis and photoautotrophy in an underground plant organ. Plant Physiol. 101: 363-371.

Fontanel A & Tabata M (1987) Production of secondary metabolites by plant tissue and cell cultures, present aspects and prospects. Nestle Research News 1986/87, Nestec Ltd.,Vevey, Switzerland, pp. 47-54.

Fujita Y, Hara Y, Suga C & Morimoto T (1981) Production of shikonin derivatives by cell suspension cultures of *Lithospermum erythrorhizon* II: A new medium for the production of shikonin derivatives. Plant Cell Rep. 1: 61-63.

Fukuda H & Komamine A (1980): Establishment of an experimental system for the study of tracheary element differentiation from single cells isolated from the mesophyll of *Zinnia elegans* Plant Physiol. 65: 57-60.

Fukuda H & Komamine A (1982) Lignin synthesis and its related enzymes as markers of tracheary-element differentiation in single cells isolated from the mesophyll of *Zinnia elegans* Planta 155: 423-430.

Fulzele D, Heble MR & Rao PS (1995) Production of terpenoids from *Artemisia annua* L. plantlet cultures in a bioreactor. J. Biotechn. 40: 139-143.

Gamborg OL & Phillips GC (eds) (1995) Fundamental Methods of Plant Cell, Tissue, and Organ Culture and Laboratory Operations. Springer, New York.

Gamborg OL, Miller RA & Ojima (1968) Nutrient requirements of suspension cultures of soybean root cells. Exptl. Cell Res. 50: 151-158.

Gautheret RJ (1959) La Culture des Tissues Végétaux. Techniques et Rélization. Masson, Paris.

Gertlowski C & Petersen M (1993) Influence of the carbon source on growth and rosmarinic acid production in suspension cultures of *Coleus blumei* Plant Cell Tissue Organ Cult. 34: 183-190.

Girod PA & Zryd JP (1991) Secondary metabolism in cultured red beet (*Beta vulgaris* L.) cells: differential regulation of betaxanthin and betacyanin biosynthesis. Plant Cell Tissue Organ Cult. 25: 1-12.

Goddijn OJM, de Kam RJ, Zanetti A, Schilperoort RA & Hoge JHC (1992) Auxin rapidly down-regulates transcription of the tryptophan decarboxylase gene from *Catharanthus roseus.* Plant Mol. Biol. 18: 1113-1120.

Gontier E, Sangwan BS & Barbotin JN (1994) Effects of calcium, alginate, and calcium alginate immobilization on growth and tropane alkaloid levels of a stable suspension cell line of *Datura innoxia* Mill.. Plant Cell Rep. 13: 533-536.

Green KD, Thomas NH & Callow JA (1992): Product enhancement and recovery from transformed root cultures of *Nicotiana glauca* Biotechnol. Bioeng. 39: 195-202.

Gränicher F, Christen P & Kapetanidis I (1995) Production of Valepotriates by hairy root cultures of *Centranthus ruber* DC. Plant Cell Rep. 14: 294-298.

Gränicher F, Christen P, Kamalaprija P & Burger U (1995) An iridoid diester from *Valeriana officinalis* var. *sambucifolia* hairy roots. Phytochemistry. 38: 103-105.
Guo ZJ & Ohta Y (1993) A synergistic effect of glutathione-depletion and elicitation on the production of 6-methoxymellein in carrot cells. Plant Cell Rep. 12: 617-620.
Guo ZJ & Ohta Y (1994) Effect of ethylene biosynthesis on the accumulation of 6-methoxymellein induced by elicitors in carrot cells. J. Plant Physiol. 144: 700-704.
Haagendoorn NJN, Zethof JLM, van Hunnik E & van der Plas LHW (1991) Regulation of anthocyanin and lignin synthesis in *Petunia hybrida* cell suspensions. Plant Cell Tissue Organ Cult. 27: 141-147.
Haberlandt G (1914) Physiological Plant Anatomy, MacMillan, London.
Hahlbrock K & Scheel D (1989) Physiology and molecular biology of phenylpropanoid metabolism. Annu. Rev. Plant Physiol. Plant Mol. Biol. 40: 347-369.
Hamill JD, Parr AJ, Robins RJ & Rhodes MJC (1986) Secondary product formation by cultures of *Beta vulgaris* and *Nicotiana rustica* transformed with *Agrobacterium rhizogenes* Plant Cell Rep. 5: 111-114.
Hamill JD, Robins RJ, Parr AJ, Evans DM, Furze JM & Rhodes MJC (1990) Overexpressing a yeast ornithine decarboxylase gene in transgenic roots of *Nicotiana rustica* can lead to enhanced nicotine accumulation. Plant Mol. Biol. 15: 27-38.
Hara M, Kitamura T, Fukui H & Tabata M (1993) Induction of berberine biosynthesis by cytokinins in *Thalictrum minus* cell suspension cultures. Plant Cell Rep. 12: 70-73.
Hara M, Morio H, Yazaki K, Tanaka S & Tabata M (1995) Separation and characterization of cytokinin-inducible (S)-tetrahydroberberine oxidases controlling berberine biosynthesis in *Thalictrum minus* cell cultures. Phytochemistry. 38: 89-95.
Hashimoto T & Yamada Y (1994) Alkaloid biogenesis: Molecular aspects. Annu Rev. Plant Physiol. Plant Mol. Biol. 45: 257-285.
Hayashi T, Gotoh K, Ohnishi K, Okamura K & Asamizu T (1995) 6-methoxy-2-benzoxazolinone in *Scoparia dulcis* and its production by cultured tissues. Phytochemis try. 37: 1611-1614.
Hirano H & Komamine A (1994) Correlation of betacyanin synthesis with cell division in cell suspension cultures of *Phytolacca americana*. Physiol. Plant. 90: 239-245.
Hirose M, Yamakawa T, Kodama T & Komamine A (1990) Accumulation of betacyanin in *Phytolacca americana* cells and of anthocyanin in *Vitis* sp. cells in relation to cell division in suspension cultures. Plant Cell Physiol. 31: 267-271.
Hirotani M, Zhou Y, Rui H & Furuya T (1994) Cycloartane triterpene glycosides from the hairy root cultures of *Astragalus membranaceus*. Phytochemistry. 37: 1403-1407.
Holden PR & Yoeman MM (1994) Variation in the growth and biosynthetic activity of cloned cell cultures of *Capsicum frutescens* and their response to an exogenously supplied elicitor. Plant Cell Tissue Organ Cul. 38, 31-37.
Hook I (1994) Secondary metabolites in hairy root cultures of *Leontopodium alpinum* Cass. Edelweiss. Plant Cell Tissue Organ Cult. 38: 321-326.
Hoshino T, Chida M, Yamaura T, Yoshizawa Y & Mizutani J (1994) Phytoalexin induction in green pepper cell cultures treated with arachidonic acid. Phytochemistry. 36: 1417-1419.
Hosokawa K & Fukunaga Y (1996) Production of essential oils by flowers of *Hyacinthus orientalis* L. regenerated *in vitro*. Plant Cell Rep. In press.
Häusler E, Petersen M & Alfermann AW (1993): Isolation of protoplasts and vacuoles from cell suspension cultures of *Coleus blumei* Benth. Plant Cell Rep. 12: 510-512.
Ikemeyer D & Barz W (1989) Comparison of secondary product accumulation in photoautotrophic, photomixotrophic and heterotrophic *Nicotiana tabacum* cell suspension cultures. Plant Cell Rep. 8: 479-482.

Ikenaga T, Oyama T & Muranaka T (1995) Growth and steroidal saponin production in hairy root cultures of *Solanum aculeatissimum*. Plant Cell Rep. 14: 413-417.

Ikuta A, Kamiya K, Satake T & Saiki Y (1995) Triterpenoids from callus cultures of *Paeonia* species. Phytochemistry. 38: 1203-1207.

Ionkova I, Witte L & Alfermann AW (1994) Spectrum of tropane alkaloids in transformed roots of *Datura innoxia* and *Hyoscyamus* x *gyoerffyi* cultivated *in vitro*. Planta Med. 60: 382-383.

Ishiguro K, Nakajima M, Fukumoto H & Isoi K (1995) Co-occurrence of prenylated xanthones and their cyclization products in cell suspension cultures of *Hypericum patulum*. Phytochemistry. 38: 867-869.

Ishimaru K, Arakawa H, Yamanaka M & Shimomura K (1994) Polyacetylenes in *Lobelia sessilifolia* hairy roots. Phytochemistry. 35: 365-369.

Jaziri M, Shimomura K, Yoshimatsu K, Fauconnier ML, Marlier M & Homes J (1995) Establishment of normal and transformed root cultures of *Artemisia annua* L. for artemisinin production. J. Plant Physiol. 145: 175-177.

Jung KH, Kwak SS, Choi CY & Liu JR (1996) An interchangeable system of hairy root and cell suspension cultures in *Catharanthus roseus* for indole alkaloid production. Plant Cell Rep. 14, in press.

Jung KH, Kwak SS, Choi CY & Liu JR (1994) Development of a two-stage culture process by optimization of inorganic salts for improving catharanthine production in hairy root cultures of *Catharanthus roseus*. J. Ferm. Bioeng. 77: 57-61.

Kanegae T, Kayija H, Amano Y, Hashimoto T & Yamada Y (1994) Species-dependent expression of the hyoscyamine 6β -hydroxylase gene in the pericycle. Plant Physiol. 105: 483-490.

Kim SW, Jung KH, Kwak SS& Liu JR (1994) Relationship between cell morphology and indole alkaloid production in cell suspension cultures of *Catharanthus roseus*. Plant Cell Rep. 14: 23-26.

Knöss W (1994) Furanic labdane diterpenes in differentiated and undifferentiated cultures of *Marrubium vulgare* and *Leonurus cardiaca*. Plant Physiol. Biochem. 32: 785-789.

Knöss W (1995) Establishment of callus, cell suspension and shoot cultures of *Leonurus cardiaca* L. and diterpene analysis. Plant Cell Rep. 14: 790-793.

Kreis W, Heinz H, Sutor R & Reinhard E (1993) Cellular organization of cardenolide biotransformation in *Digitalis grandiflora* Mill. Planta 191: 246-251.

Krombholz R, Mersinger R, Kreis W & Reinhard E (1992) Production of forskolin by axenic *Coleus forskohlii* roots cultivated in shake flasks and 20-l glass jar bioreactors. Planta Med. 58: 328-333.

Kurz WGW & Constabel F (1998) Production of secondary metabolites. In: Altman A (ed.) Agricultural Biotechnologies. Marcel Dekker Inc., New York.

Kutchan TM, Ayabe S, Krueger RJ, Coscia EM & Coscia CJ (1983) Cytodifferentiation and alkaloid accumulation in cultured cells of *Papaver bracteatum*. Plant Cell Rep. 2: 281-284.

Lindsey K & Yoeman MM (1983) The relationship between growth rate, differentiation and alkaloid accumulation in cell cultures. J. Exp. Bot. 34: 1055-1065.

Loyola -Vargas VM & Miranda-Ham ML (1996) Root culture as a source of secondary metabolites of economic importance. Rec. Adv. Phytochem.(in press).

Luckner M, Nover L & Boehm H (1977) Secondary Metabolism and Cell Differentiation. Springer Verlag, Berlin-Heidelberg.

Ma W, Park GL, Gomez GA, Nieder MH, Adams TL, Sahai OP, Smith RJ, Stahlhut RW, Hylands PJ, Bitsch F & Shackleton C (1994a) New bioactive taxoids from cell cultures of *Taxus baccata*. J. Nat. Prod. 57: 116-122.

Ma WW, Stahlhut RW, Adams TL, Park GL, Evans WA, Blumenthal SG, Gomez GA, Nieder MH & Hylands PJ (1994b) Yunnanxane and its homologous esters from cell cultures of *Taxus chinensis* var. *mairei*. J. Nat. Prod. 57: 1320-1324.

Mahady MG & Beecher CWW (1994): Quercetin-induced benzophenanthridine alkaloid production in suspension cell cultures of *Sanguinaria canadensis*. Planta Med. 60: 553-557.

Maldonado-Mendoza IE & Loyola-Vargas VM (1995) Establishment and characterization of photosynthetic hairy root cultures of *Datura stramonium*. Plant Cell Tissue Organ Cult. 40: 197-208.

Maldonado-Mendoza IE, Ayora-Talavera TR & Loyola-Vargas VM (1992: Tropane alkaloid production in *Datura stramonium* root cultures. *In vitro* Cell Dev. Biol. 28P: 67-72.

McCaskill D, Gershenzon J & Croteau R (1992) Morphology and monoterpene biosynthetic capabilities of secretory cell clusters isolated from glandular trichomes of peppermint (*Mentha piperita* L.) Planta 187: 445-454.

McClellan D, Kott L, Beversdorf W & Ellis BE (1993) Glucosinolate metabolism in zygotic and microspore-derived embryos of *Brassica napus* L. J. Plant Physiol. 141: 153-159.

Mersinger R, Dornauer H & Reinhard E (1988) Formation of forskolin by suspension cultures of *Coleus forskohlii*. Planta Med. 54: 200-204.

Meyer HJ & van Staden J (1995) The *in vitro* production of an anthocyanin from callus cultures of *Oxalis linearis*. Plant Cell Tissue Organ Cult. 40: 55-58.

Mita G, Gerardi C, Miceli A, Bollini R & DeLeo P (1994) Pigment production from *in vitro* cultures of *Alkanna tinctoria* Tausch. Plant Cell Rep. 13: 406-410.

Miura Y, Hirata K, Kurano N, Miyamoto K & Uchida K (1988) Formation of vinblastine in multiple shoot cultures of *Catharanthus roseus*. Planta Med. 54: 18-20.

Mizukasi, S, Tanabe Y, Noguchi M & Tamaki E (1973) Changes in the activities of ornithine decarboxylase, putrescine N-methyltransferase and N-methyl-putrescine oxidase in tobacco roots in relation to nicotine biosynthesis. Plant Cell Physiol. 14: 103-110.

Moreno PRH, van der Heijden R & Verpoorte R (1993) Effect of terpenoid precursor feeding and elicitation on formation of indole alkaloids in cell suspension cultures of *Catharanthus roseus*. Plant Cell Rep. 12: 702-705.

Mothes K (1972) Pflanze und Tier, ein Vergleich auf der Ebene des Sekundaerstoffwechsels. terr. Abh. Akad. Wiss. Wien, Sonderh. zu SB. Abt. I, Vol. 181.

Mukherjee S, Ghosh B & Jha S (1996) Forskolin synthesis in cultures of *Coleus forskohlii* Briq. transformed with *Agrobacterium tumefaciens*. Plant Cell Rep. 15 (in press).

Murashige T & Skoog F (1962) A revised medium for rapid growth and bioassays with tobacco tissue cultures. Physiol. Plant. 15: 473-479.

Nakagawa K, Konagai A, Fukui H & Tabata M (1984) Release and crystallization of berberine in a liquid medium of *Thalictrum minus* cell suspension cultures. Plant Cell Rep. 3: 254-257.

Nessler CL & von der Haar RA (1990) Cloning and expression analysis of DNA sequences for the major latex protein of opium poppy. Planta 180: 487-491.

Nessler CL (1994) Sequence analysis of two new members of the major latex protein gene family supports the triploid-hybrid origin of the opium poppy. Gene 139: 207-209.

Ozeki Y & Komamine A (1994) Effects of growth regulation on the induction of anthocyanin synthesis in carrot suspension cultures. Plant Cell Physiol. 27: 1361-1368.

Paech K (1950) Biochemie und Physiologie der sekundaeren Pflanzenstoffe. Springer, Berlin.

Paiva NL, Oommen A, Harrison MJ & Dixon RA (1994) Regulation of isoflavonoid metabolism in alfalfa. Plant Cell Tissue Organ Cult. 38: 213-220.

Park HH, Hakamatsuka T, Sankawa U & Ebizuka Y (1995) Rapid metabolism of isoflavonoids in elicitor-treated cell suspension cultures of *Pueraria lobata*. Phytochem istry. 38: 373-380.

Pasqua G, Monacelli B, Cuteri A, Botta B, Vitali A, delle Monache G (1991) Cell suspension cultures of *Maclura pomifera*: Optimization of growth and metabolite production. J. Plant Physiol. 139: 249-251.

Petersen M & Alfermann AW (1993) Plant cell cultures. In: Rehm HJ & Reed G (eds) Biotechnology 2nd ed. (pp. 577-614) Verlag Chemie, Weinheim.

Piatti T, Boller T & Brodelius PE (1991) Induction of ethylene biosynthesis is correlated with but not required for induction of alkaloid accumulation in elicitor-treated *Eschscholtzia* cells. Phytochemistry. 30: 2151-2154.

Razzaque A & Ellis BE (1977) Rosmarinic acid production in Coleus cell cultures. Planta 137: 287-291.

Reichling J, Becker H, Martin R & Burkhardt G (1985) Comparative studies on the production and accumulation of essential oil in the whole plant and in the cell culture of *Pimpinella anisum* L. Z. Naturforsch. 40c: 465-468.

Reichling J, Martin R & Thron U (1988) Production and accumulation of phenylpropanoids in tissue and organ cultures of *Pimpinella anisum* L. Z. Naturforsch. 43c: 42-46.

Reinert, J (1958) Untersuchungen über die Morphogenese an Gewebekulturen. Ber. Dtsch. Bot. Ges. 71: 15.

Rho D, Chauret N, Laberge N & Archambault J (1992) Growth characteristics of *Sanguinaria canadensis* L. cell suspensions and immobilized cultures for production of benzophenanthridine alkaloids. Appl. Microbiol. Biotechnol. 36: 611-617.

Rhodes MJC, Parr AJ, Guilietti A & Aird ELH (1994) Influence of exogenous hormones on the growth and secondary metabolite formation in transformed root cultures. Plant Cell Tissue Organ Cult. 38: 143-151.

Roewer I, Cloutier N, Nessler CL & DeLuca V (1992) Transient induction of tryptophan decarboxylase (TDC) and strictosidine synthase (SS) genes in cell suspension cultures of *Catharanthus roseus*. Plant Cell Rep. 11: 86-89.

Rubluo A, Flores A, Jiminez M & Brunner I (1994) *Piqueria trinervia* Cav.: In vitro culture and the production of piquerol. In: Bajaj YPS (ed.) Biotechnology in Agriculture and Forestry. Medicinal and Aromatic Plants. Vol. 8. Springer, Berlin.

Rudat A & Ehwald R (1994) Induction of betacyanin formation in *Chenopodium album* cell cultures by co-cultivation with duckweed *Wolffia arrhiza*. Plant Cell Rep. 13: 291-294.

Rush MD, Kutchan TM & Coscia CJ (1985) Correlation of the appearance of morphinan alkaloids and laticifer cells in germinating *Papaver bracteatum* seedlings. Plant Cell Rep. 4: 237-240.

Salem KMSA & Charlwood BV (1995) Accumulation of essential oils by *Agrobacterium tumefaciens*-transformed shoot cultures of *Pimpinella anisum*. Plant Cell Tissue Organ Cult. 40: 209-215.

Salisbury FB & Ross CW (1985) Plant Physiology. Wadsworth Publ. Comp.,Belmont/CA

Sanyal MAD, Bannerjee D & Datta PC (1981) In vitro hormone induced chemical and histological differentiation in stem callus of neem *Azadirachta indica* A. Juss. Indian J. Exptl. Biol. 13: 84-85.

Sato F, Takeshita N, Fujiwara H, Katagiri Y, Huan LP & Yamada Y (1994) Characterization of *Coptis japonica* cells with different alkaloid productivities. Plant Cell Tissue Organ Cult. 38: 249-256.

Sato H, Tanaka S & Tabata M (1993a) Kinetics of alkaloid uptake by cultured cells of *Coptis japonica*. Phytochemistry. 34: 697-701.

Sato K, Yamazaki T, Okuyama E, Yoshihira K & Shimomura K (1991) Anthraquinon production by transformed root cultures of *Rubia tinctorum*. Influence of phytohormones and sucrose concentration. Phytochemistry. 30: 1507-1509.

Sato Y, Sugiyama M, Gorecki RJ, Fukuda H & Komamine A (1993b) Interrelationship between lignin deposition and the activities of peroxidase isoenzymes in differentiating tracheary elements of *Zinnia*. Planta 189: 584-589.

Sauerwein M, Ishimaru K & Shimomura K (1991a) Indole alkaloids in hairy roots of *Amsonia elliptica*. Phytochemistry. 30: 1153-1155.

Sauerwein M, Yamazaki T & Shimomura K (1991b) Hernandulcin in hairy root cultures of *Lippia dulcis*. Plant Cell Rep. 9: 579-581.

Schuchmann R & Wellmann E (1983) Somatic embryogenesis in tissue cultures of *Papaver somniferum* and *Papaver orientale* and its relationship to alkaloid and lipid metabolism. Plant Cell Rep. 2: 88-91.

Schulthess B & Baumann TW (1995) Stimulation of caffeine biosynthesis in suspension-cultured coffee cells and the *in situ* existence of 7-methylxanthosine. Phytochemistry. 38: 1381-1386.

Seitz HU & Gaertner DE (1994) Enzymes in cardenolide-accumulating shoot cultures of *Digitalis purpurea* L. Plant Cell Tissue Organ Cult. 38: 337-344.

Shimomura K, Sudo H, Saga H & Kamada H (1991) Shikonin production and secretion by hairy root cultures of *Lithospermum erythrorhizon*. Plant Cell Rep. 10: 282-285.

Smith, JI, Smart NJ, Kurz WGW & Misawa M (1987) Stimulation of indole alkaloid production in cell suspension cultures of *Catharanthus roseus* by abscisic acid. Planta Med. 53: 470-474.

Smolko DD & SW Peretti (1994) Stimulation of berberine secretion and growth in cell cultures of *Thalictrum minus*. Plant Cell Rep. 14: 131-136.

Sommer S, Severin K, Camara B & Heide L (1995) Intracellular localization of geranylpyrophosphate synthase from cell cultures of *Lithospermum erythrorhizon*. Phytochemistry 38: 623-627.

Song KS, Tomoda M, Shimizu N, Sankawa U & Ebizuka Y (1995) Mannan glycopeptide elicits p-coumaroylamino acids in *Ephedra distachya* cultures. Phytochemistry. 38: 95-102.

Songstad DD, Giles KL, Park J, Novakovski D, Epp D, Friesen L & Roewer I (1989) Effect of ethylene on sanguinarine production from *Papaver somniferum* cell cultures. Plant Cell Rep. 8: 463-466.

Spencer A, Hamill JD & Rhodes MJC (1993) *In vitro* biosynthesis of monoterpenes by *Agrobacterium* transformed shoot cultures of two *Mentha* species. Phytochemistry. 32: 9111-919.

Stano J, Nemec P, Weissova K, Kovacs P, Kakoniova D & Liskova D (1995) Decarboxylation of L-tyrosine and L-DOPA by immobilized cells of *Papaver somniferum*. Phytochemistry. 38: 859-860.

Steward FC, Mapes MO & Mears K (1958) Growth and organized development of cultured cells. II. Organization in cultures grown from freely suspended cells. Am. J. Bot. 45: 705-708.

Strasburger E, Noll F, Schenck H & Schimper AFW (1991) Lehrbuch der Botanik für Hochschulen, 33rd edition by Sitte P, Ziegler H, Ehrendorfer F & Bresinsky A., Gustav Fischer Verlag, Stuttgart.

Strauss A, Spengel SM & Schaffner W (1995) Saponins from root cultures of *Phytolacca acinosa*. Phytochemistry. 38: 861-865.

Stuhlemmer U, Kreis W, Eisenbeiss M & Reinhard E (1993) Cardiac glycosides in partly submerged shoots of *Digitalis lanata*. Planta Med. 59, 539-544.

Subroto MA & Doran P (1994) Production of steroidal alkaloids by hairy roots of *Solanum aviculare* and the effect of gibberellic acid. Plant Cell Tissue Organ Cult. 38: 93-102.
Sugimoto Y, Yoshida A, Uchida S, Inanaga S & Yamada Y (1994) Dauricine production in cultured roots of *Menispermum dauricum*. Phytochemistry. 36: 679-683.
Suri SS & Ramawat KG (1995) *In vitro* hormonal regulation of laticifer differentiation in *Calotropis procera*. Ann Bot. 75: 477-480.
Suvarnalatha G, Rajendran L, Ravishankar GA & Venkatamaran LV (1994) Elicitation of anthocyanin production in cell cultures of carrot (*Daucus carota* L.) by using elicitors and abiotic stress. Biotech. Lett. 16: 1275-1280.
Suzuki K, Amino S, Takeuchi Y & Komamine A (1990) Differences in the composition of the cell walls of two morphologically different lines in suspension-cultured *Catharanthus roseus* cells. Plant Cell Physiol. 31: 7-14.
Swain T (1977) Secondary compounds as protective agents. Annu. Rev. Plant Physiol. 28: 479-501.
Swanson SM, Mahady GB & Beecher CWW (1992) Stevioside biosynthesis by callus, root, shoot and rooted-shoot cultures in vitro. Plant Cell Tissue Organ Cult. 28: 151-157.
Tada H, Shimomura K & Ishimaru K (1995) Polyacetylenes in *Platycodon grandiflorum* hairy root and campanulaceous plants. J. Plant Physiol. 145: 7-10.
Tam WHJ, Kurz WGW & Constabel F (1980) Codeine from cell suspension cultures of *Papaver somniferum*. Phytochemistry. 19: 486-487.
Tanaka N, Takao M & Matsumoto T (1995) Vincamine production in multiple shoot culture derived from hairy roots of *Vinca minor*. Plant Cell Tissue Organ Cult. 41: 61-64.
Tulecke WR & Nickell LG (1959) Production of large amounts of plant tissue by submerged culture. Science 130: 863-864.
Ushiyama K, Oda H, Miyamoto Y & Furuya T (1986) Large scale tissue culture of *Panax ginseng* root. Proc. 6th Congr. IAPTC, Minneapolis. Abstr. p. 252.
Vazquez-Flota F, Moreno-Valenzuela O, Marinda-Ham ML, Coello-Coello J & Loyola Vargas VM (1994) Catharanthine and ajmalicine synthesis in *Catharanthus roseus* hairy root cultures. Plant Cell Tissue Organ Cult. 38: 273-279.
Verpoorte R, van der Heijden R, Schripsema J, Hoge JHC & ten Hoopen HJG (1993) Plant cell biotechnology for the production of alkaloids: present status and prospects. J. Nat. Prod. 56: 186-207.
Westcott RJ, Cheetham PSJ & Barraclough AJ (1994) Use of organized viable Vanilla plant aerial roots for the production of natural vanillin. Phytochemistry. 35: 135-138.
Wetter, LR & Constabel F (1982) Plant Tissue Culture Methods, 2nd ed., National Research Council of Canada, Saskatoon.
White PR (1934) Potentially unlimited growth of excised tomato root tips in a liquid medium. Plant Physiol. 9: 585-600.
Wibberley MS, Lenton JR & Neill SJ (1994) Sesquiterpenoid phytoalexins produced by hairy roots of *Nicotiana tabacum*. Phytochemistry. 37: 349-351.
Wickremesinhe ERM & Arteca RN (1994) Taxus cell suspension cultures: optimizing growth and production of taxol. J. Plant Physiol. 144: 183-188.
Williams RD, Chauret N, Bedard C & Archambault J (1992) Effect of polymeric adsorbents on the production of sanguinarine by *Papaver somniferum* cell cultures. Biotech Bioeng. 40: 971-977.
Wink M (1994) The cell culture medium - a functional extracellular compartment of suspension-cultured cells. Plant Cell Tissue Organ Cult. 38: 307-319.
Wolters B & Eilert U (1983) Elicitoren - Ausloeser der Akkumulation von Pflanzenstoffen. Ihre Anwendung zur Produktionssteigerung in Zellkulturen. Dtsch. Apotheker Ztg 123: 659-667.

Yamamoto H, Suzuki M, Suga Y, Fukui H & Tabata M (1987) Participation of an active transport system in berberine-secreting cultured cells of *Thalictrum minus*. Plant Cell Rep. 6: 356-359.

Yoshikawa T & Furuya T (1985) Morphinan alkaloid production by tissues differentiated from cultured cells of *Papaver somniferum*. Planta Med. 110-113.

Yoshimatsu K & Shimomura K (1992) Transformation of opium poppy (*Papaver somniferum* L.) with *Agrobacterium rhizogenes* MAFF 03-01724. Plant Cell Rep. 11: 132-136.

Yukimune Y, Tabata H, Hara Y & Yamada Y (1994b) Scopolamine yield in cultured roots of *Duboisia myoporoides* improved by a novel two-stage culture method. Biosci. Biotechnol. Biochem. 58: 1820-1823.

Yukimune Y, Tabata H, Hara Y & Yamada Y (1994a) Increase of scopolamine production by high density culture of *Duboisia myoporoides* roots. Biosci. Biotechnol. Biochem. 58: 1447-1450.

Yukumine YH, Yamagata YH & Yamada Y (1994) Effects of oxygen on nicotine and tropane alkaloid production in cultured roots of *Duboisia myoporoides*. Biosci. Biotechnol. Biochem. 58: 1824-1827.

Zenk MH, El-Shagi H, Arens H, Stoeckigt J, Weiler EW & Deus B (1977) Formation of indole alkaloids serpentine and ajmalicine in cell suspension cultures of *Catharanthus roseus*. In: Barz W, et al. (eds) Plant Tissue Culture and its Bio/Technological Application. (pp. 27-43) Springer, Berlin

Zhou Y, Hirotani M, Rui H & Furuya T (1995) Two triglycosidic triterpene astragalosides from hairy root cultures of *Astragalus membranaceus*. Phytochemistry. 38: 1407-1410.

Zimmermann A & K Hahlbrock (1975) Light-induced changes of enzyme activities in parsley cell suspension cultures. Purification and some properties of phenylalanine ammonia-lyase. Arch. Biochem. Biophys. 166: 54-62.

Index

MIX
Papier aus verantwortungsvollen Quellen
Paper from responsible sources
FSC® C105338

If you have any concerns about our products,
you can contact us on
ProductSafety@springernature.com

In case Publisher is established outside the EU,
the EU authorized representative is:
Springer Nature Customer Service Center GmbH
Europaplatz 3, 69115 Heidelberg, Germany

Printed by Libri Plureos GmbH
in Hamburg, Germany